Alternative Strategies for **Desert Development** and **Management**

Related Titles in the Environmental Sciences and Applications Series

Biswas CLIMATIC FLUCTUATIONS AND AGRICULTURAL AND WATER RESOURCES PLANNING

Biswas DESERTIFICATION STUDIES

Hinckley RENEWABLE RESOURCES IN OUR FUTURE

Holy EROSION AND ENVIRONMENT

United Nations DESERTIFICATION: Its Causes and Consequences

Pergamon Titles of Related Interest

Biswas/Biswas ENVIRONMENT AND MAN

Biswas WATER AND ENVIRONMENT

Samaha et al WATER MANAGEMENT FOR ARID LANDS

Related Journals*

ATMOSPHERIC ENVIRONMENT

CONSERVATION AND RECYCLING

THE ENVIRONMENTAL PROFESSIONAL

NUCLEAR & CHEMICAL WASTE MANAGEMENT

WATER SUPPLY AND MANAGEMENT

***Free specimen copy available upon request.**

Alternative Strategies for **Desert Development** and **Management**

Proceedings of an International Conference held in Sacramento, CA, May 31-June 10, 1977

United Nations Institute for Training and Research

Desert Management volume 4

Volume 3 in the Environmental Sciences and Applications Series
Series Editors: Margaret R. Biswas
Asit K. Biswas

Pergamon Press New York Oxford Toronto Sydney Paris Frankfurt

Pergamon Press Offices:

U.S.A.	Pergamon Press Inc., Maxwell House, Fairview Park, Elmsford, New York 10523, U.S.A.
U.K.	Pergamon Press Ltd., Headington Hill Hall, Oxford OX3 0BW, England
CANADA	Pergamon Press Canada Ltd., Suite 104, 150 Consumers Road, Willowdale, Ontario M2J 1P9, Canada
AUSTRALIA	Pergamon Press (Aust.) Pty. Ltd., P.O. Box 544, Potts Point, NSW 2011, Australia
FRANCE	Pergamon Press SARL, 24 rue des Ecoles, 75240 Paris, Cedex 05, France
FEDERAL REPUBLIC OF GERMANY	Pergamon Press GmbH, Hammerweg 6 6242 Kronberg/Taunus, Federal Republic of Germany

Library of Congress Cataloging in Publication Data
Main entry under title:

Alternative strategies for desert development.

(Environmental sciences and applications ; v. 3)
Contents: v. 1. Energy and minerals -- v. 2. Agriculture -- v. 3. Water -- [etc.]
Proceedings of the Conference on Alternative Strategies for Desert Development and Management.
1. Arid regions--Congresses. 2. Arid regions agriculture--Congresses. 3. Deserts--Congresses. I. United Nations Institute for Training and Research. II. Conference on Alternative Strategies for Desert Development and Management (1977 : Sacramento, Calif.) III. Series.
GB611.A44 1982 333.73 81-23433
ISBN 0-08-022401-6 (set) AACR2
ISBN 0-08-022402-4 (Vol.1)
ISBN 0-08-022403-2 (Vol.2)
ISBN 0-08-022404-0 (Vol.3)
ISBN 0-08-022405-9 (Vol.4)

Printed in the United States of America

CONTENTS

Volume I: Energy and Minerals, includes pages 1-254
Volume II: Agriculture, includes pages 255-481
Volume III: Water, includes pages 482-946
Volume IV: Desert Management, includes pages 947-1396

Conference Papers

VOLUME I: ENERGY AND MINERALS

IMPORTANCE OF ARID LANDS IN THE GLOBAL ECONOMY

W. Gerald Matlock
University of Arizona
Tucson, Arizona

INTRODUCTION

The economic and cultural development in the arid lands includes a full spectrum of economic types, intensity and sophistication. An estimated 13 percent of the world's population resides in arid lands, most of them in the semi-arid portions.

Eight major types of activities utilize the resources of arid lands: nomadic hunting-and-gathering; nomadic livestock herding; sedentary livestock ranching; mixed livestock and crop farming; rain-fed farming; irrigation farming; mining; and urban development. The first six are forms of animal and crop husbandry, and in the order listed, exhibit an increasing intensity of land use and productivity.

Nomadic hunting and gathering is now the lifestyle of only a few thousand Bushmen and Aboriginies. Nomadic livestock herding, in contrast, is the most widespread form of economic activity for millions of people in arid regions.

Sedentary livestock ranching is found in the arid lands of North and South America, South Africa and Australia. Livestock ranching occupies lands that are either not usable, or extremely marginal, for crop production or other uses.

Farming is the principal rural occupation of most people living in arid and semi-arid countries. Mixed livestock and crop farming, and rain-fed farming are primarily found on the margins of deserts, in the semi-arid steppes and grass-lands.

Irrigation farming in arid lands, including oasis farming, is limited to places with suitable and dependable surface or ground water supplies. Great differences exist in the type of water source, water resources development technology, irrigation systems, land and water tenure arrangements, crop types, and extent of market orientation.

Fishing is of local importance on some of the oceanic desert coasts as well as on the banks of inland lakes and rivers in deserts.

A great variety of minerals are mined in arid lands. Some appear to be climatically related in that they are deposits of previously humid phases now preserved by the dry conditions. Other minerals are associated with geologic conditions not necessarily confined to arid lands.

Urban development in the arid lands has generally been highly localized, especially in the true deserts. One of the oldest functions of desert villages or cities is that of a waystop at strategic places along overland trade routes. Mining and agricultural settlements have in some cases been the catalyst for other economic activities.

In the past few decades urbanization has been stimulated by the quest of both permanent residents and tourists for some of the more widespread resources of the arid zone, especially the dry, warm (particularly in winter) climate, vacant land and desert scenery. There are now ten arid land cities with over one million population; more than 150 are in the 100,000 to one million class.

GLOBAL IMPORTANCE

The arid lands countries are of considerable importance in the global economy. They make a sizeable contribution of mineral and agricultural products to the world market. They in turn provide a market for the sale of products manufactured elsewhere. In the section to follow data will be presented on this aspect of the arid lands economy.

But there is another characteristic of the arid lands that gives them increasing importance in the global economy. There are some resource-rich nations and some resource-poor nations. The former present no particular development problems. They are or will get where they want to reach on the development ladder.

The resource-poor nations present an entirely different problem to the global economy. Although they may contribute some products (usually agricultural) to world markets and buy something in return, the value of such transactions is low. Furthermore, increasing population coupled with already declining productivity made worse by widespread and prolonged drought has brought them to the disaster level.

It is the intent of this paper to look at the arid lands and their importance to the global economy from the perspective of an outsider, one who represents the more humid countries. First, the examination will focus briefly on the value of the developing arid lands contribution to world trade. Secondly, a discussion of the problem presented by the developing arid lands to the rest of the world will be presented. The emphasis in the paper is on the latter because - at least from the viewpoint of the author - it is so much more serious.

ECONOMIC CONTRIBUTION FROM THE ARID LANDS OR WHY WE NEED THEM

Arid lands contribute agricultural and mineral products to the world market. Although the agricultural products are many, certain crops are of greater importance.

Important Crops in Arid Lands

The agriculture of the arid lands is diverse. Important crops in these regions are listed in Table 1.

Table 1. Important Crops in Arid Lands

Alfalfa	Peanuts
*Barley	*Potatoes
*Cassava (Manioc)	*Rice (Upland)
Citrus	Safflower
Common beans	Sesame
*Corn (Maize)	Sisal
Cotton	*Sorghum
Flax	Sugar beets
Lentils	Sunflower
Lima beans	*Sweet Potatoes
*Millet	*Wheat

*Among the world's twelve most important food crops

Of the twelve food crops of greatest significance to the entire world, nine play a crucial role in arid lands food production. T.Y. Canby (1975) made the following comments about the individual food crops of major importance in arid lands:

> BARLEY. Easily adapted to varied growing seasons, rainfall, and temperatures, barley flourishes in temperate climates. Animal feed and beverage malt absorb most of the yield. (Note: Barley is more salt tolerant than wheat, an important distinction in arid lands).
>
> CASSAVA. Brazilian Indians long ago learned to remove poisonous hydrocanic acid from this root, known also as manioc and yuca. Source of tapioca, it forms a low-protein staple in the tropics around the world.
>
> CORN, OR MAIZE. The only important cereal of American origin, this grain is eaten directly in many lands, but also fed to livestock. Hybrids make it the most productive of grains.
>
> MILLET. Millions of the world's poorest peoples - inhabitants of semi-arid tropics in Asia and Africa - rely for survival on millet and its equally hardy sister cereal, sorghum.
>
> POTATO. Irishmen consumed 7½ pounds of potatoes a day each before blight brought the famine of the 1840's, claiming perhaps a million lives. Native to the Andes, it reigns as the most important vegetable.

RICE. "Food" and "rice" are nearly synonymous to many Asians. And one person of every three on earth depends on it for survival. The grass does best in fields flooded to retard weed growth.

SORGHUM. This versatile African grass provides human food, feed grain, molasses, pasturage, and broom straw.

SWEET POTATO. The tropical sweet potato holds promise of genetic improvement in yield and protein content that could vastly reduce malnutrition in the hunger belt (mid latitude developing countries).

WHEAT. Monarch of cereals and the basic grain of world trade, wheat edges out corn as man's premier crop.

Native Plants

There are some plants which are native to the arid regions and which are now receiving increasing attention (Table 2). These plants have been known for centuries but now, perhaps because of energy and water shortages, their time has come.

Table 2. Potentially Important Native Plants in Arid Lands

Plant	Native to	Principal Products, Uses
ACACIA ALBIDA	Africa, Middle East	Gum arabic, forage
BUFFALO GOURD	North America	Edible oil, starch
CANADELLILA	Mexico (North America)	Wax
CASSIA STURTII	Australia	Forage
ECHINOCHLOA TURNERANA	Australia	Grain, forage
GUAR	Asia	Gum, food processing
GUAYULE	Mexico (North America)	Rubber
JOJOBA	North America	Liquid wax
PASPALUM VAGINATUM (silt grass)	Australia (seacoasts of both hemispheres)	Forage, revegetation
SALTBUSHES	Australia (throughout World)	Forage
TAMARUGO	Chile	Forage
ZOSTERA MARINA (eel grass)	Mexico (other northern hemisphere coastlands)	Grain

Source: National Academy of Sciences

They are adapted to life in the desert through a variety of mechanisms including leaf drop, dormancy, and special water storage characteristics and root systems. Present knowledge is insufficient in some cases to make them economically important, but the potential seems to be there and active research programs are underway to further develop that potential. Their principal advantage is the ability to grow and produce on limited rainfall or perhaps with some supplemental irrigation. They require much less water than traditional crops grown in the desert.

Some summary comments about the better known and seemingly more important of these crops follow:

> BUFFALO GOURD: This wild, North American desert gourd which furnishes edible seeds for the American Indians, is a source of edible oil and protein in extremely arid lands.
>
> GUAR: The high protein seeds of this Asian plant contain a gum of unusual properties used for reducing friction of water in fire hoses and in thickening ice cream.
>
> GUAYULE: A shrub of Mexican deserts, guayule contains good quantities of latex that closely resembles that from the rubber tree. Technical processing problems have limited its development.
>
> JOJOBA: This subtropical, North American desert plant is unique in the vegetable kingdom; it secretes liquid wax in its seeds instead of glyceride oils secreted by other plants. Liquid waxes are difficult to synthesize and the only other source is the sperm whale.

Livestock Production

The use of arid lands for livestock production is a logical one, but because of the widely dispersed vegetative resources the system of exploitation must include a high degree of mobility for the grazing and browsing animals. (Nomads are not irrational people!) This is not to negate the importance of having some control over their movements to avoid destruction of the resource. Cattle are of greater importance in trade with other countries but, because of their ability to utilize sparse resources, sheep and goats provide a major source of animal protein in the arid lands.

Mineral Production

The mineral products of importance in arid lands may be found as preserved deposits of a more humid era such as nitrates, salts and gypsum. Other economically important minerals are found under conditions not unique to arid lands. <u>Oil</u> in West Texas, the Middle East, North Africa and the U.S.S.R.; <u>copper</u> in the U.S., and Chile; <u>gold</u> and <u>diamonds</u> in South Africa; <u>lead</u> and <u>silver</u> in Australia; and <u>iron</u> in Mauritania are examples.

Measures of Economic Importance

The economic contribution of the arid lands to the world economy as measured by export-import values of agricultural products and mineral production is given in Table 3.

Data are presented for the total of all crops of importance in arid lands and for individual crops of significance. Major import crops are cotton, rice, wheat and maize. Major export crops are sesame, sisal, cotton, groundnuts and citrus.

Livestock represent 11 percent of world import totals and 8 percent of world export totals. Livestock value is $0.31 billion for imports, $0.22 billion for exports.

Mineral production in the arid LDC's is 43 percent of the world total for minerals listed. Of major importance are petroleum, silver, gem diamonds, copper, and iron.

Although inexact because of the fact that time did not permit separation of data for just the arid parts of some of the countries and/or such data were not available, the implication is nonetheless there - the arid lands are important in the global market.

INCREASING DEPENDENCY OF ARID LANDS OR WHY THEY NEED US

A close examination of the economies of the arid lands shows their dependency on the "rest of the world." Even in the Southwestern U.S. and Israel, long touted as models of arid lands development, the analysis would show a continuing need for close economic ties with other countries or regions. In Arizona, for example, the economy would falter badly without the outside sources of revenue brought into the state by tourists, federal subsidies and grants, and retirement income. Israel's economy has been bolstered by the infusion of vast amounts of U.S. and European capital. Less fortunate arid lands countries have had to turn to the international donor community for financial as well as technical assistance in meeting their most pressing needs.

The Problem Countries

For some years the developing nations of the world have been referred to as the Third World. This is in comparison with the First World, the countries of the industrialized West, and the Second World, the countries with centrally planned economies. A further subdivision may be warranted.

The Third World is redefined as those countries that need time rather than outside assistance to build their economies. Included are the countries in the Organization of Oil Exporting Countries (OPEC), others with sizeable key natural resources, and those which have attracted foreign investment and now can obtain capital on commercial terms.

The Fourth World countries are those with some essential raw materials, some infrastructure and educated personnel. They probably can achieve self-sustaining economic growth eventually, but will need considerable technical assistance and trade concessions to do so.

Table 3. Economic Importance of Crops, Livestock and Minerals in Arid LDC's*

	Crops		Livestock			Minerals	
Imports (% of world total)	COTTON	38 percent			Production (% of world total)	PETROLEUM	50 percent
	RICE	35				SILVER	30
	WHEAT	28				DIAMONDS (gem)	26
	ALL CROPS **	22	ALL ANIMALS***	11 percent		ALL MINERALS****	43
Exports (% of world total)	SESAME	85 percent					
	SISAL	64					
	COTTON	41					
	GROUNDNUTS	40					
	ALL CROPS	7	ALL ANIMALS	8 percent			
Imports ($ value)	WHEAT	$8.1 billion			Production ($ value)	PETROLEUM	$43 billion
	RICE	1.1				COPPER	1.5
	MAIZE	0.7				IRON	0.63
	ALL CROPS	11	ALL ANIMALS	$0.31 billion		ALL MINERALS	47
Exports ($ value)	COTTON	$1.8 billion					
	GROUNDNUTS	.38					
	CITRUS	.25					
	ALL CROPS	3.3	ALL ANIMALS	$0.22 billion			

Source: World Agricultural Production and Trade Statistical Report, USDA; FAO Production Yearbook 1975, FAO Trade Yearbook 1975; U.S. Department of Interior, 1976, 1973 Minerals Yearbook; U.S. Bureau of Census, 1976, Statistical Abstract of the United States (Table 1203); United Nations, 1975, 1973 Statistical Yearbook; American Bureau of Metal Statistics, 1974, 1973 Yearbook; Delury, ed., 1976, The World Almanac and Book of facts, 1976.

*countries used are listed in Table 4
**crops: barley, citrus, maize, cotton, flax, groundnuts, potatoes, rice, sesame, sisal, sunflower, wheat, beans, peas, lentils, and other dried legumes.
***livestock: bovine cattle, sheep, lambs, goats, swine.
****minerals: crude petroleum, copper, gold, diamonds, lead, silver, iron.

The Fifth World countries are those with few exploitable resources, and which are not, in general, self-sufficient in food production. To survive as nations, they may require permanent assistance.

The developing countries which have ten percent or more arid lands are listed in Table 4. Of great significance is the fact that 16 of 36 Third World Countries are those with substantial arid regions. Twenty-three of 52 Fourth World countries and nine of 13 Fifth World countries are those with significant amounts of arid lands. This is where the problem lies.

A study of famine and related problems can be used for emphasis. A survey of 55 major famines since World War II shows that 35 occurred in the countries listed. Twenty-five out of 27 famines in Africa and the Middle East were in the arid region countries.

Indicators of Development Status

In Table 5 a comparison of several indicators of development status is given for the arid lands developing countries. They also are compared with the same indicators for the overall region or continent of which these countries are a part, and with those for the entire world. The following conclusions can be stated:

1. Gross national product per capita is lower in the arid lands of all regions or continents. Furthermore, the rate of growth of GNP/capita is lower in the arid lands of all regions or continents except Africa.

2. Energy consumption per capita in the arid lands is only a small fraction of that in the "other" countries except in Latin America where it is about 25 percent greater. On a world basis, the developing arid lands consume only about 12 percent as much energy per capita.

3. Life expectancy in the developing arid lands is somewhat lower than in the "other" countries in all regions or continents.

4. Dietary energy supplies in the countries of arid lands are below the world average and in some cases well below those of the "other" countries.

5. Literacy rates in the arid lands are substantially lower than those in the total regions or continents and the world, except in Latin America where they are slightly higher.

Food Production

Annual food production indices provide a statistical base for evaluating individual country or regional progress in agricultural development and make it possible to compare and show trends in food production. Such a comparison of food production levels in the developed countries and the developing countries with substantial arid lands for the 1963-1974 period is given in Figure 1. The arid land countries are further subdivided into Third, Fourth and Fifth World countries as previously defined.

Table 4. Developing Countries with Ten Percent or More Arid Lands

AFRICA	ASIA	LATIN AMERICA	MIDDLE EAST
	THIRD WORLD COUNTRIES		
Algeria*		Chile	Bahrain
Libya*		Mexico	Iran*
Morocco			Iraq*
Nigeria*			Kuwait*
			Lebanon
			Oman
			Qatar*
			Saudi Arabia*
			Turkey
			United Arab Emirates**
	FOURTH WORLD COUNTRIES		
Angola	India	Bolivia	Jordan
Benin	Pakistan	Peru	Syria
Botswana	Sri Lanka		
Cameroon			
Egypt			
Ghana			
Kenya			
Malagasy Rep.			
Mauritania			
Mozambique			
Namibia			
Senegal			
Sudan			
Tanzania			
Togo			
Tunisia			
	FIFTH WORLD COUNTRIES		
Chad	Afghanistan		Yemen Arab Republic
Ethiopia			Yemen Peoples Rep.
Mali			
Niger			
Somalia			
Upper Volta			

*OPEC Countries. **Abu Dhabi, a member of UAE is an OPEC member. Other OPEC countries include Indonesia and Venezuela. Only the latter has substantial arid/semi-arid areas.

Division according to Time Magazine.

Table 5. Comparative Status of the Developing Arid Lands

Region or Continent	GNP/Capita	1965-1973 Growth GNP Per Capita	Energy Consumption	Life Expectancy	Dietary Energy	Literacy
	($U.S.)	(%)	(10^6 joules/ cap)	(years)	(kcal/ cap/day)	(% of pop. over 15)
Africa						
Arid Lands	250	3.2	4.2	43	2220	19
Other Countries	410	1.8	24.7	46	2340	30
Asia						
Arid	120	1.6	4.8	50	2090	32
Other Countries	800	4.2	41.9	60	2390	50
Latin America						
Arid Lands	790	2.4	32.5	61	2230	73
Other Countries	840	4.1	25.7	62	2540	69
Middle East						
Arid Lands	1000	5.0	21.2	53	2620	36
Other Countries	3010	6.7	76.6	71	2960	84
World						
Arid	280	2.2	3.1	49	2180	32
Other countries	1340	4.0	65.1	61	2420	56

SOURCES: IBRD, Population Reference Bureau, UN, UNESCO, Time, Inc., World Energy Conference

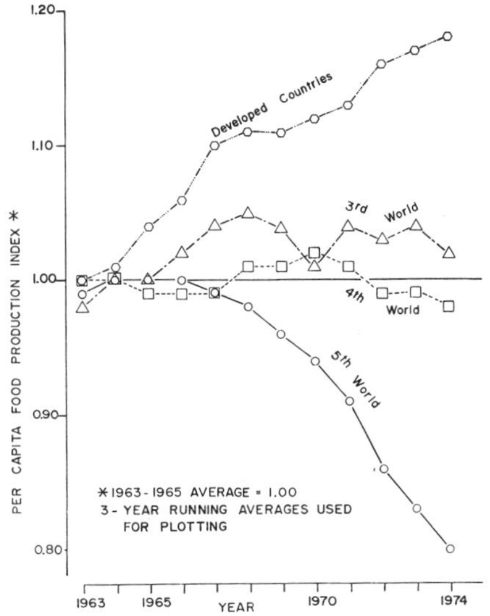

Figure 1. Per Capita Food Production in the Arid Lands.

While production per capita has been increasing modestly but steadily in the developed countries (about 18 percent in the eleven-year period), the per capita level in the arid lands developing countries shows some alarming trends. Third World countries made substantial progress in the mid and late 1960s but have decreased and even stagnated somewhat since then. Food production in the Fourth World countries has fluctuated about the same level but shows a decreasing trend in the last four years (about 5 percent). Examination of the data for the Fifth World countries, however, shows that they are in an even more precarious position. Per capita food production levels began to decline in 1966, have declined steadily since then and have declined at a disaster-portending rate in the last four years. Per capita food production in 1974 was only 80 percent of what it was ten years earlier.

Why are the arid lands in trouble? The major problem is overexploitation of natural resources in these regions under the existing systems of agricultural (and urban) use, given the present resource conditions. Drought contributes to this problem whose principal components are land degradation and water depletion associated with human activities but including, as a prime factor, the management (or mismanagement) of grazing and browsing animals.

Population in Arid Lands

Population data for the arid lands are given in Tables 6 and 7. Examination of these data show that from 1950 to 1974 population increased in the arid lands countries of Africa by 78 percent, only slightly more than in the "other" countries. In Asia the arid lands population increased by 64 percent, 8 percent more than in the "other" Asian countries.

Table 6. Population of the Developing Arid Lands by Continent or Region (millions)

Continent or Region		1950	1960	1965	1970	1974	Increase 1950-1974 (percent)
Africa	Arid	162	204	226	258	288	78
	Other Countries	55	70	77	87	96	75
Asia	Arid	419	514	562	631	689	64
	Other Countries	964	1,168	1,300	1,413	1,512	56
Latin America	Arid	43	56	67	79	89	105
	Other Countries	93	117	141	162	181	95
Middle East	Arid	56	79	88	101	115	105
	Other Countries	1.4	2.2	2.6	2.9	3.3	142
World	Arid	681	853	943	1,068	1,181	74
	Other Countries	1,805	2,129	2,340	2,547	2,723	51

In Latin America the increase was 105 percent in the arid lands or ten percent more than in the other countries. The Middle East arid lands countries increased in population by 105 percent in the same period. Israel, the only country in the Middle East "other" category increased by 142 percent, in large measure the result of immigration. On a world scale the population in the countries of arid lands increased by 74 percent as compared with a 51 percent increase in the "other" country category or, in other words, at a rate 50 percent higher.

Separation of the population data for the arid lands by development level reveals one interesting and fortunate characteristic. The increase is greatest in the Third World countries where presumably the economy is in a better position to absorb more people and least in the Fifth World countries, those least able to

Table 7. Population of the Arid Lands by Development Level (millions)

Continent or Region	1950	1960	1965	1970	1974	Increase 1950-1974 (percent)
Third World	132	175	200	231	261	98
Fourth World	500	620	679	765	841	68
Fifth World	48	57	62	70	77	62

Sources: Demographic Yearbooks of UN, FAO Production Yearbooks.

absorb the increase. Unhappily, Fifth World arid lands countries increased by 62 percent, more than all of the countries in the "other" category.

Favorable climatic cycles have been accompanied by unusual population increases in some arid lands. Then emergency relief has maintained most of the population during drought periods. Food relief in an area of frequent drought may have a counter-productive aspect. That is, supplying food to the drought area may merely make it possible for people to survive (and even continue to increase their total numbers) in an area where they probably shouldn't be living in the first place.

Livestock in the Arid Lands

The number of grazing and browsing animals in the arid lands also has been undergoing change. Data on livestock numbers in the arid lands are provided in Table 8. The numbers given are "animal units" which allows for some comparison of the impacts of different animals. One animal unit is the equivalent of a 450 kg. cow and calf. Other animals are rated as follows: 5 sheep or goats, 1 horse or mule, 2 donkeys, or 1 camel are approximately equal to one animal unit. It should be noted that this comparison is inexact because the feeding habits of the animals listed are not the same; that is, goats and camels are browsers although they also do some grazing. The other animals are essentially grazers. An animal's ability to utilize sparse rangeland also varies. For example, sheep can live on land too far depleted to support cattle, and camels can live on land the sheep have abandoned.

The data in Table 8 show that livestock in the arid lands apparently have increased at a high rate since 1950. Animal populations in the arid lands have increased at a rate from 25 to 80 percent of the rate of human population increase. The increases were greater in the developing arid lands than in the "other" countries except in Asia where the increase was less than 1/5 as much. Livestock numbers in Israel, the only country in the Middle East in the "other" category, increased by 400 percent. Of interest here, however, is the fact that livestock numbers peaked in Israel in 1960 with early recognition of overgrazing hazards. In the rest of the world animal populations peaked in the 1969 to 1971 period except in the developing arid lands of Latin America and the Middle East where they reached a maximum in 1974 and 1973, respectively.

Table 8. Livestock in the Developing Arid Lands*
(million animal units)

Continent or Region		1950	1955	1960	1965	1970	1974	Increase(%)
Africa	Arid	114	129	145	164	191	184	61
	Other Countries	35	36	39	41	42	41	17
Asia	Arid	267	277	296	318	301	307	15
	Other Countries	165	193	220	251	314	305	85
Latin America	Arid	40	47	49	51	57	59	48
	Other Countries	188	198	217	236	258	262	39
Middle East	Arid	43	47	55	56	60	60	40
	Other Countries	.1	.2	.5	.3	.3	.4	400
World	Arid	464	501	546	589	610	610	31
	Other Countries	567	624	682	751	854	772	36

*cattle, sheep, goats, horses, mules, donkeys and camels

See text for an explanation of the term, "animal unit."

Source: FAO Production Yearbooks

Causes of Increase in Livestock

Increased populations of humans have resulted in a greater need for livestock as a means of eking out a living. Health care programs for animals have been widely and effectively implemented. Relatively long periods of favorable weather have enabled herds to grow and multiply. Emergency feeding programs during drought may have contributed to the building of animal numbers or at least prevented a decrease. Furthermore, because the only way of saving available to many nomads is through cattle, they tend to increase herd sizes. The herd is their bank account.

A VICIOUS CIRCLE

There is a vicious circle of complex interactions between population, food and natural resources in the arid lands. When the system is in balance (a condition that perhaps never has occurred, at least in any sizeable area), the food supply is adequate to supply all the needs of the population, and the natural resources which supply that food are not stressed.

When the numbers of people begin to rise, more food is required to meet people's needs. This need for more food is translated into a stress on the natural resources. With continued increase in population, a point is reached where the stress on the resources exceeds first their availability and then their inherent capability to produce food. (Their inherent capability can be improved - this is what agriculture is all about - but not without limit). The cycle reverses. The decline in resource capability results in less food produced. The effect on people of less food availability is obvious - hunger and eventual starvation if the stress on the population continues.

Domestic livestock are no less affected by their own vicious circle. (Wild animals could be included; large swings in their populations are documented extensively). The same resources are involved and the same cycle of events occurs. The only difference is that the animals' "circle" is not under their own control but, instead, is dominated by people.

The cumulative effect of constantly increasing numbers of people and animals is serious damage to the limited natural resources of the arid lands. The complex interactions inherent in resource use and abuse include land degradation, water resource depletion and deforestation. Agriculture is both a perpetrator and a casualty in the process.

The problems of scarce resources in the arid lands will be exacerbated if present trends of human and animal population growth continue. There is no reason to think that it will be easier to resolve the problems when even more people and animals occupy these fragile regions.

INTERDEPENDENCE OF A MODERN WORLD OR WHY WE NEED EACH OTHER

Despite some proposed high technology schemes to transfer the world's people (and its problems) onto self-sufficient orbiting living units in space, the world remains essentially a closed system whose only input is the energy from the sun. Physical scientists and social scientists alike are beginning to recognize increasingly that no part of the globe can be isolated from any other part either in a physical sense or economically.

The developing arid lands provide us with some essential agricultural and mineral raw materials. They in turn are a market place for other agricultural products and our manufactured goods. Increasing development may change the characteristics of this international trade but will not eliminate it completely. Thus, there is an economic interdependence.

Agriculture in the developing arid lands can be improved and productivity per unit of land and per unit of labor can be increased. Effective planning and the application of appropriate technology will be required.

Minerals, where available in the developing arid lands, can be exploited. But you must remember that you can't always find a mine where and when you want it.

There is a humanitarian interdependence as well. The large populations living under harsh conditions in the arid lands unquestionably need the assistance of the rest of the world at the present time. Whether or not you hold to the position that such assistance is a "debt" of the developed countries, the fact remains that it is needed, now and in the future. And it will be given - sometimes grudgingly - but hopefully at modestly increased levels and with more careful attention to the need for well-planned and organized actions.

REFERENCES

Altschul, D. R. (1977) "The World's Arid and Semiarid Lands: An Introduction," manuscript for book in preparation, University of Arizona, Tucson.

American Bureau of Metal Statistics (1974) "Yearbook of the American Bureau of Metal Statistics, 53rd Annual issue for the year 1973," New York, N.Y. 152 p.

Anon. (1975) "Poor vs. Rich: A new global conflict," Time 106(25):34-42.

Canby, T. Y. (1975) "Can the World Feed its People?" National Geographic 148(1):2-31.

Delury, G.E., ed., (1976) "The World Almanac and Book of Facts, 1976," World Almanac, New York, N.Y. 984 p.

Department of Economic and Social Affairs (1971-1976) "Demographic Yearbooks: 1970-1975," United Nations, New York. Issues 22-27. Various pagings.

Foreign Agricultural Serivce (1975) "World Agricultural Production and Trade: Statistical Report," U.S. Department of Agriculture, Washington, D.C. 42 p.

Goeller, H. E. et al., compliers, (1974) "Survey of Energy Resources: 1974," United States National Committee of the World Energy Conference, New York, New York 400 p.

International Bank for Reconstruction and Development (1974), World Bank Atlas, Population, Per Capita Product and Growth Rates. World Bank, Washington, D.C., 22 p.

Matlock, W. G. and E. L. Cockrum (1974) "ISYALAPS, A Framework for Agricultural Development Planning," Volume 2. In: A Framework for Evaluating Long-Term Strategies for the Development of the Sahel-Sudan Region. Massachusetts Institute of Technology, Center for Policy Alternatives.

National Academy of Sciences (1975) "Underexploited Tropical Plants with Promising Economic Value," Washington, D.C. 188 p.

Population Reference Bureau, Inc. (1975) "World Population Data Sheet 1975," Washington, D.C. 1 p.

U. S. Bureau of the Census (1976) "Statistical Abstract of the United States: 1976 (97th Edition)," Washington, D.C. 1025 p.

U. S. Bureau of Mines (1976) "Minerals Yearbook 1973. Volume 3. Area Reports: International," U. S. Department of Interior, Government Printing Office, Washington, D.C. 1193 p.

Department of Economic and Social Affairs (1975-1976) "Statistical Yearbooks: 1973-1975," United Nations, New York. Volumes 25-27. Various pagings.

Economic Research (1975) "Indices of Agricultural Production for the Western Hemisphere excluding the United States and Cuba: 1965-1974," U. S. Department of Agriculture, Washington, D.C., Statistical Bulletin 540, 33 p.

Economic Research Service (1976) "Indices of Agricultural Production in Africa and the Near East: 1956-75," U. S. Department of Agriculture, Washington, D.C., Statistical Bulletin 556. 101 p.

Economic Research Service (1976) "Indices of Agricultural Production for the Far East and Oceania, Average 1961-65 and annual 1966-75," U. S. Department of Agriculture, Washington, D.C., Statistical Bulletin 555. 26 p.

Food and Agriculture Organization of the United Nations (1971-1976) "Production Yearbooks: 1970-1975," Rome. Volumes 24-29. Various pagings.

Food and Agriculture Organization of the United Nations (1976) "1975 Trade Yearbook," Rome. Volume 29, 531 p.

DESERT ECONOMIES

Susan Joekes
Hans Singer
Jeremy Swift

* * *

I. INTRODUCTION

Deserts seem to be peripheral and, apart from their oil and mineral wealth, almost useless lands. Land-use planning in desert countries concerns principally the desert edge and is concerned with stopping the extension of desert-like conditions. But this was not always so. For the specialist, deserts are not the abandoned and bleak lands of popular imagination. They have an array of highly adapted plant and animal species. Until recently, deserts formed discrete and important geographical regions, with diversified economies based on nomadic pastoralism, oasis agriculture, salt mining, and specialist long-distance trade, and were an essential link in wider economic and ecological complementarities. On this rich economic base, politically and militarily powerful societies grew up, which often dominated their sedentary neighbours. But with the growth of modern state administration and of industrial economies, desert-edge agriculture has been reoriented towards urban growth regions in wetter areas, and modern communications networks have bypassed deserts; as a result deserts have become progressively more marginal to economic and political life, a marginality often increased rather than reduced by the development of oil and mining enclaves directly linked to advanced industrial economies. Deserts now fulfil the classic role of such marginal areas, by supplying cheap raw materials, labour and commodities such as livestock for rapidly developing economic regions elsewhere. As a result, arid lands are losing biological and economic potential. The theme of this paper is how this process can be arrested, and how deserts and arid regions generally can be developed to their full potential.

Table 1 gives estimates of the size and population of the world's drylands. The categories extremely arid, arid and semi-arid are those of Meigs,(1) using Thornthwaite's aridity coefficient which measures water availability in relation to plant needs. By this criterion, 36.3 percent of the earth's land surface falls into the general dryland category; 16 percent of the world's population lives in these drylands.

Table 1. The World's Drylands

Category	Land area (million km^2)	Land area (percent)	Population (million)	Population (percent of dryland total)
Extremely arid	5.8	4.3	6	1
Arid	21.7	16.2	170	27
Semi-arid	21.2	15.8	452	72
Dryland total	48.6	36.3	628	100
World	133.9	100	3860	

Sources: Dryland percentages from Meigs 1953; population from UN Conference of Desertification 1977, Population, Society and Desertification. A/Conf. 74/8.

For some regions of the developing world the proportion of dryland, and of population and economic activity in drylands is much higher. FAO estimates that 94 percent of the land area of North Africa and the Near and Middle East (which includes Afghanistan, Iran and Pakistan) is rangeland with an annual precipitation of 400 mm or less, and that 55 percent of West, East and Southern Africa is rangeland with less than 600 mm annual rainfall. (An unknown proportion of these areas is extreme desert which does not support even grazing animals). These lands support nearly 500 million head of livestock which produce about 2.5 million tons of meat valued at 2,500 million dollars (1975) annually.(2) If production of wool, hides, and above all the value of subsistence milk production are added to this, it is clear that this land is already a major resource. For some countries, such as Somalia and some of the Sahelian countries, it is almost the only resource.

These statistics have a number of important implications:

(i) Because of the size of the arid area, even a small increase in the productivity of the rural economy will create a large total increment in output.

(ii) In spite of the low density of population the percentage of the world's population depending on arid environments for a living is by no means negligible at around 16%. The people involved are in the great majority amongst the world's poorest. Furthermore, their living is particularly precarious and vulnerable to climatic, ecological and socio-economic disasters. This was clearly demonstrated in recent years by the Sahelian, Ethiopian and Somalian droughts, but it applies also to cases of more gradual degradation as well as to the more dramatic emergencies; it is estimated for example that although pastoralists are only 6 or 7 percent of the total dryland population, they are 22 percent of the population at risk from desertification.(3)

"Dryland" is not a homogenous entity. True desert, in Meigs' sense of extremely arid conditions, comprise only a minor part (5.8 million km^2) of the total area involved. Arid and semi-arid are both much larger areas (16.2 and 15.8 million km^2 respectively).

In this paper we are concerned principally with the extremely arid and the arid zones (and use the terms arid and desert interchangeably to refer to the two zones taken together). The semi-arid zone shares many important characteristics with the two drier zones, and solutions to arid zone problems must often be closely integrated with activities in the semi-arid zone. For that reason we often refer in our discussion to the semi-arid zone also. However we do not consider in detail economic activities typical of the semi-arid zone, such as rainfed farming.

The term 'desert countries' is ambiguous. One classification of desert areas would distinguish between countries which are essentially non-desert in terms of economic activity and distribution of population but contain desert areas (often desert hinterlands) which, depending on the availability of mineral and other resources, including possibly tourist allocations, can be of great or even decisive economic importance (e.g. Iran, Algeria); countries which are almost entirely arid, with perhaps only a small fertile strip or oases (e.g. Chad, Somalia); countries which have large arid areas, but which also contain semi-arid or semi-humid settled agriculture in which there is a symbiosis between the desert population and the rest of the country (e.g. Niger).

Another important distinction is that between desert areas containing mineral resources, principally oil but also including phosphate, iron ore, uranium, etc., and desert areas so far found to be devoid of such resources (e.g. Somalia, Mali, South East Ethiopia, North East Kenya, etc.).

The economic point of such distinctions is the great variety of possible relationships between the desert areas and the rest of the country. These can range from the usual situation where the wealth of the arid areas, especially from oil and livestock, is drained off to finance the development of the non-desert areas, to the opposite (and much rarer) case where the foreign exchange earned from the export of cash crops grown in the wetter area is used to import food to keep alive the population of the arid zone in times of drought. In fact even this second, apparently advantageous, type of relationship may often turn out to the disadvantage of the arid areas if the increasing population in the cash crop areas reduces the amount of land available for the nomadic population and deprives it of essential pastures, especially those used in times of drought. This is compounded where earnings from the export of cash crops do not sufficiently accrue to the exporting country, or alternatively, are used by it for purposes not benefiting the desert population which has indirectly paid for these exports by a reduction in its living space. Also in meat production, pastoral desert dwellers may not obtain a fair share of the final proceeds. Most of the profits are made at the final stages of the distribution and marketing chain, which are almost always, in the wetter areas.

We argue in this paper that there is need to consider deserts and arid grasslands not as waste lands only useful if they have oil or minerals, but as large areas of land potentially productive in their own right, and which ought to be developed, for two reasons:

(a) equity. Arid areas are inhabited by people who are capable of making productive use of the land, and who wish to do so. In rich arid countries a part of the mineral and oil resources coming from the arid areas should be devoted to this. In poor arid countries, as is discussed later, rural development can be undertaken at relatively low cost. Rich countries can afford to undertake this sort of rural development, poor countries cannot afford not to.

(b) utility. Oil and mineral resources are limited in even the richest desert countries; long-run planning should now be thinking about sustainable uses of the land, which is an important resource for most of the countries involved. In poor arid countries, land is the major resource, and for some countries, such as the Sahel or Somalia, the products of the arid zone, especially livestock, are already the main source of national wealth. In addition, food imports in most desert countries are rising very fast. Large food imports take a substantial portion of oil (or other) exports earnings, and they create economic and strategic dependency on food-exporting countries. For both these reasons, rich and poor desert countries are well advised to increase domestic food production which means using arid lands. Increasing rural wealth from more productive pastoralism and agriculture will also provide increasing demand for domestically-manufactured goods, and will thus be a stimulus to further industrial expansion. The present expansion of arid lands also argues in favour of more attention to rural development in dryland countries. Widespread desertification threatens the livelihood of existing rural populations, and through them urban livelihoods linked to the rural economy such as livestock trading or processing irrigated industrial crops. Without rural development, rural poverty will increase, demanding greater transfer payments to relieve destitution, and existing urban livelihoods will be directly threatened. However, the water requirements for tourism should not be exaggerated. It is stated that an hotel with 300 visitors operating for six months would require no more water than 2 hectares of irrigated land.

NOTES

(1) P. Meigs, 1953, "World distribution of arid and semi-arid homoclimates." In Review of Research on Arid Zone Hydrology. Paris. UNESCO.

(2) FAO, 1975, The Ecological Management of Arid and Semi-Arid Rangelands in Africa and the Near and Middle East (EMASAR). Rome.

(3) UN Conference on Desertification, 1077, Population, Society and Desertification. A/CONF. 74/8.

* * *

II. EMPLOYMENT AND INDUSTRIALISATION: SIX CASE STUDIES

1. INTRODUCTION: DESERT ECONOMIES WITH 'EXOGENOUS' RESOURCES

Among desert economies there are a number which would be among the world's poorest economies if it were not for the valuable minerals found in their territories. Oil exporting states are the most obvious example, but countries exporting other minerals are in essentially the same position. We shall call the funds from mineral exports "exogenous" resources to reflect the fact that these funds have nothing to do with the production processes of their domestic economies. In some cases, which we will examine later, this is no longer strictly the case - the industrial structure has been modified to link up with the natural resource (e.g. refining, petrochemicals). The initial export of the resource was nevertheless an exogenous event, and in relation to other sectors of the economy it remains purely a source of finance for expansion.

The experience of mineral-rich desert economies is of interest to other desert States because they demonstrate the possibilities for development when financial constraints are removed, and no less importantly they display the difficulties of rapid development too. Various types of countries can anticipate relaxation of their financial constraint: some claim part of the aid allocations of oil-rich countries through Arab or Islamic cultural affinity; others have hopes of discovering mineral resources of their own; others still, expect to receive resources in aid from industrialised countries. All can benefit from studying their predecessors' example.

We have chosen to examine the use which six desert countries have made of their "exogenous" resources for development: Algeria, Iran, Libya, Mauritania, Niger and Saudi Arabia. (There is substantial data only for the first three.) The first two countries (and perhaps also Niger) are properly described as countries which include desert areas, whereas the others approximate more closely to a purely desert economy.(1) Algeria is important because of its relatively recent but concerted industrialisation policies, Iran because of its receipt of oil revenues over a long period. Libya, like Algeria, has only been receiving oil revenues for about fifteen years, but in proportional terms (revenues per capita) the revenues have been on a greater scale than in any other of the six countries. Saudi Arabia provides an example of a country in which the imbalance between the oil sector and the domestic economy, in the sense of the contribution of the "exogenous" oil sector to GNP, is extreme (87% in 1975 (see Table 1).(2) Finally Mauritania and Niger are important exporters of iron and uranium respectively, and yet are still very poor countries: they have both been made aware of the vulnerability of their traditional economies by the Sahelian drought. Neither of these countries has yet undertaken a large industrialisation programme (though Niger is moving in this direction). For this reason it may seem that they do not properly belong in this section; but there are signs that their employment and industrial policies are already following those of the richer countries, taking a path which precludes the optimal development of agriculture.

We concentrate, after a brief introductory description of their resource endowments, on two aspects of development in these six economies: the nature of employment and the patterns of industrialisation undertaken. The conclusion in both cases, based on quite independent reasoning, is that for the successful consolidation of industrial development agriculture must be improved. In order to avoid a wasteful allocation of labour (and a politically tightly constrained society(3)) it is preferable that it be done at the beginning of the development process. We try to explain why this is an especially important objective in desert economies.

2. RESOURCE ENDOWMENTS

(a) Current 'Exogenous' Resources

The scale of proceeds from mineral exports is illustrated in Table 1. The data are provisional and refer mostly to 1973. They are a better guide to the current situation than earlier, more certain estimates, since from about 1972 the changed conditions of trade in the world oil market are reflected in the national accounts of oil-exporting countries, though oil revenues have risen dramatically above these levels following the price rises of November 1973. For example, per capita oil revenues rose in 1974 to approximately \$558 in Iran and \$2753 in Saudi Arabia compared with \$145 and \$529 respectively in 1973.

Table 1. "Exogenous" Resources of Selected Countries US $

	Algeria	Iran	Libya	Mauritania	Niger	Saudi Arabia
GNP per capita	$ 480	$ 923[2)]	$ 2606[3)]	$ 233[3)]	$ 100[6)]	$ 986[8)]
Mineral exports as percentage of total exports	81.0%	92.0%	99.0%	84.0%	17.0%	99.0%
Mineral exports (production) as percentage of GNP	39.0%	22.0%[2)]	68.0%	44.0%[5)]	2.4%[10)]	87.0%[1)]
Mineral revenues per capita	$ 191[1)]	$ 145[2)]	$ 988[9)]	n.a.	n.a.	$ 529[2) 8)]
Mineral revenues as percentage of total government revenues	59.0%[1)]	67.0%[2)]	85.0%[4)]	n.a.[11)]	n.a.[11)]	93.0%[7)]

NOTES: Year is 1973 unless otherwise specified.

Conversions from local currency to US$ at average annual current rates of exchange.

1) Economic Intelligence Unit, Quarterly Economic Review <u>Annual Supplement 1976</u> for each country. 1975 estimate.

2) Economist Intelligence Unit <u>op. cit</u>. or calculation therefrom. 1973 data.

3) GDP per capita. GNP per capita is less.

4) R. First, <u>Libya. The Elusive Revolution</u>, Penguin 1974, p.144. 1970-1 data.

5) Mineral exports as percentage GDP.

6) Economist Intelligence Unit <u>op. cit</u>. 1972 estimate.

7) D. Wells, <u>Saudi Arabian Revenues and Expenditures</u>, Resources for the Future Inc. 1974, p. 8. 1972 data.

8) Saudi Arabian per capita estimates are unreliable. Official figures estimate 1974 population at 7.0m, but many observers estimate it at less than 4m. See EIU <u>op.cit</u>.

9) OPEC Statistics Unit <u>Annual Statistical Bulletin 1975</u>.

10) France, <u>Dossiers d'information Niger 1971-2</u>. A.2. 1972 uranium revenues; 1970 GNP only values available.

11) 'Export taxes' contributed 21.5% and 6% to government revenues in Mauritania and Niger respectively in 1973. Berg (1975) pp 198, 199.

<u>Source</u>: IMF <u>International Financial Statistics</u>, September 1975, unless otherwise specified.

The non-oil mineral exporting countries have not experienced such marked increases in export proceeds. Exports of iron ore from Mauritania were started in 1961, rose rapidly till 1964 and thereafter proceeds continued to increase but at a less rapid rate. Mauritania's economy is more dependent on mineral exports than Algeria's. Niger's exports of uranium are small by comparison. Niger has only recently emerged as a significant world producer of uranium. The first exports were in 1971,(4) but it seems likely on the basis of reserves proven so far that uranium production will incease, especially with a second mine at least as large as the first due to open in 1978-9.

The significance of these exogenous resources lies in their enormous scale in relation to the pre-existing national economy and in the fact that they are largely channelled through the central government. (They do not however guarantee freedom from balance of payments difficulties, as the Iranian example shows.) This gives extraordinary scope to the governments concerned to direct economic development. They can feasibly plan and expect to implement huge programmes of investment. We concentrate in this section on the inherent longer-term disadvantages of emphasis on industrial investment alone and on the processes by which the direct role of the government in the labour market may compound the difficulties. It follows that we recommend greater investment in agriculture; and this forms the subject of Section III.

(b) Future 'Exogenous' Resources

Rational planning for development requires information about all natural resources, with respect to future foreign exchange earnings (uncertain as such estimates must be) and to the possibility of using these resources as the material base for domestic industrialisation. Geological mapping proceeds in the case countries. Iran began a national geological survey in 1962, and since then large reserves of gas (amounting to about 30% of world reserves), coal, iron ore and copper have been discovered (see Table 2). Estimates of the iron ore and copper reserves are not available but it is known that the copper deposits at least are very large. Algeria has discovered gas in quantity and also uranium deposits almost as large as Niger's. In the other countries a wide variety of minerals has been discovered in small quantities,(5) and prospecting continues. Saudi Arabia, for example, included a geological survey in the 1971-75 Plan in order to identify suitable sites for further detailed exploration. The actual and potential diversification of mineral resources in the case countries revealed by these surveys suggests that many countries can confidently plan on the basis of long-term receipt of varied "exogenous" foreign exhange earnings notwithstanding the prospect of depletion of the primary mineral itself, in some cases (e.g. Saudi Arabia) perhaps within 25 years on pessimistic estimates.

(c) Land and Water Use

To complete the description of the case country economies we consider briefly their non-mineral resources. All are geographically very large countries. There is considerable variation in the size of their national populations, and in the population densities on agricultural land, which reflects the small proportion of the surface area which is agriculturally useful (see Table 3). "Non-cultivated(6) agricultural land" supports more or less nomadic pastoralists within a narrow range of very low population densities.

In the four properly desert economies large numbers of the population are nomads; in conjunction with the (shaky) information on land use cited, the figures though very crude, give an idea of the importance of nomadic pastoralism in these

Table 2. Minerals: Resources and Production

		Algeria	Iran	Libya	Mauritania	Niger	Saudi Arabia
COAL:							
Reserves[1] (mill.met.tons)		5(20)	193(385)	-	-	-	-
Production (1000 met.tons)	1970	13	530	-	-	-	-
	1974*	15	1100	-	-	-	-
CRUDE PETROLEUM:							
Reserves[2] (mill.met.tons)	1974	1,158	9,315	3,039	-	-	14,780
Production (1000 met.tons)	1970	47,202	191,296	159,709	-	-	188,417
	1974*	48,660	300,852	73,364	-	-	421,397
NATURAL GAS:							
Reserves[3] (100 mill.cu.m)	1974	2,837	10,602	800	-	-	1,726
Production[4] (1000 mill.cu.m)	1970	2.84	12.24	0.26*	-	-	2.26
	1974*	5.62	22.13	2.80	-	-	3.20
URANIUM:							
Resources[5] (met.tons)	1975	28,000	-	-	-	40,000	-
Production (met.tons)	1970	0	-	-	-	0	
	1974*	0	-	-	-	1,250	
IRON ORE:							
Reserves		n.a.		-		-	-
Production[6] ('ooo m.t.)	1970	1,546	3	-	5,923		
	1974*	2,064	396		7,582		

* estimates

NOTES: 1) known recoverable economic reserves in place, i.e. those which, under current economic and technological conditions actually may be extracted. Compare total known reserves in brackets. Refers to 1957 Algeria 1972 Iran.

2) amount known with reasonable certainty to be recoverable under existing economic and operating conditions.

3) reserves (thousand million cubic meters) proven, excluding liquids recoverable from gas.

4) excluding production of natural gas liquids - Algeria 1260 million m.tons 1974*; Iran 1570 million m.tons 1974*; Libya 1225 million m.tons 1974*; Saudi Arabia 4480 million m.tons 1974*.

5) reasonably assured resources; U content of ores at a market price of less than $15.00 per pound of U_3O_8.

6) iron content of marketable ores mined.

Source: UN *Statistical Yearbook 1975*

Table 3. Land Use and Population Densities, Selected Countries. 1974 or Nearest Year

	Algeria	Iran	Libya	Mauritania	Niger	Saudi Arabia
Total land area ('000 ha)	238,174	164,800	175,954	103,070	126,700	214,969
Agricultural[(1)] area as percent total area	19.1%	16.6%	5.4%	39.1%	14.2%	39.8%
Cultivated area as percent total area	3.0%	9.9%	1.5%	1.0%	11.8%	0.3%
Irrigated area as cultivated[(2)] percent area	4.3%	33.0%	5.1%	0.3%	0	24.0%
Population density persons/km^2 overall:	7	20	1	1	4	4[(4)]
on agricultural land:	36	118	25	3	25	10[(4)]
percent total population in urban areas	52.0%	43.0%	30.0%	22.0%	5.0%[(3)]	38.0%[(5)]
Total population (m)	16.3	32.1	2.3	1.3	4.5	8.7

NOTES: (1) "agricultural" land = arable land + land under permanent crops + permanent meadows and pastures.

(2) "cultivated" land = arable land + land under permanent crops.

(3) from France, Niger 1971-2, op.cit.

(4) Using official Census population 1974 in preference to (5) (7.02m). However, "many informed observers still maintain the real figure is lower than 4m." EIU, op.cit., p.3.

(5) UN mid-1974 estimate.

Sources: Calculated from FAO *Production Yearbook* figures for agricultural area and UN *Statistical Yearbook* figures for total population; urban population, UN *Demographic Yearbook*, unless otherwise stated.

N.B. Definitions of "agricultural land", especially "permanent meadows and pastures" are not standard, so above estimates are not of great value; neither is the definition of "urban place" standard.

economies. It has been estimated that 20% of the population in Libya and Niger, and 30% in Saudi Arabia are nomads (we do not attempt to define the term strictly here).(7) In Mauritania, until recently the overwhelming majority of the population was nomadic; the Sahelian drought caused a sudden nominal increase in the urban population.

Even in those economies including, rather than consisting of, desert the pastured semi-arid regions are so large (amounting to 16.1% and 6.7% of the total land area in Algeria and Iran respectively (see Table 1)), that the nomadic population is a significant proportion of the total. For Iran, official estimates put the current number at about 0.25m (less than 1% of the total population)(8) but most observers doubt these figures, putting the true number at about 2m. Whatever its exact size a considerable nomadic population remains which is of real importance to the national economy. 40% of the national herd of sheep and goats is owned by nomads,(9) and livestock products figure predominantly in exports, led by carpets, the main non-oil export item. In Algeria there is also a significant nomadic population;(10) although livestock herds provide wool, hides and dairy products, these do not make such a large contribution to exports as in Iran, though they are important in GNP.

The Mediterranean littoral and the Senegal and Niger river valleys are areas of fertile and generally well watered land.(11) On the other hand, Iran (excluding the Caspian which is not an arid area) and Saudi Arabia have sufficient water to support scattered oasis agriculture but no large river systems. A large part of the cultivated area is irrigated, often still by intricate traditional methods, from ancient water resources, and is thus more or less independent of contemporary rainfalls. The disadvantage is that such resources are not infinite, and cannot easily be replenished.

With regard to large-scale water resources, there are known to be large underground reserves in Saudi Arabia, conveniently sited in relation to current and potentially cultivable areas. But in Libya water reserves are in remote southern locations distant from present settlements - meanwhile depletion of traditional water sources is resulting in a very marked lowering of the water table in cultivated areas.(12) In Libya the emphasis should be on conservation of water resources through less wasteful distribution, and on salvage and recycling of water for use in processes in which pure water is not required. The need to avoid exhaustion of the reserves is pressing. Desalination plants are in operation in Saudi Arabia, and planned in Iran, as one way of getting round the exhaustible water resource problem. But desalination is expensive in energy use. The generation of useful energy from inexhaustiable attributes of deserts, wind and sunlight, and from tides, should be seriously considered in relation to their potential for increasing water availability.

3. EMPLOYMENT AND THE STRUCTURE OF THE LABOUR MARKET

As well as having smaller proportions of the population as nomadic pastoralists, Algeria and Iran have much larger total populations (see Table 3) and more complex and diversified economies than the other case countries. A considerable part of the labour force is engaged in non-agricultural activities, 60% in Iran, and 40% in Algeria, as against proportions of perhaps 5% in Mauritania and as little as 1% in Niger.(13) Libya and Saudi Arabia are in an anomalous position. They have been receiving substantial exogenous revenues for long enough for the distribution of their labour forces to have been profoundly altered away from agriculture towards services. The share of the labour force in non-agricultural activities in Libya was about 70% in 1972.(14) Nevertheless within that number the share of the total labour force in manufacturing is much less than before; it had fallen from 19% in 1964 to 7% in 1972.(15) In Saudi Arabia 54% of the labour force in 1966 were in non-agricultural activities, of whom only 4% were in manufacturing.(16)

It is helpful in this context to refer to the concept of a labour surplus economy, that is, essentially, one in which there is population pressure on the land.(17) This is the fundamental feature, it is reflected in high rates of rural (and urban) underemployment, overt urban unemployment, urbanisation and international emigration. Iran and Algeria are the two countries of the six generally described in these terms. "The heart of the employment problem in Iran is manifestly the sizeable volume of under-employment, mainly in rural areas."(18) The level of employment in agriculture has been constant over the past 15 years;(19) during that time all increments to the labour force have appeared in urban areas and urban unemployment has increased. In Algeria, according to the 1966 Census, 1.4m, or as many as 53% of the labour force, were recorded as unemployed or under-employed.(20)

The availability of finance should in principle be able to modify the employment situation, insofar as employment possibilities (by contrast to the term 'labour surplus') are also given by endowments other than land, i.e. capital. Surprisingly, with exogenous resources the excess supply of labour seems intensified instead of alleviated, for unemployment is <u>increasing</u> rather than the reverse. By contrast, the two other countries most obviously not in this position, Libya and Saudi Arabia, have been able to use large numbers of foreign workers to overcome their shortages of labour.

Mauritania and Niger (not classified so far) could be seen as <u>intermittent</u> labour surplus economies.(21) The urban population in these countries has been greatly increased in recent years by influx of rural population whose livelihood has been undermined by the Sahelian drought. The population of Niamey, capital of Niger, rose from 74,000 in 1970 to 105,000 in 1972; and 250,000 people moved to the towns in Mauritania over the same period.(22) There have also been large movements of population internationally, generally southwards away from the Sahara, which has caused some friction between Sahelian states, e.g. between Mali and Upper Volta in 1974.(23)

These population movements were the response to a great, but temporary disturbance in the carrying capacity of local land.(24) They represent the traditional accommodating mechanism of nomadic life in extreme form. There may be a danger that in this case the governments concerned fail to recognise this, and instead of providing temporary relief perceive the problem as one of (sudden) conventional 'labour surplus' demonstrating the futility of nomadic pastoralism, and attempt, for example, to offer employment to the nomads and remove the incentive for them to return to pastoralism. This seems particularly likely to happen in Mauritania which has increased its central resources since the nationalisation of the iron mines in 1973.(25) Later in this section we explain why this would be a damaging and destructive course for the government to take on any large scale.

The direction and composition of international labour movements are further revealing indicators of the labour supply situation in the various case countries. The long standing phenomenon of large numbers of expatriate Algerian workers is the result of a chronic labour surplus. There were about 650,000 Algerians abroad in 1973, the great majority in France,(26) amounting to 20% of the active population. Many unskilled Iranians are also working abroad in Kuwait, Bahrain and the Gulf States,(27) but the larger population movements in this case are internal: the rate of growth of urban population has been 5.3% from 1956-70 compared with total poulation growth of 2.9% over the same period.

By contrast, the large number of foreigners working in Libya and Saudi Arabia is evidence of their shortage of labour. The share of foreigners in the labour force is about 20% in Libya and 32% in non-agricultural activities in Saudi Arabia.(28)

Apart from its scale, the pattern of imported labour utilisation marks out these countries as labour shortage economies. There are foreigners working in all sectors, but the distribution is noticeably bipolar in occupationnal terms: there are many more expatriate employees who are highly skilled or totally unskilled than there are of intermediate skill levels. This is the rule throughout the Arab world, among those countries importing labour. The emergence of occupational discrimination against unskilled foreign workers is clear in Kuwait: foreigners are concentrated in construction (only 6% of that sectoral labour force is indigenous), and in general, wages and conditions of work are more favourable to Kuwaitis than to foreigners.(29) In Libya a similar pattern is emerging. In the recent past as members of the Libyan population have become economically active they have taken up employment almost exclusively in the service sector,(30) and there has been an absolute decline in the numbers of Libyans in mining, manufacturing and construction. Increase in demand for labour in those sectors has been satisfied by foreigners. "A future where no Libyans will take up unskilled and semi-skilled occupations can be envisaged."(31)

The situation of chronic labour surplus which we suggest holds in Algeria and Iran (though in neither country to such a marked extent as in, say, Egypt) does not preclude shortages of particular types of labour in relation to demand. Both countries have shortages of skilled labour. In Algeria before independence virtually all employees with high level skills were French and most left abruptly when independence was finally achieved. There was accordingly a sudden demand for technical and managerial personnel which could only be filled by re-employment of French workers, brought about on a different contractual basis as part of technical assistance programmes.(32) Thus there were estimated to be 65,000 skilled foreign workers in Algeria in 1970.(33) Iran was never faced with such a drastic shock to the labour market but there is nevertheless held to be "a serious shortage of skilled industrial and public service (mainly education) workers,"(34) in spite of the presence of 55,000 skilled foreign workers in 1976.(35) Iran also suffers from the brain drain. In both countries the lack of local skills is well illustrated by extensive foreign participation in recent turnkey industrialisation projects.(36)

The shortage of skills in the context of a general labour surplus represents one type of imbalance in the labour market of desert economies. What is more surprising is the existence of pockets of underemployment within an overall labour shortage economy. Yet it is arguable that this situation, representing a severe misallocation of labour resources, exists in the relevant case countries, notably Libya:

> "When in 1956 oil companies started exploring the desert they offered employment at wages substantially higher than the rates prevailing either in agriculture or industry... The wage-differential, expectations, the stimulus given by the oil industry to the service sector all attracted rural migrants... Labour was transferred to the towns. ... When oil revenues started to flow into the Treasury the Government found itself able [to absorb the urban surplus]"(37)

Currently unskilled wages are lower in the oil industry than in agriculture and construction. The government on the basis of taxes on oil operations has taken over from the oil industry itself as wage-leader in the economy and is by far the largest single employer, of "useless civil servants, workers, orderlies and watchmen."(38)

It has contributed thereby to a serious shortage of manpower, particularly in construction, but also in agriculture where the increased cost of labour has led to the abandonment of some arable land.(39) The oil companies were able to pay relatively high wages when they first set up operations, and they were prepared to do so in order to induce labor out of its traditional occupations and no doubt also with a view to good political relations with the host government.(40) The government is now similarly both able and willing to increase public sector employment at high wages.(41)

The pressures on the State to provide employment in Iran and Algeria relate to the more intractable situation of a fundamental labour surplus. Nevertheless the same phenomenon of coexistence of pockets of underemployment and excess demand for the same type of labour exists in these countries as well, within the narrow but crucial sub-market for skilled labour. It can be attributed again to wage and employment policy in the public sector. The under-use of labour is of two casually connected types: underemployment in the public sector and urban unemployment of skilled labour. Overexpansion of the public sector (that is, of employees in public administration rather than in health and education services) is difficult to prove but queues for public sector jobs, as in Iran,(42) indicate that skilled applicants are prepared to wait in hope of a post in the bureaucracy, highly paid; expectations have been raised by the pattern of public sector employment in the past which rose in total from 248,000 in 1956 to 474,000 in 1966.(43)

The share of the urban labour force in service employment is often held to be a catch-all employment sector in societies where there are no formal unemployment benefit systems and only members of comparatively high-income households are likely to be able to fit the statutory definition of full employment. In a sense we are suggesting a variant of this argument, that the very rapid growth of public sector employment has emerged as a latter day version of an unproductive income providing mechanism. In an overall labour surplus situation the existence of such a mechanism may _a priori_ seem justified. But it is misconceived as a distributional measure in so far as it tends to divert highly skilled labour and thus leads to bottlenecks in productive sectors, i.e. _reducing_ the present value of national income.

The socio-political repercussions of the existence of very large public sector employment in what has been called the "rentier" State,(44) that is, one that receives sizeable 'exogenous' resources in our sense, are very far-reaching. This is not the place to examine the political economy of the case countries in detail, but it is relevant to point out not only how the pattern emerges but that it has strongly self-perpetuating characteristics which make the expansion of public employment a hazardous principle of development planning even though it may seem justifiable in the short term in a labour surplus economy. It leads to the existence of a privileged class entrenched in Algeria in terms of power with the emergence of a class of 'state capitalists' corresponding to the programme of industrialisation under state capitalism; entrenched in Iran in terms of material consumption and to a lesser, but still considerable extent, in terms of power in the sense of leverage in the socio-political framework. In Libya, the emergence of indigenous workers as a privileged employment class over foreigners leads to great pressure on the State to maintain this wasteful allocation of labour by continuous expansion of the public sector. The next stage of Algerian development policy aimed at greater participation in production decisions in agriculture and industry (under 'autogestion') faces an inherent hostility to loss of influence and control on the part of public sector employees. In Iran the fall in oil revenues in 1975 led to downward revisions of expenditure in defence and productive investment rather than consumption, which illustrates the difficulty of this variant of the rentier State: the consumption demands of the rentier classes come to have a central claim on the national income, and reinforce a marked inequality of income. That inequality has itself constrained the pattern of industrialisation in the past.

The concept of a rentier economy elaborates an early well-known paper by Dudley Seers on the mechanism of open petroleum economies. He concludes:

> "Factors that [in non-petroleum economies] would express themselves in balance of payments crises, such as wage increases or inadequate initiative in developing local industry, will here cause growing unemployment."(45)

As we have seen this unemployment is formally accommodated at the first stage by expansion of the public sector. But this has the secondary effect of inducing further wasteful unemployment among groups who choose a period of unemployment in the hope of fulfilling their job expectations.

The importance of the 'rentier' model is not only that it illuminates the current situation in Algeria, Iran and Libya, but that there are reasons for supposing that <u>desert</u> economies which find sudden riches may be especially vulnerable to finding themselves on this path. The discovery of exogenous resources means that expectations of higher incomes and easier employment spread rapidly through the rural population, and act as a strong inducement to large-scale movement off the land. Furthermore, the (reserve) price of labour is closely linked to opportunities elsewhere, and so the price of rural labour is extremely likely to rise. Existing arid area ecological systems are so finely adjusted that even a small increase in the price of labour will make many previously viable agricultural activities uneconomic in regions where the land is successfully exploited to the full extent of its economic carrying capacity.(46) So, added to the common experience whereby migration to the towns takes place faster than productive jobs can be created there in the short term, it is likely in arid regions to entail deterioration in directly labour created agricultural capital as well, i.e. deterioration in the <u>quality</u> of land - <u>all</u> agricultural land. The fact that land is quantitatively reduced also, i.e. that areas are abandoned in these circumstances, has been noted in Libya.(47) The nature of deterioration of agricultural capital, primarily of course irrigation systems, and the speed with which decay becomes irreversible, are well illustrated in a paper describing the decline of two cases in Oman.(48) Traditional irrigation channels are complex structures which require regular skilled maintenance; if this is neglected they become derelict and unusable within two or three years. The drop in <u>land</u> productivity which follows the deterioration of irrigation systems, even if average <u>labour</u> productivity does rise slightly, is a very serious development.

Libya and Oman are labour shortage economies; it is not surprising that the man/land ratio in agriculture should fall as the balance of economic activities shifts to urban areas, nor indeed that it should be accompanied by a short-term fall in agricultural output.(49) Our analysis however suggests that the same process of deterioration in land quality is likely to take place in labour surplus economies as well. This is because of a further well-known feature of rural to urban migration: its selectivity,(50) which becomes crucial in the light of the specialisation of labour involved in the construction and maintenance of traditional irrigation systems. The category of labour traditionally used for these tasks is precisely that which is most likely to migrate. Birks and Letts make clear that the deterioration in Omani irrigation channels is indeed due to the absence of disproportionately large numbers of men (especially young men), not to the absence of labour power as such. There is no reason to think that labour surplus economies in arid areas will automatically escape this process. The significance of it is that the decline in the capital stock in irrigation is irreversible except at very great cost; and it cannot easily be replaced by modern types of capital intensive mechanised equipment which 'exogenous' foreign exchange earnings could buy. The experiences of Iran and Libya show that the widespread use of deep pumps etc. quickly leads to over-rapid depletion of water resources, and a generalised fall in the water table which disrupts the existing traditional arrangements for obtaining water.

Thus if the wage differential in favour of modern sector employment, which is the primary stimulant to migration out of agriculture, is allowed to emerge on a significant scale the subsequent development of agriculture is made very much more difficult, in arid lands. In desert countries expansion of agricultural output and rural employment at wages comparable with urban wages should be given the highest priority from the earliest days of receipt of 'exogenous' revenues. Agricultural fiscal and price policies must be designed and implemented in such a way that they do not contribute to a worsening of the inter-sectoral distribution of income away from agriculture. (The existence of 'exogenous' revenues reduces the pressure on the government to find ways of taxing agriculture, which must normally supply the bulk of domestic resources for development.) Furthermore, the development of agriculture in arid areas is subject to special difficulties in relation to employment, as well as to the technical matters described in section II: investment in cultivated agriculture must be done with careful reference to its impact on the nomadic pastoral economy, for experience has shown that it runs the risk of undermining that economy, and throwing a very large segment of the rural population onto the labour market.

4. INDUSTRIALISATION

In this section we examine industrialisation in Algeria and Iran. The former is remarkable for the size of its industrialisation programme and its unusual concentration on heavy industry, the latter for its longer industrial history and different intra-sectoral sequential pattern. They represent quite well two polar types of industrial development. Both countries have concentrated on industry more or less to the exclusion of agriculture until very recently. Of the other case countries, only Libya has paid much attention to agriculture - but in a number of large settlement schemes in previously uninhabited areas, rather than in a serious examination of the potential for raising productivity in currently cultivated and pastured lands.(51)

(a) Industrialisation Experience and Policy in Algeria and Iran

In Iran the contribution of manufacturing to GNP was 15% in 1974-5,(52) and in Algeria 10.7% in 1974.(53) The sector has expanded very fast in both countries and ambitious growth targets are set in current development plans.

The pattern of industrialisation in Iran is one familiar to many developing countries. Broadly speaking, it has been one of import substitution, beginning with manufacture of consumer goods; recently diversification into intermediate and capital goods has begun. The process started in the 1930s with state investment in textiles and food processing. The shift into heavy industry, petrochemicals, steel and machine tools dates from the mid 1960s, again in state-owned plants. A large privately owned complementary manufacturing sector has grown up in the context of buoyant consumer demand and under the additional stimulus of protection. Though the government has not sold its long standing interest (mainly in textiles) in the light industrial manufacturing sector, in which private capital is now dominant, public investment now tends to be confined to large scale, slow gestation projects which do not attract local private capital. Local private enterprise is in any case based on exogenous resource based finance, channelled through state owned development banks, which has resulted in a symbiotic industrial relationship between public and private sectors.

Foreign private investment has been considerable. European steel producers participate in a variety of actual and projected steel plants, and many other joint ventures have been set up (joint with public as well as private enterprise) in such activities as vehicle assembly, tractor production, and domestic appliance manufacture.

With respect to local resource based industries, there are plans to expand petrochemicals production to become the biggest in the Middle East. In effect the Iranians wish to encourage the development of substitutes to oil as an energy source, a resolve which is not undermined by the possession of large reserves, in order to conserve it for use as a feedstock to continue to provide a range of products for industrial and consumer use.

It can be argued that in Iran, as elsewhere, the pattern of protection has encouraged excessively capital intensive techniques in new production and relied on an unequal distribution of income to provide sufficient demand for the expansion of domestic consumer industries. However, these policies have not been sufficient in this case to displace an unusually large traditional small-scale manufacturing sector. The handicrafts sector is an important component of it, particularly carpet production, but it does not comprise the whole. Other small-scale workshops supply a range of mass consumption goods such as food, textiles and shoes, and of building materials and furniture.(54) In some products they may not be competing directly with imports or domestically mass-produced substitutes, but even where they are the secular expansion of consumer demand has allowed the sector to survive. It is of great strategic importance for another reason: much small-scale manufacturing takes place in small towns and rural areas, and its expansion would help to redistribute industry away from excessive concentration on Tehran and the other major cities.

The Algerian pattern of industrialisation has in one sense been the reverse of Iran's, starting with heavy and finishing with light industry. Development began in the mid 1960s with the establishment of petrochemicals and steel production, along with general austerity in individual consumption. The share of consumer goods (mostly food) inputs has been kept fairly low, and a general realignment of trade away from dependence on France has accompanied industrialisation. Contracts for the large-scale heavy industrial projects, among which the petrochemicals scheme is remarkable for its huge size have relied heavily on foreign technology, and been spread among a variety of suppliers; provision of expertise has often been separated from provision of finance.(55)

The next stage was the development of machine tool, diesel engine, vehicle, and electronics production. Food processing plants constituted the only domestic manufacturing industry at independence, and they are still very important. From the late sixties textile and shoe production was added, and further developments for manufacture of intermediate and consumer goods are planned.

What are the problems for the further industrialisation of these two economies? There is a convergence now in that both aim for a comprehensive, integrated industrial sector, internalising externalities by capitalising on intra-sectoral linkages as much as possible. Up to now the progress of industrialisation in both countries has been relatively successful, in terms of fulfilment of expenditure plans and subsequent utilisation of capacity. Iran's strategy of starting with consumer goods production has been consistent with the creation of a large high-income class through the distribution of oil revenues. However, as in other cases of import substituting industrialisation, the country is now facing the severe problem of trying to achieve international competitiveness: It has been estimated that domestically manufactured goods prices are on average 23% higher than world prices.(56) Apart from this, the problems for further development in Iran and Algeria are illustrated by considering employment and demand factors. As has been mentioned, Iran is probably in a situation of growing underlying unemployment which it is not desirable or realistic to think should be absorbed in the public sector. Algeria must plan for the possibility that the labour market in Europe will not absorb any more immigrants, and that if conditions worsen some may return to Algeria. (The 'employment problem' is not so much disguised as in abeyance, and it may materialise at short notice.)(57) Contrary to popular opinion, the production of capital goods is often fairly labor-intensive, owing to short production runs(58) and non-mechanised ancillary operations, so in Iran the move towards capital goods production may provide some employment opportunities, but certainly not enough to employ all the extra members of the labour force. The answer must lie in the expansion of the relatively labour-intensive agricultural and small-scale manufacturing sectors.

Expansion of demand for the output of the mass-consumption goods manufacturing sector requires a general raising of personal incomes among the mass of the population, i.e. in rural areas, particularly in agriculture. Agricultural expansion must be facilitated by provision of greater finance, though _not_ at cheaper rates, since that would tend to encourage the adoption of less labour-intensive techniques than hitherto. In recent years the foreign exchange spent on imports of food has risen very rapidly;(59) increase in agricultural productivity would help also to relieve this expenditure (which is otherwise likely to grow very fast) and restore self-sufficiency or balanced trade in food. In Algeria, too, the current need is for increased productivity in agriculture primarily to provide demand for consumer goods industries and by derivation (directly in the case of fertilisers etc.) for intermediate industries. On both counts then it is desirable that the terms of trade be improved for agriculture, and the currently wide differential in rural: urban wages in Algeria (estimated about 1:4 in 1973)(60) gives scope for that to be done. However, this also suggests that the state capitalist class enjoys income advantages which it will be loath to see undermined, and is in a position to resist.

For the time being, therefore, the continued expansion of the productive capacity of the domestic economy in both countries is crucially dependent on the continued receipt of exogenous revenues, and more particularly on the efficacy of investment in agriculture.(61)

(b) Peculiarities of Resource Based Industrialisation

If there is an apparent fallacy in the patterns of industrialisation carried out in the two countries, it is the emphasis on petrochemicals. It might seem that the petrochemicals industry cannot in the long run be the centre of the whole industrial sector because it relies on continued oil supplies; whereas it is the purpose of the industrialisation programme to build up a productive sector to maintain the level and growth of GNP exactly for the time when local resources are exhausted.

However, the case _for_ petrochemicals is strong. The arguments in favour are fourfold. First, there is the externalities argument of learning the art of industrial production by doing it, which is to consider the experience of production _per se_ a transferable skill independent of the nature of the product. Second, the plant will not be redundant even when local raw materials have been exhausted. It will have been the basis for the whole industrial structure, set up initially with tremendous cost advantages in terms of local fuel and feedstock, and it could continue to fulfil this functiom even using imported supplies. In a sense the true industrial transformation will then have been achieved, with the country importing raw materials and exporting finished products. Thirdly, development of processing industry is a means of increasing the share of mineral-based value added and thereby increasing the resources available for more diversified industrial and other investment. Lastly and most importantly, depletion of the physical resource is not necessarily synonymous with receipt of revenue. Resource rich countries can refuse to supply the world market with crude or refined petroleum withholding it instead for local processing plants.

Similar arguments apply to the domestic production of steel. Both countries (and Saudi Arabia) have large-scale steel projects planned or already in operation. Algeria and Iran have domestic supplies of iron, so that the combination with oil or gas as fuel makes steel production a highly profitable operation; Saudi Arabia has in fact no iron reserves, yet steel production is going ahead. The cost of extraction of Saudi oil and gas is so low and fuel costs are such a large part of toal production costs that the country is potentially the world's lowest-cost producer of steel.(62) A barter deal has been arranged whereby oil and iron ore are to be exchanged between Saudi Arabia and Brazil, and the steel products shipped to various other markets. Desalination plants are to be set up to provide water for this plant. The other attractions for a venture of this kind are said to be, first, the ample, cheap supply of unskilled labour - foreign labour in this and other labour shortage economies. By virtue of their immigrant status, it is easier for governments to prevent these workers organising for collective wage bargaining; the advantage of low labour costs may therefore be retained for a long time.

The second locationally attractive factor is the absence of pollution controls. These now involve heavy industry in considerable costs to producers in the industrialised countries, and there is every likelihood that rich societies will count the disutilities of pollution more and more highly as time goes by, and require stricter control of production processes. No such requirements exist in the suddenly oil-rich economies. There is likely, therefore, to be considerable interest among Western MNCs and public corporations in arid areas as sites for certain types of production. Desert economies may discover an important comparative advantage here, even when the extra costs of production involved in the adaptation of machinery to desert conditions are taken into account. Whether such relocation of production is desirable for desert economies must rest firstly on the global effects of such pollution, and secondly on the local impact on the desert ecology of pollutants which are physically damaging as well as socially unacceptable in other, more congested countries. If the global polluting effects are severe, but the immediate local effects weak, other nations may collectively compensate desert countries if they agree not to allow polluting activities to be carried out in their territory. If both global and local effects are positive but tolerable, and are less severe in the desert than elsewhere, desert countries may attract polluting activities and levy taxes on them. In either case the arid environment effectively provides a source of government revenue.

NOTES

(1) The same distinction is noted by F. Halliday, "Migration and the labour force in the oil producing states of the Middle East" Development and Change vol. 8 no.3 July 1977. A third category is introduced, of city states in a desert setting, viz Kuwait, Bahrain, Qatar, UAE.

(2) The tables are placed at the end of section III.

(3) The meaning of this term will become clear. It refers to the difficulties governments face in reducing high income (urban) consumption and service sector employment of a particular type, on which it is based, once they become entrenched.

(4) France. Direction de l'aide au developpement. Secteur Information Economique et Conjoncture. Dossiers d'information Niger 1971-1972, 1973.

(5) For details of small mineral discoveries in individual countries, see Economist Intelligence Unit, Quarterly Economic Reviews, Annual Supplements.

(6) This does not exactly coincide with extremely arid areas on either the climatic or soil and vegetation definition. Underground water supplies may allow some cultivation in climatically extremely arid conditions, whereas the absence of well spaced watering places will mean that some grasslands cannot be used for grazing.

(7) Economist Intelligence Unit. Quarterly Economic Reviews Annual Supplements.

(8) Compared with 2.47m at the beginning of this century. See J. Bharier: The Economic Development of Iran 1900-70, Oxford University Press, 1971, p. 31.

(9) Bharier op cit

(10) Estimated at 170,000 units. Economic Intelligence Unit. Quarterly Economic Review Annual Supplement 1976, p. 7. It is not clear whether the figure refers to individuals or households.

(11) But the Niger and Senegal rivers had fallen to 72% and 44% of their normal flow respectively by 1972/3. Berg E The Recent Economic Evolution of the Sahel University of Michigan Center for Research on Economic Development, 1975, p. 154.

(12) See Section III.

(13) Economic Intelligence Unit. Quarterly Economic Review Annual Supplements taking numbers of "wage and salary earners" as synonymous with non-agricultural employment. An alternative estimate of non-agricultural employment is 11% in Mauritania and 4% in Niger in 1965. Berg E (1975) p. 6. But see footnote 47 below.

(14) R. First: Libya, the Elusive Revolution. Penguin 1974, p. 173

(15) R. First (1974), p. 173 for 1972. ILO/ECWA Seminar on manpower and employment planning in Arab countries. Manpower and Employment in Arab Countries. Some critical issues, ILO 1976, p. 7 for 1964.

(16) UNIDO ID/167. Comparative Study of Development Plans of Arab States. UN New York 1976, p. 181.

(17) The concept of 'a labour surplus' is controversial. It is argued that the concept is invalid on two different counts:

(i) the descriptive definition of the term (that the ex-post marginal product of labour is zero) reduces to the condition that there is a constant rate of substitution between income and leisure at prevailing wages. It is very doubtful that this condition holds.

(ii) the significance of a labour surplus for development is that it implies an "unlimited supply of labour" at prevailing wages. In a system of work-sharing households, however, this cannot logically hold (as opposed to the less realistic assumption of an organised wage labour system with paternalistic centralised support for the unemployed). The force of these criticisms does not affect the argument here. We use the term 'labour surplus' as a shorthand for the set of employment conditions experienced in Algeria and Iran and countries like them, not in any rigorous sense; and in any case the conclusion of this section is that whatever their labour supply conditions, certain deleterious processes seem likely to emerge in resource-rich desert countries.

(18) ILO. Employment and Income Policies for Iran. ILO 1973, p. 27.

(19) ILO (1973), p. 27.

(20) K. Farsoun "State Capitalism in Algeria' Middle East Research and Information Project, Report no.42. 1975, p. 22. Remili in ILO/ECWA (1976) reports 1 million unemployed in rural areas in 1954.

(21) The concept of labour surplus abstracts from the seasonality of demand for labour in agriculture, of which seasonality on the Sahelian scale is the extreme case.

(22) France (1973) Appendix 6 for Niger; Berg (1975) p. 52 for Mauritania. Both estimates are very rough.

(23) Berg (1975) p. 62.

(24) It may be argued that there had been a gradual ecological overloading of the Sahel as generally good rainfall accommodated expansion of human and animal populations up to the mid or late 1960s, and that by extension the immediately previous levels of population should not be reestablished. Whatever the truth of this matter, the argument pursued in this section about appropriate employment policies outside agriculture still stands.

(25) Berg (1975) p. 52 reports that the Mauritian government is indeed setting aside considerable sums for the permanent settlement of a large number of migrants.

(26) Economist Intelligence Unit. Quarterly Economic Bulletin Annual Supplement 1976, p. 3 Farsoun (1975), p. 33, cites another source putting the figure at 700,000.

(27) Bharier (1971) p. 28. The official estimate of the number of Iranians working in the Gulf States is 100,000 in 1967, but this is thought to be an underestimate even for 1967; certainly the current numbers are much greater.

(28) A M Farrag 'Migration between Arab countries', ILO/ECWA (1976) p. 84ff. Compare the share of foreigners in the labour force of other Arab countries: about 20% in Oman, and in the City States of Kuwait, Qatar and UAE 75%, 83% and 90% respectively. Halliday (1977) p. 280.

(29) ILO/ECWA (1976), p. 88

(30) R First (1974), p. 174

(31) ILO/ECWA (1976), p. 91.

(32) ILO/ECWA (1976), p. 138.

(33) Halliday (1977), p. 276.

(34) ILO (1973), p. 63ff.

(35) Halliday (1977), p. 277.

(36) See below pp 20, 21.

(37) 'Labour Supplies and Labour Stability: a case study of the oil industry in Libya' Bulletin of the Oxford University Institute of Economics and Statistics vol. 32 no.4, November 1970, p. 327.

(38) R. First (1974), p. 152, citing R. Mabro "La Libye, un état rentier?' Projet 39, Nov. 1969.

(39) R. Mabro (1970), p. 338.

(40) This was true also of the establishment of the oil industry in Iran. J. Bharier (1971), p. 162.

(41) This remains true whether the government is in fact the wage leader or follows private firms in its wage policy, especially the mineral producing transnational corporations. Data on relative wage levels in the case countries is very scarce; but informed opinion suggests that private sector wages are probably the higher in Algeria and Libya.

(42) ILO (1973), p. 26.

(43) It is reported at second hand that "the (Iranian) Government themselves believe 25% of the public employees to be redundant," see M A Katouzian 'Some observations on the Iranian economy and its recent growth' Tahqiqat-e-Ectesadi vol. IX nos. 27 & 28, 1972, p. 87. It may of course imply that they are not thought suitable candidates by private sector employers; if so, the phenomenon also indicates the inadequate adaptation of the educational system to the composition of demand for labour, in permitting a disproportionately large number of students in 'literary' rather than technical and professional subjects.

(44) H. Mahdavy: 'The Patterns and Problems of Economic Development in Rentier States' in M A Cook (ed) Studies in the Economic History of the Middle East from the Rise of Islam to the Present Day, Oxford University Press 1970; R Mabro (1969).

(45) D. Seers: 'The Mechanism of an Open Petroleum Economy' Social and Economic Studies vol. 13 1964, p. 236.

(46) The apparently obvious solution to maintain output would be to shift towards more capital-intensive techniques, i.e. to mechanise operations. But the almost universal experience is that current methods of mechanisation (in both irrigation and cultivation) are destructive and counter-productive in practice in arid lands.

(47) See above p. 14.

(48) J S Birks and S E Letts 'Dying oases in Arabia', Tidjschrift voor Econ. en Soc. Geografie, vol. 68 no. 3, 1977.

(49) This follows conversely from the strict definition of labour surplus (reduction in the agricultural labour force does not reduce agricultural output): where there is no surplus a fall is expected.

(50) As described recently, for example, by M Todaro Internal Migration in Developing Countries, ILO, Geneva, 1976, p. 65.

(51) The Libyan manufacturing sector is still extremely small; it contributed only 2% to GNP in 1973. Of the other case countries, the industrial sector is significant only in Niger (11% of GNP in 1966 - Berg (1975), p. 6). Industrial activities consist mostly of processing of local foods and fibres and animal skins (minerals were not being extracted then, and have not since increased the size of the sector very much). The large size of the sector is therefore perhaps associated with the large cultivated area in agriculture, another respect in which Niger is distinctive among the desert case countries. Unfortunately information is too scanty for us to include a study of industrialisation in Niger in this section.

(52) Economic Intelligence Unit. Quarterly Economic Review Annual Supplement Iran, p. 3.

(53) Ibid Algeria, p. 6.

(54) ILO (1973), p. 56

(55) K. Farsoun (1975), p. 13. For example, the Annaba steel complex was built by a Western consortium but financed with a Soviet loan.

(56) F. Halliday in a seminar at the Institute of Development Studies, UK, March 1977, citing Looney's estimates.

(57) France has announced repatriation incentives for foreign workers since this paper was drafted.

(58) ILO (1973), p. 55.

(59) Between 1970 and 1973 Algerian food imports more than doubled in value, and rose from 10% to 13% of total imports; and between 1971 and 1974 Iranian food imports increased more than five times, and rose from 8% to 15% of total imports. UN Yearbook of International Trade Statistics 1975.

(60) Farsoun (1975), p. 20.

(61) Of course structural changes may be a precondition of productive investment in agriculture. Possible structural changes relevant to the pastoral economy are noted in the following section. We do not consider these and other structural changes here because both Algeria and Iran have introduced agrarian changes over a considerable period (the "autogestion" system in Algeria, and land reform and other later changes in the organisation of production and in tenurial conditions in Iran) which have not proved sufficient to promote fast growth of agricultural productivity on their own; they must be complemented by financial investment.

(62) J. Stork: 'Oil Revenues and Industialisation' Middle East Research and Information Project Report no.42 1975, p. 10.

* * *

III. PASTORALISM AND AGRICULTURE

The previous section emphasised the importance of rural development in arid zone countries. In rich/arid countries, sustainable, high productivity uses of the land should be an important goal of present development planning against the day when exogenous resources fail; rural development now will also stop leakage abroad of an increasing portion of income from oil or minerals to pay for food imports, and will ensure that rich arid countries minimise the risk of strategic dependency on food suppliers. In poor arid countries, the rural sector must still provide a livelihood for a majority of the population, and there may not be many other development possibilities.

The ecological characteristics of arid counties were briefly mentioned in the introduction. Here we examine them in greater detail. The focus is on the extremely arid and arid zones according to the criteria used in the introduction, and only parenthetically on the semi-arid zone.

PASTORALISM

The Arid Environment

In deserts and semi-deserts with less than 400-500 mm annual precipitation, rainfall is too low and scattered to permit successful unirrigated agriculture except in special circumstances. There is nevertheless extensive natural vegetation in all but the most extreme desert areas, which can be used by wild and domestic animals, and thus by hunters and nomadic pastoralists. Solar energy is converted by photosynthesis into green plant material, is consumed by domestic animals and converted into products which pastoralists can eat or sell. The first step in this process is the primary production of plant material by photosynthesis. Since in most arid areas there is adequate incoming sunlight throughout the year, plant growth is limited principally by the quantity of available water; as a result, in areas of less than 600 to 700 mm annual rainfall, net primary production is more or less directly related to annual rainfall. This relationship is shown in Fig. 1, using data from North America; the same direct correlation between annual precipitation and plant production has been recorded in other dry areas, including western and southern Africa.

Table 1 gives an estimate of primary production for the main ecological zones of the earth's land surface.

It should be noted that the table underestimates total area, primary productivity and energy fixation from arid lands since it excludes (for lack of data) dry thorn bush woodland, which amounts to perhaps as much as a third or half of total woodland, as well as arid cultivated land. If these were included total arid land area would be 36 percent of the globe, and the arid land percentage of total energy fixation would be correspondingly higher, probably in the region of 15-20 percent. Nevertheless the table does show the potential significance of arid lands as producers of food and material energy for man. Of the major biomes, arid tropical grassland and savanna is second only to tropical and raingreen forest in total net primary productivity, and in total annual energy fixation, exceeding for example cultivated land. A more direct comparison of energy available to man is not feasible since (a) energy in tropical savanna and desert scrub has to be further converted, at a substantial loss, by domestic animals; although energy inputs into traditional pastoralism are low, it is not possible on available data to estimate potential production of human food through pastoralism from these

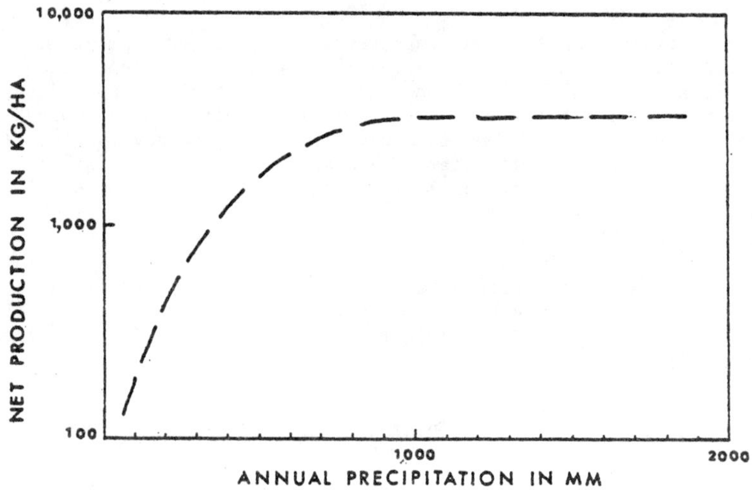

Figure 1. Relationship between rainfall and plant production

Source: F.H.Wagner cited in B.Lundholm 1976 Domestic animals in arid ecosystems. In Can Desert Encroachment be stopped? eds. A.Rapp et al. Stockholm: Ecological Bulletins No.24.

estimates of total energy fixation by plants, since study of secondary productivity in pastoral ecosystems is not yet as far advanced as that of primary production; (b) energy produced on cultivated land is more directly available to man in the form of crops, but energy inputs, particularly outside traditional agriculture, are high. It should also be noted that although total energy fixation in forest biomes is much greater than either savanna or cultivated land, little of this is available to man as food, since it consists in production of leaves, shoots and new wood. Some of this energy is used by man as fuel or materials, but most production is in the major tropical forests, relatively little used.

Thus arid lands create a huge ecological potential through energy fixation by photosynthesis. But its use is not easy.

The main problem is the great variability of energy fixation. Plant production is directly related to rainfall, and in arid lands rainfall is not only scarce but also irregular; its relative irregularity increases with declining annual total rainfall. In the centre of major deserts, in the extremely arid zone, rainfall is highly irregular and several years may pass without effective rain. Where some rain falls most years, its relative variability (mean variation relative to mean total over several decades) is very high: 127 percent at Qseir on the Red Sea Coast, 92 percent at Adrar and 79 percent at Djanet, both in the central Sahara.(1) This irregularity declines as annual totals increase, being about 60 percent annually in the north Sahel, and 40 percent on the southern edge of the Sahel in west Africa. As a result of variation in time, rainfall at any one moment is spatially highly variable: differences in annual totals in the same catchment area 5 km apart regularly vary by 40 percent. The main pattern of rainfall in desert and semi-desert regions is violent, geographically localised

Table 1. Plant Production and Energy Fixation by Major Biomes

	Area covered (million square km)	Total Net Primary Productivity (10^9 metric tons)	Total Annual Energy Fixation (10^{18} cal)
A. Major Biomes:			
Forests	50	81.6	368.6
Woodland	7	4.2	19.6
Dwarf and open scrub	26	2.6	11.0
Grassland	24	19.2	76.8
Extreme desert	24		0.1
Cultivated land	14	9.1	37.8
Total for land surface excluding fresh water	145	116.7	513.9
Of which:			
B. Major arid zone ecosystems:			
Desert scrub	18	1.3	5.4
Tropical grassland and savanna	15	12.0	48.0
Dry desert	8.5		0.1
Partial total arid ecosystems	41.5	13.3	53.5
Arid ecosystems as partial percentage of world total (B/Ax100)	28.6%	11.4%	10.4%

Source: H. Lieth 1975 Primary productivity in ecosystems: comparative analysis of global patterns. In Unifying Concepts in Ecology. eds W.H.van Dobben and R.H.Lowe-McConnell. The Hague: Junk.

storms, resulting from small-scale convective cells, that flood a small area leaving neighbouring areas dry.

Since water is the main limiting factor in plant growth, variability in rainfall leads to great variability in plant growth, and thus in energy available for domestic animal and human use. Variation in plant production within one year is very large due to seasonal water availability. Measurements in the Negev desert (with 75-150 mm rain) and neighbouring mountains (with 550 mm) showed that summer plant cover 82-95 percent less (in biomass weight) than that at the period of maximum growth in spring.(2) Variations between years is also large. Recent research at the International Biological Programme station in north Senegal, in an area of 200-300 mm rain, has defined a "period of useful rainfall", during which soil water is available to plants, and active growth takes place.(3) The period of useful rainfall, and with it plant production, in a "normally" good year is double that of a "normally" bad year; in the extreme drought of 1972 the period of useful rainfall and plant production both fell to zero. These figures are shown in Table 2, together with an estimate of the number of cattle which could safely graze the area in the different years.

Table 2. Variations in Plant Productivity and Theoretical Carrying Capacity for Cattle, North Senegal[4]

	Period of useful rainfall (days) (1)	Above ground plant production (kg dry matter/ ha) (2)	Theoretical carrying capacity for cattle (head of cattle/1000 ha) (3)
"Normal" good year	110	1300	187
"Normal" bad year	50	590	85
1972	0	0	0

This great variation in primary productivity creates problems for animal use of the environment. Noy-Meir(5) has listed five special attributes necessary for organisms (plants and animals) to survive such conditions, to which Lundholm(6) has added a sixth:

1. Tolerance of extreme conditions, especially extreme drought and heat (to which should be added cold).
2. Ability for rapid recovery when conditions are good.
3. Adaptation to the transitions between pulse (expanding) and reserve phases.
4. Flexible and opportunistic feeding habits.
5. Nomadic migrations.
6. Special distribution patterns, with concentrations in favourable areas.

Pastoral economies

Traditional pastoral societies developed economic strategies resembling these ecological strategies shared by all arid zone organisms; pastoralists are flexible and opportunistic in use of pasture resources, they move and they have integrated the grazing economy with complementary economies (desert mining and trading, savanna agriculture, and crafts) in a way that allows population and economic activity to shift beween different activities as conditions change, and allows rapid expansion of pastoralism in good rainfall years. This regional economic interdependence between complementary desert, steppe and savanna economies is well exemplified by the precolonial West African Sahara, Sahel and savanna.(7) In Central Niger and Northern Nigeria, the pastoral economy provided animal products, the Sahara salt and dates, and the savanna grain and the products of town craft industry such as cloth and leather goods. These various economies were integrated by the economic operations of Twareg nobles, who ran the pastoral and desert trading economies, and maintained a commercial infrastructure in the Hausa agricultural areas and towns in the savanna. There was exchange of goods and of labour between these economic sectors, and more important, the chain of integrated activities linking the desert to the savanna provided a safety valve for pastoralists during drought; when rain failed and plant production in the desert edge and steppe areas was very low, pastoralists migrated to the southern end of their trading chain where, because of their land holdings, they were assured of access to pasture and to non-pastoral activities. The desert and Sahel economy contracted in these drought periods and its centre of gravity shifted into the savanna; in good periods the reverse happened, and the pastoral economy expanded to take advantage of the abundant pasture. To a certain extent, these types of movement still operate. Although the degree of economic interdependence between different zones was probably exceptional in this Saharan and Sahel area, a comparable pattern probably existed along the edges of most Old World deserts.

With 20th century development of centralised state administration and the penetration of new types of production and new production relations into the arid zone pastoral areas, these traditional economies have become increasingly peripheral. Development strategy has generally been unbalanced. In Africa, governments have attempted to intensify marketed offtake of animals by improving communications, markets, and by providing animal disease control and new wells, but have rarely improved pasture; in most cases the flexibility and the interdependence of complementary economies from the desert to the savanna has been lost, and has often been restricted. These and other changes have brought many African pastoral economies and societies to the edge of disaster, so that environmental destruction is widespread, and a serious drought can precipitate famine. In resource-rich areas, on the other hand, the emigration of adult men or of whole families to a rapidly expanding extractive or urban construction centre has left pastoral economies without labour, and has led to a rapid decline in pastoral production, and to unused pasture resources.(8)

Development Strategies for Pastoral Areas

In the light of the ecological characteristics of arid lands, and the precarious state of most such areas now, it is desirable to look for new general development policies for these areas. Desert and arid grasslands are ecologically unstable, but resilient, in the useful distinction made by Holling(9) and applied specifically to arid ecosystems by Noy-Meir(10): unstable in the sense that there are quantitatively large annual variations in plant and animal growth, but resilient in that, unless there is gross misuse, they are able to recover high productivity in good years.

This resilience is the result of the stabilising mechanisms listed above, such as flexibility, opportunism and the ability to recover from drought rapidly. Human use of these lands should build on these basic ecological strategies; planning should seek to recreate at a higher level of productivity the traditional advance and withdrawal strategy, integrated with neighbouring economies, of previous desert and steppe pastoral users. In economic terms this means not aiming at constant yields through large technological and institutional investments, but rather at management which accepts fluctuations in yield, aims to make productive use of these grasslands in good years, and accepts small or nil yields in bad years.(11) The best way to achieve this is by extensive pastoralism which builds on the flexibility of existing nomadic pastoral societies, in which grazing pressure is closely adjusted to the amount of energy fixed by plants each year. Technically these calculations are simple. The problem is to combine this ecologically sound use with economically realistic and socially just forms of organisation.

The heterogeneity of arid regions can be analysed around two variables, population density on grazing land and national wealth, and this indicates the feasible developement strategies:

1. Pastoral population. The size, rate and direction of change in pastoral populations relative to pasture resources is important. Two general categories of country can be distinguished, depending on whether there are still substantial pastoral populations or not:

    (a) there are countries where the original inhabitants (for example US Indians, Australian aboriginees, Kazakh and other nomadic pastoralists of the Soviet Central Asian Republics) were removed by a central government with varying degrees of force, and either pushed into marginal reservations (USA and Australia) or organised into new types of economic and social formation (USSR). There is another group of countries where the discovery of oil or minerals (Middle East, Libya, the process is starting in Algeria and Mauritania), or ancient agricultural policies (Egypt, Iraq) have attracted many or all nomads away from pasture areas. In both cases the pasture areas have been largely emptied of their original inhabitants and are (or in the case of the industrial countries, were at the time when the original inhabitants were cleared) a tabula rasa for new development policies.

    (b) in the other countries, either because of a lack of mineral resources or other economic opportunities (e.g. most Sahelian countries including Niger where uranium mining has created few jobs, Somalia, Sudan, dry areas of eastern and southern Africa, Afghanistan, Pakistan, north west India, Mongolia), or because abundant pastures make traditional pastoralism economically competitive with other activities (Iran), there are still substantial pastoral populations on the land, although these are always of low density by comparison to agricultural populations.

The size of future populations of pastoral land-users will depend on the size of present populations, and also on their growth rate. There appear to be significant differences in rates of population growth between geographical areas. In Africa, pastoral populations apparently have low rates of natural increase compared to neighbouring agricultural people; these are the result of a combination of low fertility and mortality. In parts of south west Asia on the other hand (e.g. Iran), pastoral populations apparently have higher rates of natural increase than neighbouring sedentary agricultural populations and there is a constant movement of population from the former to the latter.(12) In Rajasthan in India, population growth rates of nomads and settled farmers are higher in the desert region than in the wetter region.(13)

2. Resources. Strategy options differ according to the resources available for development in the pastoral sector.

    In the rich arid countries of concern in this paper, these will mostly be "exogenous" resources from the export of oil and minerals. So far few of these resources have been made available for pastoral development but they may be in the future; we argue in this paper the desirability of such a policy.

Poor countries do not have this possibility, although some poor countries may have a special claim for development aid from mineral-rich countries in the name of Islamic solidarity, and most countries with large pastoral populations fall in the category of least developed countries and will thus benefit from special policies for this group. However in general poor countries will have to plan for pastoralism to generate its own resources for development. There are important differences in potential surplus generation between pastoral economies, depending on productivity of pastoralism itself (which in turn depends on environmental productivity, on the species herded and on the social organisation of production), and on available markets. Thus Iranian sheep and goat pastoralism in the Zagros mountains benefits from high environmental productivity and a ready market, and the surplus generated is considerable. Somalia and Sudan, although of low environmental productivity, have ready access to Arabian markets, and so also generate a large surplus. Chad and Afghanistan on the other hand, despite sizeable pastoral economies, are in a much less favourable position; even though the former country has large livestock exports, the distance from markets and the low ecological productivity make the potential resources pastoralism can generate for its own development less certain, although they certainly could be substantial given the right economic policies.

As for the joint effects of the population and wealth variables, most rich countries have low population densities on rangelands while most poor countries have high densities. But there are also a few exceptions: relatively rich countries with high pastoral densities, and poor countries with relatively low pastoral densities.

This indicates two types of development policies for arid grasslands:

(1) That of developed countries, which is likely to be followed by rich third world desert countries, where pastoral population densities are now low. This involves capital intensive ranch-type development (private or state owned), in which pastoral emphasis is shifted from milk production for immediate human consumption to meat or wool production; labour is displaced by capital (fenced pastures, water pumps, animals moved and watered by lorry), and free communal grazing is replaced by a tenure system in which the pasture is owned by an individual, a company or collective. Ranches coupled with a developed economic infrastructure (communications, markets, government support prices, fodder production and fattening, insurance against drought, elimination of disease) enables efficient use to be made of variable pastures; in time of drought, animals can be trucked off the range to wetter regions, or in the worst case can be slaughtered and the carcasses frozen for export or later consumption. A developed economic infrastructure allows profits from wet periods to be banked as insurance for dry periods. Hence this type of development does recreate at a higher level of productivity the traditional pulsating pastoral strategy integrated with complementary economic operations. The capital cost is high, and employment is reduced; available jobs are filled by a new rural proletariat (cowboys, gauchos or jackaroos). The ecological efficiency of commercial ranches is low; in the Amboseli area of Kenya, the traditional Masai pastoral system has a food chain efficiency (measuring the relationship between plant production and animal consumption) of 7.8 percent while that of nearby commercial ranches is only 6.5 percent, despite the fact that the Masai are sharing their range with wildlife, while the ranches are not, and that the ranches are on better, wetter land.(14) However pastoral development of this sort provides an opportunity for resource-rich aid countries to produce, from their large pasture resources, high value meat for rapidly increasing local demand, or for export.

(2) Poor arid zone countries are sometimes tempted to follow this example and move pastoralists off the range in order to set up state ranches. There are several major disadvantages to such a policy: it is expensive and replaces abundant labour with scarce capital, it creates a large number of unemployed pastoralists for whom there is usually no alternative source of livelihood, it generates foreign exchange or provides meat for middle and upper income earners but reduces food production by and for the rural poor. For these and other reasons it is likely to be resisted by the pastoralists themselves.

Instead, a policy of developing extensive pastoralism within new economic and social frameworks may be advantageous. Mongolia is most advanced in this respect, with experience of pastoral cooperatives (negdels) dating from the 1930s.(15) Currently some African countries are experimenting with similar approaches, for example Somalia and Sudan.(16) The main difference between this type of policy and the previous one is that pastoral populations are kept on the land, and the pastoral economy is developed in a labour-intensive manner, to provide in the first place an improved food supply. Production relations are cooperative, with producers retaining ownership of their livestock and acquiring decision-making power over land and water. The essential task is to provide a framework within which herdsmen's use of range and water is better controlled and directed. This may be done by attributing specific areas of land to specific groups of pastoralists. Once control of the land has been vested in the users, investments can be made by them or by the State that will not be subject to the usual problem of common property resources, that is they cannot be appropriated by outsiders.

Such investments can be simple, labour-intensive measures. Land conservation by bunds and earth dams allow the development of subsidiary agriculture or fodder production. Pasture quality can be improved at low cost by use of fodder trees and shrubs; experiments with replanting Acacia albida in the Sahel, long conserved by the local population for its fodder value, are encouraging. In Soviet Central Asia, desert shrubs and semi-shrubs (mainly Chenopods) have been used successfully to enrich pastures, and create fodder reserves and create fodder reserves and shelter belts.(17)

Livestock marketing has to be regulated so that the terms of trade for pastoral products do not decline or fluctuate too rapidly against agricultural and other products. Marketing levies can provide a stabilisation fund for this, and also resources for further pastoral development expenditure. Pastoral cooperatives or associations can also provide a framework through which the state can provide education and health services to a mobile population (tent schools and nomadic front-line health workers are already active among some Iranian pastoralists, so the techniques exist; the problem is to provide such services in the context of a more scattered pastoral nomadism as in Africa). Pastoral associations can also serve as a vehicle through which the state can consult pastoralists about planning, and inform them of decisions taken; a channel of two-way communication now generally lacking in the difficult conditions of arid and desert pastures.

In any pastoral development strategy, economic integration of pastoralism with complementary activities must provide for fattening of animals for sale or slaughter. This can either be done in small-scale feedlots created in farms in ecologically favoured places within the pastoral zone (river valleys, wadis, oases), or by a system of stratified animal production extending over several ecological and economic zones; this is now being planned for West Africa from the Sahel to the savanna. The main problem of both these systems, especially the second, is in ensuring that pastoral producers receive a fair share of the total value of the finished animal: at the moment in most such systems the producers (the nomadic pastoralists) carry most of the risk (calf mortality can be 30-40 percent in the first year) while most of the profit is made in the final stages of the fattening or marketing chain.

Policies such as these do not need to use much capital investment, and thus have an added advantage for poor countries. However the demands in skilled manpower to organise and carry through such pastoral cooperatives may be high, and such skills may be in short supply. For this reason it is important that existing trained personnel are used to maximum advantage by combining their work with much increased participation by pastoralists in decision-making at all levels.

With these sorts of development it seems likely that pastoral development can take place without great initial investments even in poor arid countries. Raising the productivity of pastoralism in the ways outlined above will broaden economic options. There will be a three-way trade-off, likely to be resolved in different ways in different regions, between using the increased economic surplus thus generated for national development in other economic sectors (attractive in countries such as Somalia, Sudan, Afghanistan and the Sahel, where livestock is a major resource of an otherwise poor country), using it for further development of the pastoral economy itself at the same population level (and settling the natural population increment into agriculture, industry or services), or supporting a higher pastoral population at the same level of technology and output.

Past neglect of pastoral development means that there are probably relatively large gains in productivity to be made in the pastoral sector with small investments. Thus in the deserts of the Turkmen SSR in the 1960s, improved range management was sufficient to make extensive karakul sheep production more profitable even than irrigated cotton growing.(18) In Africa, it should also be noted that since pastoral populations have a lower rate of population growth than agricultural or urban populations, investment in the pastoral economy is likely to provide greater per capita output than investment in agriculture, other things being equal. If, as has been suggested, the low rate of increase of African pastoral populations is the result of social control of fertility as an adaptation to some features of the pastoral economy,(19) settlement of pastoralists into agriculture or urban jobs is likely to lead to the abandon of this demographic control and to increasing rates of population growth; this has been noted when pastoralists in the Sudan settle into agriculture.(20) In this case, other things being equal, per capita output will be maximised by keeping pastoralists in the pastoral economy and by investing to increase its productivity, rather than by encouraging settlement into agriculture or industry, where total output gains will be reduced by greater population growth rates.

AGRICULTURE

Rainfed Agriculture

Rainfed agriculture becomes marginally possible on the edge of the arid zone considered in this paper; because of the irregularity of rainfall, yields are highly variable. Figure 2 shows the linear relationship between precipitation and millet yield in the Sahel, illustrating the marginal nature of this cultivation.

Rainfed cultivation has spread in recent decades into the arid zone beyond the margin of what is ecologically safe or economically satisfactory. Desertification in most of Africa and the Middle East has been attributed to ecological breakdown following the extension of marginal rainfed farming. Because of its ecological unsuitability, rainfed agriculture usually gives a lower economic return than pastoralism in these marginal lands. Thus in the Rajasthan desert of north west India, the cultivated part of the arid zone is increasing rapidly, leading to land degradation;(21) however even the drought-resistant crops which are now widely grown give a less good return than a natural grass/livestock economy.(22)

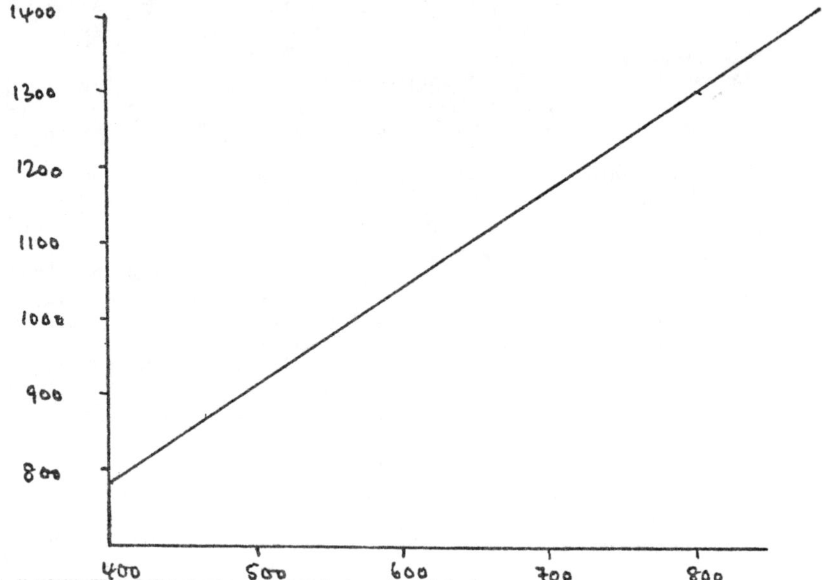

Fig. 2. Millet yield and rainfall in the Sahel

Source: J.Cocheme and P.Franquin 1968 Etude agroclimatique dans une zone semi-aride en Afrique au sud du Sahara. Geneva: World Meteorological Organization.

Irrigated Agriculture

The inhabitants of the Old World desert belt have long practised sophisticated irrigated agriculture, with techniques including the 'funnel' agriculture of the Souf region of the north eastern Algerian Sahara (funnels of a few hundreds yards diameter are dug in the sand dunes down to the water table, and date palms and vegetables grown), underground channels taking water from an underground source several miles out into otherwise dry land (the Sharan *foggara*, Omani *falaj* and Iranian *ganat*), and the many systems of diversion and use of river and *wadi* water.

Where soils are suitable and water can be supplied cheaply, irrigation can create entirely new agricultural ecosystems in the desert; the ample supply of solar energy throughout the year is an important advantage of these areas. Thus the construction of the Karakum Canal in the Turkmen SSR has in ten years created productive agricultural areas in the desert. Many of the world's major desert and arid areas are crossed by large rivers, giving rise to great irrigation potential. Many arid areas also have extensive groundwater tables. In the Rajasthan desert for example there is considerable potentially usable groundwater at present unused, and the greatest potential is in the most arid parts.(23) In the deserts of the Middle East, as Table 5 shows, potential yields from both irrigated and unirrigated land could be substantially increased by a package of appropriate technical measures, if social and institutional problems can be solved.

The problems facing improved and extended irrigation are well-known and serious. Nearly all presently irrigated and potentially irrigable lands in the Middle East and western Asia require drainage to counteract waterlogging and salination. In the Mesopotamian plain, salination has reduced yields of winter barley to as low

Table 5. Present and Potential Yields of Some Major Middle Eastern Crops

Country and crop	Yield per ha			Total production		
	Present	Potential	Increase	Present	Potential	Increase
	(metric tons)		(%)	(million metric tons)		(%)
Egypt:						
wheat	2.7	6.7	148	1.62	5.4	233
maize	3.6	6.3	75	2.36	4.4	86
rice	4.1	8.0	95	2.0	5.6	180
Syria:						
wheat	0.9	2.8	211	1.05	5.6	433
barley	0.9	2.6	189	0.59	1.3	120
Iraq:						
wheat - rainfed		2.7			2.7	
- irrigated		6.7			13.4	
Six Middle Eastern countries, all grains				10.95	49.59	353

Source: M. Clawson, 1971, "Social and institutional obstacles to agricultural development of arid lands in the Middle East." In food, fibre and the Arid Lands. eds W.G.McGinnies, B.J.Goldman, P.Paylore, Tucson: University of Arizona Press, p. 8.

as 450-550 kg/ha(24); the natural wild vegetation, without irrigation, could probably do better. Salination and waterlogging threaten or have already put out of production 50 percent of irrigable land in Iraq and the Euphrates valley in Syria, 30 percent in Egypt, 23 percent in Pakistan.(25) A large amount of water in irrigation systems or dams is lost by evaporation (especially intense in arid countries) and by seepage: evaporation in Lake Nasser behind the Aswan High Dam loses one litre out of every eight, and total losses from all sources may reach 50 percent in some irrigation systems. Where the source of irrigation water is a slowly recharged or fossil watertable, irrigation may run down resources rapidly: it is estimated for example that the 50 percent increase in irrigated agriculture in south Tunisian oases in the last 90 years has been the main cause of a substantial decline in the groundwater table.(26)

In resource-rich areas, ancient irrigated agricultural systems are also breaking down for economic and social reasons. In most cases they depended for their functioning on a hierarchic social system with a high degree of communal resource management, and a large labour force usually including low status social groups specialised in the difficult and dangerous work of tunnel cleaning. The growth of modern industrial operations with high wage rates has attracted a large proportion of the labour force away from these tasks, and the previously rigid social systems that managed communal resource allocation are breaking down under pressure from modern changes. As a result, previously cultivated land in the arid zone, with plentiful water supply, is going out of production not from overuse but from the withdrawal of labour. Thus in two Omani oases, land under cultivation has declined by more than 60 percent in a few years as a result of emigration to the Persian Gulf, and to the subsequent lack of upkeep of the traditional _falaj_ underground water channels.(27)

One important source of agricultural land in some arid areas is the bed of major rivers, such as the Niger or Senegal, which are cultivated as seasonal flood waters retreat. This cultivation can probably be extended safely, although there are often conflicts with pastoral use of flood retreat pastures as a dry season reserve. There may also be possibilities for extending the small-scale irrigation from spirng water, practised in arid mountain regions throughout the developing world.

Another possibility now being explored is irrigation with brackish or moderately saline water. The use of brackish water reduces soil evaporation and plant transpiration, and increases plant resistance to drought; thus it allows economies in water use. However, in practice this saving is usually balanced by the additional water necessary to leach the salts from the soil. Water with up to 3-4 g of salts/litre can be used for irrigation if drainage is adequate. Tunisian experiments on this are well advanced.

Although in general there is room for improvement and extension of irrigated and flood-retreat cultivation in arid lands, the most immediately profitable irrigable land has already been developed; many already irrigated areas now need drainage and other expensive investments. New irrigation schemes will be increasingly expensive, and as a result will only be economically feasible if high value (usually export) crops are grown in them, thus further increasing the imbalance in many arid zone developing countries between domestic food production and export crop production. Indeed water pumped from some deep water tables, or obtained by desalination, may be too expensive to be used economically for agriculture at all, but should instead be used for industry or mining where it brings a much greater return.

New irrigation schemes may also have high opportunity costs. Some of the land already taken for irrigation, and much of that potentially irrigible, is either the best land for rainfed agriculture growing food or is essential dry season or winter pasture for nomadic pastoralists. When this land is irrigated, rainfed farming may not be replaced but rather pushed into marginal land where yields become even more uncertain and environmental damage increases. The pastoralist finds that not only is the newly irrigated area denied him, but also a much larger area of pasture for which the now irrigated area was an essential dry season reserve; this has been a consequence of agricultural development in most dry pastoral areas, for example in the Sahel, the Awash valley in Ethiopia, and in the lowland winter quarters of some of the pastoral groups of the Iranian Zagros mountains.

Runoff Agriculture

Perhaps the greatest potential for arid zone agriculture lies in techniques of runoff agriculture, by which surface or subsurface runoff from a wide catchment area is concentrated onto a small area and used to grow a crop. With the right soil, and rain intense enough quickly to create an impermeable surface crust, from 30 to 50 percent of the rain water runs off as a flood and can be concentrated on one field or around one fruit tree; an added advantage is that the water is pure, without the high salt concentrations common in arid zone well water. Traditional techniques of runoff agriculture were well developed in several places in the arid zone, but with the economic developers' preference for large, highly technical projects, these small-scale traditional methods have, with one or two important exceptions, been neglected. It is time for a reappraisal. Such techniques are inexpensive in land, water, materials, skilled manpower and organisation, they are within the existing technical competence of land-users, and they can easily be integrated with other forms of land-use such as pastoralism, with which they rarely compete for space.

There are two main types of runoff agriculture: surface and subsurface. Subsurface water flow takes place in the alluvial bed of wadis and has long been used by nomadic pastoralists who dig shallow wells in the wadi bed to tap the flow. But there are few examples of it being successfully used for agriculture. The advantage of this type of flow is that the water is clean, having been filtered naturally, that it flows slowly, and above all, that evaporation is reduced to a fraction of that above ground. This type of flow can be tapped by subsurface dams, which raise the water to the surface. This has been done successfully in the northern Sahara. The town of Tamanrasset in the Ahaggar mountains in the central Sahara used to get all its domestic and some of its agricultural water through a _foggara_ tapping the underground flow in a major wadi.(28)

Surface runoff has traditionally been used in a number of ways. The simplest is the cultivation of dune hollows (_grair_) in the Mauritanian Sahara, or the _dayas_ of the northern Sahara. A more evolved version of this technique consists in digging shallow hollows to retain surface runoff, like the _pukios_ of the Chilca valley south of Lima (in which maize, fruit and vegetables are grown); the same technique is used for fruit trees and other crops, for example for figs and vines on Lanzarote in the Canary Isles, and for olives and fruit trees in North Africa.(29) Surface water tanks are also dug in many parts of the arid zone to capture and store surface runoff water for animals: the _r'dirs_ of the Algerian Sahara and the _barkads_ of Somalia. In Iran, water storage cisterns, _āb-anbār_, made of mudbrick with domed roofs to prevent excessive evaporation while encouraging cooling, and fed by surface runoff directed by low banks of earth and stone (often less than 1 metre high) are of ancient construction(30); although used mainly for household water and animals, there is no reason why they should not be used also for small-scale agriculture. Covered cisterns of this sort, or the ancient underground rock cisterns of the Negev, have much lower evaporation losses than open surface water storage.

A more substantial surface water flow occurs irregularly in many arid zone wadis, especially where mountains adjoin a desert basin, as in the Saharan mountains such as the Ahaggar, Ayr or Tibesti, or the Zagros and Elburz chains in Iran. Rainfall in these mountains collects into short violent floods in wadis, most of which is usually lost in the desert plains. Some use has been made of this water by farmers and pastoralists. In the northern Sahara, simple water spreading systems date from Roman times and are still in use, by which simple earth dams divert the flash flood through unlined canals onto agricultural land where wheat and barley are grown. This flood water has the added advantage that it has often collected organic matter from a wide area, so it contains its own fertiliser. In the closed basin delta of the Gash in the Sudan, cultivators belonging to the pastoral Beja group have on their own initiative since the 1930s used flood waters to cultivate cotton; in a good year the cultivated area in this and the neighbouring Tokar delta reaches 200,000 acres.

Recent attempts to develop this sort of small-scale wadi diversion, as at Tcherozerine in Niger, have proved successful. Although the irregularity of wadi flood water, with long dry periods followed by violent floods, creates technical difficulties, the problems can be overcome. The potential is enormous: in the wadi Jizan on the Red Sea coast of Saudi Arabia for example it is estimated that there is enough flood water to irrigate 50,000 acres in an average year.(31)

The most successful and systematic use of surface runoff water for agriculture is the current research in the Negev desert which has recreated ancient (950 BC to 630 AD) farming practices. Each farm consists of the farmland proper and a catchment area from which stones are removed into heaps, and where for best results the soil is compacted by rolling. Runoff water collected in this way is channelled onto the farm, in a farm: catchment proportion of 1:20 to 1:30. Microcatchments ranging in size from 16 to 1000 square metres, each with a small basin on the lowest part are also used to grow fruit trees and Atriplex salt bush as animal fodder. These methods have proved an economically feasible way of turning apparent waste land into productive agricultural land without any additional irrigation. However, much more work needs to be done on the economies of such agriculture. Labour costs of construction of runoff farms appear high, although labour required to maintain the catchments is negligible: 5 man-days per 10 years are enough to maintain a 40 ha watershed, and the labour needed in the micro-catchments is even less.(32) The Negex study area has the advantage of fertile loess soils in the depressions and plains, but the method is probably technically applicable in many deserts of the world; similar schemes are working in Niger, Tunisia and Afghanistan. Much more work is also needed on the social aspects of runoff agriculture, and how it relates to existing production technologies and social relations of production.

Perhaps the greatest advantage of runoff agriculture is that it uses low cost water which has few alternative uses. Although the economics of runoff agriculture (including pasture irrigation) have yet to be investigated in detail in a variety of arid environments, the Negev experiments suggest that the cost of installation, and more important of upkeep, are relatively low. And since this is rainwater which if not used in this way would either infiltrate the soil in too dispersed a manner to grow useful vegetation, or would evaporate, its use for agriculture is not competitive with higher yielding industrial uses.

Water and Rural Development

In some part of the arid zone water is the limiting factor for rural development, either because of absolute scarcity or because of high cost. However this is not universally so, and too great a preoccupation with water can distort development priorities. In addition, water quality is not homogenous, which complicates the task of economic analysis of alternative uses.

In many arid areas, water is not a critical development constraint; indeed water may be present in abundance. Thus in Omani villages, where the fields are watered by falaj tunnels or by wells, the decline of agriculture is due to emigration from villages, which leads to a lack of agricultural labour and to decay of water supply systems. Shortage of water itself is not the cause of agricultural decline. This must rather be sought in the unequal development of the industrial/urban and the rural sectors, so that the latter gets little research or investment; consequently productivity and wages are low relative to the modern sector, and labour moves from one to the other. In most pasture areas likewise, the problem of recent years has not been a shortage of water. Large quantities of water have been provided, at low cost, either by tapping the groundwater table as in the Sahel, Sudan and Botswana, or by collecting surface runoff as in Somalia. The critical unresolved problem of pastoral development is the organisation of grazing around these water points; pasture, not water, is the limiting resource in short supply, and it is a problem that has to be solved by new institutions, especially those of a sort that would encourage greater local community regulation of local resource use.

The cost of water supplies will be an important determinant of new agricultural enterprises. Expensive water, for example from desalination plants, is unlikely to be available for agriculture, except for high value industrial or export crops and even these are unlikely to be competitive with industrial uses. However this is not necessarily true of some of the major groundwater supplies underlying arid regions, which may be neither very expensive to tap, nor sited so as to be of use other than for agriculture or pastoralism. And perhaps more important, most of the types of agricultural development outlined above as being especially suitable for arid areas, such as river flood retreat cultivation, saline water irrigation and runoff agriculture, all use cheap water which is too irregular and scattered, is often in the wrong place and is of too low quality to have ready alternative urban or industrial uses; the opportunity cost of its use for small-scale food crop agriculture is thus low.

Wildlife Conservation

Arid lands everywhere, but especially the Old World arid belt from Mauritania to Mongolia, are an unusual and important set of habitats containing plants and animals which have adapted to the special problems of life with little water and great heat and cold. In particular, the great deserts of the Sahara, Arabia and central Asia and their borders contain a unique large mammal fauna; the most spectacular member of which is the addax antelope of the Sahara. This large animal may never drink water in its life, getting all it needs from the desert vegetation. It used to be common throughout the Sahara, but has now been hunted to the point of extinction.

Addax and other desert and semi-desert mammals such as oryx may have economic significance. The ability of these animals to go largely without water and to use scattered and sparse desert and subdesert pastures makes them potential commercial animals of use in areas where water is insufficient for cattle or camels. (Addax were apparently at least partly domesticated in the Sixth dynasty of ancient Egypt). And these animals could be important tourist attractions in desert national parks. But their significance is not primarily economic. Desert plants and animals form ecosystems of the greatest biological interest because of the variety and elegance of their adaptations to extreme aridity and heat. As such they are part of the world heritage or international commons, natural things and artifacts so valuable that they are of importance to the world as a whole, not just to the nation in which they occur; because of this the world as a whole must pay for their preservation. A first step would be the creation of new national parks and other protected areas. Deserts were left out of the great wave of national park creation in the first half of this century because it was thought that their isolation and size gave them adequate protection. But this has changed: the discovery of oil and minerals (and the search for more), and the new national frontiers guarded by underemployed soldiers have filled deserts with hunters equipped with vehicles and rifles. A few efforts have recently been made: the new Banc d'Arguin national park on the Atlantic shore of the Sahara in Mauritania, the Wadi Rime national park in Chad, new national parks in the Dasht-i-Kevir in Iran and in the Mongolian Gobi. More are warranted as part of an international effort.

NOTES

(1) UNESCO, 1978, Obstacles to Arid Land Development. In press

(2) G.Orshan, S.Diskin, 1968, Seasonal changes in productivity under desert conditions. In Functioning of terrestrial ecosystems at the primary production level. Paris: UNESCO

(3) J.C.Bille, 1974, Recherches écologiques sur une savane sahélienne du Ferlo septentrional, Sénégal: 1972, année sèche au Sahel. La Terre et la Vie 28, 1:5-20

(4) Columns (1) and (2) from Bille, Recherches écologiques; column (3) from Jeremy Swift, 1966, Sahelian pastoralists: underdevelopment, desertification and famine. Annual Revenue of Anthropology 6:457-478

(5) I.Noy-Meir, 1974, Stability in arid ecosystems and the effects of man on it. In Structure, Functioning and Management of Ecosystems. Proceedings of the First International Congress of Ecology. Wageningen: Centre for Agricultural Publishing and Documentation.

(6) B.Lundholm, 1976, Domestic animals in arid ecosystems. In Can Desert Encroachment be Stopped? eds. A.Rapp et al. Stockholm: Ecological Bulletins No. 24.

(7) P.E.Lovejoy, S.Baier, 1976, The desert-side economy of the Central Sudan. In The Politics of Natural Disaster: the case of the Sahel Drought. ed. M.H.Glantz. New York: Praeger.

(8) See for example S.Birks, 1978, The mountain pastoralists of the Sultanate of Oman; reactions to drought. Development and Change, 9, 1, in press; P.Bonte, 1975, Conditions et effets de l'implantation d'industries minières en milieu pastoral: l'exemple de la Mauritanie. In Pastoralism in Tropical Africa. ed T.Monod. London: Oxford University press; D.L.Johnson, 1973, Jabal al-Akhdar, Cyrenaica: An Historical Geography of Settlement and Livelihood. University of Chicago: Department of Geography Research Paper No.148.

(9) C.S.Holling, 1973, Resilience and stability of ecological systems. Annual Revue of Ecology and Systematics 4:1-24

(10) Noy-Meir, Stability in arid ecosystems.

(11) Noy-Meir, Stability in arid ecosystems.

(12) F.Barth, 1964, Nomads of South Persia; the Basseri Tribe of the Khamseh Confederacy. London: Allen and Unwin.

(13) H.S.Mann, S.P.Malhotra, 1976, Biotic factors and desertification in arid zone of Rajasthan. In Problems in the development and conservation of desert and semi-desert land. 23rd International Geographical Congress, Working Group on Desertification, Ashkhabad USSR, July 1976. pp 55-59

(14) D.Western, 1978, The environment and ecology of pastoralists in arid savannahs. In The Future of Nomads in Africa: Modern Change among Traditional Hunter-Gatherer and Nomadic Pastoral Societies. ed. Jeremy Swift. London: International African Institute. In press

(15) C.Humphrey, 1978, Pastoral nomadism in Mongolia; the role of the herdsmen's cooperatives in the national economy. Development and Change 9, 1, in press

(16) Republic of the Sudan, 1974, Southern Darfur Land-Use Planning Survey; Development Plan. London: ODA. International Labour Office, 1977, Economic Transformation in a Socialist Framework; An Employment and Basic Needs Oriented Development Strategy for Somalia. Addis Ababa: ILO/JASPA.

(17) Z.S.Shamsutdinov, 1976, Methods of raising the productivity of desert pastures in central Asia. In Problems in the Development and Conservation of Desert and Semi-Desert Land. 23rd International Geographical Congress, Working Group on Desertification.

(18) N.T.Nechayeva, 1971, Pastures and sheep production in the arid zone of the Soviet Union. In Food, Fibre and the Arid Lands. eds. W.G.McGinnies, B.J.Goldman, P.Paylore. Tucson: University of Arizona Press.

(19) Swift, Sahelian Pastoralists.

(20) R.A.Henin, 1969, The patterns and causes of fertility differentials in the Sudan. Population Studies 23:171-198

(21) Mann and Malhotra, Biotic factors.

(22) N.S.Jodha, 1969, Drought and scarcity in the Rajasthan desert; some basic issues. Economic and Political Weekly 4, 16, April 19: 669-703

(23) A.K.Sen, H.S.Mann, 1976, Exploration, utilization and quality of groundwater in Rajasthan. In Problems in the Development and Conservation of Desert and Semi-Desert Land. 23rd International Geographical Congress, Working Group on Desertification.

(24) M.Clawson, 1971, Social and institutional obstacles to agricultural development of arid lands in the Middle East. In Food, Fibre and the Arid Lands. eds. W.G.McGinnies, B.J.Goldamn, P.Paylore. Tucson: University of Arizona Press.

(25) UNESCO, Obstacles.

(26) H.Flohn, 1971, Investigatins on the Climatic Conditions of Advancement of the Tunisian Sahel, Geneva: World Meteorological Organization.

(27) J.S.Birks, S.E.Letts, 1977, Diqal and Muqayda: Dying Oases in Arabia. Tijdschrift voor Econ. en Soc. Geografie 68, 3:145-151

(28) G.Drouhin, 1962, The possibility of utilizing minor water resources in Algiers. In The Problems of the Arid Zone: Proceedings of the Paris Symposium. Paris: UNESCO.

(29) UNESCO, Obstacles.

(30) E.Beazley, 1977, Some vernacular buildings of the Iranian Plateau. Iran

(31) G.F.White, 1962, Alternative uses of limited water supplies. In The Problems of the Arid Zone: Proceedings of the Paris Symposium. Paris: UNESCO.

(32) M.Evenari, L.Shanan, N.Tadmore, 1971, The Negev; The Challenge of a Desert. Cambridge: Harvard University Press; and idem, 1971, Runoff agriculture in the Negev Desert of Israel. In Food, Fibre and the Arid Lands. eds. W.G.McGinnies, B.J.Goldman, P.Paylore. Tucson: University of Arizona Press.

* * *

IV. DEVELOPMENT OPTIONS OF RESOURCE-RICH DESERT ECONOMIES

Resource -- rich desert countries are faced with special problems arising from the very nature of the process which has given them a leadership and demonstration role in relation to the rest of the Third World:

(i) In so far as these resources are typically depletable resources, as distinct from the typically renewable resources of many of the non-desert areas with considerable agricultural and industrial development, the desert countries are faced with special problems regarding time factors. The problem consists in making sure that by the time the oil or other depleting resources are either physically exhausted, or possibly economically obsolete, the economy will have been put on a basis where development can be self-sustaining even without the original exogenous resource. (It is here that the previous development models retain their validity even in relation to OPEC type countries.) This will normally require agricultural and industrial development which will be easier for the countries which are largely or partly non-desert countries. In the fully desert countries, it will raise special problems of adapting agricultural and industrial technologies for permanent production activities to desert conditions. Even in the former group of countries, it will raise problems as to whether the permanent activities should be in the desert areas or outside them. The exploitation of oil or other depletable resources in desert areas will inevitably involve the creation of a certain infrastructure of productive activities and will equally inevitably attract some population to the desert areas even if capital-intensive technologies are used. The centres of activities thus created will turn into 'ghost towns' unless permanent activities are initiated in the desert areas themselves. In the US this phenomenon of ghost towns is well-known in the mining areas of some of the western States; there it was accepted without creating special problems because there was free movement with the rest of the economy and the deserted areas were only a small part of the whole economy. Some of the desert countries would be in a similar position as the US in this respect and hence could contemplate the creation of ghost towns in desert areas with some equanimity. But others are not.

On the other hand, the population centres and infrastructure being built up in resource-rich desert areas could become the nucleus of more permanent activities, particularly if some advance thinking and advance planning is given to combining present activities for resource exploitation with more lasting agricultural, industrial and other activities.

(ii) As has been pointed out above, depletable resources add a special time dimension to development programming. For this purpose, it becomes necessary first to form an estimate of the total available reserves, at various cost levels of exploitation. In the light of uncertainties and of new information and techniques becoming available, this estimate must be constantly revised and brought up to date. It is then a matter of cost/benefit calculation to determine the optimum rate for the country concerned of depleting the resources involved. A number of factors have to be taken into account in such a cost/benefit calculation:

a) The absorptive capacity of the country, i.e. the rate at which foreign exchange revenues can be currently converted into development potential. Current use and developmental use are not identical. In actual life, a part - often a very considerable part - of the oil revenue is currently used, but not for development purposes. It may be used for importing armaments, or for paying out subsidies to friendly countries, or paying debts or profits to foreign investors or supporting international aid programmes,[1] etc. Although cost/benefit analysis is geared to developmental use of revenues from the depleting resource, the introduction of non-developmental expenditures does not in a formal sense introduce any new element into the cost/benefit calculation. Other things being equal, a country which has a higher absorptive capacity of converting revenues into current developmental use will consider itself justified in increasing the rate of resource utilisation and depletion, provided the development thus financed is of a self-sustaining character once the resource is exhausted. It is not this point that developmental and non-developmental uses are different: non-developmental uses will not normally lead to self-sustaining development (quite on the contrary, armament expenditures may involve a country in continued dependence on further armament imports of maintenance, replacement, training in use, etc.). However, the reverse is not necessarily true. Industrial development, for example, is not necessarily self-sustaining if the industries built up are high-cost, inefficient capital-intensive industries depending on imported raw materials and components. Countries with depleting resources must be especially concerned about the long-term pattern of their development as well as the immediate quantitative requirements and absorptive capacity.

b) Failing immediate current use, a country with depleting recources still has an option of either leaving its resources in the ground by having a low rate of utilisation, or else investing their proceeds abroad as income-earning assets. This will not be an easy calculation to make. Arguments for leaving reserves

---

[1] In this case, the use may be defined as developmental, but not in respect of the resource-rich country itself.

in the ground would include the political, economic and strategic power which possession of scarce resources may give in the future as well as the possibility that the price of the resource may increase in the future as it gets increasingly scarce. As against these prospects, there is the risk that the resources may become obsolete or fall in price as substitutes are developed and/or competitors discover additional reserves of the same resource. This will obviously be a difficult calculation to make, requiring bold assumptions about the future. Similarly, on the other side of the account, there are conflicting considerations about the value of foreign investments of surplus revenue: If the investments are wisely made their income-earning capacity can be very high and they could also give considerable political and economic influence to the country owning these investments. The latter point also applies, if the surpluses above current use requirements are held in the form of high currency reserves. On the other hand, unwise investments may yield little income and lose their value with inflation or other eventualities. They could even provide political or economic leverage against the investing resource country. In consumption oriented societies with highly unequal distribution of income, foreign investments may be deliberately made for purposes other than maximising subsequent income for the country; e.g. the holding of Swiss banking accounts, or prestige investments in buying up real estate, etc. Even from the few factors mentioned it will be seen that the balance between using reserves or leaving them in the ground in the absence of current use, can fall in different directions depending on the assumptions and motives of decision-makers.

c) When a country moves rapidly from poverty to overall wealth as the result of the exploitation of a valuable desert resource, and where in addition incomes may be very unequally distributed - these two factors can in turn be functionally interrelated - there is a likelihood that much of the current use made of depleting resources is consumption oriented rather than development-oriented. Where this additional consumption takes the form of a higher standard of living, including better health and education provision for a previously very poor population - including the nomadic or previously nomadic desert population - such consumption expenditures can be justified on grounds of reducing poverty and can even be productively justified in terms of improving capacity to produce. However, inequality of income distribution will often result in considerable imports in items of luxury consumption, combined with building up an industrial structure - including also agricultural and services structures -geared to the needs of a limited high-income group and combined with a pattern of investment abroad which benefits individual wealth owners rather than the development of the country as a whole. In such cases there is a divergence between the results of private and social cost/benefit analysis.

d) The principle of calculating an optimum pattern of resource depletion is easily defined in the formal sense, i.e. to maximise the present value of total future output. However, this conceals very formidable difficulties because in order to calculate the present value of future outputs and revenues we must use some kind of social discount rate. If a high rate is used, i.e. a short time horizon is applied and future outputs and revenues are heavily discounted, a different result will be obtained than in the case of using a low discount rate, i.e. where decision-makers apply a long-term and intergenerational point of view. There is no firm basis on which to advise decision-makers what the appropriate rate of discount should be. Ultimately, this depends on the value attached by them to the welfare of future generations. Uncertainties about the future will normally be reflected in a high discount rate. A country with low reserves in relation to current use can presumably use a higher discount rate than a country with large reserves relative to current use. In this respect Iran and Saudi Arabia are obvious illustrations of two different situations.(1)

e) The producers of scarce, depleting resources, especially, of course, oil, when deciding on their price policy may also differ on the importance which they attach to the protection of the economic system of the consumers on whom they depend in the long run for their revenue. This has several aspects. One relates to the fact that the consumers may be very poor Third World countries importing the oil, or other scarce resource. In this case the resource countries may try in various ways to protect their poor customers, either through price rebates or dual pricing (which OPEC has so far, however, declined to do), or else through aid from the resource countries (which OPEC has accepted as an obligation), or by permitting in other ways the recycling of their revenues to poor countries through the IMF, the World Bank or the new International Fund for Agricultural Development. Another aspect relates to their wealthier customers in the industrial countries. Here also, the high price, based on scarcity and cartelisaion of a scarce resource, may create serious economic difficulties and even undermine economic viability: a country like Saudi Arabia is clearly concerned, partly because it has a political interest in not undermining the Western economies, partly because it depends on their prosperity for useful and profitable investment of surpluses. Superimposed on their differences with Iran, i.e. higher reserves and lower immediate absorption capacity, this has led to well-known divergences on price policy between these two countries (and, more general, within OPEC) in recent months.

f) Resource-rich desert countries would be naturally inclined and motivated to adopt highly capital-intensive technologies, in spite of the potential and need for more labour-intensive technologies, argued in the preceding sections for both agriculture and industry. This is due to a number of reasons:

1. The big surplus revenues mean that there is no capital constraint as there is normally with Third World countries.

2. The scarce population in the desert areas means that, normally, there is no perceived <u>rural</u> surplus population, at least not in the desert areas themselves. This perception may have some reality in countries such as Saudi Arabia, Libya or Mauritania, among the case studies presented above, but not in the cases of Algeria and Iran which have considerable surplus-population problems. Moreover, even in the arid and desert areas, there may be surplus population in the more relevant economic sense of very low marginal productivity of labour.

3. The desire to produce speedy development before the resource is depleted also favours capital-intensive investments, particularly in the form of turn-key factories. By comparison, the development of indigenous labour-intensive technology is a much slower process, and even where such technologies exist the organisation of projects on a labour-intensive basis will usually take much more time.

4. Capital-intensive technologies may also commend themselves to resource-rich desert countries as a means of increasing their imports from industrialised customers and hence of recycling their surpluses, avoiding balance of payments crises for their customers.

5. Finally, capital-intensive technologies may be needed because of low levels of training, education and skills among the local population, possibly added to by special problems applying to the nomadic population, and partly to

the tremendous distance, in cultural, psychological and social terms, which separates modern industrial production from desert civilisation.

The above five reasons combine in explaining the capital intensity of the technologies which we find in practice in the countries concerned. To this is added the fact that the processing of the resources involved is itself also usually highly capital-intensive, e.g. oil refining and petrochemicals. The adoption of large capital-intensive technologies, of course, contains considerable dangers for the desert resource countries. There may be a lack of employment and domestic linkage effects, so that the population becomes partly dependent on social distribution of revenues from resources rather than on employment. This has been described as a 'Rentier State'.(2) There is also a risk that development based on highly capital-intensive technologies will fail to be self-sustaining, lead to continued technological dependence, and leave the country in a very vulnerable situation when the resources are depleted, or when revenues fall for other reasons.

g) As the case studies for Algeria and Iran make clear, countries basing their development on depleting resources from desert areas are under particularly strong pressure to discover new resources to take the place of depleting resources. These new resources can either be additional deposits of the same resource, e.g. new oilfields, or they can be other mineral resources whether arid areas tend for geological reasons to be mineral rich or not, the large area of most desert countries makes the probability of discovering a large endowment of such resources relative to population and GNP. high.

The additional revenues now obtained as a result of higher prices, as well as the increasing rate of depletion and shortening of the expected rate of duration of presently known resources should give strong additional incentive to full surveying. For instance, it is a striking fact that in the case of Saudi Arabia, while the presently known and accessible reserves would last for 35 years if production were stabilised, they would be exhausted in less than 20 years if output increased by 10 per cent each year.

Employment and Industrialisation Problems of Desert Economies

As the case studies of employment and industrialisation show,(3) the balance of argument regarding emphasis on agricultural versus industrial development comes down strongly in favour of investment in agriculture in desert economies. (In resource-rich desert economies this does not mean that investment in agriculture precludes simultaneous industrialisation.) In the long term the improvement of agricultural productivity will leave available a greater proportion of foreign exchange resources which will otherwise, given rising personal incomes and fast rates of population growth, be spent on imported food; hence the country will be in a stronger position if and when revenue from exogenous resources comes to an end. The short-term importance is twofold: as a way of creating employment and of increasing direct and derived demand for nascent domestic industry.

The employment consideration is of great importance even in presently labour shortage economies. The development of desert mineral resources (or receipts from other exogenous resources) is likely to lead via the wage leadership initially of the extractive industry and then of the public sector, to a general rise in expectations of well-paid employment on the part of members of the rural labour force, and to large-scale movement off the land to the towns. With oil revenues of $1000 per capita and 20 per cent of the labour force of foreign origin, testifying to a severe overall shortage of labour, Libya is nevertheless described as having "a surplus of unskilled labour. Most of the workers....have migrated from the country to the city."(4) Exogenous resources are liable to transform labour-short into labour-surplus economies, and to underwrite and perpetuate misallocation of certain types of labour in this type of economy. Two measures are required to prevent the institutionalisation of continually rising expectations, over-rapid urbanisation and the nominal absorption of labour in well-paid but unproductive activities. First, and crucially, a large wage differential should not be allowed to emerge between exogenous resource based and traditional activities. But, secondly, in view of the fact that the exogenous revenues do represent a real increase in national wealth, some increase in individual consumption should be permitted in poor desert countries. In order that this should be done equitably to benefit all sections of the population, agricultural incomes must be able to rise. Unless the carrying capacity of semi-arid lands can be raised this will not be possible, and the process of dynamic misallocation of labour and absolute decline in the land resource base of the economy will set in.

Wherever the attempt is made to set up an integrated domestic industrial structure, agricultural and rural incomes have to rise to provide local demand. The case studies examine industrialisation strategies which differ in respect of the precise role of the State (though not in its massive involvement, corresponding to the fact that it is the recipient of the export income from minerals), and of the sequence in which particular industrial activities are introduced. In neither respect is one pattern obviously more or less appropriate to industrialisation in specifically desert conditions. But it is shown that in each case the time has come where full utilisation and successful further development of the industrial sector depends on mass local effective demand including demand from the agricultural and rural sector.

The comprehensive integrated industrial sector is not likely to be a universal objective in desert States. Iran and Algeria both have relatively large populations and Iran in particular has a substantial traditional small-scale manufacturing sector, out of which many of the 'modern' private enterprises have grown. Resource-rich countries without these advantages might do better to concentrate on processing locally their mineral resources and using them as fuel and raw material for the production of steel and other products for export to industrialised economies; to exploit, when the technology becomes available, their inexhaustible attributes of sun, wind and tides for energy generation to provide water and perhaps also for export; to capitalise on traditional commercial skills for international entrepot trading; to develop a high level of skills among the population; to underwrite equitable and generous welfare systems; and last but most importantly, to concentrate on improving the proven but poor traditional pastoral nomadic systems and methods of cultivation since the two are closely interdependent.

NOTES

(1) See the case studies of six desert economies (Section II).
(2) by H.Mahdavy, in his contribution to M.A.Cook (ed) *Studies in the Economic History of the Middle East from the Rise of Islam to the Present Day*, Oxford University Press 1970.
(3) Section III below.
(4) UNIDO (1976), p.174. See footnotes to Section III.

FOOD SUPPLY AS AN ESSENTIAL ELEMENT IN DESERT MANAGEMENT

Marlyn L. Shelton
Assistant Professor of Geography
University of California, Davis
Robert M. Pratt
Assistant to the Director
California Department of Food and Agriculture, Sacramento

ABSTRACT

The ability to provide an adequate food supply for the population will be the ultimate standard for measuring the success of any long-term desert management strategy. A range of alternatives for supplying food is available, but these alternatives represent different combinations of reliance on food provided by on-site production, food shipped from another part of the country, or food purchased on the world market and imported. The strategy adopted by a given country will depend on the geographic and economic characteristics of that country, but it must be clearly recognized that improved nutrition does not inherently accompany the development of an industrial economy. An improved food supply requires explicit and comprehensive treatment as an integral part of a desert management plan.

Fundamental elements that influence food supply decisions and which should be considered in planning for the development of arid lands are population size, environmental limitations of on-site food production, energy and monetary costs of importing food, and the impact of the failure of a non-agricultural resource on the food supply system. An adequate food supply for a low density population may be provided by traditional desert agricultural practices, but high density populations in the desert require the introduction of modern intensive agriculture, a reliance on imported food, or a strategy incorporating both approaches. The selection of crops that can be produced at a given site in the desert and the productivity of the land are influenced by biophysical constraints represented by climate, the water supply, soils, terrain, natural vegetation, and pests and diseases. However, deserts possess significant advantages for irrigated agriculture that can be beneficially exploited to achieve high productivity.

Shipping food produced in remote parts of the country to population centers or importing food increases the energy and monetary costs of the food supply. Fossil fuels are a signigicant energy source used directly and indirectly in agriculture, and increased energy costs increase food costs. Furthermore, the imbalance between world food demand and production is widening, and this suggests that the price of food on the world market can be expected to increase as the supply decreases. If mining, petroleum, or tourism is providing the capital to underwrite a food purchase program, failure of the resource may make food purchases impossible even if food is available on the world market. The complex interaction of factors influencing food supply decisions demands that a desert management strategy recognize the necessity for providing a continuing source of food.

INTRODUCTION

Providing an adequate food supply must be the first objective in designing strategies for desert management because the success of long-term development plans for arid lands will ultimately depend upon the success achieved in feeding the population. The precise manner in which the food supply is provided is likely to assume various forms, but each alternative will represent different combinations of reliance on food provided by on-site production, food shipped from another part of the country, or food purchased on the world market and imported. The strategy adopted by a given country will depend upon the geographic and economic characteristics of the resource endowments of that country and upon the priorities dictated by variable social and cultural considerations. The desert areas should not be regarded as space to accommodate an expanding population. It is necessary for land to support people, not merely to provide space for them. The purpose of this paper is not to advocate a particular strategy for providing food, but rather to advocate that consideration of how the population is to be fed must have high priority in desert management planning. Other papers presented during these meetings will discuss details of water conservation, drainage and other practices necessary for desert agriculture.

Present trends in economic, social, and cultural sectors in arid lands reflect a desire for higher standards of living and for sharing the world's most advanced technologies and economies. These aspirations are interrelated in that a higher standard of living assumes income growth and improved nutrition, and these goals may be achieved through adoption of advanced agricultural and industrial technology. However, accessibility to food supplies required to provide improved nutrition does not automatically accompany the adoption of new technologies or the development of industrial or mineral-based economies. Examination of the evolution of the world's industrialized economies reveals that these nations developed a strong agricultural base first. Agriculture must have the ability to support an expanding population so as to free an increasing number of people to pursue industrially oriented tasks. This historic pattern of development is clearly established, and there should be no mistake in recognizing that the availability of a food supply supports the development of industry rather than the presence of industry generates the food supply. It is clear that improved nutrition as an essential in a higher standard of living in arid lands cannot be assumed as a benefit of developing an industrial economy, but that is must be achieved through direct attention to the character of the available food supply.

The world's desert nations are characterized by diverse environmental conditions and by a wide range of alternative means for supplying food for their inhabitants. However, some common factors with respect to the food supply are identifiable which represent broad fundamental considerations, and these features merit particular attention in planning for the development of arid lands. Therefore, the following discussion will focus on population size, environmental limitations of on-site food production, energy and monetary costs of importing food, and the impact of the failure of a non-agricultural resource on the food supply system. These topics are suggested as essential considerations in any desert management plan because of their potential influence on the food supply.

POPULATION AND THE DESERT FOOD SUPPLY

In planning for the food supply required by the development of arid lands, certain assumptions are necessary concerning the population of the area. Probably the most important consideration is whether the population will remain stable or whether it is expected to increase. The significance of this assumption is that it will largely dictate whether traditional agricultural methods involving hunting and gathering, grazing, and oasis agriculture can be expected to provide the needed food supply or whether alternative sources of food must be sought. In general, traditional agricultural methods, with some modification, may provide an adequate source of food if population is held constant, but it is inconceivable that the traditional agriculture of arid lands can provide the food required to support a growing population.

The population that can be supported by traditional desert agriculture is limited because hunting and gathering, grazing, and oasis agriculture are most successful in low population density settings. The population that can be supported by these methods is variable depending upon local conditions, but, in general, as the region becomes more arid and the human community that can be supported becomes smaller and the territory occupied by the community becomes larger. The population density of the Kalahari Bushman, for example, is less than two persons per square kilometer (Peterson, 1970). Traditional desert agriculture and pastoralism appear to rule out large populations, and to require that population sparsity be an inevitable condition for arid areas where these practices persist as the dominant food production method.

A growing population will require greater amounts of food than can be produced by the limited capacity of extensive agriculture or herding. Therefore, in arid regions experiencing or anticipating an expanding population, traditional agriculture can be expected to provide a limited input only and alternative means of providing the major portion of the food supply must be identified and developed. The use of intensive agriculture, a reliance on imported food, or a strategy in which both sources are used to a varying degree will have to be developed and evaluated on the basis of the physical, human, and economic characteristics of the region. None of these basic food supply sources can be developed without incurring certain costs in the physical, human, and economic realms, and these costs make the decision-making process extremely complex. Selection of the food supply strategy appropriate for a given area should be guided by the geologic, geographic, and climatic characteristics of the region. These traits define the basic resource endowments that are available for food production, and they determine largely the relative advantage of intensive agriculture or food importation as a principal food source for a region.

BIOPHYSICAL CONSTRAINTS ON FOOD PRODUCTION IN THE DESERT

Intensive agriculture in arid lands may be practiced in immediate proximity to the population center, or it may be pursued in a distant part of the country from which the food is shipped to the population center. In either case, the desert environment imposes biophysical constraints that impair man's ability to develop and to maintain intensive agriculture, which in the desert implies irrigated agriculture. The major biophysical constraints on human food systems are climate, soils, terrain, natural vegetation, and pests and diseases. These constraints limit the types of crops that can be produced at a given site and the amount of food available for human use (Duckham, 1974). In the desert, the first two constraints are particularly apparent.

The deficiency of water, excessive evaporation, and the limitation of soils suitable for agriculture in the desert are limiting factors for large-scale irrigated agriculture. The most apparent conflict between the desert environment and irrigated agriculture is that water supplies are inadequate in the desert but irrigated agriculture is a high water-use technology. This mandates the use of water conservation measures in agriculture. Waste waters from urban areas should be reclaimed for reuse to the fullest extent possible.

Rainfall is the basic water supply, but in the desert it occurs in small amounts, is infrequent, and is unpredictable. Water provided by surface or ground water sources is often found in restricted areas, such as oases or river valleys, and these sources may experience seasonal or irregular fluctuations in the amount of available water. A total absence of water during some periods of the year, or for an extended period, is not uncommon.

The meagerness of desert precipitation is a concern for agriculture, but the uncertainty concerning when and in what amounts rainfall will occur is a problem of equal concern. It is generally recognized that as the amount of annual precipitation decreases the variability or uncertainty of precipitation increases. In humid climates the average annual deviation of precipitation does not exceed 10 to 20 percent, but in the arid zone it reaches 30 to 40 percent and actual deviations may be so large as to be devoid of any statistical meaning (Nir, 1974). Average annual precipitation in the desert may be a nearly meaningless statistic, also because of the great annual variability of precipitation. Two examples illustrate this point. In the Sonoran, average annual rainfall at Yuma, Arizona, is 85 mm, but Yuma has received as much as 235 mm and as little as 7.2 mm of precipitation in a given year. In the Atacama, the average annual precipitation from 1918 to 1924 at Trujillo, Peru, was 5.2 mm, but in March 1925 Trujillo received 350 mm of precipitation. Rainless periods of long duration in the desert further erode the meaningfulness of average annual precipitation. In the Sahara and the Atacama, dry periods have occurred with durations of eight and eighteen years, respectively, and shorter periods of extended dryness occur with great frequency.

The irregular spatial distribution of desert precipitation is an additional environmental limitation which impairs the use of rainfall as a dependable supply of moisture for intensive agriculture. Areas that receive the majority of their rainfall from summer thunderstorms are particularly afflicted by this problem. For example, twelve separate storms which occurred in the Sonoran and covered an area of 300 square km produced rain over an area of 7 square km only (Nir, 1974). Orographic influences produce significant spatial disparities, also. The 1000 m high coastal mountains in northern Cyrenaica, Libya, receive 600 mm of annual precipitation, but 200 kilometers west along the coast of the Gulf of Sirte there are no coastal mountains and annual precipitation is about 100 mm (Peterson, 1970).

Where perennial streams or rivers occur in the desert, surface water is more reliable than precipitation as a source of water for agriculture. However, intermittent streams that flow only after rain producing storms are more common than perennial streams, and the intermittent streams share the unreliability of rainfall. Perennial sources tend to be large exotic rivers, such as the Nile, the Euphrates, and the Hwang Ho, and a real accessibility to water from these sources is relatively limited. The short length of smaller perennial streams limits the area used by agriculture that might benefit from water withdrawn from these sources. In Libya, for example, the longest perennial stream is only 4 kilometers long (Jones, 1966*).

Ground water provides an important, and usually dependable, source of water in all desert areas. Ground water may represent remnants of water stored during earlier, more humid times, or it may be water that is periodically recharged by infrequent rains or streams that discharge onto coastal plains or onto interior desert floors. Significant utilization of ground water for agriculture usually requires the mining of some fossil water reserves, since few desert regions receive sufficient precipitation to provide ground water recharge each year (Peterson, 1970). However, many of the aquifers possess such vast dimensions and store such large quantities of water that yearly recharge may be a small concern (Drouhin, 1970). In the Sahara, the Continental Intercalary and the Nubian sandstone contain aquifers of tremendous importance, and the Continental Intercalary receives periodic recharge from the Saharan Atlas. The eastern side of the Arabian Peninsula is underlain by sedimentary strata of late Cambrian to Tertiary age and extensive aquifers in these formations are restored by runoff from Al Hijaz Asir (Peterson, 1970).

Springs provide natural access to ground water and wells or ganats may be dug to intercept an aquifer. Ground water reservoirs at great depths can be utilized if the necessary geologic knowledge and the capital for investing in modern drilling equipment are available. In the Algerian Sahara, wells draw water from the Continental Intercalary down to a depth of 1500 meters (Drouhin, 1970). Desert ground water may be highly mineralized, especially when drawn from deep wells, and caution should be exercised in developing these sources for agriculture until the quality of the water is known. An additional consideration in the development of ground water sources is the infrastructure required to fully utilize the resource. Even where known ground water sources occur, it may be difficult to establish agricultural systems due to an absence of transportation and marketing networks, energy supplies, and an educational system to disseminate agricultural information.

For agriculture, the problem of limited water supplied in the desert is compounded by high rates of evaporation. The climatic controls responsible for the delivery of scanty precipitation to arid lands also produce little cloudiness, intense solar radiation, high temperatures, and low relative humidity, and these conditions cause excessive evaporation. Evaporation is a complex natural process that is a function of available energy, vapor pressure differences, and wind velocity, and the combination of these elements in the desert produces evaporation potentials of extreme magnitudes. In the central Sahara, annual evaporation exceeds 5 meters when water is constantly provided to permit the process to proceed at the potential rate.

High evaporation rates in the desert mean that the benefit to agriculture derived from one cubic meter of water in arid climates is less than the benefit produced from that amount in humid lands. Water losses from conveyence systems due to evaporation and direct evaporation losses from water applied to crops are both greater in arid environments than in humid climates. The net effect of high evaporation rates in the desert is that greater amounts of water are required in arid lands than in humid lands to achieve the same irrigation objective. Technological advances, such as drip irrigation, have helped to reduce evaporation losses from irrigated lands and have achieved increased farm productivity without increasing the amount of irrigation water needed. Another area of technological advance is the use of reclaimed water from urban sewer systems and processing plants. Reuse of such water may make possible a limited agriculture near population centers in the desert. Use of reclaimed water for industrial purposes, such as cooling, could make other waters available for agriculture.

Deserts represent a land bank of 20.8 million square kilometers of which 8.10 million hectares are considered to be potentially arable land (Thorne, 1970). Soils in these areas have developed under conditions of low rainfall and high temperature, and their texture and structure are dominated by the products of mechanical weathering. Limited leaching has left arid land soils rich in most of the biologically essential nutrients, except nitrogen, but is has caused many of the soils to be affected by direct or latent salinity. In some sections of the Algerian Sahara, and possibly elsewhere, deep and well-developed fossil soils produced during more humid conditions are found under a few centimeters of gravel (Drouhin, 1970). In either case, these soils require irrigation for effective crop production, but the application of irrigation water produces drainage and soil salinity problems.

Close attention must be given to the drainage requirement of irrigated desert land to avoid the hazards of waterlogging and salinity. Excessive application of water is a common mistake in irrigation farming in arid lands, and the excessive water gives mobility to greater amounts of minerals. The salinity problem is aggrevated by waterlogging of the soil which occurs easily in the flat and poorly drained desert plains used for irrigation farming. Between irrigation applications, high evaporation rates create an increased upward capillary movement of water in the soil, and evaporation of the water concentrates the precipitated minerals in the dry soil. Waterlogging and salination can be combatted by efficient application and proper temporal spacing of irrigation water, but full-scale drainage must start concurrently with irrigation. Sufficient water must be applied to move excess salts from the root zone to the drains.

Although intensive irrigated agriculture in the desert is faced with limitations imposed by the physical environment, human obstacles may be even more difficult to overcome than the physical limitations. Replacing traditional farming methods with irrigation agriculture will require the introduction of an economic system and modern cultivation technology that may not be compatible with existing social and cultural traditions. Modern irrigation farming requires purchased inputs which are not usual in traditional farming and which introduce an economic risk factor that traditional farmers may be reluctant to assume. To achieve maximum benefit, irrigation technology cannot simply to transferred to new areas of the desert but rather it must be assimilated into the prevailing environment. To a great extent, traditional farming in the desert is based on adaption to the environment, but irrigated farming is founded on managing the environment. The transition to irrigated agriculture cannot be expected to proceed without producing significant changes in the social structure.

Noting the environmental limitations of irrigated agriculture in the desert is not intended to imply that only negative considerations are present. Deserts, in fact, possess significant advantages for irrigated agriculture that can be beneficially exploited to achieve high productivity. One irrigated hectare can produce the yield of several hectares with sufficient moisture, and a combination of irrigation and fertilizer can increase multiple cropping and bring unused land into production.

The abundant sunshine and low humidity provide a prolonged growing season and they provide unique advantages for scientific and mechanized agriculture. The low precipitation means that farm operations are not disrupted by wet ground, and precisely timed use of diversified equipment and farming practices is possible. The scientific regulation of water essential for the production of high yields and high-quality farm products can be achieved because of the limited rainfall. The chemical treatment of crop insects and diseases can be timed with accuracy and applied with assurance that the materials will not be washed off the plants before they are fully effective. Finally, harvesting schedules can be determined by crop maturity without regard for potential disruption due to wet weather (Thorne, 1970).

Using the fragile desert environment for irrigated agriculture continuously without destroying it is a complex problem which requires more comprehensive planning and management then would be required in other areas. The planning process for arid lands must recognize the limited resources of deserts and their limited range of alternatives. The conversion of deserts to intensive agriculture requires careful appraisal of regional and local environmental conditions including the social customs and competence of the population to adapt to a new lifestyle, if the project is to succeed. Initial appraisal of the environmental implications of the project must be followed by reevaluation at each stage of planning and implementation and this means that deserts can rarely be expected to lend themselves to fast development or to quick technological changes (Amiran, 1970).

TRANSPORTING THE DESERT FOOD SUPPLY

A decision to transport food to the population from the area of production, either within or outside the country, should be preceded by a thorough examination of the costs and consequences involved. In general, it can be assumed that shipping food to the population will create greater energy and monetary costs than would be experienced if the food was produced in proximity to the population. This assumption is particularly valid for deserts where habitable land is found in limited areas only, and these areas are separated by large expanses of inhospitable land that are largely unoccupied. This discontinuous settlement pattern is a serious obstacle for the creation of an efficient and economic modern regional infrastructure of roads, railroads, powerlines, and services necessary to provide economical transportation networks, storage facilities, and processing plants for distributing food in the desert.

A reliance on food transported to the population places an increased dependence on the expenditure of energy for food, but energy utilization for this purpose is staggering already. Energy expended in food processing, distribution, and cooking is much greater than the energy expended in food production in both traditional and modern human food systems (Revelle, 1976). On a worldwide basis, nearly 25 percent of all energy goes into the human food system, and the percentage increases to more than 60 percent for the developing countries (Pimental, el al., 1976).

The use of fossil energy is an integral component of the energy-cost increase associated with shipping food. Fossil energy is one of the prime resources used in agriculture for pumping irrigation water, for planting, cultivating, and harvesting crops, and for transporting and processing farm products. In addition to the energy applied directly in farming, fossil energy is used in manufacturing farm tools, machinery, and chemical fertilizers, and in transporting these farm inputs. Escalating usage of fossil energy has created dramatic increases in world prices for these commodities, and the increased energy costs have resulted in increased monetary costs for producing and shipping food. The relatively long distances that food must be shipped in the desert means that increasing transportation costs will be an especially acute factor in supplying food in arid lands. The importation of food shares the energy and cost problems related to transportating food produced within the country, but transportation costs are even greater for imported food.

When considering imported food as an alternative, a concern with potentially greater implications than the cost of transportation is the question of the availability and cost of food on the world market. Today, world food needs are greater than at any time in history, and the imbalance between food demand and production is widening. Except for the United States, Canada, Australia, New Zealand, Argentina, and Thailand, all nations are consistent net food importers (Pimentel, et al., 1976). This global demand for imported food creates keen competition on the world food market as surplus food supplies are reduced, and the food importer must be prepared to pay increasingly higher prices.

It appears unlikely that the disparity between global food demand and production will be reversed. For example, cereal grains provide a significant percentage of the dietary protein for the world's population; yet most countries import grain from North America and the degree of dependence has increased rapidly over the past decade. Outside of North America scores of countries have become important food importers during the past quarter century, but not one new country has emerged as a major cereal exporter. One impact of this pressure on the grain exporters is that reserve stocks of grain are declining. In 1961, reserve stocks were equivalent to 105 days of world grain consumption, but reserves in 1976 were reduced to 31 days of consumption (Brown, 1975). These trends suggest that a world of cheap, abundant food may now be history. Consequently, a food supply strategy that relies on imported food is obligated to contain a provision for meeting increasingly higher food costs, and it must provide an alternative food source in the event imported food is unavailable.

RESOURCE FAILURE AND THE FOOD SUPPLY

Considering food importation as an alternative food supply strategy in the desert assumes that capital is available to underwrite a food purchase program and that the purchasing power has been derived from economic development in the arid land. Economic growth which produces capital for purchasing imported food may be supported by the development of any of the natural resources that compose the resource endowment of the desert. Agriculture represents one alternative, but minerals, petroleum, weather, and vacant space may serve as the basis for area development and economic growth as well. The implications of the failure of the natural resource cannot be disregarded in desert management planning, especially as they pertain to impairment of the capacity to continue feeding the population of the developed area. A desert management strategy designed around the development of a natural resource is not complete unless it addresses the problem of what happens to the population and how will they be fed if the resource fails.

The mineral, petroleum, and amenity resources of arid lands are in great demand, and area development based on these non-agricultural resources is an attractive alternative in desert planning. However, non-agricultural resources are particularly vulnerable to the problem of resource failure due to either exhaustion of the resource or to a serious decline in the market price of the resource. Minerals and petroleum are finite resources, and they may be exhausted even though the most rigorous conservation measures are employed. In a similar fashion, amenity resources may be depleted through utilization in that extensive development of tourism may destroy or damage the natural characteristics that make the desert setting attractive for tourists.

In the desert, ghost towns and mining settlements with shrinking populations reveal the impact of resource exhaustion, the related decline in economic conditions, and the loss of ability to maintain a food supply. The desertion of mining settlements after the resource is exhausted appears to be more drastic in the arid zone than in humid climates. Good illustrations of ghost towns and dying settlements related to mining failures are found in the arid zone of the United States where gold and silver mining was widespread in the late nineteenth century and in the Chilean desert where gold, silver, copper, and nitrate mining were pursued. These examples may foretell what lies in the future for mining settlements like Kalgoorlie and Broken Hill in Australia where gold and silver, respectively, are mined; for Fort Gouraud in Mauritania where iron ore is mined; for the large town being developed in Niger at the foot of the Air for mining uranium. Contemporary mining centers, such as these, are maintained by enormous efforts in organizing food and water supply systems and are serviced by modern highways and railway systems constructed to transport the ore. The high degree of dependence between mining and the service facilities, the size of the populations involved, and the contribution of each center to regional economic growth are all reasons why allowance for failure of the resource must be included in management strategies.

Petroleum and natural gas are currently generating economic growth in arid lands along the coast of Peru and in a broad zone stretching from Iran to the Algerian Sahara. In these oil-exporting countries, the oil industry is producing employment for large numbers of the population through construction and operation of refineries, and it is creating purchasing power which is being used to support expanding food imports. However, petroleum and natural gas are being consumed at increasingly rapid rates, and the known world reserves are expected to be more than half depleted within the next 25 years (Pimentel, et al., 1975). As these arid land resources are depleted, alternatives must be available for feeding and employing the affected population.

The sunshine, vacant space, and unique landscape of the arid zone represent an amenity resource which is the basis for a potentially flourishing tourist industry in the desert. Attractive desert recreation sites occur at both coastal and inland locations, and a considerable number of recreation centers have been developed. Coastal tourist centers include the Mediterranean coasts of Libya and southern Tunisia, Eilat in Isreal, and La Paz in Baja California, Mexico. Inland resorts are found at such locations as Palm Springs, California, and at Tamanrasset and Djanet in the Algerian Sahara. Increasing affluence in many societies and improvements in air transportation have placed these desert resorts within the reach of increasing numbers of people, in terms of both travel time and the expense of the journey.

In addition to amenity resources, the tourist industry requires a dependable supply of fresh food and water and a wide range of subsidiary services and facilities, including transportation and communication networks. All of these activities generate employment opportunities, and, in this way, resorts make a significant contribution to the local economy and may serve as a basis for area development. Unfortunately settlement growth and heavy tourist usage may cause a failure of the resource by destroying the natural attractiveness of the desert setting. The development of air pollution problems, which can accompany population increases in many arid areas because of the occurrence of inversion layers, is one example of tourism creating one of the conditions in the desert which tourists are trying to escape. Also, the consumers of amenity resources are imported and this means that a resource failure may occur because of internal factors which are largely unrelated to the resource itself. Both an economic crisis and political disagreements can create a collapse of the tourist industry and the loss of income needed by the population to purchase food.

All activities associated with economic development in the desert compete for a limited water supply. Irrigated agriculture is a water intensive technology, mining uses water for ore enrichment, petroleum production requires water for drilling, and tourism demands abundant fresh water. But while these activities compete for water, they may be mutually helpful within the framework of a management plan aimed at area development and employing a strategy designed to provide a continuing source of food for the population, even if the resource fails. The interdependence of these economic activities is best illustrated by the manner in which revenue generated by mining and petroleum resources can be utilized to develop domestic agriculture.

Petroleum and mineral resources produce large blocks of natural revenue that can be invested in developing and subsidizing agriculture to provide increased food production. The new wealth can be used in two ways. One way is to initiate temporary actions that may have to be abandoned later, but which produce immediate returns in terms of an improved standard of living and a more abundant food supply. For example, the irrigation of poor soils with water of poor quality may be expedient for achieving a quick increase in agricultural productivity, but sustained productivity cannot be expected. A second way to use the revenue is to invest in the development of new irrigation projects designed to provide a sustained yield, plus investment in the support system for agriculture necessary to maintain growth in food output. The support system includes an indigenous research capacity, technical advisory services, farm credit services, roads, transportation facilities, markets, and, in many instances, reforms in land tenure (Brown, 1975). This second alternative is more likely to develop a stable agricultural base that will be self-supporting in the event that the petroleum or mineral resource fails.

Mining, petroleum, and, in some instances, tourism represent attractive opportunities for achieving relatively rapid economic growth and an improved standard of living in arid lands. An improved standard of living must include better nutrition for large numbers of the population. It is desirable that local agriculture be able to satisfy a large proportion of the food demand, but constraints on food production and agricultural development in the desert suggest that very sharp increases in food demand will have to be satisfied initially by imported food. However, consideration of the character of the food supply system over the long-term deserves special consideration.

REFERENCES

Amiran, D.H.K. (1970), Geographic Considerations in Plans for Development, in H. E. Dregne, ed., Arid Lands in Transition, American Association for the Advancement of Science, Washington, D. C.

Brown, L.R. (1975), The World Food Prospect, Science, 190, 4219, 1053-59.

Drauhin, G. (1970), Alternative Uses of Arid Regions, in H. E. Dregne, ed., Arid Lands in Transition, American Association for the Advancement of Science, Washington, D. C.

Duckham, A.N. (1974), Climate, Weather and Human Food Systems - A World View, Weather, 29, 7, 242-51.

Jones, J.R. (1966), Ground-water Exploration and Development in Libya, Water Well Journal, 20, 2, 13-16.

Nir, D. (1974), The Semi-arid World: Man on the Fringe of the Desert, Longman Group Limited, London.

Peterson, D.F. (1970), Water in the Deserts, in H. E. Dregne, ed., Arid Lands in Transition, American Association for the Advancement of Science, Washington, D. C.

Pimentel, D., et al. (1975), Energy and Land Constraints in Food Protein Production, Science, 190, 4216, 754-61.

Pimentel, D. et al., (1976), Land Degradation: Effects on Food and Energy Resources, Science, 194, 4261, 149-55.

Revelle, R. (1976), The Resources Available for Agriculture, Scientific American, 235, 3, 164-78.

Throne, W. (1970), Agricultural Production in Irrigated Areas, in H. E. Dregne, ed., Arid Lands in Transition, American Association for the Advancement of Science, Washington, D. C.

CERTAIN CONSIDERATIONS AND CRITERIA FOR POSSIBLE DEVELOPMENT OF COASTAL DESERT AND/OR ARID AREAS IN DEVELOPING COUNTRIES*

Jakov Z. Bradanovic
Centre for Natural Resources
Energy and Transport
Department of Economic and Social Affairs
United Nations, New York

SUMMARY INTRODUCTION

Coastal areas have the unique characteristic of being the focal point for human activity. They are usually defined as being located where land, sea and air meet and interact, profoundly influencing the environmental aspects of the bordering land. Human settlements, trade and industry have flourished in coastal areas since antiquity, and now approximately two-thirds of the world's population is said to live near the coast (1). Of the approximately 280,000 miles which is said to be the total length of coastline in the world, about 18,800 miles may be considered to pertain to the desert coastal area.

The limitation of industrial growth in the coastal area of a developed region is also due to the increasing social constraints caused by ecological and environmental considerations. In the light of these ad other limitations, policies concerning future locations of industries are being reviewed, including the consideration of industrial development in the desert coastal areas suitably located on maritime traffic routes.

As a result of economic and technological advances as well as political emancipation since the end of the Second World War, the geography of the natural resources has been substantially changed. The earlier centres of natural resources have been or are being exhausted, and new areas rich in natural resources are being discovered, explored and developed. Concepts concerning the location of industry have also come under consideration and review lately, not only because of the geographical changes in sources of natural resources, but also because of environmental and ecological considerations as well as social constraints in the developed regions of the world.

---

*The views and opinions expressed in this paper are those of the writer of this paper and do not imply the expression of any opinion on the part of the United Nations.

Industry always endeavours to establish itself at a place where it can obtain the most favourable production conditions; namely, the lowest costs for production as well as transport. Newly discovered sources of raw materials and energy are located in the undeveloped areas of the world, in most cases far away from the existing large industrial complexes of the developed regions.

Structural changes have occurred in transport, particularly in maritime and inland waterway transportation. Thus, the large distance "disadvantage" between the sources of natural resources and the industrial processing and manufacturing complexes have, from the economic viewpoint, been overcome by the introduction of economies of scale in the maritime transportation of bulk raw materials as well as "general cargo." Related to the economy of scale in maritime transport are the changes in maritime transport infrastructure, namely, the ports. Modern maritime transportation industry, therefore, implies on the one hand, the application of the economy of scale which simply means the employment of large-size vessels over long distances as well as, on the other hand, the concomitant influence on ports as the basic maritime transport infrastructure.

The economies of scale in transport influence the choice of port locations and development concepts which, therefore, include the technical, operation, financial, economical, and political considerations. Large-size vessels need deep-water berthing facilities and special bulk handling equipment for the fast turn-around of vessels at both terminals.

In the absence of other natural resources in the hinterland of the desert coastal area, maritime-oriented activities would seem to be decisive to the strategy for the development of the coastal areas. This paper, therefore, will endeavour to review certain basic considerations and criteria for possible desert coastal area development.

GEOGRAPHY OF COASTAL DESERTS

Geography of coastal deserts has been comprehensively covered within "Arid Zone Research" by UNESCO in 1966, and the attached maps 1-14 including key. Map 3 shows coastal deserts and/or very arid zones of the world.

Desert coastal areas have lately come into the focus of attention, thus, a review of their development potential is appropriate from the viewpoint of realistic development strategies to be considered within the broad programmes of the yet to be defined new economic order. Table No. 1 identifies the geographical regions and attempts to estimate the length of the respective country's desert or very arid coastal area.

In the absence of hinterland rich in natural resources in the desert and/or arid coastal areas, the industrial development orientation of these areas has to be primarily maritime oriented.

Advancements in transportation technology, as well as in industrial production technology, have made a decisive contribution toward the identification of tangible advantages which the siting of industrial complexes in coastal areas may have on the socio-economic development of the region. Needless to say, the geographical location, physical dimension, hydrography and topography of the coastal area - along with availability of human resources and a favourable balance of other natural resources - are essential prerequisites for development.

Table 1. Approximate Length of Desert and/or Very Arid Coastal Areas in the World

COUNTRY	Estimated desert and/or very arid coastal area			REMARKS
	Nautical miles	Statute miles	kilometers	
Mauritania	360	414	666	
Namibia	748	861	1,385	
W. Sahara *	490	564	907	
Somalia	1,596	1,837	2,955	
Afars and Issas	132	152	244	
Libya	910	1,047	1,685	
Egypt	1,307*	1,505	2,420	*Mediterranean 538 n/m Red Sea 769 n/m
Sudan	387	445	716	
Ethiopia	546	628	1,011	
Tunis	269	312	500	
Morocco	242	281	450	
Estimated total AFRICA	6,987	8,046	12,939	
Saudi Arabia	1,316*	1,515	2,437	*P. Gulf 296 n/m Red Sea 1,020
Iran	990*	1,140	1,833	*P. Gulf 635 n/m Arabian Gulf 355 n/m
Yemen Arab Rep.	244	280	451	
Democratic Yemen	654	753	1,211	
Pakistan	404	468	750	
India	593	687	1,100	
Estimated total ASIA	4,201	4,843	7,782	
Peru	1,258*	1,448	2,329	*Partly desert and very arid
Chile	850*	978	1,574	*about 850 north of Valparaiso very arid
Mexico *	2,051	2,375	3,800	* Baja California
Argentina *	917	1,062	1,700	* Patagonia
Estimated total SOUTH & CENTRAL AMERICA	5,076	5,863	9,403	
Australia	2,537*	2,921	4,700	* Very arid

The maritime orientation of specific coastal areas is greatly enhanced if its geographical location is along the world's major ocean trading routes, i.e., in the mainstream of the maritime traffic flows. Without entering into detailed explanations, the attractiveness of desert and/or very arid coastal areas is likely, if ever, to be along the basic east/west and north/south direction trading routes.

The east/west direction may be defined as primarily using the Mediterranean/Red Sea maritime transport route. This route has Gibraltar on the western end and Bab el Mandeb Strait on the eastern end, with the man-made Suez Canal linking the Mediterranean with the Red Sea.

The alternate east/west direction is, in fact, the north/south trading route around the Cape of Good Hope and along the West African coast.

In relation to these routes, the desert and/or very arid areas can geographically be identified as being located in the southwestern and western parts of Africa, a few Mediterranean countries on the African continent, and practically all countries riparian to the Red Sea, Gulf of Aden and the Persian Gulf, Pakistan and part of India.

In north/south trading route on both west coasts of North and South America may, but need not have only hemispheric maritime traffic connotation.

MARITIME DEVELOPMENT ORIENTATION OF COASTAL DESERTS

In the absence of hinterland rich in natural resources, marine and maritime orientation in the development of coastal deserts seem to be meaningful alternatives. With the exception of coastal recreation (tourism), fishing and offshore mineral resources as well as maritime transport and port-related activities appear to be viable catalysts for developmental activities.

Fishing

is a spontaneous and traditional human activity in the coastal areas primarily oriented to the exploitation of living marine resources. There is no statistical record readily available on fish catches in desert and/or arid coastal areas. It is known, however, that during the 1948-1970 period the world fish catch increased by over 250 percent. In recent years the growth rate has been slow, prompting reevaluation of forecasts for future levels of the world catch. Thus, while the present annual catch of fish and shell fish is about 70 million metric tons today, many experts now feel that we will be fortunate to reach 100 million metric tons of the more familiar types by 1980 (2).

An increase in the fish catch could possibly be achieved in the future by an expansion into marine aquaculture (mariculture) operations in the coastal desert areas of the developing countries. Marine aquaculture, however, has not developed as rapidly as anticipated. Meanwhile, with regard to the development of fishing and the fishing industry infrastructure, and as a result of the Conference on "Fishing Ports and Markets" (3), a most comprehensive treatise on the subject was published in 1970 by the Food and Agriculture Organization (FAO) of the United Nations and the main lesson conveyed was the:

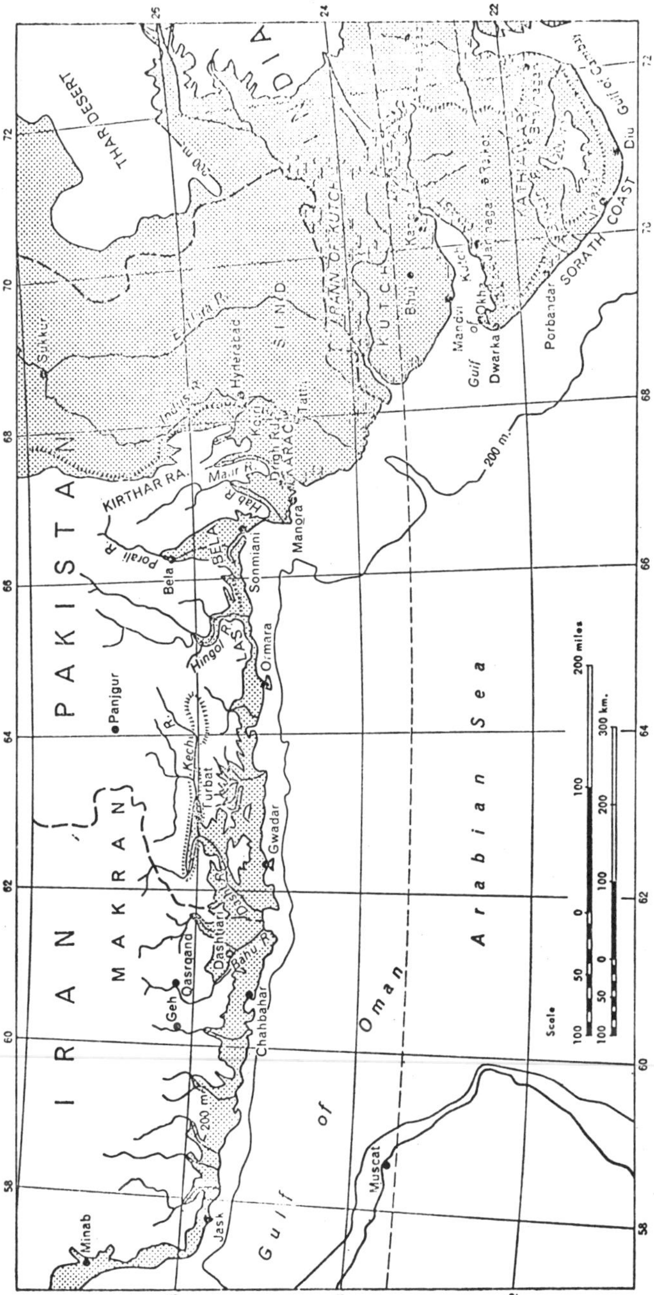

Source: Geography of Coastal Deserts by Peveril Meigs UNESCO - 1966

Map 1. Kathiawar-Makran.

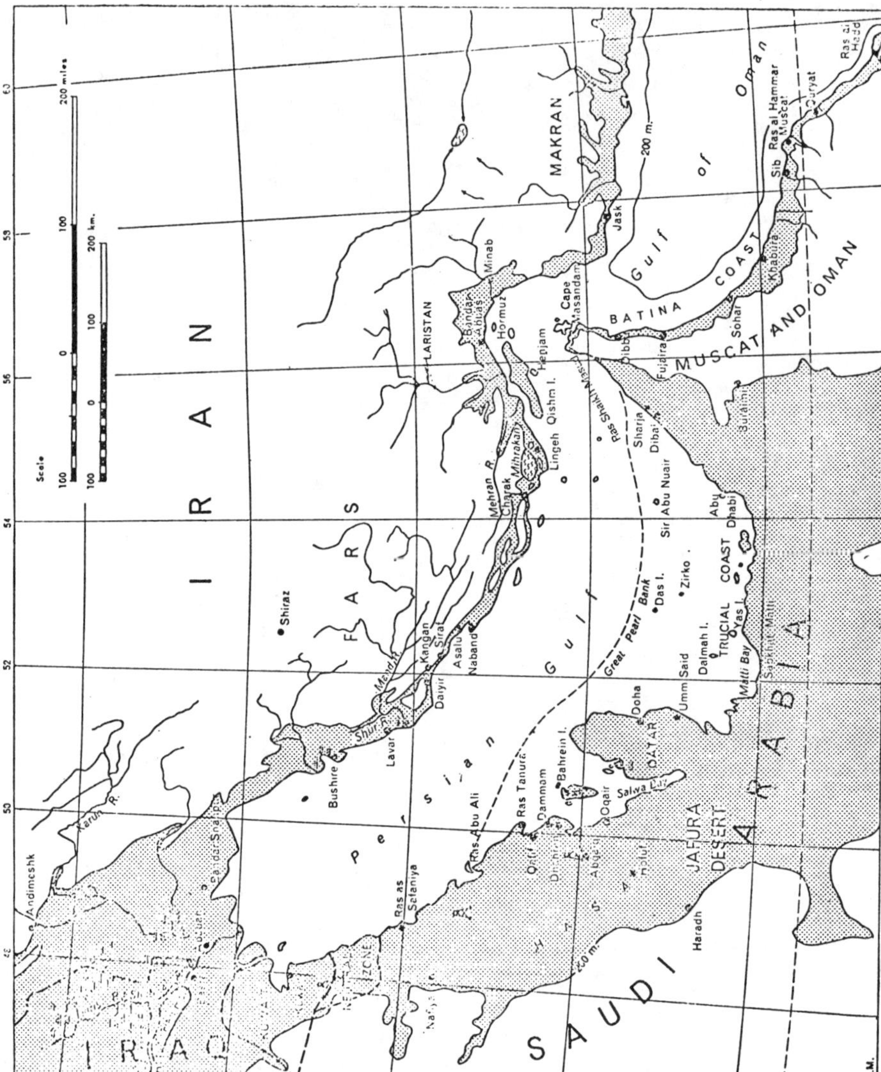

Source: Geography of Coastal Deserts by Peveril Meigs UNESCO - 1966

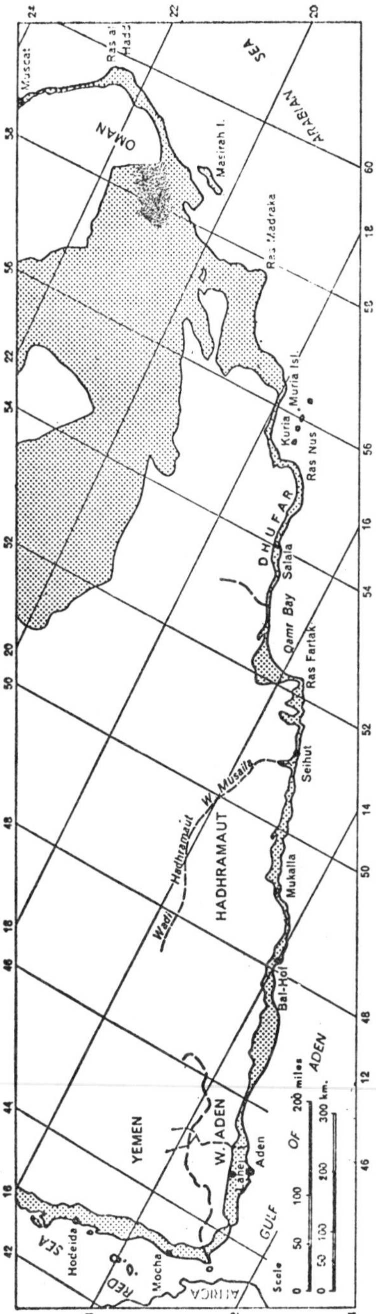

Map 3. South-East Arabia.

Source: Geography of Coastal Deserts
by Peveril Meigs
UNESCO - 1966

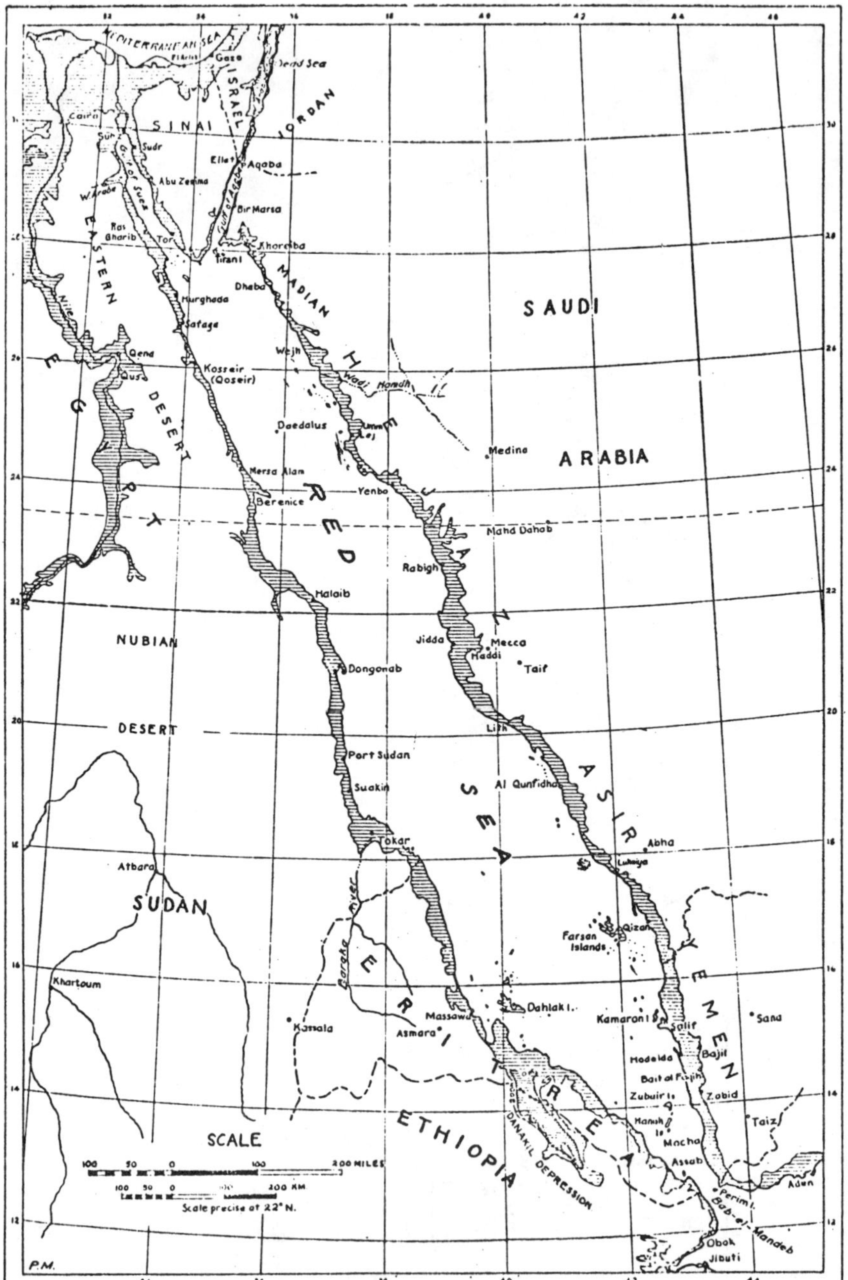

Map 4. Red Sea.

Source: Geography of Coastal Deserts
by Peveril Meigs
UNESCO - 1966

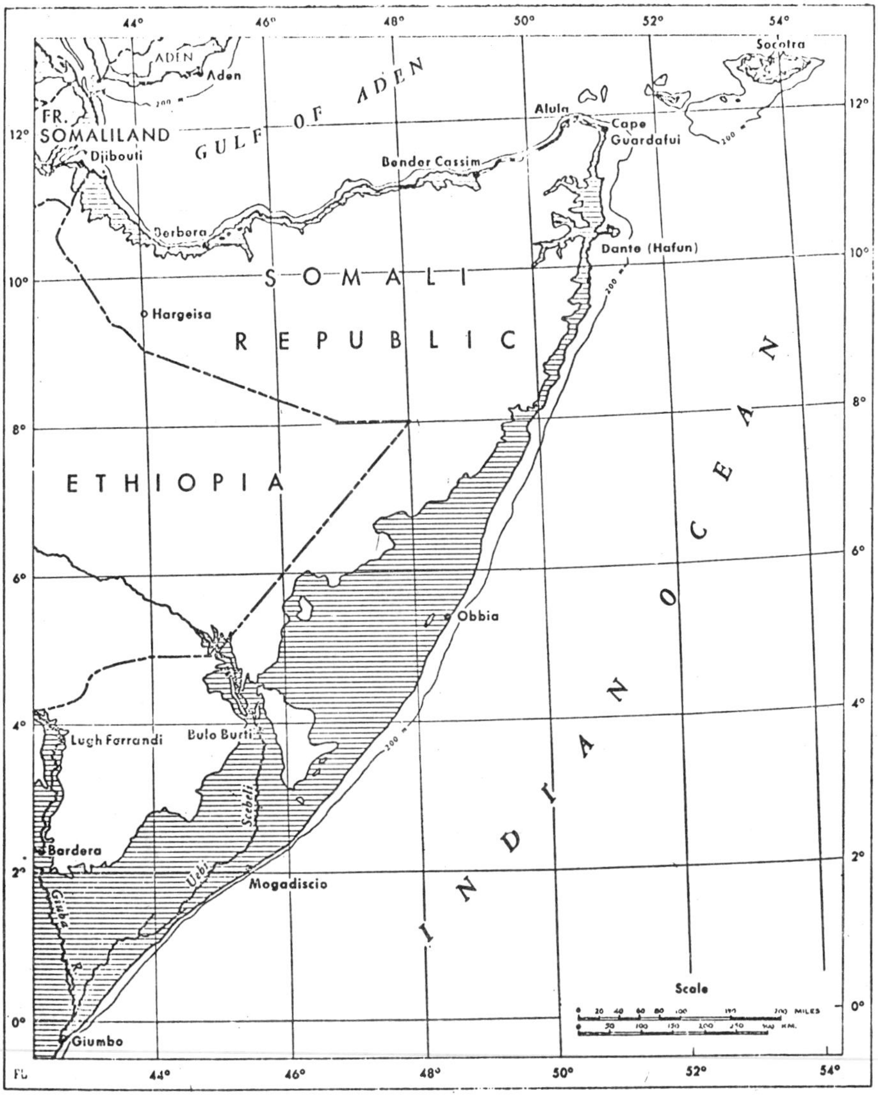

Map 5. Somalia.

Source: Geography of Coastal Deserts
by Peveril Meigs
UNESCO - 1966

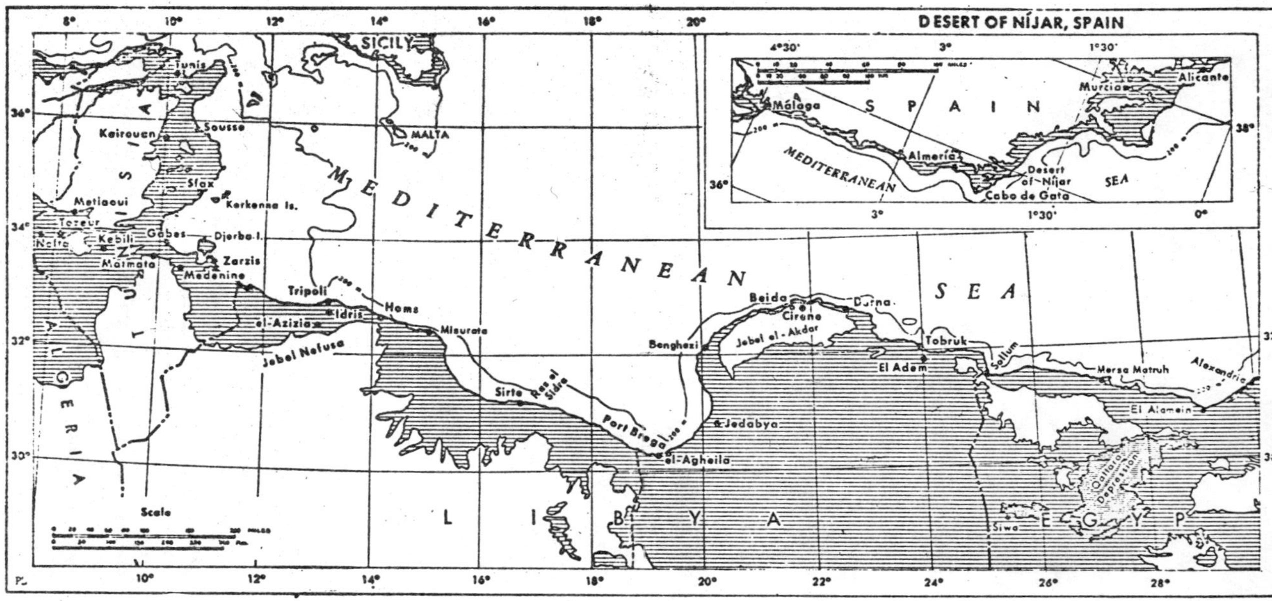

Map 6. Mediterranean.

Map 6ª. Nile delta.

Source: Geography of Coastal Deserts
by Peveril Meigs
UNESCO - 1966

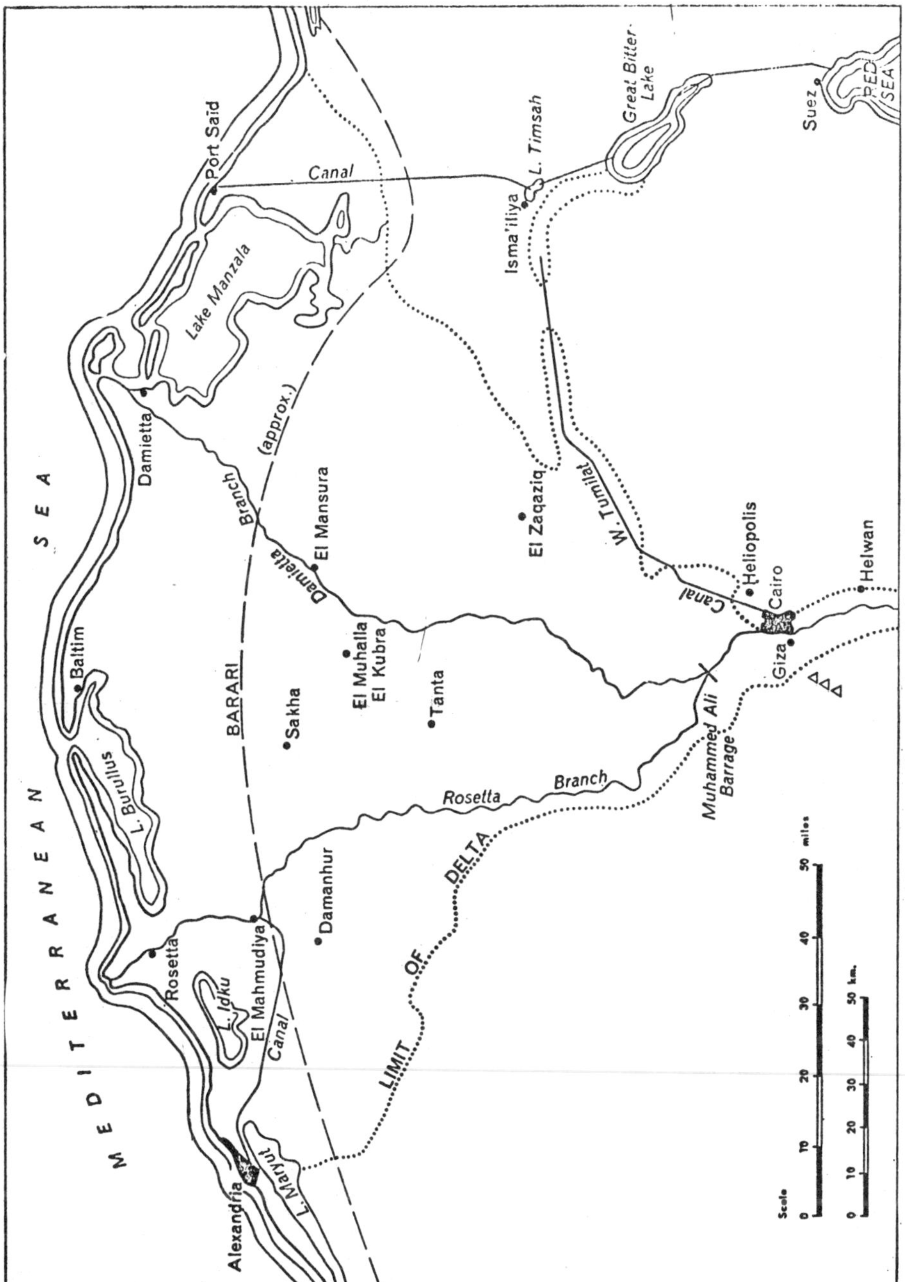

Source: Geography of Coastal Deserts
by Peverill Meigs
UNESCO - 1966

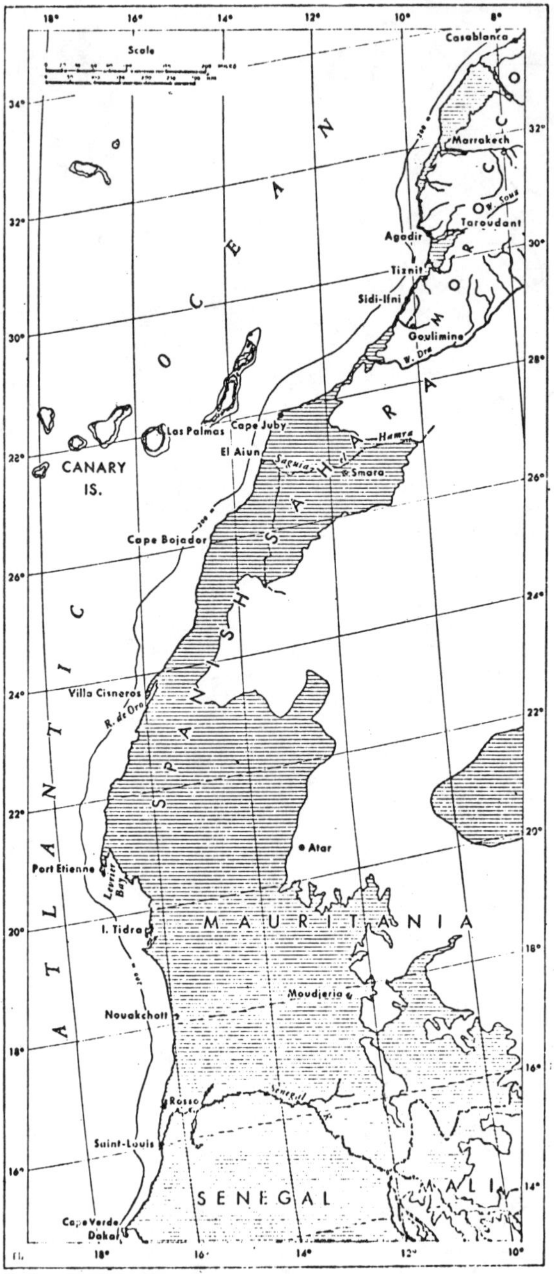

Map 7. Atlantic Sahara.

Source: Geography of
Coastal Deserts by
Peveril Meigs
UNESCO - 1966

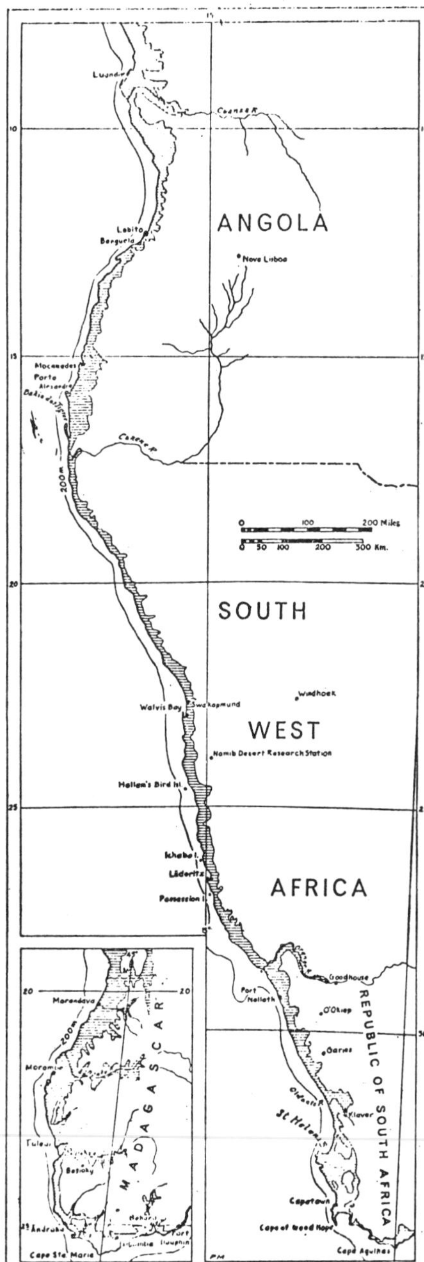

Map 8. Southern Africa.

Source: Geography of Coastal
Deserts by Peveril Meigs
UNESCO - 1966

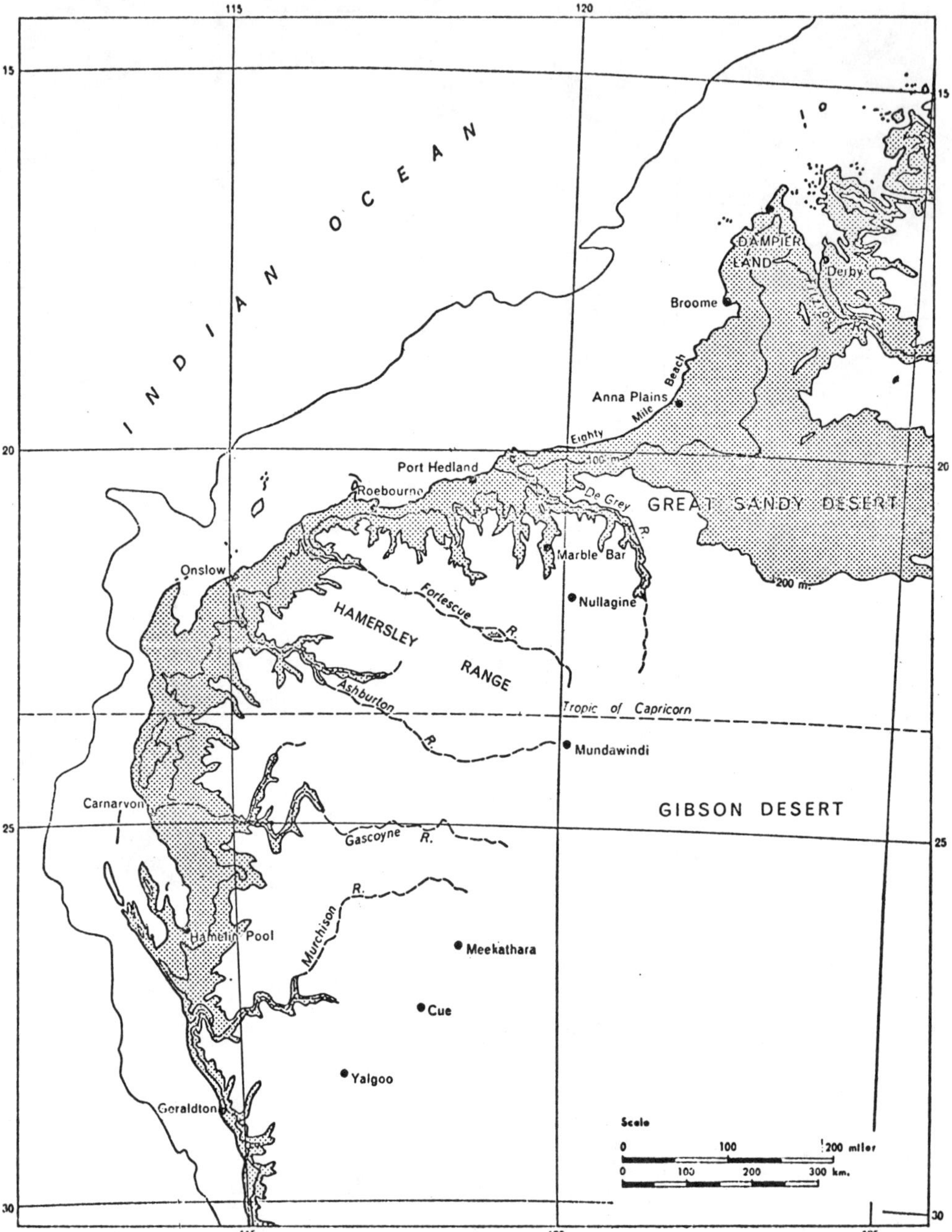

Map 9. Western Australia.

Source: Geography of Coastal Deserts by
Peveril Meigs
UNESCO - 1966

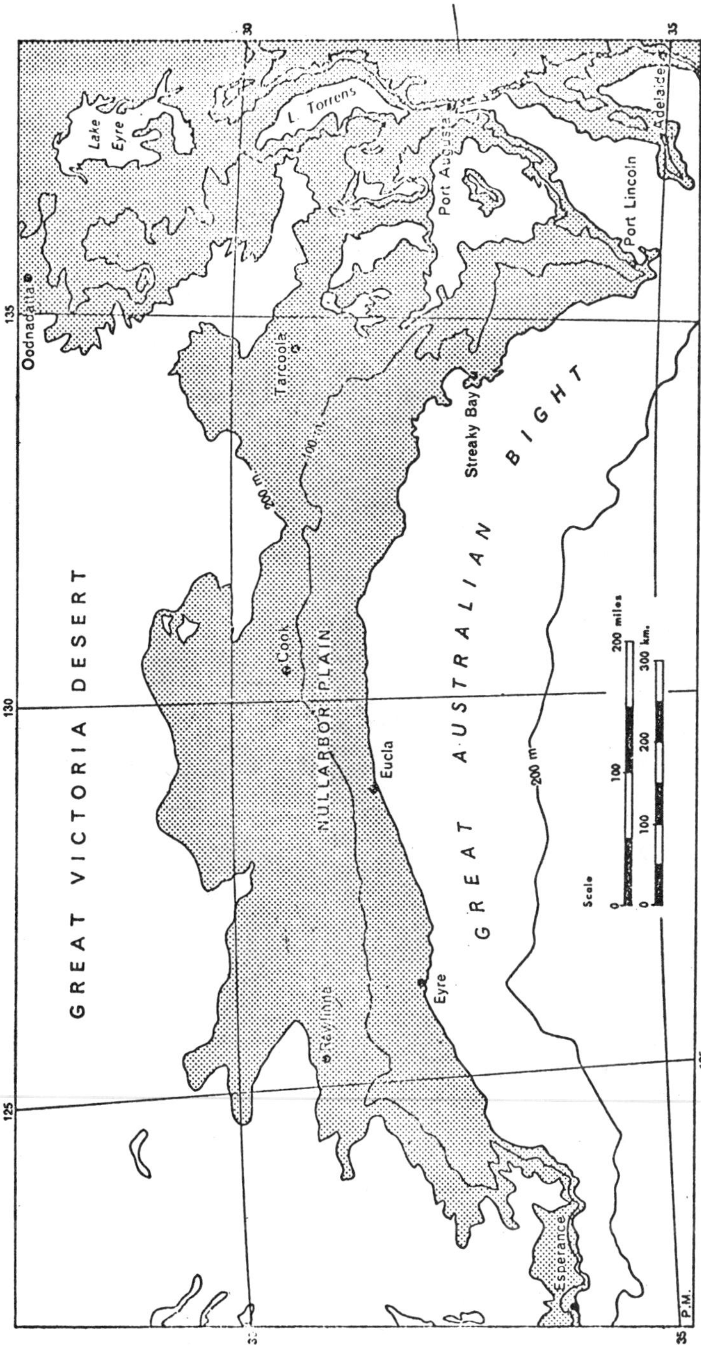

Source: Geography of Coastal Deserts by Peveril Meigs UNESCO - 1966

Map 10. Southern Australia.

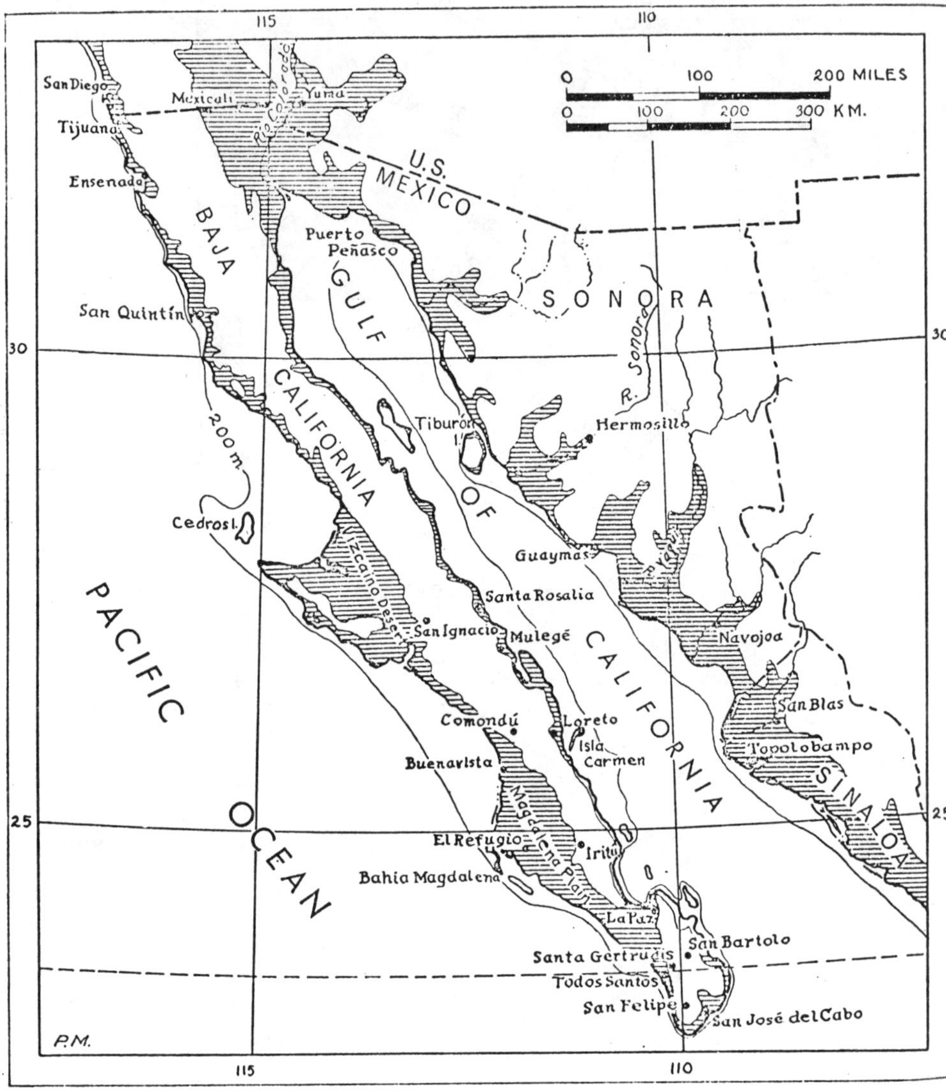

Map 11. North-West Mexico.

Source: Geography of Coastal Deserts
by Peveril Meigs
UNESCO - 1966

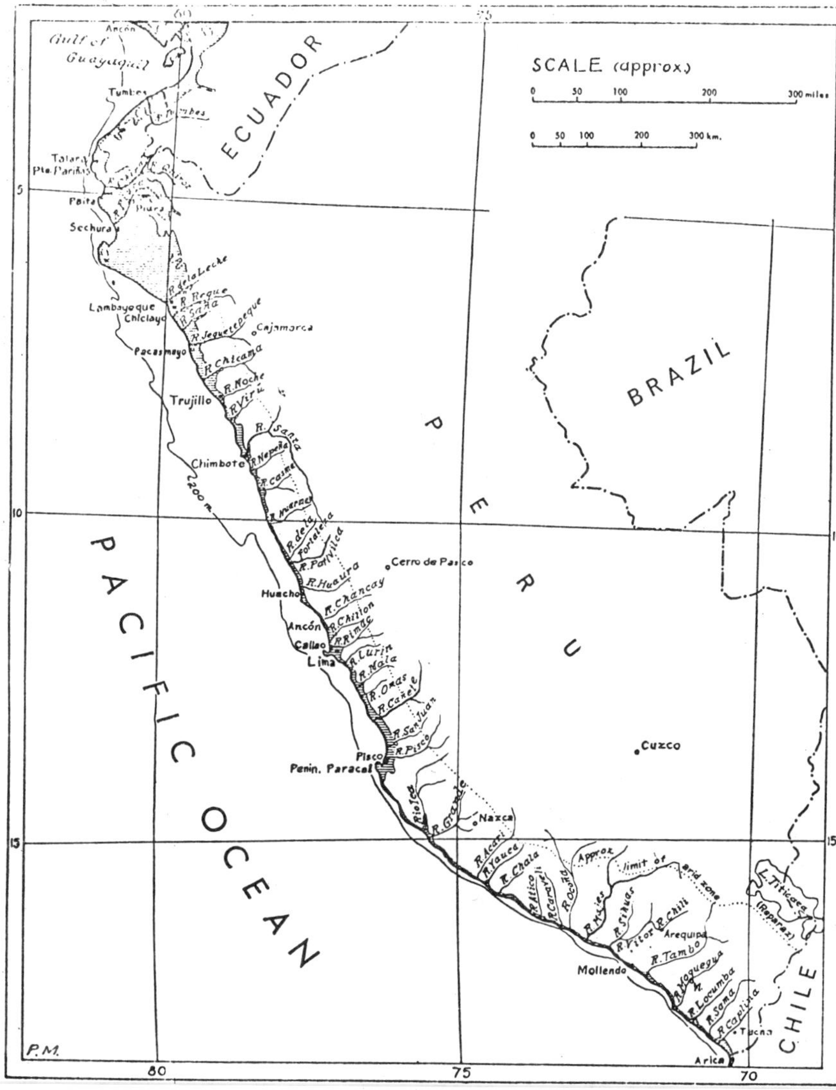

Map 12. Peruvian coast.

Source: Geography of Coastal Deserts
by Peveril Meigs
UNESCO - 1966

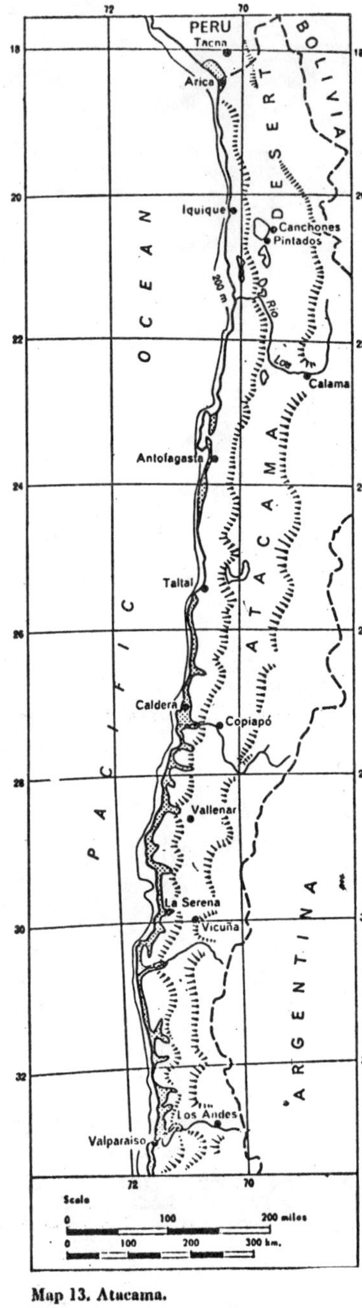

Map 13. Atacama.

Source: Geography of Coastal Deserts
by Peveril Meigs
UNESCO - 1966

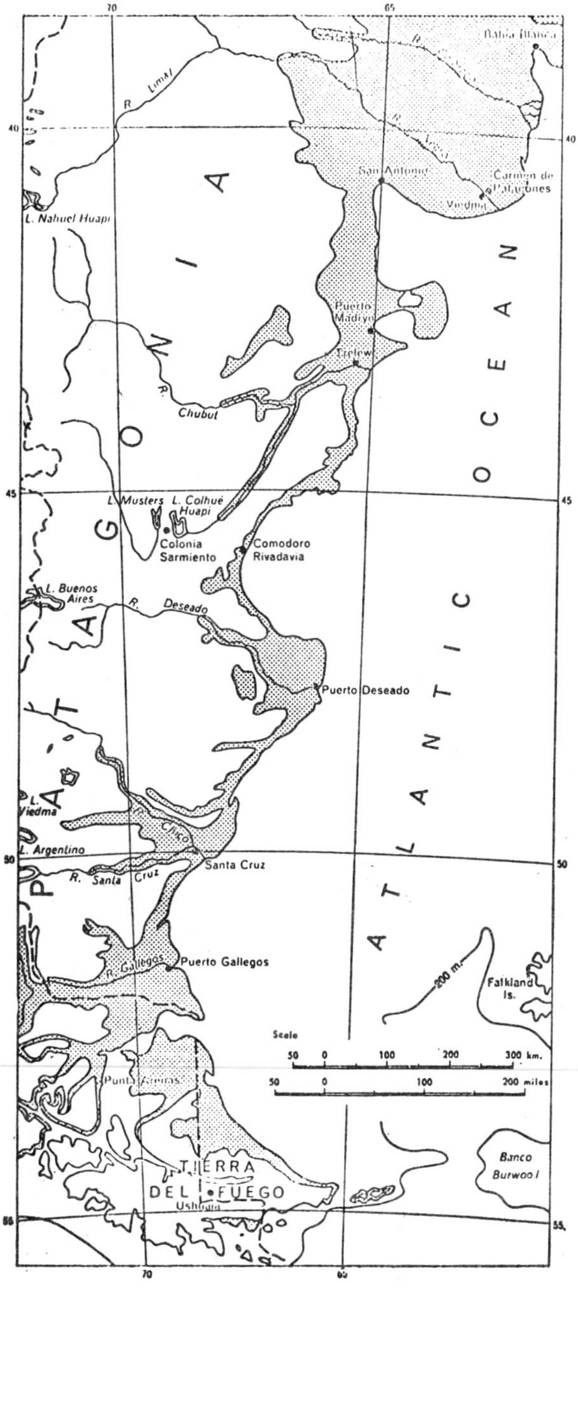

Map 14. Patagonia.

Source: Geography of Coastal Deserts
by Peverill Meigs
UNESCO - 1966

- multi-disciplinary nature of harbor development and the consequent need for coordinated planning and a team approach.

- need for realistic foresight and long-term planning, in view of the gestation period of most harbor developments, and the pace of change in fishing technology.

- essential role to be played by national and international governmental agencies in providing financing for harbor development, due to the high capital costs, the inadequacy of private financial sources and the essentially public nature of harbor facilities.

Off-shore Mineral Resources

and the feasibility of their exploitation for the benefit of mankind have been demonstrated in the off-shore oil industrial ventures in many areas of the world, including developing countries. Various technologies which would make mineral resources from the sea bed exploitable are said to have been developed. Anticipating potential benefits from marine resources, a number of countries have already declared the extension of their economic zone to 200 miles off-shore. In accordance with General Assembly Resolution 2750 (XXVI) of 1970, the United Nations Conference on the Law of the Sea is expected to adopt a convention dealing with all matters relating to the law of the sea. In the absence of the hinterland's natural resources, the development of coastal desert and/or arid areas may be induced and stimulated by the discovery of off-shore exploitable marine resources. The deficiency of water could, among other alternatives, be solved by desalination or other unconventional methods (4).

Maritime Transport

and ports as its infrastructure would seem to be a viable catalyst for the development of a variety of economic activities, including industry, especially if the coastal desert and/or arid area is geographically located along the mainstream of the marine traffic routes. The contemporary industrial development concepts revolve, in essence, around large-scale industrial production capacities. Maritime transport ensures - better than any other mode of transport - the largest economies of scale and, as a consequence, the lowest transport cost. We have all witnessed the spectacular growth in the tanker size since 1960, which is exhibited in Table 2 (5).

The dramatic growth in tankers has also been followed by the development of dry cargo carriers, the so-called ore/oil carriers, the biggest of which is said to be about 290,000 tons. The economies of scale of very large crude carriers (VLCC) are shown in Figure 1 and pertain to the building price and typical voyage cost.

The economies of scale, as applied in maritime transport and ports may, in many instances, prove to be a decisive factor in the development and exploitation of lower quality mineral resources.

The ultimate benefit, however, can only be derived from the economies of scale in maritime transportation if the corresponding port infrastructure can physically receive a VLCC, and provide adequate cargo storage facilities and sophisticated installations for fast cargo handling at both the loading and discharging terminals. All these call for capital-intensive facilities and the mechanical

Table 2. Growth in Tanker Sizes

Delivery date				Deadweight (tonnes)	Beam (meters)	Draught (meters)
1960	Universe Daphne	..	..	117,000	41.3	16.3
1966	Tokyo Maru	..	..	154,000	47.5	16.0
Dec. 1966	Idemitsu Maru	..	..	209,000	49.8	17.7
Sept. 1968	Universe Ireland	..	..	332,000	53.4	24.8
Sept. 1971	Nisseki Maru	..	..	373,000	54.6	27.1
Feb. 1973	Globtik Tokyo	..	..	484,000	62.1	28.2
1977/78	"Shell order"	..	..	540,000	63.0	28.5

handling of cargo for which, of course, a variety of technologies have been successfully developed and applied. Accordingly, this concept, in relation to port development, requires important technical as well as economic and financial solutions.

Figure 1. VLCC building price & typical voyage cost 1974/75

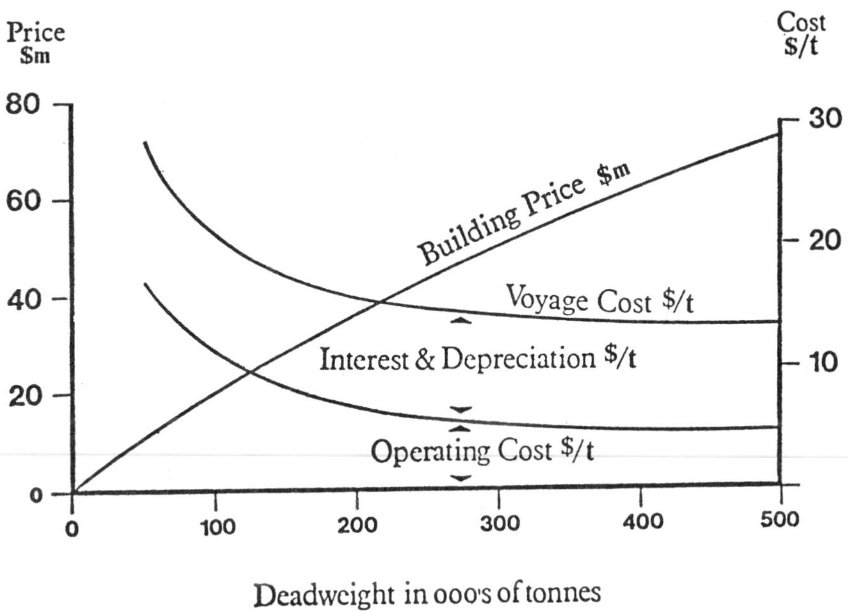

Source: A.B. Marchall - Living with scale - WEMT 1974. The Chartered Institute of Transport Journal. July, 1976.

The prerequisite of basic conditions and essential resources for human settlements in the coastal desert and/or arid region, along with the unison and harmony of both the technical and economic considerations, will provide the elements for the development of ports and the criteria for port industrial development, all of which are subject to both macroeconomic and microeconomic considerations.

Maritime Ports and Industry

Any port, including those situated in coastal desert and/or arid regions, may obviously have both national as well as international significance. The topography and hydrography, as well as other limitations in the region, will be of decisive importance in the selection of the site for the maritime port. This will include the role which such a port may play in the development process, as an important focal point for other modes of transport. A favorable geographical location of the port in relation to the mainstream of the world's maritime trading routes and traffic flows is likely to enhance their developmental opportunities. Accordingly, as an initial step for a possible broader developmental process in the coastal desert and/or arid region, a maritime port development scheme should be based on a sound analysis of both macroeconomic and microeconomic factors as well as the possible ecological and environmental impact.

Among the microeconomic considerations, water as an input into industrial production is a basic sinequanon prerequisite for certain industries. For instance, a contemporary optimal capacity for a steel mill is considered to be in the neighborhood of 4 million tons annual production capacity. While a conventional steel mill might require for its sustained production between 12 and 15 cubic meters of the so-called technological water, the latest technology in the steel production process considerably reduces water requirements. Thus, the application of modern technology in steel production makes the siting of the steel mill in the coastal desert and/or arid region a feasible consideration.

Another microeconomic consideration is the energy required as an input into industrial production. Unlike water, energy need not be an excluding factor in the coastal desert and/or arid region development concept. An energy generating plant must, therefore, be considered as an infrastructural element in the development process of the industrial zone, like the maritime port itself. Should there be a major scientific and technological breakthrough in the more efficient generation and use of solar energy, the coastal deserts and/or arid region developmental opportunities may be enhanced.

Still another microeconomic consideration in the coastal desert and/or arid area development is the transport network. When no link is needed with the hinterland, only an internal transport network between the industrial zone and that of the maritime port needs to be developed. Coastal shipping, ferry and feeder services can use the existing or specially developed facilities within the maritime port.

Shipbuilding as well as dockyard, ship repair and maintenance facilities are the traditional industrial activities found in a maritime port. Because no water is needed in its production process, it may be regarded as a compatible development scheme for coastal desert and/or arid areas. Here again, a favorable geographical location and relation to world trade routes may especially stimulate dry dock activities as well as ship repair and maintenance facilities.

Since 1974 there have been, without precedent, extensive industrial developments in the coastal desert and/or arid areas of the Persian Gulf countries. The OPEC countries in the region have launched an industrial diversification in non-hydrocarbon industries as well as in refineries and petrochemical industries. In addition, investments for the future have been made in:

- fertilizer plants
- steel mills, pipe and steel products
- copper refineries
- desalination and power plants
- cement plants
- plastics plants
- tire factories
- natural gas liquification plants
- shipyard and dry dock complexes
- sugar refineries
- maritime port and off-shore oil terminals
- assembly plants
- aluminum fitting plants
- paint factories

A very preliminary estimate of the investments in industry, infrastructure and social facilities along the entire Arabian desert coastline from Ras-Al-Khaimah in the United Arab Emirates to Shatt-Al-Arab in Iraq, including the islands of Abu Dhabi and Bahrain, is said to cost about $40 million per kilometer of coastline. A very preliminary corresponding investment estimate along the entire Iranian desert coastline between Shatt-Al-Arab and the Strait of Hormuz is said to be about $20 million per kilometer of coastline.

The growth of the Middle East maritime ports, according to the Engineering News-Record report of 28 April 1977, is shown in the following table.

Growth of Middle East Ports Projected Completion of Berths*

	1976	1978	1980	1985
Iran	37	51	70	81
Saudi Arabia	31	56	123	150
Iraq	19	26	36	36
Kuwait	23	23	29	33
Qatar	4	11	11	21
UAE**	33	43	53	147

*Does not include facilities for tankers

**A major 74 berth port is expected to be completed in 1981, thereby causing the big jump in figures between 1980 & 1985.

Source: Maritime Administration

Industrial development schemes along the desert and/or arid coastline of the Persian Gulf need not, but may be indicative as to what types of industrial plants might be compatible with similar environments of the other coastal desert and/or arid areas in the world. The spectacular growth in the berthing facilities in maritime ports also indicates the extensive import/export trading prospects for the entire area.

The rich off-shore fishing grounds of the coastal desert area may stimulate development of fishing harbor and related ancillary facilities as well as the fishing industry within or adjacent to maritime port. In a lecture entitled "The Planning and Preparatory Work for A Fishery Harbour Development Project" (6), the author exhibited a fishery harbor chain of activities which illustrated some activities to be catered to in the ideal fishing harbor. (Figure 2)

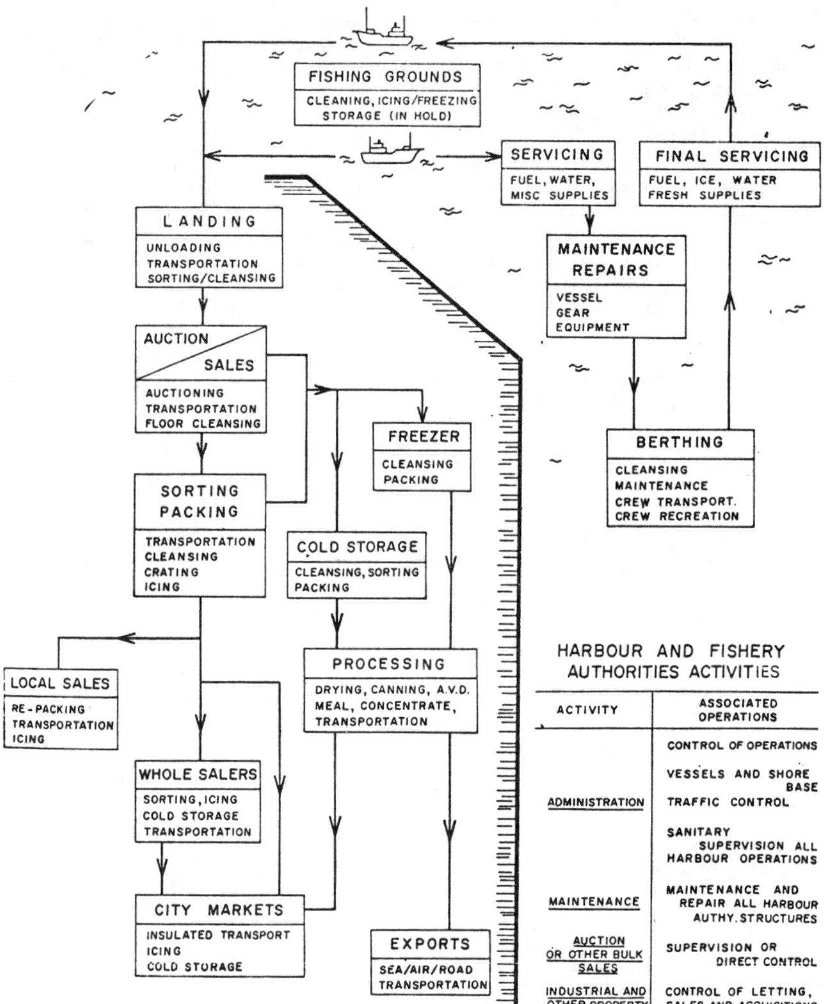

HARBOUR AND FISHERY AUTHORITIES ACTIVITIES

ACTIVITY	ASSOCIATED OPERATIONS
ADMINISTRATION	CONTROL OF OPERATIONS VESSELS AND SHORE BASE TRAFFIC CONTROL SANITARY SUPERVISION ALL HARBOUR OPERATIONS
MAINTENANCE	MAINTENANCE AND REPAIR ALL HARBOUR AUTHY. STRUCTURES
AUCTION OR OTHER BULK SALES	SUPERVISION OR DIRECT CONTROL
INDUSTRIAL AND OTHER PROPERTY	CONTROL OF LETTING, SALES AND ACQUISITIONS
TAXES/ DUTIES	COLLECTION ALL DUES
NEW DEVELOPMENT	CONTROL OF INDUSTRIAL GROWTH IN HARBOUR AREA AND RECOMMENDATIONS FOR NEW WORKS

The fishery harbour: chain of activities

Source: "Fishing Ports and Markets" published by Fishing News (Books) Ltd.

CONCLUDING REMARKS

In the absence of other natural resources in the hinterland of the desert coastal area, maritime-oriented activities, including off-shore resource explorations, would seem to be decisive to the strategy for the development of the coastal desert and/or arid regions of the world.

A favorable geographical location of the maritime port on the coastal desert and/or arid region in relation to the mainstream of the world's maritime trading routes and traffic flows is likely to enhance developmental opportunities in the region. A major scientific and technological breakthrough in the more efficient generation and use of solar energy may further stimulate the developmental process in the coastal desert and/or arid regions.

Spontaneous and traditional human activities oriented towards exploitation of living marine resources, in the coastal desert and/or arid areas would be enhanced if successful mariculture were applied, modern fishing methods and fleets were introduced and rational fish processing and marketing schemes were developed.

Traditional maritime industries such as shipbuilding, dry dock and ship repair and maintenance facilities should be regarded as compatible development schemes for coastal desert and/or arid region. The industrial development projects along the desert and/or arid coastline of the Persian Gulf need not, but may be indicative of the types of industrial plants that might be compatible with similar environments of other coastal desert and/or arid areas of the world.

In the context of the emerging environmental constraints and other limitations to further industrial growth of the coastal areas in developed regions of the world, on the one hand, and the need for continued and sustained economic growth for the benefit of mankind, on the other, it would seem reasonable that institutional as well as economic policy guidelines for long-term development of coastal desert and/or arid regions should be compatible with environmental considerations and consistent with yet to be defined policies for a new economic order.

NOTES

(1) Coastal area management and development - Report of the Secretary-General of the United Nations - document E/5648.

(2) Food and Agriculture Organization of the United Nations, International Technical Conference on Fishery Management and Development, Vancouver, 1973, Report No. 134, p. 44 - ECOSOC Document E/5650. 30/4/75, p. 4.

(3) Published by Fishing News (Books) Ltd. 110 Fleet St., London, E.C. 4, England.

(4) It has been reported that a study has been undertaken in order to determine technico-economic feasibility of towing an iceberg from Antarctica to the Red Sea port of Jeddah in Saudi Arabia.

(5) A.B. Marchall - Living with scale. The Chartered Institute of Transport Journal. July 1976.

(6) By: Wm. J. Guckian, Chief, Fishery Engineering Section, FAO - Rome, in "Fishing Ports and Markets" published by Fishing News (Books) Ltd.

NONRENEWABLE RESOURCES AND THE DEVELOPMENT OF ARID LANDS: A PLANNING APPROACH

Adam Rose
Michael Sheehan
Dale Hurd
University of California
Riverside

ABSTRACT

This paper examines the strategic role that non-renewable resources can play in the development of arid lands. An investigation into past experience reveals a pattern of sharp upswings in regional economic activity as resources are extracted, followed by downswings (often quite severe) in the wake of depletion. The extent of the downswing appears related to the strength of the economy and its dependence on mining.

These results point up to the need for a comprehensive planning framework which integrates mining with an overall economic development plan. Such a plan calls for (1) an emphasis on mining as a growth stimulating primary sector, (2) an accelerated development of secondary industries integrated with mining and (3) a more gradual diversification of the economy before depletion draws close. Such a strategy economizes on water, stretches the life of the resource, speeds development and offers stability after mining is phased out.

A multi-sector planning model has been formulated in order to test the worth of this strategy. The model incorporates several important features of arid lands such as limitations of water, population, labor force skills, and capital. The proper application of the model calls for the extension of the planning horizon so as to take into account the post-depletion consequences of the development plan.

An ideal setting for an application of the model is the Mexican State of Chihuahua. Aside from tourism and U.S. border zone industries, Chihuahua's economy is reliant upon mining and agriculture in an arid climate. The water resources of the State are not well developed, thus severely limiting the agricultural sector. On the other hand, mining is quite prominent, with Chihuahua being the leading Mexican producer of both ferrous and non-ferrous ores. Development to date, however, has not proceeded farther than the actual production of the ores; ore reduction and smelting as well as secondary manufacturing industries dependent upon the refined output have not located in Chihuahua. Given this economic structure, the exhaustion of the mines would spell disaster for the State economy. Once the resources are depleted, future development may be impossible without fundamental improvements in mining technology or the infusion of a large amount of outside aid. It is important, in this context, that the exploitation of Chihuahua's non-renewable resources provide an impetus to development and a legacy to the population that will see them through the post-depletion period.

INTRODUCTION

The primary economic feature of arid lands is their lack of development. This is not surprising in light of their locational disadvantages historically. Four factors should, however, mitigate this situation. Foremost is the drain on resources elsewhere which has focused attention on more remote sources. Second, new technologies have been devised which should facilitate the recovery and use of resources in arid lands. Third, world population pressures have placed a premium on space. Finally, there is the drive by wealthier nations and international organizations to promote development in the third world.

This paper will focus on the role of non-renewal resources in the development of arid lands. We will stress the role that planning can play in accelerating development and avoiding problems, specifically those relating to employment, arising from the unique character of arid lands. In planning there rests the possibility of overcoming the "boom and bust" phenomenon which has all too often characterized resource exploitation in the desert. Rather than rely on unrestrained market forces, which work even more imperfectly in lesser developed settings, planning may guide the policymaker toward the attainment of broader social objectives.

The unique features of development in desert areas call for some modifications of traditional planning frameworks as well. The most important of these is the extension of the planning horizon so as to minimize the true net social costs of depletion. This has already been done in a few instances--e.g., oil shale development in the U.S. where viability also includes a calculation of the costs of returning the land to its predevelopment form. A strong case can also be made for pursuing a resource development policy which insures that the economy will be at no less than its predevelopment level. Planning should take into account high post-depletion unemployment levels, transition costs and especially the possibility that a halt in mining production may feed on itself and result in a general economic deterioration. Also since the arid zone is not homogeneous, allowances should be made for variability of rainfall, population density, preponderance of resources, and availability of skilled labor and capital.

THE IMPORTANCE OF RESOURCE DEVELOPMENT IN THE ARID ZONE

The prevalence and quality of some mineral resources are enhanced by aridity (Amiran, 1965). Major resources of this type are nitrates, phosphates, gypsum, oil, various salts, copper and opal.

The important thing about most of these resources from the standpoint of development is that they have a high export demand and, therefore, can serve as a source of much needed foreign exchange. Also, in the case of certain strategic resources such as oil they may yield substantial economic profits which can be rechanneled into investment funds. Even their exploitation by foreign-owned enterprise is not the enclave situation often depicted, since substantial tax payments may be garnered by the host country (Hall, 1970).

Other factors help substantiate the case for mining as a strategic primary sector in the development of many arid lands. Paramount among these is mining's low demand for the desert's most precious resource--water. Otherwise, both mining and the other major primary sector have unique advantages and disadvantages in desert settings. The major natural advantage for agriculture is the abundance of sunshine. However, arid lands are also chacterized by poor soil, and semi-arid lands by a high degree of uncertainty as to the availability of water. The major drawback of mining is the exhaustibility of the resources involved; the only advantage unique to deserts is the richness of the deposits.

More general characteristics of these two sectors, however, may tip the balance in favor of mining. The most important of these is the relatively higher income elasticity of demand for the products of this sector and its more direct link to manufacturing. In fact the latter point may substantially temper the disadvantages of exhaustion. Most lesser developed countries directly export their primary products. As economic development proceeds, plans usually call for industrial diversification as opposed to integration of the primary with secondary activities, such as refining and smelting in the wake of mining. Such a strategy is often marked by caution and input constraints, thus slowing the pace of development. Akhdar (1974) and others have noted the advantages of vertical integration. For one thing the accelerated shift from primary to secondary industries which this may bring about will lengthen the life of the resource. Second, there may be a comparative cost advantage vis-a-vis transportation, for example, in the production of the finished products. On the other hand, the main argument against integration--the increased risk brought about by reliance on a single resource--is easily countered. There is a major difference between dependence on oil as a primary product and dependence on a multitude of separate products (gasoline, chemicals, drugs) which can be processed from it.

Note that this is not an indictment of diversification since as we shall note below it becomes crucial at a later date.

PAST EXPERIENCE OF NONRENEWABLE RESOURCE DEVELOPMENT: THE BOOM AND BUST CYCLE

Resource exploitation is desert lands can often be characterized as a boom and bust phenomenon. Development of the resource provides a healthy stimulus to the economy in question through direct, induced and agglomeration effects. Infrastructure is built, labor trained and industry put in place, but once depletion is neared the situation is characterized by a drop in expectations, dilapidation and outmigration. Of course, price fluctuations, whether caused by world trade cycles or the development of synthetic substitutes, could bring about a similar downswing.

At the root of this phenomenon is what Aschman (1970) calls the natural cycle of the mine, which consists of four stages. Initially the discovery and surveying of the proposed mine employs relatively low levels of inputs, representing primarily the prospector's stake or the modern survey team's salary. As the prospects of economical extraction increase, more intensive efforts to determine the extent and concentration of the resource stock lead to higher levels of capital and labor being employed. These prospecting and surveying costs are the components of stage one.

With the decision to begin mining operations, the development of the infrastructure needed to support the mine, the capital investment, and the recruitment of the labor force proceed rapidly. The investment in infrastructure is necessitated by the isolation of the mining location and includes not only transportation facilities but also housing and support services for the labor force. Stage two continues until the profit level reaches its peak.

As the mine deepens and the costs of operation increase while the ore quality declines, profits fall off and the mine begins its slow decline to depletion. Aggravating the situation faced by the mining concern are the demands made upon its revenues by society in general and the mining labor force in particular. The rapid increase in the mine's profitability in stage two leads the state and laborers to demand a higher share of the proceeds from its operation, usually with a realization lag. The critical point for the operators occurs near the end of stage three when the end of profitability is in sight and the decision to pull out must be made.

However, as the mine becomes unprofitable at stage four it may still continue in operation for some time. At first its continued existence is made possible by the running down of the fixed capital. Then comes the possibility of offering the depleted mine for operation to indigent laborers for a nominal fee. In this case labor with minimal capital scratches out a meager subsistence.

The phase-out of the mine is likely to be sufficiently distressing by itself; yet it usually is exacerbated by two additional factors. One is the disturbance of the hypersensitive ecological balance which, though not necessarily irreversible, is long-lasting due to the lack of regenerative capacity of dry lands.

The second is the likelihood of a general economic downswing as a result of depletion. This refers to more than just the closing of the mine itself, but to the reverberations which will spread to the mining town, region or nation. Desert ghost towns and depressed arid regions stand as bleak testimonials to this chain of events. Examples abound in the Western United States and the Norte Grande of Chile.

The cycle of the mine is a portent to the "natural cycle of the mining region." As long as the solitary prospector or isolated survey team is working in a potential mining district they may provide for their needs from sources outside the region. But as soon as a deposit is discovered and the mine begins to be developed, the inmigration of investment and service industries leads to a seedling business community dependent upon the mining operations. Typically geological deposits do not occur in isolation but are part of the broader geographic region's unique characteristics. Thus the discovery of a deposit is apt to lead to widespread surveying to determine the location and extent of any other mine sites in the area. If the area does indeed contain a number of deposits it is then possible to develop a common infrastructure. Also important is the possibility of centralizing the mining support industries in a township of growing size which is able to offer social services not available in the mine site camp. Economies of scale permit the specialization and division of labor necessary to provide the services desired by the mining community, and may even be so general as the attract industries not related to mining.

Through the multiplier/accelerator interaction the community's well being responds in an exaggerated manner to the mines fortunes both on the upswing and the downswing to follow. As the mine enters stage three investment and labor demand begin falling off. The lessened demand for equipment and workers ripples throughout the region as housing begins to be vacated, stores close their doors for lack of customers, mining equipment firms cut back operations to a low level, and a further decline begins as suppliers to these activities must also cut back on their operations. The continuing depletion of the mine results in progressively lower output levels and declining economic activity. The extent of the decline depends upon the strength and diversity the economy may have obtained on the upswing. Any rivival is contingent upon the region's strengths (other resources, usefulness as a transport link, etc.). Clearly, a recovery from a depressed position is much more difficult than a smooth transition out of mining dependence.*

---

* In some cases the region is able to exert political suasion to exact a subsidy from the rest of the economy to allow it to continue in unprofitable operation. Given the commitment people have made to developing the region they may be unwilling to leave when the resource deposits are worked out. If new profitable opportunities are overlooked elsewhere in the effort to maintain the mining operations, the region becomes a drag on the rest of the economy. Usually further investment in the infrastructure or direct investment in the mine succeeds in sustaining extraction for only a few years. Then this further investment must also be abandoned.

Three general postdepletion trends are worth distinguishing and are depicted in Figure 1. They can be used to represent situations in countries or regions at various levels of development. Case I shows a continuous upward movement in economic activity even after depletion. It might represent the case of a rather underdeveloped country where absolute levels of economic activity are extremely low and any stimulus can be absorbed in a manner conducive to development, thus leading to some sustained growth. Such an upward trendline might also be considered as a norm which proper planning might achieve for a country at any level of development. (This paper presents a planning model to facilitate such a result.) Case II shows a downturn following depletion, but a tapering off at a level above the pre-extraction phase. It is representative of an advanced nation or region with a broad enough base to withstand a major downturn in a single sector. Ironically, the case can also depict a previously uninhabited, isolated area which peaks and then returns to zero. Case III represents the least desirable situation, that of a downswing below pre-extraction levels. It depicts the situation in a semi-developed country where the extraction of the resource does not lead to any sustained growth and where a weak economy is not able to withstand the "shock" of depletion. Here a downswing in one sector or region may transmit extensively and cause severe harm to the entire economy.

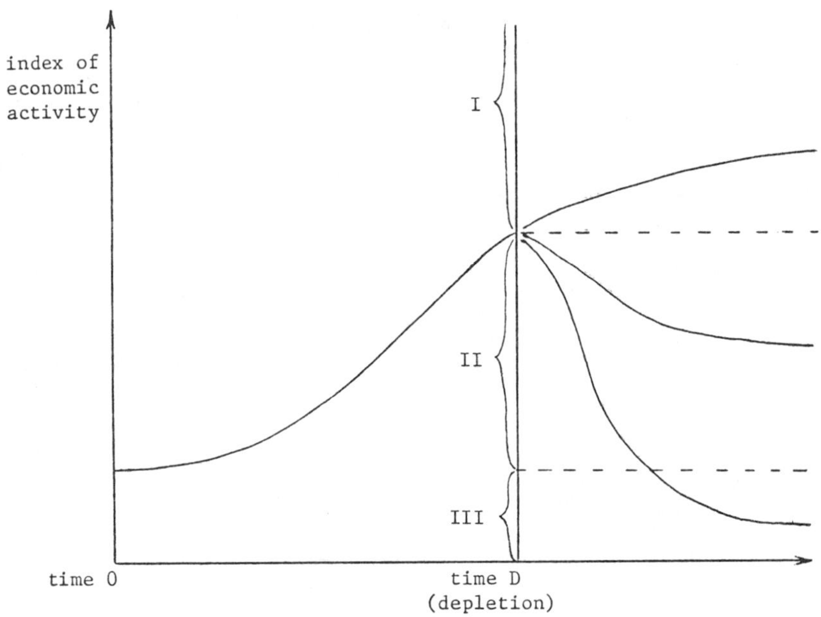

EMPLOYMENT ASPECTS

Employment trends generally coincide with the general economic tendencies displayed above. Due to the relative remoteness of deposits there is likely to be a shortage of manpower (or training skills) in the early stages, requiring importation of workers. Wage differentials or other inducements may have to be especially high in order to attract the necessary personnel to such relatively inhospital areas. In sparsely populated countries and/or where cultural conditions are resistant to such sedentary occupations this has involved going to outside sources. Mining areas were thus often inhabited by only the mine-workers and a meager support population, many of both groups having stronger ties elsewhere. The resultant single-purpose towns were readily vulnerable to abandonment as mining phased out. The end of the miners' employment in many cases meant the end of employment for all inhabitants.

This points to the strategic role of the support population in the mining region. First, a large population is crucial to specialization and the broad base needed for self-sustained growth. Mining towns can thus serve as "growth poles" and may eventually turn into the urban centers needed for development.* Also, once a mining town has been formed and the necessary infrastructure put in place, abandonment represents a terrible waste. A broader base for the town would give it a reason for existing apart from its primary resource and would make it less susceptible to cycles and more capable of absorbing unemployed miners. Third, a large population may mitigate some of the severe personnel problems which arise out of the isolation of mining in dry lands. Webb (1969) documents the psychological stress which leads to low productivity and high turnover rates in these areas.

We do not suggest that mining towns be maintained at all costs. Aschman (1970) and others have made the point that it is better to relocate workers than run an unprofitable mine. But neither should depletion immediately suggest abandonment. There are many cases where desert towns have survived depletion or the exodus of their primary industry and become crucial commercial centers in the process of regional development. The towns of Kalgoorlie and Broken Hill in Western Australia are two excellent examples. In other cases the situation may warrant some government stimulus for a short duration until the town is self sustaining. The ideal solution, however, is a proper sequencing of resources and a development strategy which avoids such a situation in the first place.

A PLANNING MODEL

In order to escape the natural cycle of the mining district it is necessary to digress from myopic market forces and move toward broader social planning. In most arid lands this calls for the integration of resource exploitation into the overall development planning effort. Attention to several important features can lead to a smoothing out of the cycle and the furtherance of a higher standard of living. Such features include the time path of extraction, sequencing of resources, location of towns, and allocation of capital between sectors.

Below we will outline a planning model for this purpose. There is no pretense about its ability to capture all of the essentail features of the problem. We do intend to focus on the more important ones, however, within the context of a computationally manageable model with modest data requirements.

The model is a linear programming version of a dynamic input-output model. There have been several multisector optimizing models applied to development, most notably those of Chenery and MacEwan (1966), Manne (1966), Bruno, Fraenkel and Dougherty (1970), Sandoval and Golladay (1972), Goreux and Manne (1973). The model presented below resembles that of Bruno *et al.* (1970), who applied their work to development planning for Israel. The major modifications we have made relate to the more important natural features of arid lands: nonrenewable resources, water, population

* The emphasis on towns reflects the usual pattern of concentrated development in the desert brought about by the scarcity of sources of water and sparseness of the transport network.

and skill levels; and also include a much more sophisticated set of investment equations to more accurately capture cyclical effects.

The model is presented below. The notation used and sector designations are contained in the appendix. Note that while the model is presented mathematically in a rigorous manner, the absence of testing means that it is still only heuristic at this point.

Objective Function--maximization of discounted per capita consumption (with due acknowledgement to its limitations as an indicator of overall well being or development):

$$\sum_{t=1}^{T} \frac{(C^t/N^t)}{(1+\delta)^{t-1}}$$

Unemployment Constraint. Major policy variable used to adjust the amplitude of the economic cycle:

$$L_S = \sum_{s=1}^{3} (1=\xi) L_s^{t-1} + \sum_{s=1}^{3} L_{fs}^t$$

$$L_D = \sum_{s=1}^{3} \sum_{j=1}^{15} \ell_{sj} X_j^t$$

$$U^t = \sum_{s=1}^{3} U_s^t = \sum_{s=1}^{3} (1-\xi) L_s^{t-1} + \sum_{s=1}^{3} L_{fs}^t - \sum_{s=1}^{3} \sum_{j=1}^{15} \ell_{sj} X_j^t$$

$$U^t \leq \lambda^t \rho^t N^t + \lambda_f^t \rho_f^t N_f^t$$

Resource Constraint--presence of one or more non-renewable resources in the region:

$$\sum_{t=1}^{T} X_r^t \leq \overline{R}_r \qquad r = 2A. . .2G$$

Water Constraint--maximum annual natural fresh water use which can be attained without permanent damage to the supply plus the volume of the supply increase in t:

$$W_D^t = \sum_{j=1}^{15} w_j^t X_j^t$$

$$W_S^t = W^{t-1} + W_e^t$$

$$\sum_{j=1}^{15} w_j X_j \leq W^{t-1} + W_e^t$$

Population Constraints--historical trend (exogenous) plus subsidized in-migration:

$$N^t = (1+\gamma_1)N^{t-1} + N_f^t$$

$$N_f^t \leq \gamma_2 N^{t-1}$$

Skill Level Constraints. Base year skill levels (exogenous) with provision for upgrading:

$$(1-\xi_s)L_s^{t-1} + L_{fs}^t = \sum_{j=1}^{15} \ell_{sj}^t X_j^t + U_s^t$$

$$L_{fs}^t \leq \rho^t N^t - (1-\xi_s)L_s^{t-1}$$

Other less unique constraints and relations are presented notationally below without elaboration.

Output Determination:*

$$X_i^t = \sum_{j=1}^{15} a_{ij}X_j^t + \Gamma^t(b_{ij}^t X_j^t - b_{ij}^{t-1}X^{t-1}) + \Pi^t d_{ij}^t X_j^t + C_i^t + G_i^t + S_i^t + E_i^t$$

Consumption Relations:

$$C_i^t = \beta_i C^t + \alpha_i$$

$$C_m^t = \beta_m C^t + \alpha_m$$

$$C^t = \sum_{i=1}^{15} C_i^t + C_m^t + C_v^t$$

Savings-Investment Relation:

$$I \leq sV + F_p$$

Government Spending Constraints:

$$G \leq tV + F_a$$

$$G_f = W_f^t + \sum_{s=1}^{3} c_L L_{fs}^t + c_N N_f^t$$

$$G_f \leq \omega G$$

* The investment equation depicted in equation (7) is of the more general form of those presented in equation set (11).

Investment Constraints:

$$I_r^t = \Gamma^t(b_{ir}^t X_r^t - b_{ir}^{t-1} X_r^{t-1}) + \Pi^t(d_{ir} X_{ir}^t)$$

$$I_j = b_{ij}(X_j^t - X_j^{t-1}) + d_{ij} X_j^t$$

$$\Gamma^t = f\left(\frac{R_r^t}{\overline{R}_r}\right) \qquad \Pi^t = q\left(\frac{R_r^t}{\overline{R}_r}\right)$$

$$I = \sum_{r=2A}^{2J} I_r + \sum_{\substack{j=1 \\ j \neq r}}^{15} I_j + I_m$$

Export Constraints:

$$E_i \leq \overline{E}_i$$

Imports and Import Substitution:*

$$S^t \leq S^{t-1}$$

$$M + \sum_{j=1}^{15} S_j = \sum_{j=1}^{15} m_j X_j + C_m + I_m + G_m$$

---

* Specific industries are not singled out for import substitution so as to avoid biasing the results in favor of any given development plan.

Foreign Exchange Constraint:

$$\sum_{i=1}^{15} \mu_i E_i + \mu_v F_v + F \geq \mu_m M$$

Capital Stock Constraint:

$$K^t \geq K^{t-1}$$

GNP Identity:

$$V^t = C^t + G^t + I^t + E^t - M$$

There are two special features of the model which merit further discussion. The first is the incorporation of "resource enhancing industries." These do not affect the nonrenewable mineral resources, but affect water, population and skill levels. (1) There is a desalination industry which transforms water with a high brine content into potable irrigation, or industrial water supplies. Though desalination techniques are not usually economically viable, a tight constraint translating into a high shadow price may lead to a positive output in this industry. It might also serve as a proxy for a more general "water project industry" as well. (2) There is a "population recruitment industry" which represents subsidies, travel costs or wage differentials to induce people to the region. (3) Finally there is an education activity which upgrades the skill levels demanded by the optimal production program. All of these are assumed undertaken by the state as part of its development program and transferred to private industry at the price of their natural counterparts (where appropriate).

The second feature relates to the investment equations of the model and their usefulness in generating the cyclical effects of resource development. The investment equation for the mining sector is distinct from that of the rest of the economy. The mining sector's investment equation (11A) reflects the strategic role this sector plays as a finite economic base upon which to build. The unique aspect of this equation is the inclusion of a capital stock adjustment factor which is a function of the resource remaining in the ground (see eq. 11C above). The prospect of holding a large stock of specialized capital at the time of depletion must be balanced against the gains made from the increased extraction rate. As depletion nears, the model calls for a slowdown of both net and replacement investment and more intensive utilization of the existing capital stock.

The time path of the model and the cycles inherent in resource exploitation are derived from the reduced form of equation (5').* That is:

$$X_j^t = A\left(\frac{-\Gamma b_{ij}^{t-1}}{a + \Gamma^t b_{ij}^t + \Pi^t d_{ij}^t - 1}\right)^t \qquad (5')$$

The improvement wrought by resource development planning within an overall economic development planning context can be determined by performing three simulations. The first is a baseline projection consisting of a private optimum depletion path (calculated from any conventional model of the mine, e.g. Scott, 1967) and the time path of the rest of the economy (based upon an extrapolation of present conditions via an input-output model). The second simulation would involve an integration of the resource sector into an overall planning effort (calculated with the model presented in this section). The final simulation would call for the consideration of post-depletion conditions (calculated by extending the time horizon in the above model). Thus the difference between simulation 1 and 2 would reveal the improvement due to integrated development planning. The difference between models 2 and 3 would represent the improvement attributable to a consideration of post-depletion factors.

The important questions which can be addressed by the model include:

(1) the optimal rate of extraction of a series of nonrenewable resources,
(2) the identifications of "bottlenecks" to development,
(3) the optimal allocation of capital between sectors; likewise the identification of key industries to induce development,
(4) the optimal sequencing of non-renewable resource extraction and construction of infrastructure,
(5) the cyclical pattern of resource development,
(6) the sensitivity of an economy to exogenous factors* relating to non-renewable resources,
(7) the policy mix needed to adjust to such factors in an optimal manner.

---

* Exogenous factors include technological displacement and price fluctuations as well as depletion.

AN APPLICATION OF THE MODEL

The Mexican State of Chihuahua provides a good example of a large arid region with an economy based primarily on dry land agriculture and mining. Located between Texas, Sonora, Coahuila and Durango, Chihuahua is made up of a dry central plain with rainfall in 250-400mm range, bordered by a fringe of the Sierra Madre Occidental on the west.

The population of the area is involved in three major economic pursuits. First, tourism, the bracero program, and the border industries, have attracted large numbers of workers from throughout Mexico. Wages are relatively high when work is available but the un- and under-employment rates are substantial. Second, about 60,000 men find employment working in Chihuahua's 59 mines (Bonine, 1970). Employment in this area varies sharply with the substantial changes in the levels of production (see the Maximum Variation column of Table I) which themselves vary with the vicissitudes of the export markets. Finally, about one-half million people (about 43% of the total state population) work in agriculture where disguised unemployment is a major problem.

TABLE I
Output of Major Minerals in Chihuahua, 1970-1974

Resource	Unit	1970	1971	1972	1973	1974	Maximum Variation
Zinc	1000 Metric Tons	266	265	262	271	263	263/372
Molybdenum	tons	85	79	78	41	43	41/85
Mercury	tons	1043	1220	776	700	894	700/1200
Manganese	tons	99	96	106	131	145	96/145
Lead	1000 tons	177	157	161	179	218	157/218
Gold	Kg.	6166	4692	4543	4117	4182	4117/6166
Silver	tons	1332	1140	1166	1206	1168	1140/1332
Tungsten	tons	288	408	362	348	309	262/408
Copper	1000 tons	61	63	79	81	83	81/83
Iron	1000 tons	2612	2819	3053	3113	3338	2612/3338

Source: United Nations (1974).

As might be imagined in this context, agriculture is primarily based on dry land farming of wheat and other suitable field crops along with a substantial amount of stock grazing. Unfortunately only 53,000 hectares (127,000 acres) are actually irraged of a total land area of 247,000 Km^2 (Bonine, 1970). Much of the limitation in this regard is due to the non-strategic location of the rivers, which generally run along the periphery of the extensive arid plains (the Bólson de Mapimi). However, another contributing factor has been the failure of the government to find funds for more than 3 projects totalling a meager 405 million m^3 in impounded surface water capacity in 61 years in all of Chihuahua. Additionally, money for surveying and development of groundwater resources has been severely limited by budgetary constraints.

Mining, on the other hand, has had a long and interesting history in the area.* Gold and silver have been extracted for many years in large amounts, and more recently zinc, copper, tungsten, manganese, molybdenum and mercury have been extracted in worthwhile but relatively small amounts. Iron output is also significant but the leading producer in the area, in relation to world production, is lead, with the mine network at Hidalgo de Parral being one of the world's largest current producers (see Table I). Ranked in value of ore produced Chihuahua is first among all the Mexican states in both ferrous metals (value of output in 1965 = 62 million pesos) and non-ferrous metals (at 1.3 million pesos in 1965). Approximately 11,000 miners were employed in the ferrous metal mines and 50,000 in the non-ferrous (Bonine, 1970).

Investigating the forward linkages on the non-ferrous side of the industry we find that only one smelter handling these ores is available in all of Chihuahua. What is to be found is that two major rail lines run from the Chihuahua mining areas: one to El Paso and the second to the major ore producing-smelting areas of Arizona.

Of equal negative importance for the local economy is that no steel making facilities exist within the state. All the iron ore must either be shipped to mills in Coahuila, Neuvo Leon or Central Mexico. This means that it is unlikely that Chihuahua will have an opportunity to develop a manufacturing industry based on its raw materials. The data bear this out:

TABLE II

Value of Production of Primary Iron and Steel Industries, 1965
(thousand pesos)

1. Nuevo Leon	2,875,000
2. Coahuila	2,548,000
3. Mexico (State of)	1,589,000
4. Vera Cruz	814,000
5. Distrito Federal	709,000
6. Chihuahua	145,000

Source: Bonine (1970).

* The emphasis here is on mineral resources. Other non-renewable resources such as geothermal brines, which can be used in the generation of electricity, process heat for industry, or desalination, are also located in the Northern deserts and may play a prominent role in the region's development.

The situation in the non-ferrous area is even worse: of 74 establishments in all Mexico for the fabrication of primary non-ferrous metals not one is located in Chihuahua and only four are located in adjacent states.

We find that the forward linkages within Chihuahua are virtually non-existent. This means that not only are the "terms of trade" going to be against Chihuahua in relation to other Mexican states and the U.S. but that an important avenue for reducing unemployment both in agriculture and in the border zones has been blocked.

In combination these factors make Chihuahua an ideal example for the examination of policies for the mitigation of problems of unemployment, business cycles and optimal patterns of resource extraction. Our study builds on the works of King (1970), Barkin and King (1970), Manne (1966), and Manne and Goreux (1973). King's first work provides an overview of the Mexican economy with special attention to trade and industry in the years 1940-65. Barkin and King are concerned with the development of Mexican surface water resources generally, but focus attention on a case study of the Tepalcatepac River basin in Michoacan-Jalisco, where tropical-temperate conditions prevail and where little mining exists. The studies by Manne and Goreux have emphasized the modeling aspects of development planning in Mexico. Unfortunately, they have not emphasized either the problems associated with the depletion of exhaustible resources nor the complications of their exploitation in arid regions. These studies have been helpful, however, in the determination of Mexico's ample capacity for integrating new capital goods industries and the provision of the input-output tables we are currently utilizing in regionalized form for our policy simulations.

CONCLUSION

This paper has emphasized the strategic and ever more promising role of non-renewable resources in the development of arid lands. It has suggested that mining can be a key primary sector and has several advantages over agriculture in that it is less water intensive and a more direct link to manufacturing. We have suggested a strategy of vertical integration at first in order to speed development and extend the life of the resource. Diversification is suggested at later stages in order to protect against possible cyclical downswings in the wake of depletion.

Conditions vary greatly between countries even in the arid zone so that the optimal pattern of development may warrant more or less emphasis on the propositions presented here. It is hoped that the model we have sketched will be useful in testing some of these propositions and thus help further the development of arid lands.

APPENDIX I
Variable List

Endogenous	Exogenous
C ≡ total private consumption	
C_i ≡ private consumption of commodity i	
C_m ≡ private consumption of imports	
C_v ≡ residual consumption	
E_i ≡ exports of commodity i	$\bar{E}_i$ ≡ limitation on the export of commodity i
F ≡ total foreign capital inflows	
F_a ≡ foreign aid	
F_p ≡ private foreign investment	
F_v ≡ inflows due to residual export sales	G ≡ total government spending
G_F ≡ spending for augmenting resources	G_M ≡ government spending on exports
I ≡ total investment	
I_m ≡ investment in imports	
I_r ≡ investment in mining industry r	
K ≡ total capital	
K_r ≡ capital in mining industry r	
L_D ≡ demand for labor	
L_S ≡ supply of labor	
L_{fS} ≡ training of workforce to skill s	
L_s ≡ workforce with skill s	
M ≡ imports	
N ≡ total population	
N_f ≡ immigration	
R_r^t ≡ amount of ore reserves remaining in period t in industry r	$\bar{R}_r$ ≡ total volume of resource r at t=0
S ≡ import substitution	
U_s ≡ unemployment of workers of skill s	
U ≡ aggregate unemployment	
V ≡ GNP	
W_D ≡ demand for water	W^t ≡ amount of existing water supply in period t
W_S ≡ supply of water	
W_e ≡ water augmentation	
X_j ≡ aggregate output of industry j	Π ≡ capital stock adjustment mechanism (replacement I)
X_r ≡ output of the mining sector	Γ ≡ capital stock adjustment mechanism (net I)

Parameters

a_{ij} ≡ input-output coefficients
b_{ir} ≡ capital coefficient for mining industry r
b_{ij} ≡ investment coefficient for all other industries
c_L ≡ unit cost of producing an extra unit of labor with skill s
c_N ≡ unit cost of importing a unit of new labor N_f
d_{ij} ≡ replacement investment coefficient
ℓ_{sj} ≡ labor input coefficient
m_j ≡ import coefficient
w^t ≡ water input coefficient
$\alpha_{i,m}$ ≡ autonomous consumption
$\beta_{i,m}$ ≡ marginal propensities to consume
γ_1 ≡ population growth rate
γ_2 ≡ maximum absorption rate for immigrants
δ ≡ rate of social time preference
ξ_s ≡ attrition rate for each skill
λ^t ≡ acceptable unemployment rate in year t
λ_f^t ≡ acceptable unemployment rate for immigrant workers
μ_i ≡ export value of a unit of commodity i
μ_v ≡ export value of a unit of residual exports
μ_m ≡ average value of imports
ρ^t ≡ labor force participation rate for local workers
ρ_f^t ≡ labor force participation rate for immigrant workers
ω ≡ coefficient of limitation on government spending for resource enhancement

APPENDIX II
Sectors of the Mexican Model

1. Agriculture
2A-J. Mining (disaggregated by mineral)
3. Petroleum
4. Food
5. Textiles
6. Wood
7. Chemicals
8. Non-metals Processing
9. Basic Metals Processing
10. Machinery
11. Construction
12. Electricity
13. Commerce
14. Transportation
15. Services

REFERENCES

Akhdar, F.H. (1974), Multinational Firms and Developing Countries: A Case Study of the Impact of the Arabian American Oil Company "ARAMCO" on the Development of the Saudi Arabian Economy, Unpublished Doctoral Dissertation, University of California, Riverside.

Amiran, D.H.K. (1965), Arid Zone Development: A Reappraisal Under Modern Technical Conditions, Economic Geography, 41, 5, 189-210.

Aschman, H. (1970), A Natural History of the Mine, Economic Geography, 46, 2, 172-189.

Barkin, D. and T. King (1970), Regional Economic Development: The River Basin Approach in Mexico, Cambridge University Press, London.

Bonine, M.E., et al (1970), Atlas of Mexico, University of Texas, Austin.

Bruno, M., C. Dougherty and M. Fraenkel (1970), Dynamic Input-Output, Trade and Development, in A. Carter and A. Brody, eds., Applications of 1-0 Analysis, North-Holland Publishing Co., Amsterdam.

Chenery, H.B. and A. MacEwan (1966), Optimal Patterns of Growth and Aid: The Case of Pakistan, in I. Adelman and E. Thorbecke, eds., The Theory and Design of Development, John Hopkins Press, Baltimore.

Cummings, R. (1972), Water Resource Management in Northern Mexico, Resources for the Future, Baltimore.

Golladay, F.L. and A.D. Sandoval (1972), Optimal Development Policy in an Open Regional Economy, Journal of Regional Science, 12, 2, 185-198.

Goreux, L.M. and A.S. Manne (1973), Multilevel Planning Case Studies in Mexico, North-Holland, Amsterdam.

Hall, M. (1969), New Dimensions in Natural Resources--Discussion, American Economic Review, 55, 2, 129-131.

Kelso, M.M. (1970), Economic Growth Opportunities and Constraints, in H.E. Dregne, Arid Lands in Transition, American Association for the Advancement of Science, Washington.

King, T. (1970), Mexico: Industrialization and Trade Policies Since 1940, Oxford University Press, London.

Manne, A.S. (1966), Key Sectors of the Mexican Economy, in I. Adelman and E. Thorbecke, eds., The Theory and Design of Economic Development, John Hopkins Press, Baltimore.

Murakami, Y., et al (1970), Efficient Paths of Accumulation and the Turnpike of the Japanese Economy, in A. Carter and A. Brody, eds., Applications in I-O Analysis, North-Holland Publishing Co., Amsterdam.

Scott, A. (1967), The Theory of the Mine Under Conditions of Certainty, in M. Gaffney, ed., Extractive Resources and Taxation, University of Wisconsin Press, Madison.

Webb, M.J. (1969), The Impact of Capital Intensive Industries on Arid Australia, in R.O. Slatyer and R.A. Perry, Arid Lands of Australia, Australian National University Press, Canberra.

THE ROLE OF THE UNIVERSITY IN APPLIED TECHNICAL ASSISTANCE

Michael E. Norvelle, Research Associate;
Donald L. Percious, Research Associate;
William H. Brooks, Research Associate;
Kennith E. Foster, Associate Director;
Office of Arid Lands Studies - University of Arizona
Tucson, Arizona

ABSTRACT

Many of the developing nations occupy the arid and desert regions of the world and when considering the development of these areas, the application of technical assistance and the transfer of technology is of central importance. At the same time that strategies are being formulated to reckon with the development and management of these area resources, the means must be provided to transfer these technical solutions to those who will put them to work - often in less technologically advanced societies than those in which they were devised.

This paper outlines the philosophy and structure of the Laboratory of Native Development, Systems Analysis and Applied Technology hereafter called (NADSAT) established at the University of Arizona. Although based in the U.S. and designed to assist the economic development of the Indian tribes of the arid and semi-arid Southwest portion of this country, the program's approach is viewed as potentially transferable to universities of developing areas throughout the world. NADSAT has been designed to attack certain problems in the technical assistance process. Its initiators believe that this program is a departure from the normal pattern of university sponsored or affiliated technical assistance and a great departure from the patterns of private consultants.

Universities have been actively engaged in international technical assistance on many levels; on home ground, however, their role in development has been of an essentially passive nature. In mid-1976 the Laboratory of Native Development, Systems Analysis and Applied Technology (NADSAT) was initiated through a grant from the Office of Technical Assistance, Economic Development Administration (EDA), U. S. Department of Commerce. The program as originally conceived as two primary objectives: 1) to actively provide technical assistance to American Indian tribes of the arid and semi-arid Southwestern United States for more effective utilization and economic development of their reservations' natural resources and; 2) to transfer current technology to the tribes to further assist them in their expressed goal of greater self-determination. The program has since been broadened to include administrative management assistance through contracted funds from the Bureau of Indian Affairs in response to the technical assistance requirements of Public Law 93-638.

NADSAT is organized to operate with a small core staff of broad experience in the natural resources development area which works in conjunction with other programs and departments within the university community so that it has the ability to tap the skills of experts in a wide variety of disciplines. When the expertise needed is unavailable within the university, the program has the flexibility to subcontract with qualified organizations outside that community in order to fulfill program needs. Temporary staff are added on the basis of project requirements to supplement permanent staff capabilities and to increase the number of projects capable of being accomplished at any one time.

Conceptually, the program operates in response to the specified needs of the tribes providing much greater assurance that the tribe concerned has a definite stake in the results and on the premise that the University, as one of its educational functions, will continue to serve the recipient after the formal conclusion of the project providing explication of problems that inevitably crop up in any undertaking.

An endeavor is made to transfer technology during project operation through:

1. direct involvement to tribal personnel throughout the duration of the project.
2. appropriate methods of information presentation.
3. specific technical training in project related fields.

Technical assistance provided by colleges and universities can fill important gaps which currently exist in the linkages connecting science and the technological advances with the user in developing communities concerned with realizing their economic potential. The university as a source of scientific research and development of technology can actively participate in the application of methods and systems best suited to particular local conditions due to their intimate knowledge of the immediate environment. And, they may act as an intermediary between imported technology which can need modification and its application.

INTRODUCTION

Many developing nations occupy the world's arid and desert regions. When considering developments of these areas the application of technical assistance and the transfer of technology is of central importance. At the same time that strategies are formulated to reckon with development and management of resouces in these areas, plans need to be made with respect to implementing them: the means must be provided to transfer technical solutions to those persons who will employ them, persons often working in less technologically advanced societies than those in which the technologies were devised.

This paper discusses the philosophy and structure of the Laboratory of Native Development, Systems Analysis and Applied Technology (NADSAT) established at the University of Arizona as a working example of one alternative to provide technical assistance and accomplish technology transfer. Although based in the United States and designed to assist the economic development efforts undertaken by the Indian Tribes of the arid and semi-arid U.S. Southwest, program's approach is potentially transferable to universities in developing areas throughout the world. NADSAT is designed to attack certain problems inherent in standard technical assistance processes. Its initiators believe that the NADSAT approach is a departure from the normal pattern of university-sponsored or affiliated technical assistance and a great departure from the pattern of private consultants.

In the evolution of standard technical assistance systems and as they now operate, development has often been approached as though it were possible to provide meaningful assistance through relatively brief contacts with representatives of recipient societies. Commonly, specific problems are studied in varying degrees of depth, conclusions drawn, recommendations made, reports filed and the consultants, most of whom are well-meaning depart, because their commitment to the project then is fulfilled. This process is often completed with less than optimal contact with those persons ultimately responsible for implementing recommendations at the local level. Within this framework rarely do consultants remain long enough to stimulate and to follow the implementation process (Solo, 1975). Stacks of unused feasibility reports gather dust on bookshelves throughout developing regions of the world.

Although individuals providing assistance usually are competent, their technical expertise frequently is most suitable for application within the geoclimatic and socio-political environment in which they were trained. The consultant's normally short advisory capacity tenure often does not afford the opportunity to adjust suggested technology to specific local conditions (Solo, 1975). In addition, large numbers of technical advisers are not trained in the social change processes; therefore they have little learned feeling for the magnitude of social adjustment required by the recipient of technical assistance. Here, it is not the accuracy of technology problem solution in question, but the appropriateness of the solution to specific local conditions which can tend to diminish its feasibility: the recipients of technical assistance may not be able societally to appreciate and supplement the technical applications recommended.

This problem is compounded by lack of direct communication between adviser and user, as well as the sociological factor of hesitancy on the part of recipients to trade a "secure" status quo method for an advanced system to whom the reliability of which has not been thoroughly demonstrated in most cases. In these situations experience points to the tendency among recipients to let adviser-consultants "do it for them" and when the adviser-consultant departs, the application of advanced technology well may terminate.

An example may illustrate this point. During a program to increase agricultural production in the southwest of Saudi Arabia, an area with limited groundwater, members of a western advisory group suggested that a trickle system was technically the best method for a certain farmer to irrigate his orchard. In view of the farmers limited water supply and his practice of using unlined ditches in a sandy, well-drained soil which provided less than an adequate supply of irrigation water, the advisers were undoubtedly technically correct in their suggestion. But this traditional farmer would not conceptualize how his orchard could gain sufficient water from the low-volume output of the emitter no matter how many disgrams were presented to him, and refused to use the system after it was installed and operational. On the other hand, an intermediate system utilizing pipes to convey water to secondary and tertiary ditches and/or tree basins was readily accepted by other farmers in the region. It was a system to which they related and yet one which provided sizable improvements in water-use efficiency. Perhaps it was not the technical optimum, but was certainly the cultural optimum at this early stage of Saudi Arabia's agricultural development program.

Under the normal mechanism of technical assistance, technology transfers, the keys to technical assistance processes are difficult to accomplish, often because of the inherently short tenures of the operations. Technical assistance programs normally are designed to provide reports or to construct facilities without involvement in training functions. These training functions often are left to schools, colleges, and universities and there is little assurance that graduates will implement programs or operate facilities.

Universities have been actively engaged in international technical assistance on many levels through efforts of individual faculty members, direct contact with representatives of recipient countries government-financed programs sponsored by agencies such as the U.S. Agency for International (AID) and consortia such as the Consortium for International Development (CID). On home ground, however, university roles in development have been essentially passive; for example, these universities often do not motivate foreign students who are expected later to apply university training in their jobs. Universities have been among the most important sources of research in new technologies or better applications of current knowledge and certainly some portion of new or improved techniques will eventually reach developing communities. But, the application of science and technology in field situations, whether in industry, commerce, agriculture or, in this case, technical assistance generally has been left to others. Notable exceptions to this pattern include the Agricultural Extension Services, a direct link between the developer and user of technology, and the sometimes semi-institutionalized connection between various colleges of medicine and education and rural development programs in the United States.

DISCUSSION

NADSAT, established in mid-1976, is funded through a grant from the Office of Technical Assistance, Economic Development Administration (EDA), U.S. Department of Commerce. The program administered by the University of Arizona Office of Arid Lands Studies (OALS), has two primary objectives: 1) to provide technical assistance to American Indian Tribes of the arid and semi-arid Southwest United States for more effective utilization and economic development of their reservations' natural resources and; 2) to transfer current technology to the tribes to further assist them in their quest for greater self-determination. EDA is particularly supportive of projects designed to provide tangible results in advancing tribal economic development.

The three conditions necessary for increased tribal economic development offered by Anderson (1976) are: 1) jurisdictional control over their resources, 2) financial capital, and 3) management assistance. The NADSAT program seeks to supply tools for more effective natural resources management and, as a result, to contribute to American Indian economic development. The program was expanded recently to include administrative and management assistance through contracted funds from the U.S. Bureau of Indian Affairs (BIA) in response to the technical assistance requirements of Public Law 93-638, the 1975 Indian Self-Determination and Education Assistance Act.

NADSAT is organized to operate with a small, core staff who have broad experience in natural resources development. The core staff undertakes projects within its scope of expertise -- hydrology and water resources management: agricultural development, management and economics; and, range and forest ecology and management. In projects involving other areas of specialty such as geological engineering and geophysics, or industrial development, cooperative programs are conducted by other OALS staff and University departments under NADSAT auspices. For example, close coordination is maintained by NADSAT with the Applied Remote Sensing Program of OALS. When the expertise needed is unavailable within the University, NADSAT has the flexibility to subcontract with qualified organizations outside the University to fulfill program needs. Temporary staff are added on the basis of project requirements to supplement permanent staff capabilities and to increase the number of projects capable of being accomplished at any one time. This structure enables the program to employ the multidisciplinary approaches in providing technical assistance -- an acute need when dealing with natural resources development and management.

Projects which the program can address include inventories using remote sensing and conventional ground water survey methods to assess land, water, and mineral resources; operations research and modeling; management systems development; economic feasibility analyses; technical training; social service assistance; community development assistance; and industrial development assistance.

Conceptually, the program operates in response to the specified needs as identified by the tribes. It is not NADSAT's function to define tribal development requirements and goals. When a request for technical assistance is issued by a tribal governing body, rather than an outside agency, there is much greater assurance that the tribe concerned has a definite commitment in getting and using results.

NADSAT operates on the premise that the University, as one of its educational responsibilities, will continue to serve recipient tribes after the formal conclusion of projects providing further explication of problems that inevitably arise in any undertaking. Ties of an informal nature also are established between members of the University community and the tribes. These ties are mutually beneficial and can be as simple as a telephone number of a University staff member able to direct Indian questions to the best source of information. Such a beginning is of no small importance for Indians beginning to cope with the complex and often anonymous university organization.

Another important factor is the geographic area with which NADSAT functions. Resource development problems are regionally specific, in varying degrees, depending upon the development problems involved. For example, agricultural development involves geoclimatic crop suitability, variance in compatibility of irrigation methods, soil management, water resources, etc. The University's resources are also regionally specific, in varying degrees, depending upon the department involved. Although general informational and study bases are provided, advanced students and faculty, particularly those in earth sciences and agriculture, often become regionalized because they work with regional examples in the study environment. The University of Arizona NADSAT program operates within the arid and semi-arid Southwest United States, with the OALS providing one of the foci for natural resource studies on campus. Alternative technological applications recommended by the program may require some modification before application is made within Indian socio-institutional frameworks, but we can be confident that those involved with NADSAT are specialists in the physical environment. In addition, coordinating the phases of various projects, from formulation through presentation of the final report, with members of the tribe, University staff and others closely associated with the tribe, assures that projects conform to the needs, objectives, and cultural framework of a particular tribe. For this reason, close alignment is maintained with the University of Arizona Office of Indian Programs.

During project operation endeavors are made to transfer technology through:

1. direct involvement of tribal personnel throughout the project, including specification of its constituent elements, assessment techniques, and analyses methods;

2. appropriate methods of information presentation. The method by which information is presented is as important to the accomplishment of technology transfer as the information itself is to the total technical assistance function. Therefore, reporting is accomplished to provide the recipient tribe with an appreciation of the types of data acquired and their relationship to the conclusions drawn. For example, although several hydrologic and agricultural development studies have been completed concerning a river basin on one reservation, no one had effectively articulated to tribal planners and decision makers the relationship between soil and crop potentials and the poor quality of the

water in that basin. Consequently, during a preliminary reporting session tribal members still believed that all that was necessary to farm the area was to divert or pump the available water onto the land. The problem of paternalistic agencies having "done it for them" in the past undoubtedly played a role in this instance of misunderstanding. A primary step in technology transfer is to convey an appreication for the relevance of the technology to the situation at hand in language and form that will be understood and accepted by the recipient tribe. Therefore, emphasis is given to oral presentation with graphic illustration in support of narrative reports; and

3. specific technical training related to inventory and assessment methods, information systems, irrigation technology, accounting procedures, and other subjects as they arise.

It has been the experience of NADSAT that an appreciation of the scientific method of conceptualization and analysis is not generally a feature of American Indian culture. The scientific method is, in fact, a transplant from an alien culture. This factor can be a hinderance in the application of useful technology by the tribe. American Indian culture, with few exceptions, traditionally has been rooted in the concept of gaining sustenance and well-being from that which nature provides - intrinsically a non-manipulative perception of the natural environment. The American Indian has not been engaged in a contest with nature. The 18th Century Seneca Chief, Red Jacket, said it this way:

> There was a time when our fore fathers owned this great island. Their seats extended from the rising to the setting of the sun. The Great Spirit has made it for the use of Indians. He created the buffalo, the deer, and other animals for food. He made the bear and the deer, and their skins served us for clothing. He has scattered them over the country, and had taught us how to take them. He caused the earth to produce corn for bread. All this he had done for his red children because he loved them...
> (Anderson, 1976).

Therefore, the conveyance of non-Indian perceptions without seeking diminution of Indian cultural views, is an essential task to be performed by this program under the scope of its technology transfer objectives, i.e., not only to transfer the method of doing something in a certain way, but also to transfer an appreciation of why the method is being utilized. Effecting this double transfer will not only lay the ground work for increased applications of technological methods to aid in attaining the tribe's economic development objectives, but also will enhance communication between advisers, analysts, and the tribes. As a result, the tribe gains substantive knowledge from the advisers work. Over a period of time, reliance on the services of outsiders is reduced, consequently advancing the tribe's goals of self-determination in controlling and managing their land resources. In effect, the aim of the University's program is to go out of business - an objective that those engaged in private enterprise dare not seek.

To accomplish this aim, NADSAT's technology transfer emphases has been placed upon:

1) the general approach of technical analysis;
2) specific analytical methods;

3) specific operational management requirements;
4) specific operational methods; and
5) operational management assistance in an operational environment.

In addition, the development of a framework for decision making, based on cost-effectiveness analysis also have received considerable attention. However, since this is the subject covered by another paper as it applies to water resources management, it is only mentioned here. (Percious, *et al*, 1977).

CONCLUSION

Technical assistance provided by colleges and universities can fill important gaps which currently exist in the linkages between science and technological advances and users in developing areas who are concerned with realizing the economic potentials of their lands (Hawthorne, 1970). The university as a source of scientific research and technology development can actively participate in the application of methods and systems best suited to particular local conditions because of its intimate knowledge of the immediate environment. And, it may act as an intermediary between imported technology which can need modification and its application. Universities and colleges are here defined narrowly as those institutions which are: (1) situated among or near recipient groups, and (2) involved, in varying degrees, in an educational relationship with memebers of those groups. Assistance on an international level, or perhaps even on a domestic interregional level, is included only peripherally. This is technical assistance provided by "insiders" - optimally those representatives of the institution best able to transfer technology according to the requirements of a particular local, physical environment. Thses individuals should have or acquire a reasonably intimate knowledge of the cultural milieu and institutional framework. Indeed, the ideal technical adviser would be found from within the recipient group itself and hopefully will be more frequently in the future. Understanding and trust are often still basic to working relationships in traditional communities to establish effective dialogues. Two-way channels of communication greatly enhance the ability to go beyond the provision of a product and to effect a transfer of technology, or at least to achieve an appreciation of the technology's relationship to the product and the relationship of the product to desired goals.

NADSAT intends to approach the barrier of technology transfer in very practical terms, with the first principle being the certain knowledge that each situation is unique. For each group of people, whether they are American Indians or the inhabitants of developing countries, has very individualized requirements if proposals for development and change are to be adopted in concrete terms. Concrete is here defined as resulting at some point in an ongoing developing activity carried on by the recipients of technical assistance.

ACKNOWLEDGEMENT

The Operation of the Laboratory of Native Development, Systems Analysis and Applied Technology (NADSAT) is supported by a grant (No. 99-6-09555) from the Office of Technical Assistance, Economic Development Administration, U. S. Department of Commerce.

REFERENCES

Anderson, A.T., Nations Within A Nation, The American Indian and the Government of the United States, Union Carbide Corporation, 1976.

Domergue, Maurice, Technical Assistance, Theory, Practice and Policies, New York: Frederick A. Praeger, 1968.

Gruker, William H. and Marquis, Donald G., Factors in the Transfer of Technology, Cambridge, Mass.: The M.I.T. Press, 1969.

Hawthorne, Edward P., The Transfer of Technology, Istanbul: Organization for Economic Corporation and Development, 1970.

Percious, D., Norvelle, M., Wright, N., Foster, K., Mouat, D., "An Appropriate Multidisciplinary Methodology for Water Resources Planning and Decision Making:" Paper Presented at UNITAR Conference on Alternate Strategies for Desert Development and Management, Sacramento, California, U.S.A., June, 1977.

Solo, Robert A., Organizing Science for Technology Transfer in Economic Development, East Lansing: Michigan State University Press, 1975.

"HUMAN SETTLEMENTS AND DESERT DEVELOPMENT"

The Centre for Housing, Building and Planning
Department of Economic and Social Affairs
United Nations

INTRODUCTION

As the title of the Conference suggests, the theme is largely exploratory -- its concern is with the less widely studies aspects of desert development, particularly those which relate to the integration of arid and semi-arid areas into the mainstream of national development planning. While problems of water supplies and related hydrological and hydrogeological matters predominate in programme planning for the development of desert regions, the lessons of experience point clearly to the need to approach the issue on a broader front. One such area of the broad front is that of human settlements, that is, the network of communities ranging from the smallest village to the largest metropolis.

The most significant event in recent years in the field of human settlements was the convening of Habitat: United Nations Conference on Human Settlements which was held from 31 May to 11 June 1976 at Vancouver, Canada. When dealing with the subject of "Human Settlements and desert development" it is perhaps instructive to recall the very first General Principle of the Habitat Conference which stated that "the improvement of the quality of life of human beings is the first and most important objective of every human settlement policy."

The very term "development" in a national or regional context connotes policies and actions for the betterment of the social and economic life of people. It is hard to envisage development -- desert development or development in any other environment where man is the main actor -- without human settlements. From another viewpoint, human settlements themselves constitute an instrument of development, with interdependence, both spatial and economic, with various development sectors.

In view of the oft-cited fact that arid and semi-arid lands cover one-third of the globe's land area, and that they are the habitat of one-sixth of the world's population, prospects for physical, social and economic development of these regions are inextricably linked with human settlements policies and planning.

The purpose of this paper is to outline some of the issues that could provide the framework for considering the role of human settlements in the context of desert development. In order to further define the scope of the paper a distinction should be made between arid and semi-arid areas, particularly areas undergoing the process of desertification as a result of human actions. Discussion of the human settlements implication of the desertification process due to soil erosion and deforestation in areas such as the Himalayan Arc (comprising the Indus, Ganges and Brahmaputra rivers and their tributaries) lies beyond the scope of the present paper. In short, the focus here is on human settlements in the context of existing desert areas rather than areas subject to the encroachment of desertification through such causes as the burning of vegetation, overgrazing, woodcutting and the overcultivation of marginal lands and other similar phenomena associated with human settlements development.

THREE TYPES OF DESERT SETTLEMENT PLANNING

Three separate but related settlement situations will be considered:

(i) New desert settlements

(ii) Existing desert settlements

(iii) Settlements in the context of nomadism

The crucial common feature is the fragility of the eco-system and the shortage of water. Indeed it is precisely the latter which underscores the difference between settlements planning in a dynamic sense in the context of a hot-humid environment and in the desert environment. Though in different ways, the eco-system in the humid environment is just as delicate as in a desert environment and both share other similarities regarding heat exchange and high temperature. But the scarcity of water in deserts necessitates an entirely different approach.

Desert areas have become in relatively recent years the focus of interest and concern, particularly in the developing regions of the world. This has come about for different reasons in different countries and regions -- population factors and social change, new wealth in some parts, the need to exploit the desert's richness of mineral deposits -- but whatever the motivating forces, the one factor that makes any such development immediately possible is the availability of powerful technology.

In the past the location, form and socio-economic functions of desert communities followed the common sense and survival instincts of their inhabitants. These processes evolved over a very long time span, usually centuries, and reflected, in the purest sense, man's constant battle and accomodation with the forces of nature. Until recent times, in the extreme conditions of desert regions, the environmental forces pressed hard on the evolution and development of society and human settlements, making progress -- though beneficial to communities as a whole -- a slow process.

In the present-day context the need for development is accelerated, as are the means. Using new technologies, the age-old balance between the terms of man's settlement and the natural elements is dramatically altered. The time frame of implementation is also shortened. As compared with the historical situation, change can be sudden and fundamental rather than gradual and marginal.

What man learned in generations past as a result of the slow and frequently painful lessons of experience must now be anticipated and acted upon before the fact through careful planning. By comparison with the past, present and future desert area development initiatives are too big in scale and too costly in resources to be easily attenuated or reversed if mistakes are made. While this condition is broadly true of many facets of modern industrial and social life, it is particularly valid in the present context where both the fragility of the eco-system and the physical and financial resources available to the countries in question do not permit a wide tolerance for error.

The same technology which in the space of a generation has so vastly expanded the horizons of development has also increased the temptation for expedient and short-term solutions to major problems. Lack of experience in dealing with the phenomenon of rapid growth can create new problems leading to adverse social, cultural, political and environmental consequences.

Planning human settlements, since it touches so fundamentally on the quality of life in all its dimensions must be integrated at the outset with planning for social and economic development. In the sparse and inhospitable environment of the desert, such integrative planning is especially critical because the margin for error, in both physical and social terms, is slight.

On the positive side the availability of new technology, combined with the wealth of accumulated local experience offers challenges and opportunities for creating new systems of human settlements in a relatively unspoiled physical environment.

NEW DESERT SETTLEMENTS

While considering the development of new desert settlements a distinction should be made between human settlements as a dynamic element in social and economic development and human settlements as a service element in the same context. Stated briefly, the term "service element" here refers to the circumstance in which the existence of the settlement is predicated by other objectives, for example, supporting mining activity or servicing a tourist resort. This is not to say that such a settlement may not play a dynamic role at a later stage, or that the activity involved in its actual construction may not induce other activities and so play a dynamic role by virtue of the linkages forged with other sectors. But its initial conception is in a service sense to support other basic activities. Thus, a settlement of this type plays a "residual" role.

In the dynamic sense, however, the settlement plays a pioneering and catalytic role in opening up a hinterland, usually as a staging-point for radial exploration and development. In this situation, the settlement is the "leader" and the economic base is the "follower." Though it represents, probably history's most dramatic example, Brasilia is essentially a recent manifestation of the age-old process of establishing settlements in the van of opening up a new frontier.

There is, however, an important difference between the Brasilia model and its historical antecedents: whereas former outpost settlements frequently originated from military garrisons in one form or another (that being their fundamental "economic base," and raison d'etre), the Brasilia model is predicated not on a military, politico-expansionist drive but rather on a base of central government administration. Thus, the economic base of the settlement is ready-made and large enough to support the requisite diversity of service activities, leaping over traditional evolutionary stages. While this kind of settlement adventure constitutes the extreme form of deliberate settlement planning, the important lesson in the present context is the degree to which national governments have the means to direct resources, physical, human and financial to deliberate development of hitherto undeveloped or underdeveloped regions. The issue is definitely pertinent to developing countries that might be planning to develop new settlements in desert areas.

The question is whether this model of settlement development is viable or desirable in the context of desert development -- even allowing for considerable reduction in scale. If it is not, what other approaches are feasible. Preliminary consideration of this question raises immediately a number of others, such as: (1) what natural resources are available for exploitation and what is the spatial distribution of such resources, (2) the location and volume of water resources, (3) the type and scale of technological possibilities for the provision of water and energy,

(4) the possibilities for reclaiming sizeable tracts of potentially suitable soil for agricultural production, (5) proximity of market outlets and the availability of necessary inputs for production activities, (6) the social problems arising out of isolation, a consequence of settlement dispersal made necessary by the low support capacity of desert areas.

No settlement can sustain itself for long without a viable economic base. In the case of new desert settlements in developing regions of the world it seems reasonable in the light of present-day knowledge to suppose that they would favour the "resource follower" model. The supposition is that any such settlement would be contemplated because it is essential for the servicing of some economic activity which is determined to have potential for national or regional economic and social development. The likelihood is that the economic activity would be extractive or agricultural in nature. It is logical, therefore, that the settlements to be built would be geared in size and form to complement these functions and to play an appropriate role as a part of the broader national network of human settlements.

Another issue relates to the time horizon for the activity. If the resources in question would be exhausted in a given area in relatively few years then in all probability a temporary settlement would be best. In this kind of situation and where the physical conditions are critical it is reasonable that the form and structure of the settlement should correspond to the requirements of the economic activity, provided however that minimal environmental conditions are met. The optimum specification for the settlement in this case will depend on the scale of the economic activity itself -- how many people need to be housed at the site and what on-site services are required (such as recreation, health, education). The experience of a number of countries with just this kind of situation suggests the advisability of building a settlement which is largely pre-fabricated and which can be removed when the resource is exhausted, so as to leave the area (hopefully) intact. In the case of the desert environment such a policy would take careful planning due to the very slow recuperative capacity of the terrain to even the slightest intrusion.

If the scale of the economic activity is large and/or if the predictable time horizon is lengthy, the appropriate settlement planning becomes considerably more complicated. For example, in the case above, the needed water supply has been taken for granted. If bore-hole supplies are inadequate, water can always be delivered to the settlement. In the instance under discussion, a make-shift solution, however, may not be satisfactory. Since the settlement derives from the objective of resource exploitation, its structure (and rate of growth) must first be evaluated in terms of the physical constraints.

In this connexion there are a number of planning principles which may be cited as examples to be applied to each new desert settlement project:

(1) Each such development effort should from the beginning be viewed as a generating force for a range of future activities, even if those future activities cannot be defined precisely at the outset. In other words, the aim should be to create a diversified economic base through opportunities presented in every phase of settlement development.

(2) In the planning sequence, the new settlement should be developed from the centre out, that is the amenities of a central core should be provided at an early stage. Apart from creating a focal point for the new community such a central core can also provide services and facilities for the surrounding residential communities as soon as these communities are formed.

(3) As implied by the above, planning for single enterprise communities, as in the case of mining towns, may -- unless the activity is clearly short-term in nature -- include the possibility of future expansion for other purposes. In the long-run, and provided that the settlement development from its inception does not adversely impinge on the environmental and social conditions of the residents, the generation of varied activities and opportunities for the inhabitants (present and future) contributes to the growth and prosperity of desert settlements, enriching the quality of life of their inhabtants.

These principles have been cited merely as examples of the type of issues that need to be raised in the development of new desert settlements. Likewise, it is possible to give examples of specific urban location and urban design guidelines that may be developed, adapted and applied as appropriate in the context of new desert settlements:

(1) The location of the new settlement in relation to the location of the primary activity itself should be determined on the basis of environmental and site conditions. Site selection criteria can be based on two main groups of elements, environmental and functional. The former category would include topography, soil characteristics, direction of prevailing winds and cultivatable land while the latter category would cover such considerations as commuting distance, access to the main highway, access to water supply and problems of sewage disposal.

(2) The form of the settlement should derive from the climate and the conditions related to the directions of the prevailing winds and orientation to the sun.

(3) Regarding urban design, the settlement plan should consider the interrelationship of land-use, modes of transportation, the practicality of the utilities network, the suitability of housing types reflecting climate and traditional characteristics best suited for the desert.

A major question in connexion with the development of any new settlement is that of estimating its demographic make-up, both for the short-term and the long-run. This issue carries enormous implications for the very structure of the settlement, both in its initial phases and as it grows. This would include the appropriate mix of housing for single workers, dwelling units for families of various sizes, schools, health and community facilities and so on.

In short, the conception of a new settlement must take a specific shape, even at the initial stages, if it is felt that a wider, more diversified economic base can be developed so that in time the settlement's dependence on the basic resource is diminished. Clearly, examination of the options in this context can only be undertaken in the wider regional, national or inter-regional framework. There must be, from the very beginning, a solid appraisal of the economic and spatial linkages which could be forged as the settlement grows. On the economic side the normal progression of events is from primary activities (agricultural or extractive) to secondary (manufacturing) to tertiary (services). (The Brasilia model is characterized by the appearance in the first instance of substantial tertiary activity -- true also of a number of science and defense oriented and tourism based desert settlements in the United States). In general, it is not easy to see diversification and expansion of the economic base occurring in the conventional manner under the physically constrained conditions of arid regions. The large water supply and waste disposal requirements of most manufacturing and processing activities constitute just one of many such constraints.

Apart from considerations of economic base, water supply and other elements mentioned above, perhaps the most important issue in the case of large scale desert settlements is their envisaged role in the process of overall socio-economic development. A clear definition of this role would also determine the priority accorded to the new settlement and the commitments that the government is prepared to make in terms of resource allocation for this type of enterprise. A closely related issue is the integration of settlement planning with overall development planning in terms of the national objectives to be achieved, the formulation of plans, and the working out of detailed programmes and their implementation.

EXISTING DESERT SETTLEMENTS

A natural consequence of the life support system of desert regions, and in particular the scarcity of water, is the tendency for settlements to be small and widely dispersed. To the extent that desert settlements are unplanned, at least by a central authority, their relative location and the fluctuations in their density reflects a continuing and fluid, (though not frictionless) self-regulating mechanism in the struggle for survival. The mechanism is, indeed, in the Malthusian sense classically simple: where there is an abundant supply of life support resources (water, arable land, etc.) in an area in relation to the human and animal population in that area, the standard of living of the settlement's inhabitants rises. This induces a greater concentration of people and animals in the area which in turn causes a reversal of the temporary gain. The inability of the rangelands in some parts of Africa and the Middle East, for example, to support the inhabitants of these areas under the pressure of human and livestock population growth is a case in point. The Sahelian tragedy is itself but an extreme example of this phenomenon.

The above scenario applies with greatest force to semi-arid areas where the potential productivity of the land is much greater than in the arid areas. A temporary beneficence of nature in the semi-arid areas is more likely to generate the illusion of long-term gains. It is precisely this cycle which must be considered if planned human settlements development in the desert is to be sustained. Where a population concentration builds up, or can be expected to build up as a result, for example, of the provision of new water supplies, the settlement structure which evolves must be spatially and socio-economically compatible with the eco-system. The planning approach must involve safeguards against the natural tendency to over-exploit the resources in the short-run. For the long-run it should ensure that the growth of the settlement is based on a sufficiently wide range of economic activities to minimize the danger of the exhaustion of critical resources. To be fully effective, such a policy should be fashioned in a national or regional context.

Apart from the points noted above, the major issue relating to existing desert settlements is the quality of life of the communities. On the one hand there are the physical requirements of comfort and security which can be addressed directly through government action in such areas as the delivery of health and education services, basic utilities and so on. There is, however, a more pervasive problem stemming from the environmentally-dictated dispersal of settlements -- namely, the isolation of communities one from another. This major problem, which probably more than any other colours the social fabric and value structure of desert communities, should be accorded high priority in social and economic development planning. As has been stressed earlier, the technological means (particularly in communications and transport) are now at hand; what is needed is a commitment and will to use them effectively.

NOMADIC DESERT SETTLEMENTS

Nomadic migrations can essentially be thought of as mobile human settlements. The lifestyle and the quality of life of nomads has been the subject of considerable study, and to a lesser extent, governmental action. From the standpoint of this paper, it is accepted that in certain environments nomadism may be the only feasible "settlements" structure consistent with physical, economic and social realities. The relevant question then is whether the conditions of living of nomadic peoples can be improved by introducing some of the "modern sector" amenities of settled communities, and if so how can these elements best be delivered.

The technology now exists to enable substantial improvements in the conditions of life of nomads. Particularly as a result of "remote sensing" capability, it is now possible to generate the information upon which the migratory pattern of nomads can be rendered more "efficient." Thus, areas of rainfall, as well as the level of precipitation, can be detected. This information could then be conveyed to nomadic groups to be of immediate practical value. At certain locations, to which nomads would come regularly in relatively predictable numbers, it would be possible to establish minimum site and service community facilities. These could include clinics, schools, trading facilities and so on. In this way, and without prescribing change in the nomadic lifestyle, some of the advantages of sedentary life can be delivered to the nomadic populations. In areas where the permanence of the resting point is doubtful, much the same services can be provided on a mobile basis, an example being the mobile schools in Africa proposed by UNESCO. Other mobile services may include health units, agricultural advisory services, libraries and cultural activities.

Particularly in the case of communities which are on the move, the potential of audio-visual communication techniques can be exploited to great advantage. The first and most obvious benefit to be derived is the one cited above, namely communication via satellite concerning rainfall, grazing areas, and other such features of interest. Audio-visual techniques can also be used to convey programmes of education, emergency assistance, medical information and so on. It must be recognized, however, that the introduction of these technologies will have a far reaching, though possibly slow, impact on the social fabric and value structure of the nomadic societies.

NEW DIMENSIONS IN DESERT SETTLEMENT PLANNING

While the foregoing has focused on specific issues related to desert settlements, it is worth concluding this paper with some remarks on the wider context in which such development can now be perceived. Three principal areas come to mind: (a) relatively stabilized political structures and recognition of the positive role of government, (b) the widespread and growing concern for the social goals in national development objectives, and (c) the many technological breakthroughs of recent years in the areas of water, energy, building and communications.

In a fundamental way, the key to planned development, for desert areas or any other region, stems from the central role that governments have assumed in shaping social and economic policy and in implementing programmes to achieve stated goals. The elaboration of human settlements policies and plans for desert area development, or the intent to elaborate such policies, is a reflection of a sense of national sovereignty. This is spearheaded by the desire and ability of governments to mobilize resources nationally (and internationally), to promote national growth and development and to distribute the benefits of development to all areas and segments of the population on an equitable basis. In large measure, therefore, desert area development will occur, subject to environmental and resource constraints, as a result of political will and national commitment.

The second element referred to -- the social norm expressed in the desire to help improve the conditions of living of the relatively disadvantaged sectors and communities -- is strongly associated with the first. Both these elements are crystallized into programmes and action plans, rather than remaining simply unattainable desiderata, on the basis of present-day technological feasibility.

The application of one technological breakthrough, remote sensing, has already been mentioned in the case of nomadic settlements. In this case, as in others, the key concept is "adaptation", since existing technology has to be adjusted and modified to the existing conditions -- social, cultural, physical and political -- if it is ultimately to produce the desired results. In the context of human settlements a number of areas for technological adaptation can be cited. The application of hydroponic and drip-root agricultural techniques is one such area. Clearly, where such techniques can be applied, and the necessary investment made, then human settlements appropriate scale can be supported. Likewise, it is clear that the size of any desert settlement will depend, other things being equal, on the availability of water, both for human and animal consumption as well as for industrial processes (however minimal), and for the growth of some crops, garden cultivation and the like.

One relevant issue in the present context concerns the different solutions to the water question that are possible depending upon whether the desert settlement is located near the sea (coastal desert areas) or far inland. In coastal settlements it is more readily conceivable that the water needs of even a large settlement could be met with relative efficiency by the installation of a suitable desalinization plant. Where a major plant is not feasible, because of the dispersion of population or small size of communities, smaller individual family scale solar stills for water distillation might be appropriate. Such units have been used with success in a number of Caribbean countries and might well be adapted for use in certain coastal desert areas.

Another technique that can be employed involves using the day/night temperature differential or some cooling device to squeeze water out of the air. Some desert areas of the United Arab Emirates, for example, have relatively high air humidity, and therefore there exists the potential for generating, if not all, at least a part of a settlement's water supply from this source.

In the field of energy, considerable attention has been paid to the concept and implications of "energy-intensive" development for arid areas -- a reflection of relative abundance in certain regions of oil reserves vs. the relative (and absolute) shortage of water. At the margin (in both an economic sense and technologically) there is a degree of substitutability between the energy and "conventionally available" water in as much as high-energy absorbing techniques can, for instance, be applied to recycling processes and other water harnessing methods.

Though costly, cloud-seeding and evaporation-suppression technologies can also be adapted and implemented. Many other techniques are today available that are capable of easy and relatively inexpensive installation for capturing the freely available solar and wind resources.

In a very fundamental sense the quality and character of human settlements depends on the quality of the physical structures in which inhabitants are housed. This in turn is intimately related to (1) traditional construction techniques and building materials and (2) new materials and techniques that reflect the best use of available resources and are appropriate to local manpower skills. Admittedly much research and experimentation still remains to be done in this regard. However, a number of general observations can be made. First, building techniques which have been successfully developed over a period of centuries in one area may be unknown elsewhere, despite the existence of many similarities in the situation of the respective communities. This kind of concentration of, or lack of diffusion of knowledge suggests that substantial benefits could be derived on a regional or even interregional basis from the systematic promulgation of proven construction techniques. Indeed, such an effort lends itself ideally to programmes of technical co-operation among the respective developing countries themselves -- starting perhaps on a pilot or demonstration project scale, and growing on the basis of experience into larger scale mutual co-operation programmes.

As an example to illustrate the need for developing suitable building materials for desert settlements, it may be noted that the most abundant building material, namely sand, in some desert areas contains concentrations of various salts that disintegrate concrete structures in a very short period of time, in some cases six months. Suitable stone for coarse aggregate in concrete is also difficult to obtain in those areas. Thus, much work remains to be done to further the use of desert sands and soils as a viable building material. Through the provision of such techniques and the development of parallel on-site construction methods, major improvements can be made in the quality of desert settlements.

In conclusion, this paper has raised a number of issues that should be part of the framework of alternative strategies for desert development. It should be clearly understood, however, that these issues cannot be regarded as a satisfactory substitute for a well-considered, comprehensive national policy on human settlements embodying the distribution of population and related economic and social activities over the national territory, as recommended by the Habitat Conference.

PLANNING FOR LAND MANAGEMENT ON THE CALIFORNIA DESERT

Ed Hastey, California State Director
Bureau of Land Management
U.S. Department of the Interior

ABSTRACT

The California Desert comprises 40,000 square miles (104,000 square kilometers) in southeastern California and constitutes one-fourth the total land surface of the state. The California Desert is part of the North American Desert, fifth in size among the 12 principal deserts of the world, and includes three of the four major desert subtypes. Approximately one-half the California Desert is National Resource Land administered by the Bureau of Land Management (BLM), U.S. Department of the Interior.

There are indications that man has occupied the Desert for more than 10,000 years, but during the American settlement period the desert was regarded as a hostile obstacle to be crossed, except to a few who found gold, silver and other mineral wealth there. It has been used extensively for military maneuvers. The desert remains an area to be crossed, but now by powerlines, pipelines and canals moving energy, fuel and water to the densely populated coastal area of South California. The California Desert is unique because of its proximity to metropolitan areas with a combined population of some 12 million.

More recently the desert has become a playgound for many of the people in the adjacent urban areas. Uncontrolled recreational use and conflicts with other uses, such as livestock grazing, mining, and wildlife habitat, led the BLM to initiate the California Desert Study in 1967.

From the desert study came the Bureau's California Desert Program consisting of concurrent elements of long-range planning and interim management. Interim management includes a desert ranger force, visitor assistance and management, recreational facilities, and public information and education. The rangers have not had enforcement powers, but have sought voluntary compliance with laws and regulations. Much of the early planning effort was devoted to a management program for off-road vehicle use.

Basic resource data were scant in an area that historically had been overlooked by most of the population in spite of the presence of important natural and economic resources, including minerals, irrigated agriculture, wildlife and grazing.

Passage of the Federal Land Policy and Management Act of 1976 (Public Law 94-579) established the California Desert Conservation Area and required preparation of a comprehensive, long-range plan for management, use, development and protection of the California Desert by September 30, 1980. The initial two years will be devoted to intensive resource inventory and the final two to plan preparation. A concurrent program of interim management will include development of a ranger force authorized to carry arms, make arrests and issue citations; additional facilities including campgrounds and interpretive centers; and resource protection.

Inventories will be in the areas of geology-energy-minerals; soil, water and air; vegetation and range; wildlife; cultural resources; and recreation. Socio-economic data also will be gathered, and projections of future uses made.

Public participation will be an integral part of the program throughout the planning process.

All elements will be synthesized in the planning process to produce a viable land use allocation plan which will provide a framework for preserving the rare and endangered life forms and cultural resources of the California Desert while allowing resource development and use to meet the needs of man.

INTRODUCTION

The California Desert is part of the North American Desert, which ranks fifth in size among the 12 principal deserts of the world. The North American Desert includes four major subtypes: the Great Basin, Mojave, Sonoran or Colorado, and Chihuahuan Deserts. All except the latter are found in the California Desert. The California Desert contains 40,000 square miles (10,000,000 hectares), or one-fourth of the total area of the State. Approximately one-half of this is National Resource Land administered by the Bureau of Land Management (BLM), U.S. Department of the Interior. Approximately 12 percent of the desert is in military reservations, and 10 percent is administered by the National Park Service.

Rainfall ranges from less than two inches (5 cm) annually at Furnace Creek in Death Valley National Monument to 20 inches (50 cm) in the higher elevations of the mountains in the Mojave Desert such as the Clark Mountains. Summer temperatures of more than 120 degrees F. (48.9 degrees C.) are common, while freezing temperatures are not uncommon in winter. Elevations range from 286 feet (85.5 meters) below sea level at Badwater in Death Valley to more than 11,000 feet (3,300 meters).

Remnants of earlier cultures indicate man adapted to the inhospitable climate of the desert, surviving because he lived in harmony with his surroundings. Unlike modern man, he had no other choice.

During the American settlement period, the desert was a hot, hostile, dangerous obstacle to be crossed, except to a few people who found gold, silver and other mineral wealth there. During World War II the desert was used extensively for military maneuvers by General George S. Patton's troops. Today the desert remains an area to be crossed, but now by power lines, pipelines and canals moving energy, fuel and water to the densely populated coastal area of Southern California. The California Desert is unique because of its proximity to metropolitan areas with a combined population of some 12 million.

People's attitudes toward the desert remained basically unchanged until the era following World War II, when the desert increasingly became a playground for many of the people in the adjacent urban areas. Increasing leisure time and discretionary income started a revolution spurred by the development of motor vehicles that could go with comparative ease where vehicles had not ventured before.

Concurrently, public attitudes toward the remaining federal lands were shifting. The public domain lands left over from the great westward movement were increasingly recognized for their values for open space, natural resource and recreation.

The Bureau of Land Management initiated a study of the California Desert in 1967. The initial phase focused on the recreational resources, use and potential of the desert. A second phase considered all the resources of the desert and their interrelationships. From that study came the Calfironia Desert program designed to manage the explosive recreational use of the desert as one of many uses in a total multiple use program. The program consisted of concurrent elements of comprehensive, long-range planning and interim management to protect resources that were threatened by the rapidly growing, uncontrolled use of the desert. Interim management includes a desert ranger force, visitor assistance and management, and programs of public information and education. The rangers are natural resource specialists who seek compliance through information, education and cooperation rather than legal sanctions. They have been unarmed and without law enforcement authority.

The Desert Planning Staff was established in 1972 to begin development of a long-range plan of land uses and management of desert resources.

Basic resource data were scant in an area that historically had been overlooked by most of the population in spite of the presence of important natural and economic resources. The value of mineral production from the desert was estimated at $160 million in 1968, and is expected to double by 1980, but comprehensive mineral surveys are lacking.

Private farm lands in the Imperial and Coachella valleys account for approximately 10 percent of California's fruit, vegetable and meat production - a significant amount when it is noted that California produces about one-fourth of the entire agricultural production of the United States. Additional large acreages of the desert have soil types which indicate they are suitable for agriculture if water can be obtained for irrigation.

Public lands historically have provided forage for sheep and cattle.

Similarly, the California Desert contains some 200 species of animal life and more than 1,000 plant species, many of which are endemic, but comprehensive data are lacking.

There are approximately 3,500 miles (5,600 km) of powerline corridors criss-crossing the desert; more than 1,200 miles (1,900 km) of petroleum and fuel pipe-lines; 340 miles (550 km) of underground telephone cables. There are more than 100 microwave repeater sites. There is one large known geothermal resource area and several smaller potential areas. There are thousands of archaeological and historical sites.

Approximately 450,000 people live in 100 cities, towns and smaller settlements in the desert. Tourism and outdoor recreation use of the desert are of major importance to the economy of these communities.

The rapid increase in recreational use, particularly that involving off-road recreational vehicles, required that initial efforts of the Desert Planning Staff be directed toward developing a management plan for this use. The result was the Interim Critical Management Program (ICMP) for off-road vehicle use of the desert.

From 1972 to present planning proceeded on a unit-by-unit basis, with resource inventories of four of thirty-one planning units containing four million acres (1.62 million ha). It was estimated that this unitized planning following the Bureau of Land Management's planning system would have taken 15 years to produce a comprehensive desert plan. Much planning would come after irreversible commitments had been made of land and resources.

CALIFORNIA DESERT CONSERVATION AREA

A dramatic turning point came in 1976, when the Federal Land Policy and Management Act of 1976 (Public Law 94-579) was enacted by Congress. Title VI of this Act established the California Desert Conservation Area, consisting of 25 million acres (10 million ha). While only half of this is National Resource Land administered by the Bureau of Land Management, the Bureau was given the lead in planning for the area. The law adopted the two-pronged approach developed earlier by the Bureau of Land Management with concurrent elements of long-range planning and interim management.

The Secretary of the Interior was directed to prepare and implement, by September 30, 1980, a "comprehensive, long-range plan for the management, use, development and protection of public lands within the California Desert Conservation Area." It further directed that such plan take into account the principles of multiple use and sustained yield in providing for resource use and development, including, but not limited to, maintenance of environmental quality, rights-of-way and mineral development.

"Multiple use" is defined as the management of the public lands and their various resource values so that they are utilized in a combination that will best meet the present and future needs of the American people, a combination of balanced and diverse uses that takes into account the long-term needs of future generations for renewable and non-renewable resources.

"Sustained yield" is defined as achievement and maintenance in perpetuity of a high-level annual or periodic output of various renewable resources consistent with multiple use.

The Act also directs the Bureau to use a systematic interdisciplinary approach to achieve integrated consideration of physical, biological, economic and other sciences; to give priority to designation and protection of areas of critical environmental concern; to consider present and potential uses of these lands; to consider the relative scarcity of the values involved and the availability of alternative means and sites for realization of those values; and to weigh long-term benefits to the public against short-term benefits.

During the period beginning with approval of the Act by the President and ending on the effective date of implementation of the comprehensive, long-range plan, the Secretary of the Interior through the Bureau of Land Management is to "execute an interim program to manage, use and protect the public lands and their resources now in danger of destruction, in the California Desert Conservation Area, to provide for the public use of such lands in an orderly and reasonable manner such as through the development of campgrounds and visitor centers, and to provide for a uniformed desert ranger force."

The rangers will be trained law enforcement officers authorized to carry arms, make arrests and issue citations.

The law also authorizes the appropriation of up to $40 million - $30 million for interim management and $10 million for planning - but did not actually appropriate funds. The funding needed to complete the program must be justified each year through the budgeting process of the federal government and the Congress.

Finally, the Act requires that the Bureau of Land Management involve the public heavily in the planning process. This is done in a number of ways, but a key element is the California Desert Conservation Area Advisory Committee. The committee consists of a group of distinguished citizens representing a broad spectrum of disciplines, interests and uses.

The four-year timetable presents the Bureau of Land Management with a planning challenge that is unique in size, scope and importance. It will be necessary to utilize all available techniques, including many spinoffs of the Space Age, to achieve the objective of a comprehensive plan based on sound resource inventories. Many of the findings will be applicable to other desert lands, and the knowledge gained here will be shared with other nations.

RESOURCE INVENTORY

We have learned through experience what many natural resource managers have long held: A land use plan will be only as good as the resource information on which it is based. On the California Desert, we have found surprisingly little systematically collected information on the resources and trends of this vast region. Some of our basic tasks are to identify critical areas, plan for development of economically valuable resources, prescribe areas for special management (including areas for scientific study) and the foundation for all resource decision - to determine the carrying capacities of the desert's resources.

Thus, an extensive inventory is beginning on a desert area for which we must produce a plan in just four years. Two years will be devoted to completing this inventory. Following are brief descriptions by resource groups to provide an idea of the current known resource situation and the objective and procedures for gathering new knowledge.

Geology-Energy-Minerals

The Resources: Among the many natural resources present in the California Desert, geology, energy and minerals are considered as forming a closely related group known in our studies as G-E-M Resources. Our knowledge about the members of this group varies. We know more about the geology of the area than we know about the energy resources. And we know more about the latter than we know about the minerals of this area.

Our paucity of knowledge about the minerals is surprising when we consider that this area constitutes one of the primary mineral sources in California. Although a considerable amount of data is available for some world-famous mineral localities such as the rare-earth carbonative deposit at Mountain Pass, or the salines and borates of Searles Lake, these are found in only a very small portion of the total area.

The value of raw minerals produced in this area in 1968 was estimated at $160 million. The estimated value in 1980 is $300 million. Minerals which are leasable by the Bureau of Land Management include coal, petroleum, phosphate and sodium salts. Minerals which the Bureau sells include sand and gravel, building stone and fill material. Minerals upon which claims may be filed for private development include gold, zinc and iron.

The desert contains one large known geothermal resource area and several smaller ones with active development taking place. Numerous power generating station sites - thermal and nuclear - are being analyzed for possible construction during the next 10 years. Many of these sites are on public lands and all of the sites involve transmission lines across public lands. In addition, because of our growing energy shortages, increasing attention is being given to possible development of solar and wind energy sources in the area.

The gamut of rock formations range in age from early Precambrian metamorphic and intrusive rocks to the recent sand dunes and volcanic rocks. The geology of the area is complex lithologically and structurally.

Based on physical features, the area is subdivided into four major geomorphic provinces. The largest is the Mojave province occupying the middle and southeast part of the area; followed in size by the Great Basin province in the northern third of the area; the Colorado Desert province in the south and southeast; and the eastern extension of the Transverse Range province.

The complex geology, numerous exposures and relatively easy access from centers of high population make the CDCA an open classroom for studying and/or researching this geologic environment.

The potential for all G-E-M resources in the area appears to be impressive.

The Planning Approach: Developing a data base for G-E-M resources in such a large area, and in such a short time frame, obviously is not simple. We estimate that 300 man years of effort would be required to do a reconnaissance survey, using conventional data gathering methods. The conventional approach is not possible, so we have developed a conceptual systems approach which combines concept, working hypothesis and modern methods.

The inventory part of the program has been divided into two major phases which will be sequential and tentatively are planned to be completed by the end of 1977 and 1978, respectively.

Phase I is synoptic-desertwide and includes:

1. Lineament study of satellite imagery compared to available data on gravity, aeromagnetics, lithology, structure and mineralization. This will provide the basis for an interpretive overlay map of comparatively low confidence.

2. Geostatistical study. This technique will use mathematical modeling and statistical analyses of available geological and mineral data. This will give us a classification of areas within the CDCA based on probability of mineral deposits at different levels of confidence.

3. Lithology/Fracture Study. A reconnaissance field study in preselected areas.

4. Field Verification Project. This will consist of two types of varification. One type is simply verification of maps and/or literature data; the second type is verification in the field of results from numbers 1 and 2.

Satellite imagery will be studied in direct relationship to known geology and unusual mineral concentrations. There will be studies to interpret the geology and structure as seen in larger scale than obtained from satellite. When correlated with known occurrences, we expect to be able to identify geologic environments with relative potential.

The following methods are likely to be used over some of the selected areas: airborne magnetometer, gamma-ray spectrometry and infra-red (Fals color and/or thermal).

Soil, Water, Air

<u>The Resources</u>: The entire California Desert has been mapped at a general level as to soil types. The broad groupings, suitable for regional planning, identify 31 soil associations made up of approximately 70 soil series and miscellaneous land types. The U.S. Soil Conservation Service has inventoried about 3 million acres (1.2 million ha) within the desert. This land is mainly concentrated within private lands and areas where agriculture has been established, such as Imperial Valley, Coachella Valley, Antelope Valley and the Palo Verde area.

The Bureau of Land Management has inventoried about 4 million acres (1.6 million ha) within the desert. The inventory shows approximately 60 to 70 soil series and miscellaneous land types. These soils have been evaluated to determine their susceptibility to erosion and surface disturbance. Other determinations were the potential for agriculture, sand and gravel extraction, road location and building sites. About 20 percent of the soils were considered suitable for agriculture. Studies will be undertaken to determine the impact of man, animals and off-road vehicles on the soil. Also needed are studies of the effects of dust on people. Pollen which can cause problems of allergy and spores of different genera of fungi which can cause human disease have been found in dust in parts of Arizona.

Little is known about water resources of the total California Desert. A few areas have been studied in detail but the over-all evaluation of water resources of all the basins and the interrelationships of the basins has not been done. There needs to be an inventory of water resources, including surface-water runoff, ground water basins, ground water supply and availability, ground water recharge and discharge, water quality, location of wells and springs, well yields and spring flows, water temperature, geothermal resources, hazardous waste disposal potential and other water related problems. Any future use of water in the desert will have to be evaluated with great care. Ground water levels have been built up over a long period of time and any rapid use of water, because of limited recharge, could lower water tables drastically.

Planning Approach: The soil-water inventory for the CDCA will be done at two levels, the first at a level of medium intensity with accompanying water data; and the second at a very general level using data obtained from the first. Satellite information and general soil-association data will be evaluated. Outside the inventoried areas, soils will be evaluated by projecting known data. These areas will be spot checked in the field and limited laboratory tests performed. There will be studies to determine actual amounts of soil loss through erosion. Climatic and soil conditions will be evaluated to determine their relation to soil loss. There also will be studies to determine the carrying capacity of soil to sustain certain types of recreational activity. There will be a desertwide evaluation of impacted areas to learn the degree and kinds of impact. We will coordinate our water quality evaluations with the State of California's two Regional Water Quality Control Boards within the planning area, as well as with the State Water Quality Control Board in Sacramento. The planning process on BLM lands will focus on identification of ground and surface water sources and their quantity and quality. We will identify existing land uses and existing and potential threats to water quality. Water pollution control recommendations and programs will be established.

The U.S. Environmental Protection Agency has established a program to maintain the air quality of areas where the air is cleaner than the national standards. The State of California is developing a similar but more comprehensive program. Our planning efforts will be coordinated with EPA, California's Air Resources Board and the Southeast Desert Air Basin Coorindating Council, consisting of county governments in the Basin. Candidate "air conservation areas" located on non-BLM lands will be reviewed and their impact on the Desert Plan determined. These could be areas designated by the National Park Service, National Forest Service or State Department of Parks and Recreation. In addition, significant resource areas on BLM land may be selected as "air conservation areas." These resource areas may have particular values such as scenic quality, vegetation, wildlife, cultural or recreational.

Vegetation and Range

Resources: The flora of the area has strong representation from three of the four regional deserts which occur in North America. They are the Great Basin, generally to the north; the Sonoran, generally to the south; and the smaller Mojave in the middle. More than half of the Mojave Desert falls within the planning area. The flora of the area is moderately rich. There are approximately 1,200 species belonging to 110 families of vascular plants. The Mojave portion is particularly rich in endemism. Notable pockets of apparently newly derived species exhibiting very limited distributions occur in the southwestern Mojave and in the Death Valley region. At this time 170 species are identified as rare and endangered. Eureka dune grass (*Swallenia alexandrae*), represented by only two populations in a valley at the northern limit of the planning area, is an example of the type of narrow endemism encountered in the area.

The vegetation of the area ranges from sparse grasslands to scrub or shrublands to woodlands. The respective transition in vegetational structure is associated predominantly with increasing precipitation along an increasing elevational gradient. Sparse grasslands usually are associated with sandy substrate and occur rather infrequently at low elevations. Shrublands occupy 70 to 80 percent of the entire area and exhibit a diversity of types. Plant ground cover ranges from less than one percent to more than 30 percent. Creosote bush (Larrea tridentata) is dominant throughout most of the area. As a species its importance extends well beyond the area. At intermediate elevations succulents (Cactacene and Agavacene) contribute heavily to the vegetation cover. Woodlands occur in flood plains and at higher elevations. They exhibit the greatest amount of plant productivity per unit area. In the Sonora portion the woodlands are found mostly in the flood plain at dry washes and are composed of small legumenous trees. In the Mojave, woodlands are characterized by pine, juniper and Joshua.

The vegetation has a distinctive temporal component. Annual species grow abundantly with favorable moisture but may be essentially absent for periods of years when the moisture is lacking. Annuals make up just under 50 percent of the species for the area in general but in specific habitats may consist of 80 to 90 percent of the number of species present. Use by domestic livestock, primarily cattle and sheep, has occurred throughout most the area for more than 100 years. There are about 6 million acres (2,400,000 ha) of federal rangeland suitable for livestock grazing. Approximately 25,000 cattle and 140,000 sheep graze these lands during a year of good growth, using 180,000 animal unit months of forage. (An animal-unit-month is the amount of forage one cow will eat in a month). Feral burros of African origin have developed highly successful populations throughout the area. The total population is estimated at 3,500 to 4,000 burros.

Some species of native plants have sparked economic interest. Jojoba (Simmondsea chinensis) from the Sonoran desert is being studied as a potential oil and wax producing cultivar (an acceptable substitute for whale oil) by several universities throughout the world. Joshua tree (Yucca brevifolia) fruits are being considered as a source of steroid drugs by a major pharmaceutical firm. These drugs include estrogen, cortizone and synthetic hormones. The extensive natural stands of Joshua tree throughout the Mojave appear to be an attractive potential source of basic raw materials for the production of these drugs. Mojave yucca (Yucca schidigera) is being exploited commercially on private lands for such uses as fertilizer and shampoo. Much of the area also could be considered a potential habitat for arid land cultivars that are native to other regions.

Planning Approach: The planning objective is to provide information on the vegetation and flora that will assist in developing a long-range comprehensive multiple use plan with a commitment to the concept of sustained yield, particularly as it pertains to vegetation and flora. Special considerations include the amount of forage available for livestock and wildlife and the protection of endangered and threatened species.

We will learn through inventory the present dimensions of the botanical resource in terms of both vegetation and flora. We must also learn how the productivity and composition of the vegetation and survival of the flora will be affected by varying intensities of imminent alternative uses.

A general vegetation map will be prepared of the entire area and a series of more detailed maps will be developed to show site specific information. To prepare the first draft of the general map, we will use satellite imagery plus specific data obtained from completed planning unit resource analyses, i.e., Yuha, El Paso-Red Mountain, East Mojave and Saline Valley; other literature and reports. The procedure will be primarily a deductive process of extrapolating from known areas to areas not yet inventoried.

Site specific inventory, now underway, is applied to areas of critical concern, such as grazing allotments, off-road vehicle open areas, construction sites and rights-of-way. In these areas vegetation stands are delimited on air photos to form polygons, are ground checked and inventoried by transect to determine species composition and ground cover. Utilization by foraging animals also is estimated.

In depth, localized annotated species lists will be prepared for flora. This will be followed by analysis and interpretation. The annotation will characterize the species with regard to important structural, functional and distributional properties. Important characteristics such as habitat, life form, palability to livestock and degree of endemism will be considered.

The initial studies will cover two growing seasons, in 1977 and 1978. After this, interpretive study will begin in earnest. Special attention will be given to building ecological profiles for rare and endangered species in order to obtain a predictive base for locating similar species in other places.

Vegetation dynamics will be evaluated both with regard to impacts by nature and man. The relationship between plant productivity and use intensity by different groups of users also will be studied.

Wildlife

Resources: The California Desert is composed of parts of three major desert systems: the Great Basin; Mojave; and Sonoran or Colorado. In addition, it contains elements of floras and faunas from the Sierra Nevada, coastal or San Diegan, Rocky Mountain and San Joaquin or Central Valley regions. This diverse region supports nearly 200 vertebrate species, ranging from the bighorn sheep to the desert tortoise. Many species are found only in this habitat type. These include such rare and endangered species as the Mojave chub, desert pupfish and desert salamander.

Basically, there is little information available for most species of vertebrates. In fact, new species are still being discovered. The situation is even more acute for invertebrates. Even the most elementary data, such as distribution and abundance by habitat type for each species, is lacking. The amount of information acquired varies by species and location. For several species listed by state and federal governments as rare and endangered, such as the black rail, Yuma clapper rail and yellow billed cuckoo, there is not even basic distributional information.

Wildlife can be found in almost every habitat type, from the playas to the mountain slopes. The tiny fairy shrimp found in some dry lakes lay dormant until the scarce desert rain brings them to life. Small relictual pupfish exist in water 112 degrees F. (44 degrees C.) and saltier than the ocean. A great variety of small animals, reptiles and birds inhabit the desert. Each of these creatures has a purpose and function in the balance of nature.

Desert plants provide shelter and food for animals, stabilize soils and protect watersheds. Because the desert is fragile and heals slowly, damage to the natural vegetation limits the ability of habitats to support life. With each successive impact, the balance of nature is tilted a little more. Due to this slow healing ability, frequent environmental impacts may eventually push the desert beyond the point of recovery. In some place this already has occurred because of heavy off-road vehicle use and ill conceived human habitation.

Planning Approach: We are now taking existing data in the form of literature and museum records for the several hundred species and analyzing it by habitat or community type, and by location and distribution. New observations are being made and studies undertaken both on individual species and on sensitive sites. In addition, studies are planned on effects of livestock grazing, off-road vehicle use, camping, mining and other human uses on wildlife habitat and species. In some areas of the desert, livestock grazing, feral burro use and recreational activities have seriously damaged wildlife populations and habitat. This is readily apparent in riparian areas, sand dunes and other fragile habitat types. Some of these problems will receive high priority for corrective measures.

During the summer and fall of 1977, wildlife habitat types will be mapped, using satellite imagery and both high flight and lower level aerial photography. This mapping will be on a broad scale. At the same time, we will prepare a preliminary outline of sensitive, unique and critical wildlife areas.

Phase 2 will begin in late 1977. This will include preparation of contracts, design of field studies and reconnaissance. In particular, there will be focus on water source studies and endangered, threatened, rare, sensitive, fully protected and other significant species.

Field work began in late winter 1976 and will continue through October 1978 on joint studies of sensitive or unique areas in conjunction with soils and vegetation specialists. There will be general studies on approximately eight major habitat types and areas that have previously been unstudied.

Wildlife habitats will be redefined and delineated at a more detailed level in late 1978 after data from all sources have been analyzed. Once habitats have been delineated, analyses will begin on distribution of vertebrate species by habitat type. Hopefully, advanced computer programs will be available to contribute to and simplify the analysis procedure.

A final definition and analysis of critical or sensitive wildlife habitats and populations can be accomplished after reports and field data on endangered, threatened, rare and other important species are recieved and after the second phase of habitat analysis is completed.

Phase 3 will be the period of final analysis. It will begin when all reports are at hand as to species and effects of other resource uses on wildlife habitats and populations.

Cultural

Resource: Cultural resources are the fragile, non-renewable evidences of human activity, occupation and endeavor which are reflected by structures, artifacts, objects, ruins and works of art. There is evidence that man lived in this area as long as 10,000 years ago but little is known about *how* he lived. Much of what he left for us to see - his rock etchings or petroglyphs - are not yet comprehensible to modern man.

The value of known data is miniscule insofar as it can help us to predict the location of other cultural resources, for understanding the peoples' adaptation in various desert regions and for assessing the significance of the sites for preservation, protection, interpretation and study.

Approximately 7,300 archaeological and historical sites have been identified in the study area by various individuals and institutions. An additional 1,000 have been recorded by BLM in the past four years. There are indications that there could easily be 100,000 cultural resource locations in the planning area. Many of these are significant enough to be nominated to the National Register of Historic Places. To date, fewer than 25 sites or site districts have been nominated or determined to be eligible for nomination.

Each year hundreds of sites are vandalized, disturbed or destroyed. Adequate protection and/or signing has been afforded to less than five percent of the known sites. Protection recommendations for several hundred additional sites are expected in the comprehensive desert plan.

Planning Approach: The California Desert will be divided into four sampling universes which correspond to clusters of planning units. One of the clusters will be surveyed by the Desert Planning Staff and the other three by contractors. The inventory reports are to include discussions of sensitivity, present and future impacts and recommended mitigation of impacts.

Each sampling universe will be stratified into environmental zones corresponding to vegetation domains, geomorphic units and hydrology. Units will be randomly selected and surveyed in detail. A prehistoric and historic land use model will be developed.

Evaluations of potential impacts to cultural resources from other activities will be assessed.

The Desert Planning Staff will undertake systematic intensive cultural resource inventories for about 250,000 acres (101,000 ha). The findings, coupled with other information, will permit 10 to 15 percent of the resources to be known by the time the plan is completed. This will help identify the sensitivity of desert lands (i.e., the density, diversity, uniqueness and significance of cultural values) to other activities. In addition, about one-third of the desert has been assessed through cultural resource overviews. These provide a review and synthesis of the existing cultural resource information, including both historic and prehistoric data. Within the next two years, all public lands in the desert will be covered in this manner.

Recreation

Resources: Our planners describe the California Desert as "the world's largest, contiguous area of dispersed high intensity recreation use under the management of one agency."

The desert is within two hours driving time of 12 million highly mobile people. Recreation use in some parts of the area have increased 1,000 percent since 1949. Recreation is by far the most intensive use of the desert and the activity that has propelled us - more than any other - into a planning job that, to the best of our knowledge, is unprecedented in size and cost in the United States.

Visitor use days on lands administered by the Bureau of Land Management total approximately 14 million a year. The visitors come to collect rocks, ride motorcycles, jeeps and dune buggies, to sightsee, take pictures, look at wildlife, rock climb, sandsail, soar, camp, hike, backpack, hunt, paint, birdwatch, and, above all, to find escape from the smog, congestion, noise and clutter that characterize the Southern California megalopolis.

Additionally, recent studies show that the California Desert is much used for research and education. Researchers come from around the world for studies in life sciences, earth sciences and applied arid land investigations. Students from all levels engage in field trips and on-site study projects.

Beginning in the late 1960s, large numbers of off-road vehicle enthusiasts began to use the desert. Now, on a weekend, you can expect to see thousands of people on motorcycles, dune buggies, jeeps and all-terrain vehicles. In 1973, the Bureau produced a recreation vehicle use plan, still in effect, which identifies areas for specific kinds of vehicle use as well as closed areas. Areas which are open to off-road vehicle use total more than one million acres (405,000 ha).

The desert has beauty that is surprising to those who have not ventured from the busy interstate highways that cross it. Many travel at 50 or 60 miles per hour (80-100 kph) across monotonous creosote flats and fail to discover the treasures of solitude and scenic delight so near at hand. The desert has cinder cones and lava fields less than 1,000 years old. Its sand dunes are among the highest and most visually spectacular in North America. There are forests of pine and juniper - even relict stands of white fir - sprinkled on slopes of desert peaks. And the quality of vastness - the opportunity to look for great distances, as far as the eye can see, without the sight of man or his effects has a special value.

We have set aside 19 areas of special beauty and designated them as National Recreation Lands. These include primitive areas, natural areas, campsites and other areas with unusual recreation value.

The Bureau maintains a few developed campgrounds on desert public lands. Some are intensively used. Undoubtedly we will need additional recreation facilities, but our policy is to keep facility development in balance with demonstrated needs.

There are numerous conflicts among users of the desert and a considerable amount of damage to resources, both deliberate and inadvertent. Law enforcement authority has been granted and is gradually being implemented. Our main efforts in reducing conflicts and in minimizing damage, however, is to provide information and assistance to users. We find that educating the visitor as to the value and sensitivity of desert resources is the key to considerate use for all but a small percentage of users.

Inventory-Planning Approach: Recreation planning will include studies and analyses on many fronts. The potential of the land to attract and sustain various forms of outdoor recreation will be identified and evaluated.

The planning effort will include preliminary field reconnaissance of potential for various activities, scenic quality evaluations, location of principal areas with notable wildland values, identification of significant recreation areas, and contacts and data exchange with users and other agencies.

Direct identification of recreation use will be established by on-site studies, aerial surveys and studies to determine use levels, recreation preferences and preferred areas.

A desertwide interpretive resource analysis will be developed which will define specific interpretive themes, identify features of significance and potential popular interest, and suggest interpretive media.

There will be a study of primitive and wilderness values. Roadless areas will be identified through analyses of aerial photographs, map study and field reconnaissance. Other studies will deal with recreation, economics and carrying capacity. Intro-recreational conflicts will be identified, as well as conflicts which derive from other activities such as mining and grazing.

SOCIO-ECONOMIC FUTURES

All of the foregoing resource information, as vital as it is for effective land use planning, is out of context until man's needs and influences can be forecast, in this case for a 20-year planning horizon to the year 2000. We must learn what will be the demographic trends in terms of desert use as in the case of recreation, or even settlement, as well as what the demands will be for the resources of the desert. Will a world food crisis call for an all-out effort to bring more arable desert lands under irrigation to grow crops? Will the water be available to meet such a demand? Will the energy supply/demand balance call for a major shift to solar energy with breakthroughs in massive applications bringing extensive areas of the desert into solar energy production to supply the nearby megalopolis?

Such questions will be addressed in the California Desert Plan by turning to a "futurist." By consulting with many futurists through contracts, we plan to employ such methods as the "Delphi Technique" in which many respected experts will be independently queried as to their views of the future in their fields on factors affecting use of the California Desert. These responses will be analyzed, correlated and follow-up inquiries run. We believe the final composite picture will be as informed a prediction of our desert planning horizon as we can obtain.

SYNTHESIS THROUGH PLANNING

Analyzing and synthesizing the vast amounts of resource data that will be collected during the two-year inventory phase into a useful land use allocation and management plan will be an enormous task. Our planners will be aided by automatic data processing programs out of necessity. Resource data will be tabulated and stored in a map-retrievable format. Yet, the essential process requires skilled analysis by individual resource specialists and social scientists working to relate resource capabilities to social and economic demands.

We will be using the Bureau of Land Management's planning system as our conceptual planning model. It has been in use and well tested over the past decade. In brief, it begins with a "unit resource analysis," assessing resource inventory information for current condition, carrying capacities, and long-term trends. An unconstrained optimum development for each resource, subject only to technical feasibility, is then projected to avoid becoming "locked in" too early on traditional paths and policies without considering full potentials. Ecological and socio-economic profiles of the region being planned for are also prepared to assure consideration of these multi-dimensional aspects of planning. Then, preparation of a "management framework plan" is begun. This is the broad general plan of land use allocation which will serve as a foundation for specific designations, decisions and management prescriptions. Our resource professionals begin with objectives for maximizing productivity of their individual resource. This time they are constrained, however, by national and Bureau policies as well as economic feasibility requirements. Through this process, of course, conflicting recommendations are surfaced for the same land areas. For example, wildlife protection and mineral development recommendations may clash. Conversley, complementary relationships may also be identified. A cultural protection proposal may well mesh with a recreation management recommendation. At this point conflicts are resolved on an interdisciplinary basis to provide alternative land use allocation and management prescriptions for consideration by decision makers. Throughout this process, public inputs and participation are incorporated and weighed in developing the final plan. This ranges from formal meetings with our

California Desert Conservation Area Advisory Committee of distinguished professionals and community leaders, to open town meetings in various communities. It also involves preparation and public scrutiny of an environmental impact assessment under terms of the National Environmental Policy Act.

Finally, when all this is completed, our plan for the California Desert will be forwarded to the Secretary of the Interior for review and decisions on implementation by September 30, 1980. Meanwhile, the use of the desert and the demands on its resources continue. The Bureau of Land Management will carry out critical interim management to the best of our ability during this period, using planning information on a continuing basis as it is gathered.

CONCLUSION

In the Bureau of Land Management, we feel an especially great responsibility for the planning of land uses and management on the California Desert that transcends even our tightly deadlined congressional mandate under the Federal Land Policy and Management Act.

We are planning for man's wise use of an area only recently recognized as having many valuable but often fragile resources. We cannot remove man from this environment to preserve a Pleistocene era museum display. Our archaeologists and anthropologists have shown us that man has long had a place in the desert, often apparently living more harmonously with it than today.

We are working, as we must, to preserve this desert's rare and endangered plants and animals. But in some places, desert resources must be developed and used - minerals, geothermal and solar energy, and possibly additional arable land and its underlying water for agriculture - to meet the needs of man.

Beyond all we recognize that we must resolve present conflicts, reserving future options for future generations.

THE ROLE OF OASES IN DESERT DEVELOPMENT AND CONSERVATION

Ahmed Gamal Abdel Samie
Vice President
The Academy of Scientific Research and Technology
Cairo, Egypt

ABSTRACT

The paper expresses in general terms the author's views concerning the recent agricultural development achievements in the hyper-arid deserts which constitute about 97% of Egypt's surface area.

It mentions the prevailing characteristics of the physical elements involved in these schemes, the lessons learned and the technical, social and organizational aspects which governed them.

In view of the accumulated knowledge and experiences, and in order to guard against ill-consequences of over-enthusiasm that often leads to large development ventures which are not based on sufficient data or well adapted and properly tested technologies, the author suggests an oases-type approach for the safe employment of the arid desert areas.

This concept includes, as a first task, an action plan for reviving the existing oases and establishing new ones. A well selected and distributed network of these focal points can provide permanent grounds for executing programmes of investigating and monitoring of the natural resources, adapting and testing new technologies in addition to education and training of desert technologists. They may also support campaigns for arousing both public and private interests through recreation and tourism.

INTRODUCTION

The utilization of some desert areas for agriculture is an old practice in many parts of the world. Ancient records and remaining signs of these practices reveal the skills of old settlers in harmonizing their life with the scarce resources and difficult climatic conditions which prevail in these areas.

In recent years, this utilization has, in some countries, taken the form of big projects. In many cases, however, agricultural techniques were adopted, rather than adapted, without fully understanding the differences in ecosystems of deserts. The accumulated results of the last few decades provide ample background for a comprehensive revision of the plans, programs, procedures, and technologies within a frame of defined and clear strategies.

The current presentation concerns mainly the true, rainless deserts. It is based on the experiences of a developing country in which 95 percent of the population live on only 3 percent of the total surface area. Its agriculture, which dates back to more than 5000 years B.C., has been confined to the Nile delta and a very narrow strip of its valley. The present and future agricultural expansion of irrigated areas is a matter of life, and the only direction where this expansion may be achieved is in the rainless deserts which occupy about 97 percent of the country's surface area. The lessons gained provide technical, social and economic examples which should be exposed for the benefit of those involved in developing areas of similar environment.

DEFINITION

Employing or re-employing deserts by modifying them into agricultural and industrial production areas is neither an easy nor a simple task. A time plan for achieving such goals requires 100 to 200 years or even more if a new stable culture is to be established. The logical first step is to reach an acceptable differentiation between aridity and semi-aridity since strategies applicable to one can be greatly different from those suitable for the other.

The present definitions of areas into climatological zones take into consideration the impact of various meteorological elements, particularly those which are directly related to potential moisture surplus or deficit. In many cases, these definitions are short of reflecting the true nature or degree of aridity especially towards its borderlines. Moreover, agricultural development practices have shown that in addition to the biological aridity of an area the lithological characteristics of its surface are important. The latter features provide useful information which aid in appraising the practical impacts of the former.

Looking at deserts from an environmental point of view, their definition should describe them as "areas which constitute potential sources of pollution, mainly physical yet often chemical, through the agent of wind during all or the major part of the year."

RESTRICTIONS

Deserts are still being looked at as mysterious, dangerous stretches which are often given as symbol of loneliness, isolation and death. In some dictionaries, they are referred to as territories of toil and gloom.

Wherever deserts are shared by neighboring countries they constitute zones of more enforced restrictions. Such measures have lead, over the years, to passiveness from ordinary citizens and attraction for smugglers and outlaws. A transfer of any civil servent to a post in the desert was taken as punishment or exile. Mass media communications have contributed to a large extent towards generating these anti-desert impressions.

Certain positive steps are being taken in recent years to encourage government employees to take more interest in serving these areas. Incentives are given as desert allowances either in salary raise or extra years of pension privileges. In addition, provisions are available for either low rent or totally free accommodation. Yet, deserts remain zones of repellence rather than of attraction even for those who were born there and, to a great extent, the educated ones of the latter group.

PREVAILING CONDITIONS

The causes of desertization are mainly climatological but they are often aggravated by technological and socio-economic activities. The environmental conditions have exerted and will continue to exert their influence on the characteristics of the natural resources. Some of these characteristics and influences may be enumerated as follows:

- the landscape has, throughout the years, reached some state of equilibrium which if critically disturbed would cause damage that is hard or even impossible to remedy.

- almost all soils are problem soils which exhibit inferior qualities either physical, chemical or biological.

- sand drift is a continuous phenomenon resulting from normal geological erosion processes that are difficult or impractical, if not impossible, to control.

- underground water resources are of comparatively limited capacities, non or very slowly replenishable, fair to medium in quality and expensive to mine.

- relative humidity of the air is low and day temperatures reach very high values. The range of daily temperature fluctuation is wide, reaching more than 40 degrees centigrade (104 degrees F).

EGYPT'S EXPERIENCE, A GENERAL VIEW

The activities involved in developing arid desert areas for agriculture in Egypt are varied. Although they started on a small scale in the thirties, yet the big development projects in these areas began in the early fifties. During the last 25 years, more than 300,000 acres were put to irrigated agriculture.

The many positive results which were obtained did not overshadow the negative ones which could be corrected to a great extent. Most of the problems are mainly the result of insufficient pre-investment studies or coordination, in addition to the lack of efficient administration and follow-up. The details of the achievements and problems were presented in many reports and publications. The following is a brief overview of some of the conditions which governed these activities, in the field of land reclamation.

The reclaimed areas covered various types of soils, ranging from pure sands to impervious, highly saline gypseous clays. Water sources were purely artesian in places and shallow underground in others. The large development schemes depended, however, on surface freshwaters and brackish waters or a mixture of both. Pumping of water for irrigation reached up to 180 feet and drainage was not entirely by gravity. Surface flooding was the basic method of irrigation-water application, yet sprinkler systems were used on a large scale.

Problems of soil and water salinization, lateral contamination to adjacent low-lying areas, water-logging, surface crusting and sand deposition are common.

Artesian waters have their own problems of quantity, quality, erosiveness and, in particular, proper means of regulating their flow and designing effective rotations for rationalizing the controlled flow.

Land tenure varies between State farms and government controlled farms although some of the latter are now sold to the private sector giving first priority to cooperative societies.

The administration of desert development projects has also changed according to stage, size and geographical location of these projects. As an example, the responsibilities were given first to a Committee which developed after a few years into an Authority. The latter grew larger and became a Ministry subsequently attached to the Ministry of Agriculture as a compact executive organization. The management of reclaimed desert areas is divided at present among geographically distributed Companies, all of which are still governmental.

Desert research is distributed among various specialized ministerial departments, specialized research centres or institutes in addition to concerned national committees.

The relationship between research, planning, execution and follow-up departments connected with development projects in deserts can be easily attained if the role, responsibilities and authority of each are clearly defined.

LESSONS AND PRECAUTIONS

In seeking alternative strategies for desert development it is necessary to benefit from past experiences in order to avoid ill-consequences. Realizing that the economic uses of such areas need large investments which are almost prohibitive to many of the developing countries in which desert development is needed most, it becomes necessary to look for a step-by-step approach. In this way, the present-day tendencies for big ventures in the deserts can be checked, at least until more safe and realistic procedures are tested.

In addition to the foregoing exposition of problems which need proper management, the following clear issues deserve specific consideration:

- an effective strategy for desert development and management means the involvement by the best of us and the contribution of the best in us. It is necessary, therefore, to foster programs which develop in individuals, groups and communities the qualities which encourage this involvement, readily.

- the complexity and interrelationships of the desert's various facets are not fully recognized. Some of the disciplines were neglected when studying and planning for large size irrigation projects in deserts, namely: health, administration and social aspects. The impact of the social element is of particular importance especially where new technologies are to be introduced.

- documentation of all activities and exchange of experiences at all levels were not always exhibited. Many data and agreements on objectives and programs were neglected due to the absence of efficient administrative infrastructure.

- there are signs to indicate that modern technologies and experts' views were not enough to ensure continued long time success. The cooperation and interest of the local inhabitants had not been enlisted on a sufficient scale to insure their participation in the execution and beneficial utilization of these costly efforts.

- the publicity on the advantages of some new desert reclamation or irrigation technologies has, in some cases, lead to over-enthusiasm which resulted in the misuse and, consequently, deterioration of both water and soil resources.

- the change in desert ecosystems due to large irrigation projects showed itself in the spread of rodents, birds, insects and plant diseases.

- the utilization of water for either agricultural, industrial, or civil purposes results in redistribution of salts in soils, waters and surroundings. The resultant effect is always clear as salt concentrates at or near the soil surface. Due to the confinement of drainage water disposal locations, these salts, particularly as surface accumulations, create active foci or pollution.

- the climatic and edaphic conditions of deserts tend to develop gradually a state in which saline agriculture becomes a normal acceptable state. When this stage is reached, any attempt to reduce salinity in soil to suit a specific crop will be uneconomical. This tendency requires more investigations in selecting, breeding and adapting salt tolerant crops which can maintain good yields under more saline root zone conditions.

- sandy soils and highly calcareous ones prevail in many of the desert areas. More experience in managing irrigation and fertilization of these soils is needed.

- closed systems of water conveyance and distribution seem to protect against many negative consequences, particularly the rising of water-table, water logging, soil salinity and above-all the inefficient usages of water.

THE ROLE OF OASES

The aforementioned brief analysis reveals that our knowledge of arid deserts are still below the minimum required to plan for their development and conservation. Hence, it is necessary to base the strategy for the coming years on discovering more about the resources, dynamics and ecosystems of these areas.

It is true that more is becoming available through modern technologies such as remote sensing and aircraft surveys. Yet, this information is available to small groups of scientists and official departments concerned. To generate more public interest in arid areas, it is necessary to open new roads and create multi-purpose centers of attraction.

This can best be achieved through a long-term action plan establishing a network of focal points throughout the arid deserts starting with the old oases and continuing with establishing new ones. A well selected and distributed number of revived oases can serve the following purposes:

- establish permenent test sites where task forces of multi- and interdisciplinary groups can collect real data on desert conditions affecting their development and management.

- permit in-depth assessment of specific problems and existing experiences.

- record, process and disseminate lacking information particularly that regarding natural resources and meteorological elements.

- provide training, extension, and education in desert management and, therefore, act as breeding places for new generations of desert technologists.

- establish germplasm of plant and animal species tolerant to the abnormal desert conditions.

- test prototypes for the uses of unconventional sources of energy, solar and wind.

- attract tourists, investors and other interested groups.

CHINESE PERCEPTION AND TRANSFORMATION OF ARID LANDS

Sen-dou Chang
Department of Geography
University of Hawaii
Honolulu, Hawaii 96822

ABSTRACT

Lying in the southeast of the largest continent, China has a substantial share of arid lands in the world. Approximately a half of China's territory has an annual precipitation below 500 mm. Within this arid or semi-arid zone in the north and northwest, more than 1,095,000 square kilometers of land are covered with deserts, accounting for about 11.4 percent of the total area of China, a size nearly equivalent to the total cultivated area in the country (Fig. 1). In the past 28 years, the desert areas in the frontier regions of the north and the northwest have become integrated, both politically and economically, into China proper, probably for the first time in modern Chinese history. As the developmental strategy of China is very much land-oriented (vs. maritime orientation), the desert lands in China have also experienced rapid modification and transformation. This paper attempts to throw light on Chinese perception and attitudes toward their vast desert environment, to discuss basic techniques in ameliorating these harsh lands, and, finally, to make a brief evaluation of the problems and achievements of China's desert policy.

CHINESE VIEW OF ARID LANDS

The attitudes in contemporary China towards the arid environments have the following implications:

Arid Lands as Strategic Borderlands

Through the long span of Chinese history, the arid lands in the north and northwest have always been considered as a zone of demarcation between the high civilization of China and the rest of the world. No Western power was able to challenge Chinese claims of sovereignty or suzerainty over these lands until the late 17th century. Except for the borderlands along the Amur and the Ussuri in eastern Manchuria, the present Sino-Soviet border, the longest border in the world,

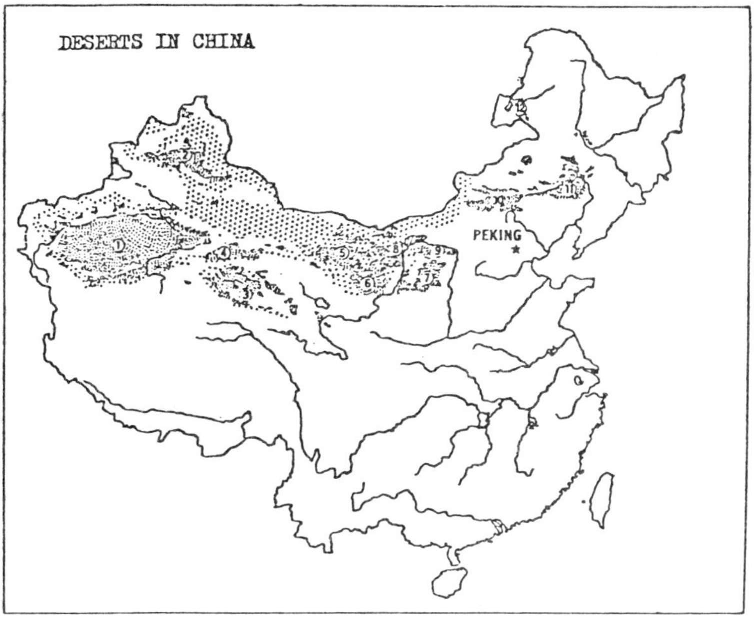

1. Taklamakan
2. Kurban-Tungut
3. Tsaidam
4. Kumutako
5. Patanchilin
6. Tengri
7. Ordos
8. Ulanbuh
9. Kupuchi
10. Lesser Tengri
11. Khorchin Sandy Land
12. Hulunbuir Sandy Land

largely passes through the arid or semi-arid areas of Mongolia and Central Asia. With subsidized colonization throughout both the Tsarist and the Soviet periods for almost a century, the Soviet side of the border has been far more developed and settled than the Chinese side. The recently available satellite picture taken by Landsats A and B have revealed the remarkable contrast between the two sides of the border. As the border tension increased in the past 17 years, the Chinese have also set up a priority in developing and settling the frontier regions along the border. The great military significance of developing these frontier lands in China is reflected in the fact that almost all these arid lands have been resettled by pioneers composed mainly of the People's Liberation Army, and organized as military colonization units of the so-called "production divisions" or "production battalions". Even though, since November, 1974, these "military colonization farms" have changed to "state farms" in name, the members of the People's Liberation Army still constitute important segments of frontier settlements in these areas. Furthermore, the employment of military spirit and tactics to combat the adverse physical elements in the arid region mirrors a fundamental attitude of contemporary Chinese leaders toward their hostile environment: "declaring war with nature".

Arid Lands as Underdeveloped Areas Where Minority Groups Live

China has more than forty million people who have various ethnic identities and who are considered as minority groups. Due to the very sparse but widespread distribution of minority groups in the arid and semi-arid regions of the north and the northwest, Chinese official publications often state that China's minority groups constitute only 6 percent of her total population but occupy 60 percent of the total land area. This statement implies that the 60 percent of the land occupied by minority group is the most thinly populated area in a very crowded country. It is believed that this 60 percent of the land contains large quantities of unworked mineral deposits and over 80 percent of her meat, milk, and wool-supplying animals. These rich resources should be developed and utilized for the benefit of the whole country.

A characteristic phenomenon of minority problems in China is that several identical groups live across the present international boundaries and pose a potentially dangerous identity problem in the political sense. From the Chinese official viewpoint, a prosperous, content minority population should be a living proof of the successes the Chinese model of socialism can have for non-Han peoples. Thus, the development of arid lands in Inner Mongolia and in Sinkiang, where the Mongols and the Uighurs live, should be on the priority list of China's regional development.

Arid Lands Where Ancient Chinese Civilization Flourished and Expanded

The earliest Chinese civilization essentially evolved and matured in the semi-arid region of the middle Yellow River valley. From this "cradle," Chinese people and culture gradually diffused and spread to various parts of present China. While the subsequent colonization in the humid subtropical areas of South China gave rise to some of the most productive key economic areas in East Asia, the military colonization and caravan trades in the arid regions of the north and the west have long been a symbol of Chinese cultural superiority and dynamism. One of the most important research tasks in historical geography in contemporary China has been the reconstruction of physical and cultural landscapes of the prosperous dynasties such as the Han and the T'ang. Archaeologists have been busy excavating sites of old settlements along the ancient Silk Road, unearthing artifacts, and estimating the agricultural productivity and population sizes of ancient ruins in past times. The task of tapping ground waters, reclaiming arable land, and laying out new transport routes seems to have aimed, spiritually and morally, to restore the prosperity and glory of the past dynasties as a first step to transform and ameliorate the adverse arid environments. Historical and archaeological research in arid lands is not entirely for pure academic purposes only. The accurate charting of ancient river beds, for example, often means the discovery of ample ground water resources.

Arid Lands as a Target of the "Transforming Nature" Campaign in the New Ideology

Perhaps one of the most dramatic changes occurring on mainland China in the past eighteen years has been the alteration in the conception of man's relation to his physical environment. The orthodox traditional view saw man as an integral part of a cosmos dominated by nature. Contentment as well as material success could come only through acceptance of the rightness of man's adjusting himself to the greater natural world to which he belonged. To question or attack nature was to contradict a broader natural order, and hence was potentially disturbing to the profoundly hierarchical social orders of traditional China. The 1949 revolution has completely changed this traditional attitude toward nature. Nature is now explicitly seen as an enemy, against which man must fight an unending war. Such a war against the environment stems at least partly from two other factors: the need which the leaders see to maintain full mobilization and employment for the entire population, and the need, further, to dramatize the revolution by a break with the traditional society and its acceptance of things as they are. The immensity and the messianic quality of the crusade against nature, suspect in many instances from a Western cost-accounting point of view, do in fact level off literally not only mountains and sand dunes, and hence produce some significant economic gains, but move the mountains of apathy and indifference which are often considered as greater obstacles in most developing societies, and thus give China what most scholars agree is a better prospect than most of breaking through the barrier of economic backwardness.

One aspect of Chinese perception toward nature which might still be quite traditional is that, in accordance with mythological beliefs of ancient China, the environment is often personalized. Thus, a campaign for dredging the silt deposits in the lower course of the Yellow River is called "clear the throat of the Yellow River". The menacing drifting sand dunes are so humanized that they have two "likings" (for strong winds and for drought) and three "fears" (fear of grass, fear of trees, and fear of water). The personalization of natural phenomena is intended to facilitate easier comprehension of the character of various hostile environmental parameters for the populace, so that man, in the context of Chinese socialist ideology, can be a better designer and architect of his own physical environment.

CHINESE EXPERIENCE IN DESERT DEVELOPMENT AND MANAGEMENT

Since the early 1950's, the Chinese Academy of Sciences has sent a number of "sand control teams" to survey and study arid lands with respect to their geomorphology, micro-climatology, vegetation, soils and salinity, hydrology and ground water, and engineering aspects of development and management. Local counties in dry areas have also set up their own office of "sand control" research, taking into consideration regionalized unique conditions. Extensive reports of these survey projects have been published. However, except for a few selected items translated in the Joint Publication and Research Service in the United States, the bulk of these technical reports on deserts is still largely in Chinese. The Academy's Research Institute for Glaciers, Frozen Soils, and Deserts at Lanchow appears to be the largest center for desert studies in China, and often gives consultations to various communes and counties dealing with desert management. The following discussion on technical aspects of stabilization of sand dunes, reclamation of arid lands, and utilization of natural resources in deserts is mainly derived from publications of the research institute at Lanchow.

Fixation of Drifting Sand Dunes

The most significant achievement in desert studies in China in the past 28 years is very likely the methods used in stabilizing the active sand dunes, which have long been a menace to agricultural lands and villages along the Great Wall in Inner Mongolia, in the Kansu Corridor, in the margins of the Taklamakan desert, and in the western edges of the Sung-Liao plain in Manchuria. One of the distinctive characteristics of the deserts in Chinese Central Asia is the high frequency of duststorms in winter and spring, with wind velocity often reaching force 5 and 6 or above in the Beaufort scale. The number of days with duststorm ranges between 120 and 150 days in the year with an average storm lasting more than ten hours. These strong winds, generally from northwest and northeast, have caused extensive saltation of sand particles and advancement of active sand drifts. As a result, the areas immediately south of arid lands have experienced various degrees of desertification in modern history.

In order to reduce the wind velocity near the surface and to slow down the speed of sand drifting, the Chinese have employed a variety of materials to build wind and sand breaks. In the area where the wind directions are changeable, gridiron-shaped wind and sand breaks with 1-2 m interval are most commonly built, with the major breaks aligned perpendicular to the most prevailing wind. In the area where a particular wind direction dominates, the breaks are in parallel alignment, 1-2 m apart, and are all perpendicular to the wind direction. Almost all kinds of bulky material which can be collected are used to build sand breaks, the most common being grass, wheat straw, rice straw, reeds, branches and twigs, pebbles and gravel, and liquid asphalt. The so-called "grass gridiron" breaks, used most extensively, are built with grass or straws partially buried under the sand 15 cm deep, and stand above the surface about 25 cm. Local conditions permitting, grass or shrubs are planted in the centers of grids. These breaks are reportedly capable of increasing the roughness of the surface, reducing the wind velocity near the surface, and thus stabilizing the position of sand dunes. They are also able to raise the moisture content of surface materials, creating a better environment for plant growth. A perennial grass which is planted widely on sand dunes in Inner Mongolia as sand breaks is Artemisia desertorum. Extensive "grass gridirons" have also covered the slopes on the west side of the Paotow-Lanchow Railroad for road protection.

A combined vegetation of trees, shrubs, and grass is generally considered the most effective measure against the drifting of active sand dunes in China. The general strategy is, first, to surround individual dunes with trees planted in the lowland area between dunes, the favorable species being Elaeagnus augustifolia, Populus simonii, and Salix matsudana. Trees are planted at the foot of the leeward side of the dune and are at least 2.5 m in height with a diameter of about 5 cm so that they are tall enough to stand above the drifting sand. Spring is the season preferred for tree planting. The roots of the trees are one meter beneath the surface. As the sand dunes are gradually moving in the leeward direction, trees are planted each year at the foot of the leeward side in a new location. In the fourth year or so, when the profile of sand dunes has been eventually lowered by wind and the micro-climate of the area has been ameliorated with the surrounding trees, grass (such as Artemisia desertorum) and shrubs (such as Caragana Korshinskii), in such order, are subsequently planted. On smaller sand dunes, it appears sufficient to plant grass and shrubs only along the lower one-third slope on the windward side to stabilize the whole formation (Fig. 2).

In protecting agricultural land from drifting sand dunes, a grid-network of trees becomes a common scene in the oasis along the margins of the Taklamakan desert, the largest desert in China. The network is composed of tree belts aligned in a gridiron formation. Each tree belt is composed of 4-8 rows of trees, one meter between rows. Populus bolleana and Fraxinus sogdiana are the favored species at a height of about 14.5 m. Tree belts are spaced with an interval distance 15-20 times the average tree height. Within this interval distance at the leeward side of the tree belt, the wind velocity is reduced to not more than 60 percent (Fig. 3). In most cases one land unit of trees is capable of protecting more than two land units of crops surrounded by sand dunes.

In Minchin county of the Kansu Corridor where cases are encroached on both sides by the Tengri desert and the Patanchilin desert respectively, the maintenance of agricultural land between drifting sands is a most challenging task to the people in the county. The only less adverse physical condition of the area is probably the relatively high ground water level, an average depth of 1-2 m, along the intermittent stream of the Shihyang Ho. Since 1952, the people in Minchin have started a campaign of afforestation. Using the hardy desert trees, the sand date, they have reportedly succeeded in getting 37,300 hectacres of land planted along roads and canals. In the past decade, every year nearly 3,000 hectares of sandy

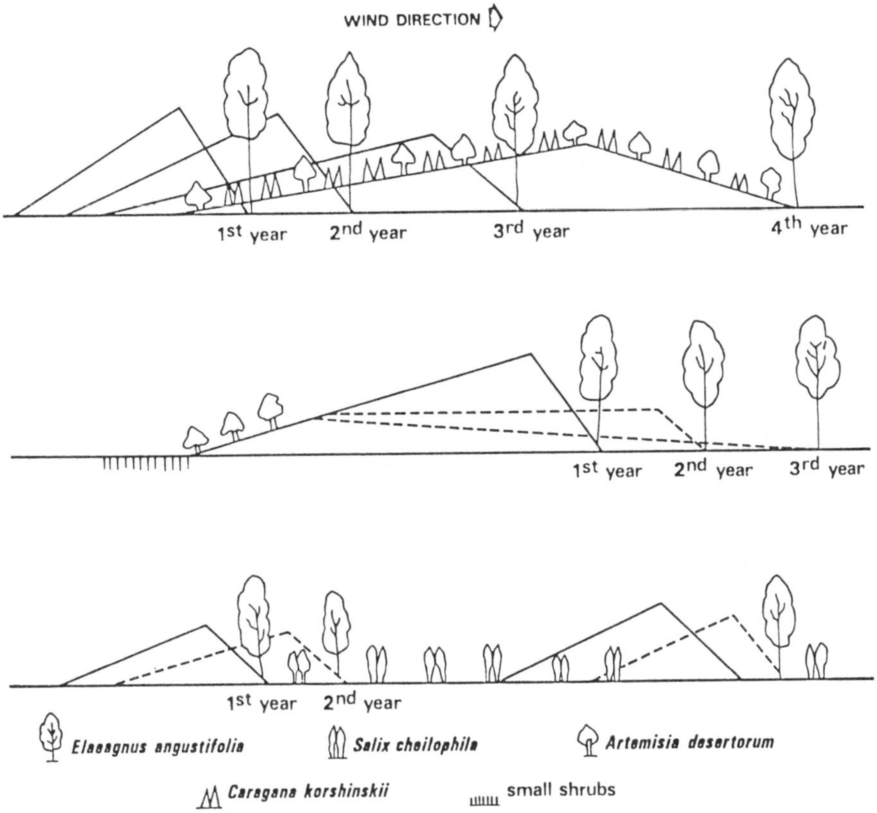

Fig. 2. Stabilization of Sand-dunes: Trees, Shrubs, Grass, Combination.

land is put into sand date trees. More than 2,000 hectares of sand dunes in the county have been flattened, surfaced with clay, and planted with bushes and grass (Fig. 4). Today, a county with a population of 220,000 in nineteen communes possesses more than 50,000 hectares of irrigated cropland, an enviable man-land ratio in the arid regions of the northwest.

Reclamation of Sandy Lands

In the sandy regions where either surface or underground drainage provides sufficient water, reclamation projects through "water washing" to level off sand dunes have been launched in certain selected areas, such as the three counties of Yulin, Hangshan, and Chinpien of northern Shansi where desertification on the southern edge of the Ordos desert has caused both depopulation and shrinkage of arable land in modern history.

The campaign of "using water to wash sand dunes" in northern Shensi is basically aimed at bringing relatively clear water from streams, lakes, or reservoirs to the lowland areas of sand dunes. When the area has been inundated, water is used to loosen up the structure of sand dunes, to destroy the vertical profiles, and to moisten the surface materials. It is reported that the "washing effect" on the

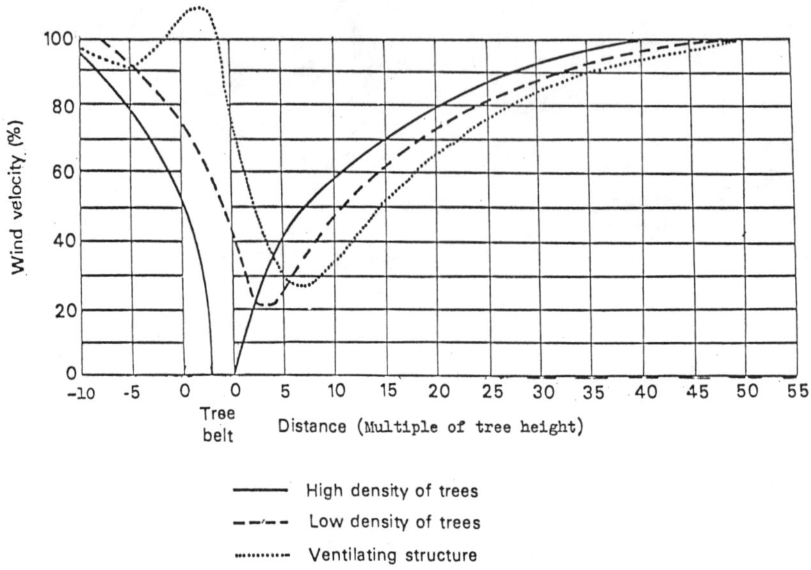

Fig. 3. Effects of Tree Belts on Wind Velocity.

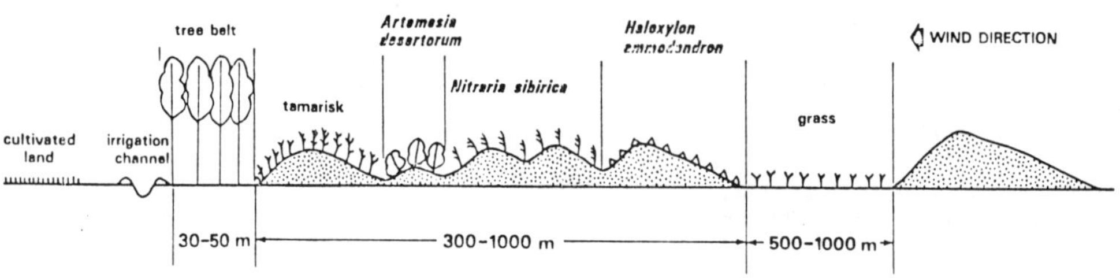

Fig. 4. Stabilization of Sand-dunes in Min-chin Hsien Kansu Province.

sand is best with flow velocity of 0.5-1.2 m/sec and a discharge volume of 1-2 cubic m/sec. The gradient of land which is suitable in using this method is preferred to be between 1/200 and 1/500. On the average, it takes roughly 2-2.5 cubic meters of water to process each cubic meter of sand. The sandy land levelled with "water washing" is not yet ready for cultivation and still needs improvement in soils structure for a number of years. A common way of soil improvement is to flood the land with stream or gully waters high in silt content in summer for several years, or, in some cases, to convert the land to paddy fields with the necessary application of green manure and irrigation waters. It often takes about ten years to really improve the humus content of the soils to over one percent, and to attain the so-called "high and stable yield" by Chinese standards.

Tapping Water Resources in the Arid Environment

Though the inventory of water resources in arid lands of China is still not complete, the preliminary estimate of the total stream flow alone is more than 120 billion cubic meters. With this amount of flow, approximately 13 million hectares of land could be reasonably reclaimed for cultivation in the oases along river valleys, near shores of ancient and present lakes and on marsh lands. Thirteen million hectares of land account for more than ten percent of the total cultivated area in China today.

As in any other desert regions, a large volume of stream flow in northwest China is lost in evaporation and seepage. In many river valleys in Sinkiang, with rudimentary irrigation, the amount of water which could be channelled into irrigation ditches accounts only 50 percent of the total stream flow. Within an irrigation system, water which could be effectively utilized for crops accounts for only 50 percent in relatively good quality projects, and between 20 and 30 percent in the average projects. A practical way to minimize the amount of seepage in Sinkiang is to pave irrigation channels with pebbles and gravel plastered with cement.

In tapping the ground waters in the arid lands in Sinkiang, a prevailing method is still the time-honored khanat (Fig. 5). According to a Chinese classic historical record (Chien Han Shu), as early as the reign of Han Wu Ti (141-87 B.C.), khanats had been constructed in the Lo Ho valley in Shensi to tap ground water for irrigation. Subsequently, the khanat was introduced to the Turfan basin. The method of khanat was not widely adopted in Sinkiang until Tso Tsung-tung started a program of water conservancy in Sinkiang in the late 1870's. In the past few decades, the construction of khanats and wells has been accelerated, so that by 1965, there were more than 2,000 khanats in the autonomous region, 1,300 of them in the Turfan Basin, a structural basin with an elevation more than 100 meters below the mean sea level. The khanats in the basin irrigate nearly 24,000 hectares of crop land and contribute 1.6 billion cubic meters of water annually, accounting for 80 percent of total water supply in the basin. It is also reported that water seeping out from springs and tapped from khanats along the foothills of Sinkiang totals more than 5.5 billion cubic meters. There are 33 major springs and khanats, each of which has an average discharge volume over one cubic meter per second, and is the potential site for the development of irrigated agriculture in this most arid region in China.

UTILIZATION OF OTHER NATURAL RESOURCES IN ARID LANDS

There are more than 1,000 different species of natural vegetation in the deserts of China, approximately one-third of them being utilized as traditional herb medicines, such as Ephedra intermedia, Glycyrrhiza uraleusis, Cistanche deserticola, and Cynomorium songaricum. Several shurbs and grass have also been utilized as raw material for making textile fibers, the most widely used being Poacynum hendersonii. On the river terraces and in the surrounding areas of salt lakes and marshes, much land is covered with wild grass, reeds, and small bushes, and has good potential for developing pastural economy. In the desert areas of Sinkiang, wild animals, such as antelopes, hares, and wild boars, abound. The red deer, probably the most valuable animal in the Chinese desert, and has been recently domesticated, as the deer antlers are always in large demand in Chinese traditional medicine.

In conclusion, this paper aims only to provide a very general overview of Chinese policy and management of arid lands based on Chinese source materials published in recent years. For the time being, it is not easy to make a reasonable evaluation of the achievement of these programs in remote Central Asia, nor is it possible to

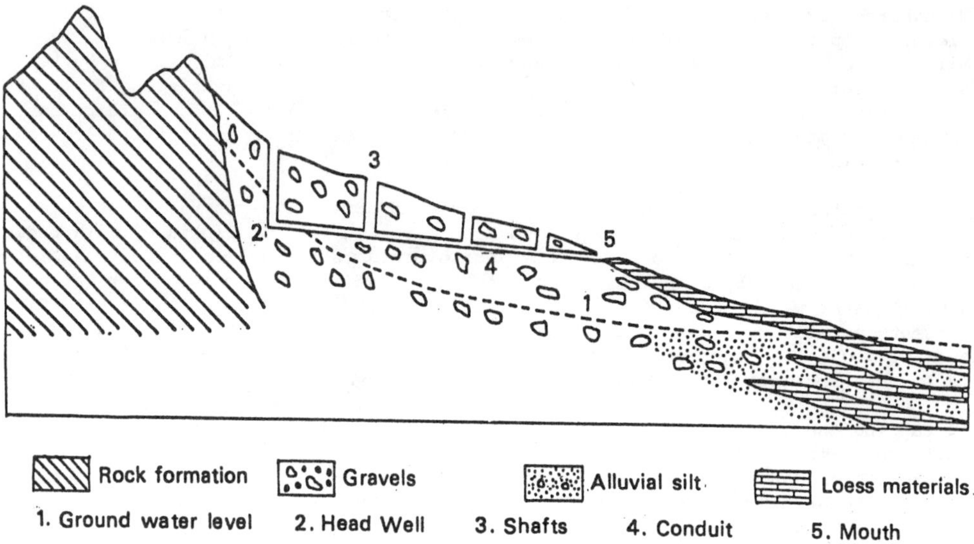

Fig. 5. Khanats in Sinkiang

make an assessment on the transferability of Chinese techniques in desert development to other developing countries with different sets of environmental parameters. In the past few years, however, there have been a number of satellite pictures of China available, taken by Landsat A and B. As the interpretation of these pictures is best in the arid regions, the satellite pictures, taken with a proper time interval, may provide a useful tool in studying desert programs in an inaccessible country like China. A preliminary visual interpretation of false color composite imagery on China testifies to significant changes in arid landscapes, such as the sizes of oasis and irrigated agricultural lands, and the increasing number of reservoirs and canals, which provide support of Chinese official claims about their achievements. A more complete picture of the impact of Chinese policy on arid lands over a longer term, however, still has to wait until multi-temporal imagery with a longer interval, together with better ground truth data published by scientific journals in China, is available in the next few years.

REFERENCES

(1) Chiao-kuan-ts'ao chi-ho chih-li liu-sha ti chu-pu chin-nien (1974) (Preliminary experiences of controlling sand-dune movements by integrated planting of trees, shrubs, and grass), Ko-hsueh tungpao (Science Bulletin), 3, 117-124.

(2) Chinese Academy of Sciences (1974), Chung-kuo sha-mo kai-lun (General Discussion on China's Deserts) Science Publications, Peking.

(3) Chinese Academy of Sciences (1976), O-kuo ti sha-mo chi ch'i chihli (our country's deserts and controlling methods) Chung-kuo Kohsueh (Chinese Sciences) 5, 492-503.

(4) Chu, cheng-ta (1965) sha-mo ti-chu yu nung-yeh kai-k'en yu kuan ti jo-kan feng-sha ti-mao wen-ti (Problems on eolian landscapes related to agricultural reclamation in the desert areas) papers presented in the 1965 annual meetings of the Institute of Geography on Geomorphology, Chinese Academy of Sciences, Institute of Geography.

(5) Golab, L. Wawrzyn (1951), A Study of Irrigation in East Turkistan, Authropos 46, 1-2, 187-199.

(6) Hopei province, Institute of Geographical Research (1976), Hsin-chao ku-ho-tao, kai-fa ti-hsia-shiu (Scarching for old river channels and exploiting underground water). Ti-li chih-shih 7, 27-30. (Geographical knowledge)

(7) Li, Te-yung (1975), Li yung tung-chih-wu hsin-chao ti-hsia shui-yuan (Finding underground water by observing animals and plants), Ti-li chih-shih (Geographical knowledge) 3, 31-32.

(8) Murphey, Rhoads (1967). Man and nature in China, Modern Asian Studies, 1, 4, 313-333.

(9) USSR Academy of Sciences (1969), Physical Geography of China, Praeger, New York.

(10) Wang, ping-ch'en, et al, (ed.) (1957) Ti-hsia-shui (underground water) Hsiu chih shih Publication, Shanghai.

BOTSWANA'S DEVELOPMENT STRATEGY FOR ITS INDIGENOUS DESERT PEOPLE, THE KALAHARI BUSHMEN

Liz Wily
Bushmen Development Officer
U. N. Development Program
Botswana

ABSTRACT

This paper describes the development strategy Botswana has adopted towards its indigenous desert people, the Kalahari Bushmen. The sole - and fairly numerous - occupants of all of Southern Africa for some thousands of years, they are today a minority, and one which is confined to the most inhospitable area of Southern Africa, the Kalahari Desert. As a hunting and gathering people, in a predominately pastoral and rapidly developing society, Bushmen are today facing traumatic change. Once a people adequately subsisting, albeit by a so called primitive life style, and in an extremely harsh environment, the Bushmen are today entering the underdevelopment of improvishment of landless dependency upon more successful strata of the rural society. It is to intervene in this process and facilitate their better integration into the wider society that is the aim of Botswana's strategy of action.

This paper is divided into three sections. The first provides a general background, first placing the Kalahara Bushmen in a historical perspective, and then describing the nature of the changes these hitherto isolated desert dwellers have faced in the last 150 years. The section ends with a brief outline of their situation today.

The major part of the paper outlines the strategy Botswana is following in recognition of their circumstances. A background to the Botswana government's explicity dealing with the programs is given, followed by a countdown of the working principles of the ensuing development programme that evolved. Following a brief description of that programme's status today, is a more detailed look at the critical access-to-resources issue. The section ends with a note on problems the current strategy is encountering.

A final shorter section looks briefly at the involvement of the international community of Bushmen affairs, real and potential.

BACKGROUND

Introduction

The Bushmen, San or the Basarwa (1), as they are known in Botswana, are one of the few remaining hunter-gatherer people in the world today. As the original residents of Southern Africa they peopled the area from the Zambezi Valley to the Cape of Good Hope. Together with their pastoral brothers, the Hottenton people, the Bushmen comprise a small but distinct race of people, known as the Khoisan. Conflicts with the incoming Bantu-speaking Negroid peoples moving down from east Africa led to a supposed severe reduction in their numbers, further decimated when the Dutch began to colonise Southern Africa in the 17th century. Today an estimated one tenth of their number remain, some 60,000 individuals restricted to the more arid areas of Southern Africa - the Kalahari Desert. The majority reside within what is the modern state of the Republic of Botswana, (est. 40,000) (2) with some 15,000 residents in Reserves in Namibia, and up to 5,000 estimated in Southern Angola. Isolated individual Bushmen do live in the more fertile regions of South Africa today, and it is generally supposed that the so-called 'Coloured' people of the South African Cape area are the result of a mixture between Bushmen and Dutch settlers. Interestingly enough the lone survival of Bushmen society in the most inhospitable region of Southern Africa, the Kalahari, is now understood to have come about, not by Bushmen being pushed further and further inland by recent Bantu or European expansion, but by virtue of the isolation from such contacts that desert environment provided.

It seems Bushmen have lived in the Kalahari not for tens or hundreds of years, but thousands. Archeological and linguistic evidence supports this (3), while on a more simple and subjective level one must ask how a society so finely balanced and attuned to the environment, and so specialized, could possibly have evolved without generation upon generation of trial and error?

I will not elaborate in this paper upon the complex social and economic dynamics of this Bushmen society, some of the more dramatic aspects of which have received almost popular treatment. One may, for example, read in many languages of their amazing ability to live without permanent water sources, surviving for half the year by extracting moisture from fruits and water-bearing roots and the rumen of wild animals. The socio-ecology of the Kalahari Bushmen, analyzing exactly how they do survive in the arid desert has received particularly thorough treatment. Suffice it to say here that such works tend to refute the impression of life of the primitive hunting and gathering Bushmen as being necessarily nasty, brutish and short. Rather we discover a life style of relative ease, leisure and abundance and a social organization somehow more workable and peaceful than our own.

In the course of these studies the Kalahari environment itself has been somewhat exonerated: one finds this 450,000 square mile sandy basin 64,000 of which is hidden amongst its varying flora of open grasslands, shrub and dunes, a rich fauna of over 200 species including some 46 mammals. It also conceals a hundred or more edible fruits, leafy vegetables, nuts, roots, tubers and berries. However, in order to not swing from an impression of Hobbes' ignoble primitive to Rousseau's most noble savage, one need only recall that Bushmen alone managed to make the Kalahari habitable. No amount of investigation refutes and only reinforces amazement at the delicate balance of the unique interrelationship the Bushmen desert dweller has evolved with the land about him.

It is the interference with this balance by recent changes that have taken place around the Bushmen in his otherwise long unchanging history, and the resulting situation he finds himself in today, that I have chosen to describe as background to this paper.

Changes and Acculturation

These changes have occurred comparatively recently, gradually and peacefully. The Kalahari-dweller, like their brothers in the more fertile southern regions outside the Kalahari are confronted with the wars of extermination or enslavement. The Tswana tribes who eventually did enter the area now known as modern Botswana, were not warlike people; and anyway, the hostile nature of the desert meant their expansion beyond its fringes and only occurred gradually and unevenly. Moreover, they were often necessarily dependent for their very survival on the Bushmen in the initial stage of their movement into the harsh Kalahari. These movements, which came from the East, and mainly by the Bakgalagadi tribe itself pushed by stronger tribes in the East, began to take on the shape of a permanent settlement between the mid and late eighteenth century. More dramatic incursions came at two points from the West: along the Ghanzi limestone ridge where (1898) Cecil Rhodes sent Boer farmers to settle and act as a buffer to German expansion from Southwest Africa, and further north, a decade or so later, Herero pastoralists came to settle, fleeing from their bloody battles with the German colonizers. Large areas of the Kalahari today remain unoccupied by any other people than Bushmen.

If the displacement of the Kalahari Bushmen has been uniquely recent, peaceful and incomplete, it has also been essentially unintentional on the part of these incoming agriculturalists. Characteristically both hunter-gatherer and cattle-owner are unaware of just how critical the continued existence of large tracts of undisturbed land is to the hunter-gatherer economy, until that resource is lost. It has only been in very recent years that an understanding of just how the Bushmen people have been 'dispossessed' has generally come about. This process seems to work in the following ways:

A Bushmen band has the use of a large area of veld (land) for centuries. (4) No other bands use that area. Other groups of Bushmen have access to other areas. Each group respects the rights of the owner band: if they ever hunt, gather food, travel into, or stay in the territory ("/nolli") of another group, permission is sought from the "owners". Eventually agricultural people enter the area visiting increasingly and eventually settling. At first the Bushmen welcome the newcomers, regarding them as another 'resource' - especially as they can provide new and exciting material goods. If they bring cattle the milk is seen as an exceptionally useful new 'resource'. Trade of tobacco, metal and other 'modern' goods for skins, ivory and ostrich feathers usually takes place.

For a while the two ethnic groups live relatively easily "together". It is only after a number of years that the negative effects begin to set in: the Bushmen find that the water that was in their pans, or in natural waterholes is not able to meet the needs of all the new inhabitants, especially where livestock have been introduced. They find the presence of cattle discourages the presence of game, and hunting is bad: they find the bush - food becomes scarce through excessive human habitation and because the cattle eat the young shoots so that edible underground roots are difficult to find. Increasingly the Bushmen are forced to focus themselves in other areas, away from the pastoral group. However, this is not satisfactory, as they must still go back to the best places for water (where they alone once lived) in the very dry months. The critical features sustaining their economy, of a low density of population and mobility are lost.

Over the years a "dual" economy is set up. For part of the year the Bushmen are able to live much off the same hunting and gathering life they lived before. For the rest of the year they must go back to that pan or place now dominated by Bakgalagadi or other pastoral peoples. Legal tenure by the newcomers is increasingly established by the drilling of boreholes, digging of wells, and so on. To survive in that area now the Bushmen must beg, or be dependent one way of another upon the pastoralist. Meanwhile the aspirations of the Bushmen exceed the returns of his old no-surplus economy and certain modern commodities become necessities.

Over the years this pattern consolidates - some Bushmen take up "employment" all year round with a cattle-owning family, some Bushmen families "attach" themselves permanently to a pastoral family much in the manner of "master and client", others maintain an uneasy coming and going from the bush to the cattle-post (5), reluctant to be permanently held to the squatter life.

THE SITUATION TODAY

The majority of the 40,000 or so Bushmen are caught somewhere along the continuum of change and acculturation as described above. The majority (50%) may be referred to as semi-independent in their life style. These are the Bushmen who usually live at cattle-posts or Government boreholes for the drier parts of the year. However, they move back to the veld when the rains allow their disposal in ungraded areas, and live by the old way.

The more social aspects of their society remain intact: band formation is still evident, and there is little integration between the Bushmen and other racial groups. They frequently express the desire to continue to hunt and gather, and their complaints tend to crystallize rapidly into pleas for "water in our nolli (place) in the bush". Essentially the skills of traditional hunter-gatherer society remain, but the opportunities to pursue it are diminished. For these Bushmen, while access to veld areas exist they are usually areas adjacent to, like, or near original 'territories' (nollis, place), which have since lost their suitability for one reason or another. The aspirations and needs of these Bushmen become more defined as areas of dependency become increasingly evident.

There is still a small percentage of Bushmen who live the traditional nomadic life by hunting wild animals and gathering bush-food, moving in accordance with the availability of both, neither planting nor storing. They have retained the band social organization, disbanding and uniting as necessary. The band continues to lack formal leadership or internal specialization, each family being self-sufficient in the skills of survival. Hunting methods are traditional (bow and poisoned arrow, spears and traps); and the gathering - stick remains the most important tool. Those few aspects of their life they have modified are essentially superficial and mostly material - for example, the use of iron pots, woolen blankets, metal spearheads, and buckets and tobacco.

Groups of such Bushmen, who number up to 5,000, live in areas within the Central Kalahari Game Reserve in the middle of the country and in odd isolated spots of the Kgalagadi, Chanzi, Northwest and Western Central Districts. The majority of these Bushmen are using the areas of their forefathers which is the key to their stability. They are the few (lucky) Bushmen who were/are living in areas over which there has been minimal or no changing land use, occupation or establishment of legal tenure by other peoples. It is for these Bushmen that the establishing of rights of tenure is the least complex, and the most clear cut direction in which development action is required - and the most possible.

A final group of Bushmen are known as sedentary 'village' or dependent Bushmen. They live for most or all of the year within or near small villages and cattle posts or freehold farms. They are not cattle owners themselves, but rather earn a living as the herders of those cattle. For this work they are usually paid one head of cattle each year, or receive gifts. Cash wages are uncommon. Their diet is heavily dependent upon bought ground meal, and the milk from the cows they herd. However, in the wettest time of the year especially, a certain amount of veld food is still gathered. Hunting is also sometimes carried out by these Bushmen, but usually with modern (and therefore illegal if no license is held) weapons, such as (borrowed) guns and steel traps. Some of the Bushmen in these circumstances have taken up horticulture, making small gardens around the boreholes, or even in some areas establishing large fields for growing maize, beans and melons. It is these more settled Bushmen moreover who begin to own small stock (especially goats) of their own, and in some individual (rare) cases, cattle. In general, the major bias is towards agricultural activity with the traditional hunter-gatherer subsistence rapidly fading. For these Bushmen dependency upon permanent water sources has consolidated, and dependent relationships with the water owners has correspondingly stabilizing - in most cases into a static 'client-master' pattern. While these Bushmen have lost the skills, the mobility, and the social organization of traditional Bushmen society, they are not yet fully acculturised into the cattle society to which they have attached themselves.

A process of change and acculturation, not to the favor of the Bushmen, is thus, well underway. In order to put this process in its widest perspective, I would like, briefly, to draw attention to one aspect of the process which derives from the Bushmen themselves, and by its nature mitigates, if only to a small extent, the incoming agriculturist as the dispossessor. This aspect is the passivity of this desert people in the face of change, and their reluctance, to this very day, to aggressively defend their territories, their resources, their culture or life-style generally.

Perhaps this is not surprising: their way of life, their society, their environment remained essentially unchanged for thousands of years. They knew of no other people or way of life in their desert isolation. Spread out in small bands over vast areas, neither was there competition amongst themselves, nor need to defend territories or develop a strong cultural identity or definsiveness.

Moving in accordance with the seasons, with game, water and plant availability, the land did not require exploiting or developing. A satisfactory living could be gained from merely reaping or gathering off the land. Neither was there need to store food for long periods. Food, if differing in species at different times of the year, was always there, and as one had to move frequently to follow the game or find water, it was clearly senseless anyway to store food.

In general, the attributes of the less attractive modern societies are lacking. In particular, progressive attitudes, which expect and deal positively with changes, were not necessary, and have therefore not been entrenched in their society unlike agricultural, pastoral and later societies, constantly confronted with changing circumstances.

Certainly today, despite the fact that only a small percentage (12%) of their number live by an unchanged hunting and gathering subsistence, the Bushmen people, in general, retain the attitudes and philosophy of a hunter-gatherer society. Its most critical aspect is a certain unprogressiveness.

Whatever its cause, it clearly remains a factor to be considered in their situation today, having facilitated if not determined their general failure to 'do well' in the face of changes in the Kalahari region. More important, it has laid them open to intimidation and exploitation on the one hand, and prone to entering heavily dependent relationships with more progressive citizens, on the other. The problems of these resulting situations merely reinforce the apathy and general demoralization that characteristically accompanies their loss of self-reliance as independent hunter-gatherers.

A viscious circle of impotence is rapidly established. The majority of Bushmen in Botswana have entered or are entering that circle. While the word 'intervention' has lost favour in the international community in the assisting of primitive minorities, the Botswana Government essentially aims to just so 'intervene' in this process of social breakdown and impoverishment, in its strategy of action.

STRATEGY OF ACTION BY GOVERNMENT TODAY

Background Leading to Government Action

The Botswana Government has since mid-1974, evolved a fairly specific strategy towards its desert-dwelling Bushmen citizens. The essence of the policy underlying this strategy lies explicity and simply, in the fact of their full and equal citizenship in the modern nation-state of Botswana.

Before describing the strategy, I will briefly outline the background to the Botswana Government's programme of action.

As a primitive, hunter-gatherer minority, and one which has perfected a remarkable and unique adaption to an extremely harsh and arid dessert environment, otherwise considered uninhabitable, the Bushmen of Botswana had always attracted a considerable amount of attention. Underlying this largely scientific interest has been a concern for their future, it being all too well appreciated that hunter-gatherer minorities hardly fare well under rapidly changing circumstances, so-called progress, or modernization.

Throughout the Colonial period of Botswana's history (1885-1966) this concern was largely limited to a recurrent discussion of whether or not Basarwa, as Bushmen are known in Botswana, were slaves or not to the Tswana tribesmen (6). It reached a peak in 1934 with a famous London Missionary Society Report which concluded that Bushmen were in a 'serf-like' relationship with the Tswana tribes, and no real evidence for the practice of slavery existed. Besides dealing with "incidents", which ususally involved maltreatment of Basarwa or cattle-theft, the British Colonial Administration generally ignored the situation and problems of Bushmen, until the mid-fifties, when it became clear that the Protectorate was moving towards Independence. At that time (1958) a Bushman Survey Officer was appointed to look into the situation of the Bushmen poeple, with a view to seeing how best they might be included within the national life of the future, independent Botswana. It is interesting for our purposes to note that one of his more specific terms of reference was to recommend where reservations should be located. In the absence of any administrative infrastructure for action, all but one of his recommendations were actually acted upon, which was the establishment of the Central Kalahari Game Reserve.

Until 1972 no other development action took place on behalf of Bushmen citizens, besides the relatively unsuccessful efforts of a small private mission in the far west of the country (7). Statements of concern were expressed, however, from time to time, both by politicians and administrators in the post-Independence Government. Meanwhile scientific research amongst Bushmen of Botswana -- who had become citizens at Independence along with all other original residents had accelerated. This work focused, however, to the point of repetition, on the small number of the Bushmen living in the western Ngamiland area. Basic demographic data on the numbers, locations and circumstances of the Bushmen population remained lacking. This was not ameliorated by the National Census carried out in 1971, because of a political decision to exclude questions on race and tribe.

Nevertheless in these years (1966-1972), pressure, sporadic as it was, and essentially from individuals, led the Ministry of Local Government and Lands which had become one of the senior bodies responsible for rural development, to set aside, in 1972, a Grant of several thousands of Rands with which it could 'assist Bushmen'. Before 1974 only R5983 was spend. Most of this went toward supporting the first real development effort to be undertaken for Bushmen in Botswana, the Bere Ranch Settlement Scheme (8). This scheme aimed to provide a small group of Bushmen with their own water point and facilities in the area where they had always lived, but were finding difficult to remain in without such assistance. It was, however, a privately-initiated and administered scheme, and Government's involvement was in practice limited to giving the scheme its blessing, and paying the salary of the teacher for the school established there. But it did prompt the Ministry of Local Government and Lands to establish a Bushmen Affairs Committee which recommended (1973) that a full-time post of Bushmen Development Officer should be established within the Ministry of Local Government and Lands, thus, making some kind of administrative commitment to assist Bushmen citizens.

The terms of reference of this post were undefined, the officer being merely directed to do something about what had become the "Bushmen problem" and to act as an administrative focus for all matters relating to Bushment Citizens. The Bere teacher who was the only applicant with development experience working with Bushmen, was appointed to the post in May 1974.

The establishment of this permanent and pensionable post within the civil service of the Botswana Government clearly marks an important turning point in the history of Bushmen citizens in independent Botswana. It indicated not only a growing awareness of their unhappy situation (if at that point still largely by individuals rather than Government as a body), but that Government had some responsibility to attend to their needs; that Government was willing and had at last the administrative capacity to tackle the "Bushmen problem". More significantly, the creation of a special post tacitly indicated that the problems and situation of Bushmen were somehow special or different, and accordingly required special attention.

At the same time it is important to note the extreme vagueness that otherwise surrounded this move by Government. The post does not seem to have been established as the result of a defined policy or plan of action, nor as a result of international, or internal political or popular pressure -- and most certainly not as a result of pressure from the Bushmen themselves, who remain even today singularly unvocal. Rather, the post was established if I may say so, in an atmosphere of liberal, if somewhat paternalistic benevolence "we must do something about our poor Bushmen", prompted by the concern expressed by increasing numbers of individuals including visiting anthropologists.

Far from knowing exactly what should positively be done for the Bushmen, Government's involvement was based on knowing what should NOT be done -- i.e., that the present reality of "laissez-faire" based on a 'leave them alone' argument was felt to be inappropriate, unsatisfactory and should end.

A Government official at the time pointed out that a 'leaving the Bushmen alone' policy had not meant that the Bushmen were so left alone. Certainly their lives had not gone on unchanged. Neither did leaving them alone to their own fate mean that they were, by so-called historical inevitability, dying out. Nor were they finding satisfactory solutions to the changes taking place in their lives, on their own accord. It was further clear that inaction because of uncertainty of what should be done was equally unsatisfactory. The situation of the Bushmen minority was merely continuing to get worse. It was felt that something had to be tried, some intervention had to be made, and to define this was to be the job of the Bushmen Development Officer.

Underlying Policy of Action

If there was no stated specific policy that led to appointing a Bushmen Development Officer, the move was nevertheless quickly rationalized. To quote and early working paper on the subject the Botswana Government's concern for its Bushmen citizens is in no way inconsistent with its general development aims and policies. Two of our most frequent-declared national principles are that we are a nonracial state, and one which lays great emphasis on the best future of the rural population. Therefore, the development attention Bushmen are receiving is not viewed in Botswana as a special privilege, but their basic, undisputed right as full and equal rural citizens. With the rapid development that has been seen in the last few years in the rural sector, the particular problems of the Bushmen have crystallized, and it is clear our general programmes are not reaching them; therefore their problems must be attended to through a specific programme of action, which takes full account of the unique historical social, economic and cultural circumstances which have in many cases, led to their under-development today (9).

Once plans began to be worked out (1974) certain working principles equally quickly evolved. Summarily these were: as citizens of Botswana the Bushmen people have the same rights and privileges as others in the society. They have these not as a result of the fact they are a primitive minority which needs protection, or assistance, nor becuase they were the first occupiers of the land, but simply because they are citizens.

It is recognized that these citizens have not always exercised or benefited from, these rights and privileges, for a number of reasons such as their ignorance of their status and rights as citizens, that they lack the knowledge, confidence and/ or resources to take full advantage of their citizenship, their isolation in the sand-veld, and their different history and culture.

It is appreciated that Bushmen citizens will require assistance to participate fully in the life of the nation, and to benefit along with the rest of the population, and that it is the responsibility of the Government of Botswana to ensure that these citizens are given that assistance.

The Government of Botswana has thus committed itself to an interventionist programme of action on behalf of Bushmen citizens.

This programme will intervene in and halt the process of under-development into which the majority of Bushmen have entered. Where once these Bushmen were adequately subsisting they are now living below subsistence level. Where once they were self-reliant and independent, many are now impoverished and dependent upon other citizens of Botswana. The short-term aim of assisting Bushmen citizens is to set them upon a course of self-reliance to enable their development and integration to take place in the best way possible.

The long-term aim of assisting Bushmen is to foster their optimal integration with the wider Botswana society, through their own choice and greater self-reliance.

An underlying fundamental right of different life-styles to co-exist is recognized. Bushmen will not be forced to adopt agricultural, pastoral or other economic activity. Nor will they be forced to move into villages. Rather, development will be extended out to them.

The hunting and gathering way of life is recognized as a viable subsistence. It is felt that for the majority of Bushmen, however, the continuation of a traditional hunter-gatherer way of life today is neither possible nor viable, nor necessarily what they desire.

The Programme of Action

By the end of 1974 an outline plan of action had been drawn up and was formally incorporated into Botswana's National Development Plan in early 1975. (Project Ls 32)

From the outset, it was endeavored to keep its policy and workshop as integrated as possible with general rural development. The reason for this was to make it clear that Botswana was in no way practicing 'separate development' as brain in the neighboring Republic of South Africa by focusing on the problems of one ethnically definable sector of the society. That Botswana is embracing a minority into national life and progress not separating it, that there are no laws or regulations specifically or exclusively relevant to Bushmen is constantly reiterated at all levels of administration.

Thus, no Bushmen Department or Bushmen Affairs office exists or will exist. There is a Bushmen officer, but the strategy of the programme is for it to be entirely implemented through existing channels.

A less controversial element of the programme's strategy is in its nature as an integrated research and development programme, and what is called an investigation--liaison strategy is followed. The aim of such investigation -- liaison work is simply to find out enough about the different situations of Bushmen citizens, and to work out -- IN CONJUNCTION WITH THEM -- appropriate and realistic development plans. While there was initially a certain amount of impatience with this approach, the feeling within Government being frequently expressed that "surely we know enough about Bushmen: we must just get on and DO something", it has proved an absolute prerequisite to action. In particular, it has provided a frame-work for helping the group of Bushmen involved to become more critically aware of their situation and to think constructively about how they would like it altered and how the programme objective could be achieved with assistance.

Today, some three years after its initiation the Bushmen Development Programme is made up of over 30 individual subprojects, each dealing with a specific situation or problem involving different groups of Bushmen all over the country. This is in addition to the general extension of services like primary schools and health centres which local district authorities are being encouraged to include in their general district rural plans. Capital funding for the entire programme (half a million dollars have been spent so far) derive from foreign donors, although Botswana domestic funds bear the running costs for projects about to reach P100,000 per year.

Over half of the projects implemented to date have been investigation -- liaison projects as described above. The findings of each give rise to action projects which almost invariably focus around water development. Assistance in educational, agricultural and handicraft development are also well underway. Over 20 officers are involved in one or other of areas or schemes.

The Resource Issue

Underlying the programme has been a concern for what must be viewed in the broadest and long-term sense as the most critical of issues, Bushmen and their land rights. However, this concern has deliberately not received attention as an isolated or particularly important issue on its own.

The reasons for adopting such a low-key strategy on this critical issue focuses around the fact that it is taken for granted that Bushmen, as rural citizens, have land rights -- i.e., their actual right to land is not disputed. Further, the social and political climate of Botswana today, while favouring assistance, even special assistance to Bushmen, and recognition of their rights over area of land they use today, is not such, that areas of land now being used by livestock -- owners could ever be "returned" to Bushmen. It should also be noted that the land used by cattle is now unlikely to be of any use to them after intensive cattle-grazing anyway. Meanwhile, a national feeling of guilt at the situation of Bushmen simply does not exist. It is impossible to imagine the Ghanzi Freehold Farms in the west, where 5,000 Bushmen are now squatters for example being "passed back" to the Bushmen. It is possible, however, and in fact already committed, that the Ghanzi Farm Bushmen have priority rights over the areas adjacent to the Farms (10).

More importantly Botswana is presently undertaking a massive land -- use evaluation exercise embodied in perhaps one of the most critical national policies since Independence (11). This policy implicitly recognizes the land needs of hunter-gatherers and their land rights. Hunting and gathering, has since the initiation of the

Bushmen Development Programme, become recognized as a valid land-use and even an optimal land-use in certain extremely marginal areas of the Kalahari.

The actual implementation of the Tribal grazing lands policy as it is called is most definitely considering the situation and needs of noncattle owners in a country where 50 percent of the rural population own cattle and where the predominant land-use (and land-abuse) is in the grazing of livestock. While fencing is the strategy this policy indicates as a means of controlling livestock overgrazing, the effects of such land enclosure on citizens like Bushmen are not being ignored. Most briefly no more fencing will go ahead until their needs in those areas designated for such development have been thoroughly investigated and accounted for.

Despite this positive policy there are certain less happy realities to be accounted for:

Firstly, although as citizens, Bushmen have 'rights' in theory; in practice they currently have no already legally established access to land in Botswana. This is -- all social factors aside -- because they do not have rights over or own for themselves water resources, the development of which has predictably become the demarcator of land in a waterless country, predominantly occupied by an agricultural-pastoral people. Furthermore, no more than an estimated 12-15 percent Bushmen in Botswana are today still living in areas where their forefathers lived and in which no other peoples are actively claiming or have established water (hence, land rights). A further estimated 45 percent are living in areas where other people have established rights, but they still have the possibility to move to adjacent areas or nearby areas where no such rights have been established.

The remaining 40 precent, however, are in situations where the possibility to establish rights to their old area or areas adjacent to it have long ago been removed, other (cattle-owning) people having already established rights over those areas.

HOW this has happened varies from case to case, but conforms with a typical and general pattern of hunter-gatherer being displaced by more self-conscious, aggressive pastoral peoples, as described in the background section of this paper. An extreme case is found in the Ghanzi Farm area where an estimated 5,000 Bushmen are unwanted squatters within fenced farms over which the owners have freehold title, but which was their original territory. Thus, for many Baswana the availability of suitable land, and in an area they know and would be prepared to live in, no longer exists.

But the problem does not stop anymore at legal access to land, but integrally involves land-use. Much of the land on which Basarwa today live is USELESS to them. The gaining by pastoralists of legal access to land meant, for the Bushmen not just "historical dispossession" or competition for resources, but also the destruction of that resource, as pastoralism is presently practiced.

The traditional hunter-gatherer economy, demands vast tracts of undisturbed land; the uncontrolled introduction of cattle into such an area has uniformly meant the eventual paralysing of hunter-gatherer practice: this in turn has resulted in cultural, social and economic breakdown, loss of traditional skills, and a typical entry into immobile dependency upon the pastoralist.

As a result, the land needs of Bushmen cannot be defined as only the need for areas to hunt and gather: in numbers of situations the skill and desire to hunt and gather is lost. Bushmen needs today arise as much out of the need for a stable rural development base, for an underprivileged demoralized sub-economic group, as from, in some cases, simply a specific need for a 'place' to hunt and gather.

Furthermore, Bushmen citizens, in whatever circumstances, are not going to obtain access to land without considerable assistance, guidance and education. They are a singularly unvocal minority, of an ancient society that does not recognize ownership over land, but only the resources upon the land, culturally unselfconscious, and nonaggressive, typically unaware of the critical bases of their livelihood until they are lost, ignorant of modern processes of obtaining legal access to land-water, and even if aware, financially unable to effect it, and until very recently, socially discouraged to aspire to such rights.

Taking all these factors into account, what is the strategy the Botswana Government has adopted in meeting the resources needs of this hunter-gatherer desert people?

The first feature to note is that assistance given in this, as most other aspects, is through existing procedures and channels. No land is being declared Bushmen land, on their behalf. No land is merely "granted".

At this time over 30 different Bushmen communities are at some point in the process of applying to the local land boards to use a particular piece of land, to register a water source, such as a sip well, as their own, to get permission to sink a borehole or to dig a well. They make these applications not as Bushmen, but as citizens and tribesmen of whatever tribal area they live in. (12)

Bushmen must actively apply for land-water and participate in the normal process of allocation, not merely be handed out allocations. The superficial justification for this is simply that if they are citizens and getting land because they are citizens, they must channel their applications and be granted rights in the same manner and through the same channels as all other citizens. However, it is also an important way of educating Bushmen of the critical role of legal access to land and water, and equipping them with the knowledge of just how to go about getting such access -- which has a lot more to do with their progress than land 'handouts' and will, it is felt, inevitably make their tenure over a secured well or area -- after the lengthy, complex process to get such access -- considerably more self-conscious and lasting.

Consistent with this is the implicit policy that there is not and never will be a Reservation system introduced as a means of providing resources to Bushmen. The political reason for this is as above. As importantly, however, it has sound sociological support in the reality that Bushmen in Botswana live in widely scattered groups ALL OVER Botswana, not in one, two or even several well-defined areas which could viably be demarcated as Reservations. Not only are the Bushmen communities of today far from each other, they are interdispersed with nonhunter-gatherer communities. The Central Kalahari Game Reserve is today perhaps the only area where a large and definable area is inhabited only by hunter-gathering people, albeit Bakgalagad and Bushmen. It is true that it is so because it was made a Reserve and thus prevented pastoral migration into it, and that other areas could have been also so 'reserved', and that there was a time -- in fact even up until the end of the Colonial administration, when assistance to Bushmen appears to have been thought of purely in terms of establishing Reservations.

However, then as much as today, a system of land allocation which required numbers of Bushmen to move far from their own "/nolli's" (territories) in order to get secure access to land-water, and to have to live with 'other' groups of Bushmen is anathema, both on social, and traditional resource distribution grounds. Contrary to the popular belief that the Kalahari nomadic Bushmen wandered randomly all over the vast Kalahari, the reality is that this desert-dwelling people exercise considerable territoriality, both in terms of where they move and their attitudes to those areas. As the background described each band moves indeed, but within a well defined area that they consider their own, and is so respected by other Bushmen groups, who have their own areas also.

Certainly within these respective areas considerable movement takes place, in accordance with seasons and the different resources available at one or other time of the year. A few hundred square kilometers is about the average size of a territory known. It is a mistake to think of the Kalahari Bushmen as a united and cohesive people. One need only note that amongst a population of 60,000 there are some eleven district languages and 33 dialects for evidence of intra-ethnic divisions. One of the major development schemes we are presently implementing is entirely conditioned by one language group's refusal to live with another, and even further the reluctance of related bands within each of the language groups to settle in close proximity to each other.

The framework for meeting the land needs of Bushmen under the programme has been through water development projects. As I noted earlier water exploration has largely been the demarcator of expansion into the Kalahari region.

Assisting Bushmen groups to dig wells, and to register sip wells they have always depended on, the establishing of cattle-less zones around pans which hold water, and the sinking of deep boreholes where wells cannot successfully be dug account for over half of the time, finance and projects being implemented. The structure for such land and water developments is usually in the form of settlement schemes.

'Settlement' and 're-settlement' are terms which have taken on negative connotation, usually deserved, when used in relation to indigenous minorities. This is particularly so when a nomadic people are concerned, the implication being that these people are being forced to become sedentary.

In particular reference to Bushmen, it is worth recalling their restricted nomadism, their movement and exploitation of the land and its resources within limited areas. It is also important to know that today all but a few communities are futher restricted in their movement by their evolved dependence upon fixed permanent water sources (wells, boreholes).

Settlement schemes should therefore not be viewed as a means of forcing Bushmen to stop moving, to 'settle down' or to give up hunting and gathering activity. Paradoxically they are designed rather to enable the latter activity to be continued in the majority of cases. Most simply 'settlement' means the establishing of the presence of a particular group of Bushmen in a particular area of land. Without such security of tenure, the process of displacement will continue and the possibility to not only hunt and gather but to even eventually take up agriculture if they so wish will also be eliminated.

Another major means of assisting Bushmen retain access to land and thus the possibility to continue their way of life (if so desired) is the imminent establishment in Botswana of wildlife Management Areas. Several local land boards are already in the process of demarcating such areas, where wildlife will have precedence over livestock. This means developments in these areas will focus on the preservation of the wildlife FOR ITS UTILIZATION. People resident in these areas, many of whom will be Bushmen, will be assisted to move and subsist off wildlife. Handicraft development, skin processing, and improvement for sale, game culling and possibly ranching will be fostered. The possibilities this represents for Bushmen is to be able to pursue a subsistence other than agriculture, but in a way which enables them to meet their own growing demands for a surplus economy, is timely and cannot be over-estimated. Several major development schemes are designed around this endorsed possibility for Bushmen (and others) to continue hunting.

Problems

The preceding outline of Botswana Government's strategy toward the Bushmen is believed by some as facing some fairly serious problems which may in time divert what is a fairly progressive policy for a recently independent developing nation to take towards a primitive minority.

These problems focus on the issue of separate development and integration. The prevalent attitude amongst the people of Botswana is that successful integration requires Bushmen to live in close proximity to the others (tswara). They should, that is, move out of the veld (bush) into the villages. Underlying this is, of course, an inherent ordering of the different societies: the hunter-gatherer life is seen as considerably less attractive or valid than Tswara pastoral life, and it can only be to the benefit of Bushmen to give it up for a more modern agricultural or village life. Integration and one way acculturation are thus confused; integration is only beginning to be appreciated as deriving from a parity in standard of living, of equal peoples, interacting, or the possibility of different, but equal lifestyles to coexist. The blatant stratification of rural society where Bushmen and Tswara live together at cattle-post or village, not to the favour of the former, is seen as the natural order of things and is interpreted as "integration". Concommitant with this general wish to see the enbetterment of Bushmen by their becoming civilized, i.e., abandoning their present way of life and "living like us" is an interpretation of Government's assisting Bushmen where they live presently, i.e., "out in the bush" in many cases, as preposterously like separate development, of which Botswana have a healthy fear living next to South Africa.

The implication of the above is simply that the Botswana Government's programme of action, enlightened as it is, did not arise out of popular demand, nor as yet enjoys much popular support within Botswana. This may help explain how the Government came to the kind of policy and strategy described above, with its heavy bias toward the assistance of Bushmen as their right as citizens, not as a struggling hunter-gatherer minority, and the integration of the programme of action firmly with Botswana's general rural development policy, plans, programmes, and channels of administration and implementation.

THE ROLE OF THE INTERNATIONAL COMMUNITY

Bearing the above in mind it will come as no surprise that Botswana does not look with much favour upon the international minority rights adherents who, at one extreme (13) advocated that the United Nations take over the administration of primitive minorities, which would become isolated independent entities within the countries they are today found. Such an idea is obviously preposterous politically, socially and economically to Botswana. On logistical grounds alone how could the U.N. or one of its agencies administer a some 40,000 Bushmen scattered throughout almost the entire area of Botswana?

At a more acceptable level of concern and involvement the international community does already play a part in the current assistance to Botswana's Bushmen citizens. I have mentioned earlier that the entire capital funding of the programme derives from foreign sources among them the funding of the longest land and water development project from a United Nations agency, UNCDF. Over half of the personnel involved in the programme are expatriate, including numbers of academic researchers who tailor their topics to meet our research for development needs. Most tangibly a group of Harvard and University of New Mexico researchers are presently joining forces to establish a Resource Centre to meet our needs in Botswana.

Support for this kind of assistance is in Botswana's interest, and the extension of such a Resource Centre at a more international level, would not go amiss. Such a body could not only provide appropriate personnel but could make available to Botswana and other interested states, information and documentation on past and present experiences, failures, successes, and strategies, being followed elsewhere. Perhaps such a Documentation and Resource Centre for all desert-related development could be discussed at this Conference an emerge as one of our major recommendations.

NOTES

1. Basawa is the official term used in Botswana for Bushmen. San is the preferred usage by academic researchers.

2. Until very recently it was estimated that there are 30,000 Bushmen in Botswana. Demographic work in the last two years puts the estimate now at a minimum of 40,000.

3. J. Yeller, archeologist, personal comm. A. Traill, linguist, History of the Koon, University of Witswatersarel, (1975).

4. Band - the name for the basic social unit of Bushmen society, comprising 10-50 individuals, half of whom are kin. There is no leader of this exogamous group, which has a certain fluidity with related bands and itself changes in size seasonally in accordance with the changing resource distribution and availability.

5. Cattle post - refers to the place where a farmer grazes his cattle, beyond the villages and fields and usually worked by a borehole or well.

6. An excellent discussion paper on this topic is Mavgo Russell's Slaves or Workers: Relations Between Bushmen, Boers and Tswaras (University of East Anglia) 1976.

7. Dekev Mission of the Sereformeede Kole of Aranos of Namibia.

8. Bve Ranch Settlement Scheme - begun by a private researcher, Dr. M. J. Heinz, 1972, who had worked on and off for 15 years with this! x 5 Bushmen Group of Bve.

9. Working paper by the Bushmen Development Office, August 1974, in the planning of the Bushmen Development Programme.

10. Sub-projects (v) and (z) of LG 32.

11. The Tribal Grazing Lands Policy, as per White Government Paper No. 2 of 1975.

12. 73 percent of Botswana's land is owned by the eight major Tawana tribes, who administer its allocation through Autonenjores Tribal Land Boards. Only 23 percent of the land is owned by the State; 6 percent is under Freehold Tenure. Bushmen resident in the Ngwato Tribal Area, for example, are considered as Balangwato tribesmen for all legal purposes, these resident in the Kweneng, as Bakwena tribesmen and so on, and accordingly have all the rights of privileges as of tribesmen of the territory.

13. J. H. Bodley, Victims of Progress, Chapter 9 (1975).

THE DROUGHT IN THE SEMI-DESERT OF SOMALIA

Mohamed J. Omar
Director of Planning
Settlement Development Agency
Mogadishu, Somalia

Pastoral nomadism is a life system dependent on the availability of fodder for the nomad's animals (mainly camels, sheep and goats) and water for both the animals and human beings. The shortage or lack of it in one area prompts migration to seek it in other areas. Over the years the migration becomes routine and is adapted into a yearly cycle of movement from some areas to others.

In the earlier days, despite physical hardships for human and animal, the system was a good adaptation to the particular climate prevailing in the arid regions where nomadism is practiced. Year in and year out, a given group of nomads would make the rounds of their traditional grazing grounds. The change in vegetation afforded variety of diet to the animals. Rest and recovery was afforded to the vegetation during the absence of the nomads and their animals on the seasonal migration.

Arid areas are not only burdened with the disadvantages of low levels of precipitation but, in addition, with the frequent failure of precipitation at the expected seasons. In time past, this failure was always cushioned by the availability of fodder (even in dry form) and resulted only in extended stays around permanent water points instead of moving away from them in the rainy season. Actual starvation of nomadic stock and consequently danger to the dependent human population was rare.

With the passing of time, livestock and humans increased without a compensating expansion of the resources on which all nomadic life depends, i.e. water and fodder. Nomadic groups gradually discovered that grazing grounds that would be unoccupied and to which they migrated at particular times of the year were being occupied by others who preceded them to these areas. While this became the cause of tremendous social conflicts, it was obvious that grazing space was shrinking as a direct result of the increase.

Societies react to imbalances or disruptions in their environment. The Somali nomads endeavored to minimize the aggrevation of their situation. The dependency on permenent water points with the resultant reduction of available fodder in their vicinity while more fodder was available at places removed from these points was solved by the construction of sunken cement water tanks which stored runoff water during the rainy season. This water would then be used in the dry season removing the need to migrate to the proximity of permanent water points.

Livestock itself, however, was transformed from subsistence and prestige-serving property to a market commodity. The commercialization process introduced veterinary medicine to the nomadic stock holders. The provision of veterinary care and the removal of long treks for water resulted in accelerated stock/growth. In a short time, the productive capacities of the land were exceeded and overgrazing resulted to be shortly followed by erosion.

The situation which developed as a result of the above was very surprising to the nomads. The animals' ability to withstand the hazards of a normal dry season was more and more reduced while the severity of the drought years seemingly increased. This was due to the fact that overgrazing reduced the vegetative mother stock each successive year and successive rains produced less and less fodder.

In 1973, the main rains (September-November) failed. This meant a long dry winter during which animal mortality reached the highest record in the country. When the spring rains of 1974 also failed, it was evident that a significant number of the nomad's livestock would perish and many human lives would be in danger. By September of that year, that was indeed the case, and in that month the Government started to assist the nomads. Very soon it became obvious that more than simple assistance was needed and on the 20th of November, the President declared a national emergency and along with committing the whole nation to combating the effects of the drought, he informed the international community about the drought situation and appealed for its help.

The drought effort was launched on two fronts:

a) The establishment and operation of relief camps for those who lost all or almost all their livestock and could not support themselves.

b) Provision of water and food to humans and stock.

The table below shows the rate of arrival at the relief camps of people disposessed of their livestock by the drought.

Nov-Dec	74	109 086
Jan	75	169 465
Feb	75	183 873
March	75	213 325
April	75	260 722
May	75	268 597

Generous and timely assistance was provided in different volume and in different forms by many countries and organizations. The mobilizing of all national resources and this external assistance enabled the Government to achieve the first objective which was to save the lives of the people. The next question that arose for the Somali people was: How will the people live after being saved?

After considering all alternatives (and indeed there were not many) it was decided to relocate those victims who wished it into the southern river basins of the county, providing them with assistance so as to become settled farming communities, while tiding them over the initial establishment period. In May 1975, people in the relief camps were asked to choose between:

a) A return to nomadism and self-support if they could.

b) Or accept relocation to the river basins of the country where -- i) they would be assisted to become farmers, or ii) fishermen, depending on their choice.

Their choice was made as follows:

a) Return to nomadism 148 000

b) Relocation to:
Agricultural areas -- 106 000
Fishermen villages -- 15 000

From this developed what is truly the most ambitious programme for a country of Somalia's means and stage of development, i.e. The Nomad Settlement Programme.

For the agricultural settlement, two sites on the Shebelle River basin (Sablaale and Kurtun Waarey) and one on the Juba River basin (Dujuma) were selected, all of which were uninhabited. For the fishermen settlements, the already existing towns of Brava, Adale and Eyl were chosen. The distribution of the people located in these settlements is as below:

Agriculture		
Dujuma	49 914	
Sablaale	29 937	
Kurtun Waarey	26 622	
Subtotal		106 473
Fisheries		
Brava	6 370	
Adale	5 029	
Eyl	3 156	
Subtotal		14 555
Total		121 028

The logistics problems entailed in the translocation of that number of people, supplying them with food, shelter, clothing, health facilities, etc. were staggering, but the drought effort was a unique experience in Somalia and showed very clearly that with national resolve, seemingly impossible tasks are attainable.

In the agricultural areas, land clearing started a few days after the option of the people in the relief camps was known. Three activities had to run simultaneously:

a) Operation of the relief camps,

b) Preparation of the people for the journey south, and

c) Preparation of the settlement sites for receiving them.

People with their belongings, which sometimes included a few goats, were airlifted to staging points near the settlement sites and then conveyed to the latter by lorry convoys. This effort made it possible for the last group of drought victims to arrive at their designated sites by August 15, 1975.

The transformation of nomads into productive self-sustaining agricultural communities, which is the objective of the present settlements, is not an easy undertaking nor is it one that can be achieved in a short period. Moving them to areas with the potential to produce does not represent the realization of the potential. The Government of Somalia saw this from the beginning and to ensure that this effort was not hampered by the usual red tape, associated with standard civil service organizations it created an autonomous authority and named it SETTLEMENT DEVELOPMENT AGENCY to be charged with the following responsibilities:

1. Service and operate the new agricultural settlement areas with the objective of transforming the previously nomadic population into self-sufficient farming communities.

2. In the process of carrying out the above, accumulate experience for the purpose of settling the nomadic population of the country.

3. Provide information on such experiences to national and international organizations, interested in nomad settlement.

In view of the relatively small number of people placed in the fishermen settlements and as it would have been very difficult for one agency to service simultaneously two areas as diverse as fisheries and agriculture, direct responsibility for the fishermen villages was entrusted to the Ministry of Fisheries. The Settlement Development Agency is, therefore, responsible for the Agricultural Settlements only. The Agency divided its task into the following categories:

a) Attending to the immediate needs of the new communities in its charge such as meeting their needs for food, housing, water, health, education, social and civic services.

b) Preparing, and, after due consultations with the relevant organs of the Government, implementing a development programme that would ultimately make these communities self-supporting.

c) Devising and putting into operation a training programme that would equip the people for their new occupations.

In the discharge of the above duties, the Agency is empowered by its charter to enter into agreements with organizations within or outside the country for financial, managerial, or institutional assistance. The basic operations of the Agency are financed through an annual budgetary grant from the Government. For essential services such as health, education, water, etc. the Agency established a working relationship with the appropriate national organizations so that they supply these services, in a manner approved by and paid for by the Agency.

Recently, the Agency has completed the formulation of a six-year programme for the three settlement areas under its responsibility. The main features of the programme which will be implemented in two 3-year periods are:

1. Land to be developed

Settlement Area	Rain-fed ha	Irrigated ha	Total ha
Dujuma	15.233	21.567	36.800
Sablaale	8.950	12.650	21.600
Kurtun Waarey	8.950	11.050	20.000

2. Health facilities

Settlement Area	Hospitals No. of beds	Dispensaries Nos.
Dujuma	230	9
Sablaale	140	5
Kurtun Waarey	130	5

3. Schools

Settlement Area	Number of Schools	Number of Classrooms	Number of Teachers
Dujuma	9	216	324
Sablaale	5	196	294
Kurtun Waarey	5	185	278

4. Housing and water

Settlement Area	Number of family houses	Number of wells
Dujuma	4500	10
Sablaale	1500	4
Kurtun Waarey	1500	4

The cost of the plan is distributed as follows:

Division	Cost Thousand Shillings*	%
Agriculture	768.108	68.2
Poultry	16.262	1.4
Education	27.012	2.4
Health	179.686	16.0
Housing	105.102	9.3
Food Storage	6.753	0.6
Utilities	24.607	2.1
Total	1.127.530	100.

* 1 US $ = 6.3 Som. Shillings

This programme is designed to help the victims of the effects of droughts which have been aggravated by the condition of the overgrazed range lands. A wiser utilization of these lands would not have stopped the drought from the climatic point of view, but the livestock and hence the people would have been better able to withstand its effects. While the pressure of 120.000 people and the majority of their stock have been removed from the northern range lands--which reduced the pressure on these lands, there are going to be more droughts even if at irregular intervals. To meet this problem the government had created a national range authority with jurisdiction over the rangelands of the country at the time when the Settlement Development Agency was established.

The Settlement Programme represents a very bold experiment for the Somali nation; the effects of its success or failure will be very dramatic. It needed nerve, resolution, and commitment to the country's needs. The biggest of the needs is the settlement of the nomads, who represent close to seventy percent of the population, so that they can attain a better life for themselves and contribute more to nation-building.

THE INDIAN ARID ZONE - STRATEGIES FOR MANAGEMENT

Ishwar Prakash
Coordinator and Principal Animal Ecologist
Central Arid Zone Research Institute
Jodhpur - India

INTRODUCTION

The Indian desert is unique among deserts of the world, neither a single oasis nor a native cactus breaks the monotony of the vast expanse of the ocean of sand. It is most thickly populated in terms of arid land standards having an average density of 46 persons per km^2 as against 3 persons per km^2 in other desert tracts. It also maintains a livestock population of over 23 million head, besides a large number of wild animals. Geographically, the Rajasthan desert holds a rather debatable position as it exhibits an admixture of peninsular, extra-peninsular and Indo-Gangetic features. Geologically, the desert has resisted orogenic forces and has been subjected to marine transgressions in the sub-recent past. Archaeologically, it was the seat of the Mohenjodaro-Harappa-Kalibanga civilizations. Historically the region was dominated by the Princely States and had borne the brunt of a successive series of invasions by Muslim Chieftains and emperors. Biogeographically, the majority of the desert biota exhibited "Western" affinities.

The Indian or the Thar desert is spread over four Indian States, the Punjab, Haryana, Rajasthan and Gujarat - situated in the west of India. The average annual rainfall of the region varies from 150 to 500 mm with a coefficient of variation as high as 60-70%. The distribution of rainfall is also erratic occurring mostly in the period July to September.

During the summer, the mean daily maximum temperature is generally 40°C and 22 to 28°C during the winter. The highest temperatures range from 44 to 45°C during summer and 30°C during winter. But the data on the highest extremes indicate that occasionally, the day temperature can reach 48-50°C in summer and 38°C during winter. The mean minimum temperature varies from 24 to 26°C during summer 4 to 10°C during winter. Occasionally the temperature drops to -5°C. The mean diurnal temperature variation ranges from 12 to 17°C. The mean relative humidity during summer varies from 36 to 59 per cent and from 66 to 78 per cent during monsoons in the morning hours. In the afternoon, however, there is an expected drop and the mean values range from 20 to 35 per cent during summer and 48 to 60 per cent in the monsoon season. The mean evaporation during summer exceeds 10 mm per day. The potential evapotranspiration values during summer (April to June) vary from 7 to 9 mm per day. The mean daily wind speed is recorded to be highest during summer and monsoon season, 8-20 km/h.

DETERIORATION OF DESERT CONDITIONS

Climatic changes

Rainfall: The five year running average of annual rainfall in respect of Jodhpur expressed as percentage of normal rainfall for the period 1901-1975 indicates that prior to 1956 there were fluctuations but generally above normal rainfall conditions existed at Jodhpure more frequently. The below normal conditions that occurred during certain years were rather of short duration. But from 1956, annual rainfall showed a decreasing trend and the below normal rainfall conditions that existed during 1962-1971 is of the longest duration within the period 1901-1975. Hence, in order to examine the rainfall conditions during 1951-1975, more closely, the actual rainfall (expressed as percentage of normal) recorded during this period was examined and the same also confirmed the decreasing pattern of rainfall from 1955 to 1971 in spite of some fluctuations and a short increasing trend thereafter.

Aridity index: It has been observed that the aridity index line of 80 shifted eastward between the first and second decades of this century, especially in the Sriganganagar, Bikaner, Churu and Jodhpur districts. The deterioration was not noticed in the Barmer and Jalore districts during that period. During 1920 to 1970 to and fro oscillations of aridity index were much more pronounced in the Churu district and Jalore district. The implication of this eastward shift of the aridity index line is that desert conditions have shifted slightly towards the east during the last 70 years. Prevalence and gradual shift of such abnormal arid conditions in recent decades result in poor agricultural production in this region, meager grass cover, resulting in higher mortality rate of livestock population and loss of fertility of soils due to increased wind erosion.

Human population: In most of the arid zones, the productivity of the land is rather poor mainly due to paucity of water, extreme temperatures and low fertility of soil. In the Indian arid zone also, 70 per cent of the area is classified under Class IV-VI land types indicating a deteriorated condition. To worsen the situation the Indian desert is one of the most thickly populated deserts of the world and has an average density of 46 persons per km^2 as against 3 persons per km^2 in most of the other desertic tracts. The daily needs of such a large population are the main causes of desertification. The fuel needs of desert dwellers are never-ending; the requirement of hutment construction material and that of thorny branches for protecting crops from wandering livestock, take such a large toll of the desert shrubs and trees that it is often difficult to find them near the well scattered villages.

The human population in the desert appears to be relatively more fertile than that of the adjoining areas. During the period 1901 to 1971, the growth rate of population in the arid zone of Rajasthan has been of the order of 158% as compared to 125% increase in its five bordering administrative districts. Within the desert, the percent increase of population exhibits an inverse relationship with the amount of annual rainfall. In the regions receiving an annual rainfall of 30 cm, between 30 and 40 cm, and more than 40 cm percent increase in population has been recorded during 1931-1971 of the order of 143.8, 132.1 and 107.2% respectively. A comparison of the 143.8% population increase in the 30 cm rainfall region with that of entire Rajasthan state (119.0%), indicate a disturbing situation, as this will result in deterioration and depletion of the natural resources, consequently still more acute desertic conditions.

Livestock pressure: In spite of the low productivity of the arid lands the Rajasthan desert sustains a fairly high population of livestock. Paradoxically, along with the reduction in the grazing area during the last two decades, the livestock population has registered an alarming increase, from 9.4 million in 1951 to 18.1 million, almost double. The livestock density per 100 hectare rose from 72 in 1951 to 175 in 1971 in the desert districts but in the adjoining districts, this enhancement was not so spectacular, 25% as against 293% in desert districts. Considering these trends, it is of utmost importance that the unproductive livestock is removed at appropriate time to relieve the pressure on the grazing lands while enhancing animal production.

Rodent pests - a factor of desertification: The population of rodents in the Indian arid zone is fairly dense as compared to that in the adjoining areas. The dietary demand of the rodent population, which fluctuates from 7.4 to 523 per hectare, is so insatiable that it maintains an appreciable pressure on the already sparse vegetation and rangelands. The total food requirement of the rodent population (477/ha) in the Rajasthan desert was found to be 1044 kg. per hectare (valued at Rs. 225/ha). The annual production of edible (for sheep) grass species was estimated to be, on an average, 865 kg/ha and the total forage production to be the order of 1100 kg/ha. The data reveal that in a rodent infested grassland, the rodents can consume over a year the entire production of edible fodder, leaving virtually nothing for the livestock.

Rodent activities expose the soil to action of weathering agents and increase erosion hazard. Rodent species exhibit a habitat specificity and are fairly abundant in the rocky, sandy habitats and in the rural environment, inclusive of the crop fields.

Besides maintaining a perpetual pressure on rangelands, the rodents are severe agents of soil erosion (Prakash, 1976). Out of 17 species found, 10 are of economic importance. Their fossorial propensity is injurious for soil conservation. Very successful strategies for rodent pest management have been evolved and standardised by the Central Arid Zone Research Institute.

Changes in land use pattern: In the sandy tracts, quite large areas are ploughed and millet and pulses, which have a relatively lower water requirement, are sown soon after the first effective rainfall. Soon after the rainy season, the crop is harvested and the land is left exposed to high velocity winds. This annual disturbance of the soil not only depletes it of the sand-binding vegetation but increases the soil erosion, particularly the semi-stabilised sand dunes. In fact, in certain areas of the Indian arid zone, rain-fed crops should be substituted for by perennial ranges, yet contrarily as a consequence of the high growth rate of population, the sub-marginal and marginal lands have been brought under agriculture instead. In 1951, the net sown area in the arid zone of Rajasthan was 26.1%. It increased to 39.0% in 1961, and to 42.2% in 1971. This increase in the net sown area has caused a reduction of about 25% in the pasture and rangelands. This substantial shrinkage of rangelands is adversely affecting the food and fodder supply, besides creating soil erosion hazards and incuding desertification.

STRATEGIES FOR DESERT MANAGEMENT

Resource Inventory through Integrated Surveys

Some of the technologies like the Integrated Survey directed towards the identification of the problems and the preparation of resource inventories of a defined region are, however, a responsibility of the government. Such a multi-disciplinary survey, as undertaken by this Institute, adopts aerial photo interpretation techniques and has so far provided precise interpretations of soil, vegetation, land capability, water resources, extent of erosion, settlement patterns, transportation etc., of a 45,269 km^2 located in the Rajasthan, Haryana, Gujarat and Mysore

States of India.

Sand Dune Stabilization

On the basis of geomorphological investigations, the sand dunes have been classified as parabolic, longitudinal, transverse, Barchan and shrub-coppice. Unstabilised sand dunes shift their location not only to encroach on the roads, railway lines, villages and clog irrigation canals but the drifting sand also covers the seedlings of food crops, occasionally in vast stretches of cultivated desert lands. Techniques for sand dune stabilization have been evolved. The rehabilitation of marching sand dunes through revegetating involves three distinct processes, a) protection against biotic influences by fencing the area, b) treatment of the shifting sands by fixing brushwood micro wind breaks, from the crest of the dune to the heel of the dune, and c) revegetating by reforestation and establishment of grasses, shrub and tree species. This technology has been successfully demonstrated in the Bikaner, Jhunjhunu and Barmer districts of Rajasthan. The cost of such technology is on an average of about Rs. 860/per hectare with an anticipated return of Rs. 1,000/ha. The cash flow analysis revealed that cumulative streams of receipts from fuel, fodder, seeds and top feed will start 'outwitting' costs from the 13th year and onwards, yielding net benefit cost ratio to the tune of 31.65% for the 13 year period.

Unstabilized sand dunes contain some moisture that permits growth of adapted vegetation. Properly exploited, this moisture could be used for establishing vegetation on the dunes for their stabilization and to provide fodder for livestock.

For the afforestation work a large number of species of *Eucalyptus*, *Acacia* and other genera from similar isoclimatic zones of the world were introduced. Out of different exotic tree species, *Acacia tortilis* *A. aneura*, *A. radiana*, *Eucalyptus camaldulensis*, *E. melanopholia*, *E. terminalis*, *Cassia phyllodenea* etc. have been found to be promising for planting in the Indian arid zone. For raising for plantation in large numbers, nursery techniques have been evolved.

Grasses and Soil Conservation

However, most of the evils of land deterioration and accentuation of desert like conditions arise due to the pressure of human population and livestock. It has been experimentally found that simple enclosure can double the forage yield of the desertic rangeland over a period of three years. Such closure followed by management practices such as bush control with weedicides and nominal use of fertilizer (22.5 kg N/ha) can give an additional increase in forage yield of 20 to 70 percent. The forage yield can be further increased on shallow soils in undulating topography by adoption of contour furrows. It has been unambiguously proved by experiments conducted at this Institute that a grazing schedule with correct carrying capacity and utilization thresholds in various condition-classes of grasslands, such as excellent, good, fair and poor (based on botanical composition) can provide a sustained production of forage over an indefinite period. However, unrestricted grazing has brought in an attentuation of vegetal cover leading to the exposure of the soil mantle and acceleration of erosion largely by wind action.

Agriculture and Erosion Problem

While avoidance of indiscriminate grazing is a responsibility to be shouldered by the desert-dwellers, it is also their responsibility to restrain the greedy impulse to bring all lands under the plough regardless of the land capability. About 40% of the land in Western Rajasthan is being cultivated today, even the dunes have not been spared human greed. While it is well known that marginal and submarginal lands should not be subjected to cultivation, the technologies related to conservation farming need widespread adoption in the arid and semi-arid areas. It has been found that pearl stubbles of 45 cm or higher when left in the field, reduce the soil loss by about 30 to 50% and often this is associated with marginal increase (ca. 10%) of yield of the subsequent pearl millet crop. Similarly cropping between the strips of grass plots (crop : grass 6:1) increase the crop yield by 6 to 17%. Again legumes can be grown with advantage as an intercrop with grasses. The introduction of grasses in the cropping pattern in arid ecosystems accrues diverse benefits such as, restricted erosion, improved soil conditions and a reasonable return when the crop fails in the event of drought.

Amelioration of Environmental Constraints

Among the diverse impediments for plant production in the arid areas frequent drought is the most serious problem. In some years when the rains are very much delayed (beyond 20th July) it has been found advantageous to grow certain legumes like *Cyamopsis tetragonoloba* (guar) which use about half the water as compared to pearl millet, which is grown most extensively in Western Rajasthan. While this technology has been adopted by many farmers, the simple drought evasion technique of transplantation of pearl millet has not gained a foothold. Raising of plants in nurseries at the beginning of July followed by transplantation at the time of suitable rains even up to the middle of August, has produced yields of around 26 quintals/ha in HB 4 in experimental fields. In drought years, certain minor millets like *Panicum miliaceum* and *Setaria italica* which complete their life cycles within 40-45 days have been found to have considerable promise. Similarly, in rain-fed semi-arid areas, adequate soil fertility conducive for optimum plant vigour helps to avoid the adverse effects of short duration drought. These gains can be further consolidated by the adoption of control measures for rodents, locusts, white grub and other insect pests for which all the techniques have been standardized and perfected.

Water Harvesting and Dryland Agriculture

While ample scope remains for popularizing these technologies, water harvesting method still remain to be adopted by farmers on a large scale. Except in the Jaisalmer area, where age old catchment cultivation enables the local farmers to grow excellent wheat and mustard crops in a zone having about 100-150 mm precipitation, adoption of this technique is hardly ever encountered elsewhere in the Indian arid zone. It has been again experimentally shown that man-made micro-water sheds having 0.5 ratio of catchment : cultivated area in sandy flat lands can also give remunerative yields in pearl millet, sesame moong, grain sorghum etc., avoiding the risks of crop failure in a drought year. In the sub-stations of this Institute located in the remote desert areas, rain water is being harvested since the early '60s from natural slopes (3 to 6%) in covered cement tanks or 'Tanka' (200 x 10^3 cu m) for human and animal consumption besides sufficient scope to recycle the water for irrigation through sprinkler systems. Although the state government has widely adopted this technique in many places, local inhabitants have not shown much interest in these technologies.

It is appreciated that simple rather than sophisticated methods stand a chance of quick adoption in the arid areas. But the success of such operations pivots around the motivation and will of the people. It has been found that s sub-surface barrier of clay (at 60 cm, 1 cm thick) in sandy soil, can significantly increase the yield of pearl millet and certain vegetable crops like Tinda (*Citrullus vulgaris* var. *fistulossus*). Water harvesting coupled with such sub-surface barrier increased the yield by about 200 per cent. Desilting of village tanks (located mostly in depressions) which is essential to maintain their capacity, may yield sufficient silt to implement these technologies at least in limited areas. The recharge of these village tanks can again be improved by lining the runoff channels with janta emulsion.

Forestry and Energy Needs

Despite the foregoing merits of agriculture, there is serious reservation among contemporary scientists as regards its adoption below the 200 mm precipitation zone. In view of the population load as mentioned before, a serious search is necessary for an alternate vocation, in case land-ploughing is restricted. Deterioration of the arid areas due to human activity is not a myth. In the Fort at Jaisalmer thousands of *Tecomella undulata* (commonly known as the "Teak of the desert") logs have been used as reinforcement for the roofs, but a visitor today driving within hundreds of miles around Jaisalmer may hardly detect a few trees of this species. Human need for fuel has also led to indiscriminate cutting of trees and shrubs. Charcoal making from *Prosopis juliflora* in Pali and use of *Calligonum polygonoides* for fuel in Bikaner area is another illustration of the seriousness of the problem. Since India is blessed with abundant solar energy (daily value of 5.5 KW/metre) development of solar energy has gained considerable momentum. This Institute has developed miscellaneous low cost solar energy appliances like solar water heaters, solar ovens, solar cabinet driers etc., which are expected to meet energy needs of local inhabitants. Thus the present practice of denuding the vegetation for fuel may be reduced over a period of time. But the crux of the problem is the popularization of the technology.

Arid Horticulture

The only fruit trees encountered in abundance in the desert tract is ber (*Zizyphus nummularia*) which has small size red berries. Looking to its potential for improvement, budding techniques were perfected by which it has been possible to transform these *Zizyphus* bushes into ber varieties such as *Gola* and *Seb* which are characterized by large size, luscious and tasty fruits with a production range of about 40-60 kg/tree.

It thus appears that diverse technologies are now available to put the land to the best use depending on its capability. But the success of these technologies largely depend on the active participation of the human beings who live in this special ecosystem. If the challenge of habitat improvement is accepted by the masses and the government, a massive effort has to be made through education and extension so that people become receptive to the technologies which have been evolved.*

REFERENCES

Bryson, R.A. and Baerreis, D.A. 1967. Possibilities of major climatic modification and their implications : Northwest India, a case for study. *Bull. Amer. Met. Soc.* 48, 136-142.

CAZRI, Jodhpur 1964. Recent Development in Rajasthan. Souvenir Volume.

CAZRI, Jodhpur 1974. Summer Institute on Desert Ecosystem and its improvement, Jodhpur (In press).

Das Gupta, K., and Khilanani, V.B. and Bhandari, U.M. 1973. An approximate assessment of exploitable ground water potential of Rajasthan. Proc. Winter School Dev. Raj. Desert, *India Nat. Sci. Acad., India*, Jodphur. Feb. 1973.

---

* Adapted from 'Case Study Reports' on Desertification, CAZRI, Jodhpur (India).

Garg, H.P., 1975. Solar energy utilization research. Central Arid Zone Research Institute, Jodhpur (Tech. Monograph No. 3).

Garg, H.P. and Krishnan, A. 1974. Solar energy utilization potential in Jodhpur and New Delhi. Indian J. Met. Geophys. 25, 473-478.

Goor, A.Y. 1955. Report to the Government of India on Control of the Arid Zone of Rajasthan, F.A.O., Rome.

Gupta, R.K. and Prakash, I. 1975. Environmental Analysis of Thar Desert. English Book Depot, Dehradun, 1-484.

Jacob, K. 1972. Some important aspects of reclamation and development of the western Indian Desert, Indian Natl. Sci. Acad. 1:45.

Kaul, R.N.1970. Afforestation in Arid Zone. Dr. W. Junk N.V. Publishers, The Hague.

Krishnan, A. 1968. Delineation of different climatic zones in Rajasthan and their variability. Ind. Jour. of Geog., 3:33-40.

Krishnan, M.S. 1952. Geological history of Rajasthan to present day conditions. Bull. Natl. Instt. India 1:19-31.

Lukose, N.C. 1972. Palynology of the subsurface sediment of Manfera Tibba structure, Jaisalmer, Western Rajasthan. The Paleobotanist Vol. 21, No. 3, pp. 285-297

Mann, H.S., Malhotra, S.P. and Jagdeesh C. Kalla. 1974. Desert spread - A quantitative analysis in the arid zone of Rajasthan. Ann. Arid Zone. 13(2) : 103-113

Ministry of Education. 1964. Proc. Symp. on Problems of Indian Arid Zone, Jodhpur 23 Nov-2nd Dec.

Ministry of Irrigation and Power, India. 1971. Dynamics of shifting sand dunes in western Rajasthan and their rehabilitation. Central Water & Power Commission, New Delhi.

Misra, V.M. and Rajaguru, S.N. 1975. Some aspects of Quaternary environment and early man in western India. Workshop on the Problems of desert in India, Jaipur.

National Commission on Agriculture, India. 1974. Interim Report on Desert Development, Govt. of India, New Delhi.

National Institute of Science. 1952. Proc. Symposium on Rajputana Desert, New Delhi.

National Institute of Sciences, India. 1969. Symposium on Planning for Drought areas, New Delhi.

Prakash, Satya. 1964. Archaeology. Recent Deve. Raj. Souv. Vol. CAZRI, Jodhpur, India 134-148.

Prakash, I. 1976. Rodent Pest Management, Principles and Practices. Central Arid Zone Research Institute, Jodhpur (Monograph No. 4) 1-28.

Roy, B.B. and Pandey, S. 1970. Expansion or contraction of Great Indian Desert. Proc. Indian Nat. Sci. Academy. 36(6) : 331-343.

Sankalia, H.D. 1952. The condition of Rajputana in the past as deduced from Archaeological Evidence. Proc. Symp. Rajputant Desert. Bull. Natl. Inst. Sc. India : 43-49.

Satyanarayan, Y. 1963. Integrated regional land survey in Rajasthan. UNESCO Sci. Conf. Arid Zone Latin Amr. Buenos Aires.

Saxena, R.K., Sharma, M.L. and Jodha, H.R. 1966. Quality of ground water of irrigation in Ahor development block, Jalore district, western Rajasthan. Ann. Arid Zone 5(2) : 215.

Sen, A.K. 1972. Agricultural Atlas of Rajasthan, ICAR, New Delhi.

Sen, A.K. 1972. Agroclimatic regions of Rajasthan. Ann. Arid Zone Vol. 11 (1&2) : 31-40.

Sen, A.K. and H.S. Mann. 1975. Exploration utilization and quality of ground water in Rajasthan. I.G.U. Commission on Arid Zone, Proc. Congress Symposium, Akashabad USSR.

Singh, G. 1952. The Rajputana Desert : Geography. Proc. Rajputana Desert. Natl. Inst. Sci. India. Bull. I. 151-152.

Singh, Gurdip. 1972. Stratigraphic and radiocarbon evidence for the age and development of three salt lake deposits in Rajasthan. India. Quarternary Res., 2 : 496-505.

Tod, J. 1929 and 1932. Annals and Antiquities of Rajasthan Vols. I & II Routleclze and Kegan Paul Ltd., London.

University of Jodhpur 1972-73. Symposium on Natural Resources of Rajasthan (Abstract).

Verstappen, H. TH. 1968. Geomorphology at the CAZRI, Jodhpur. Paris, UNESCO (Report).

Vishnu-Mittre 1975. Problems and Prospects of Palaeobotanical approval towards the investigation of the history of the Rajasthan desert. Workshop on the Problems of desert in India, Jodhpur.

Whyte, R.O. 1961. Evolution of land use in south western Asia Arid Zone Research XVII A., History of land use in Arid Region Edited by L. Dudly stamp P. 57-142, UNESCO, Paris.

MANAGEMENT ALTERNATIVES FOR DESERTS IN THE WESTERN UNITED STATES

Curt Berklund
Director
George L. Turcott
Associate Director
Bureau of Land Management
U.S. Department of the Interior

ABSTRACT

Except for irrigable acreage in the West, the arid and semi-arid lands of the Western United States are largely congruent with the public lands of the United States.

The Bureau of Land Management, Department of the Interior, has the administrative jurisdiction. The Federal Land Policy and Management Act of 1976 dedicates these public lands to multiple-use and sustained-yield management with special management attention directed to areas of critical environmental concern.

Multiple-use and sustained-yield principles are particularly suited to fragile desert environments. Multiple-use emphasizes combinations of uses without maximizing any one use. Sustained-yield requires use that enhances or maintains the capability of land to produce a perpetual flow of renewable products.

Under the Federal Land Policy and Management Act these public lands will remain in Federal ownership for multiple-use and sustained-yield management unless, as a result of a rigorous land use planning process which the Act requires, it is demonstrated that individual parcels should be transferred.

The arid and semi-arid public lands were virtually unused, when the United States first acquired them in the 1840s. They were increasingly overused from the 1900s to the 1930s when public management was added to public ownership under the Taylor Act.

Except for Alaska this public management of the public lands ended the frontier process of effecting land use adjustment by occupying relatively unoccupied lands. The open end land use adjustment remained in Alaska until the 1960s. With no large blocks of land legally open for unfettered use and occupancy, land use adjustments must now be made through political and administrative processes.

The result of the first forty years of public management of public lands has been to halt the deterioration of the public lands. The Federal Land Policy and Management Act is a clear charter for more intensive integrated coordinated management that could start an overall raising of the sustained-yield level.

Multiple-use, sustained-yield, and areas of critical environmental concern largely preclude maximized use, particularly maximized economic use. This almost necessitates public ownership and public management. Private ownership of land inherently requires maximized economic use to cover taxes and interest on investment.

Single maximized use of arid and semi-arid lands can pose a threat to environmental values. Recent studies indicate that recent high grain prices have induced short cuts on soil conservation practices in attempts to maximize economic return but drought is exposing the environmental costs of such practices.

Multiple-use and sustained-yield are particularly useful for deserts and semi-deserts. They emphasize professionally trained human intelligence and user cooperation in combination of uses instead of major alterations in the environment which tend to cost more and run higher ecological dangers.

The core of the arid and semi-arid lands of the West, i.e. the public lands of the United States, have been dedicated to multiple-use and sustained-yield. With drought once again hovering over them the time may be at hand to consider extending multiple-use and sustained-yield to nonfederal lands which can't sustain maximized economic use. Because of the non-maximizing nature of multiple-use and sustained-yield this would require public ownership in some form and probably public management in some form. Outright acquisition, planning and zoning, acquisition of maximizing rights are among the means of extending multiple-use.

Management arrangements include: continued individual management, use of quasi-public corporations, creation of a new Federal agency or extension of the jurisdiction of existing Federal agency.

POST FRONTIER LAND USE ADJUSTMENTS

The most repeated lesson of history is that those nations that allow continued abuse of land and other basic resources soon slump into decline and eventually fall. Despite the autopsies of the downfall of past great civilizations such as Babylon, the Egyptian Kingdoms, and Imperial Rome, many people continue to believe, indeed take comfort in the feeling that technology will always rescue their lands from abuse. It may be that in America we may have learned in time that there is no technology that can compensate for the abuse of our lands. In fact misapplied technology can be dangerous.

The effectiveness of our political and administrative processes related to land use will eventually decide our national fate.

Americans learned land use in the humid East. The customs, traditions, and laws on land holding and land use rested on a humid triad of soil-moisture-vegetation all energized by the sun. Somewhere west of the 100th meridian there was a marked change. The sun's rays remained, of course, but they energized an arid soil-moisture-vegetation-triad so changed that an entire new set of institutional arrangements were needed. Obviously there have been many new arrangements but the need to have this conference suggests that the task is not completed.

The passage of the Taylor Grazing Act in 1934 (Public Law 73-482; 48 Stat. 1269; 43 U.S.C. 315 et. seg.) closed the frontier in America and was the start of a new era. Before 1934, use and occupancy of the public lands were usually at the initiative of the individual citizen. If he wanted land for agriculture, he filed a homestead or similar application for a tract of his own choosing. If he wanted to graze cattle or sheep, he just turned them out on any area of public land when he could assert his right to use them. The result of such unregulated occupancy and use was the deterioration of the land.

Since 1934 we have been in a new era of land use. We can no longer effect land use adjustments by occupying relatively unoccupied land. The social and political ease of frontier adjustment is gone and political and administrative processes have taken its place as the medium for land use adjustments.

Before getting down to looking at an alternative I'd like to make a quick overview of major situations that will materially affect desert alternatives: (1) the vulnerability of the public lands; (2) the loss of soil; (3) the development of energy; (4) the relentless demand for water; and (5) urban sprawl.

Vulnerability of Public Lands

At the time the Taylor Grazing Act was passed in 1934, a drought cycle had reached its peak. This amplified the evils of past abuse. This condition was not peculiar to the public lands but was general throughout the western range country. The Western Range, which the U.S. Senate published in 1936 as Senate Doc. 199, 74th Cong., 2nd Sess., an in depth analysis of the conditions of western range lands reported that 84 percent of the public lands had lost more than half of their forage production capacity and that the entire capacity of the western range had been depleted by an average of 67 percent.

Under these conditions the Taylor Grazing Act brought management to the public lands of the United States. What has been accomplished to date under the Taylor Act?

For one thing it brought an end to the range wars. It stopped the killing, but after 43 years of public management, range conditions on these lands are still substantially below their productive potential. As of December 1, 1974, only 17 percent were classed as being in good to excellent condition. Thirty-three percent were in poor or bad condition. The fifty percent considered to be in fair condition were really unsatisfactory since both soil and plant cover are in a deteriorated state.

The latest data indicates that 19 percent of the public lands are improving, 65 percent are in a static condition, and 16 percent are declining. (Condition and Trend figures from Bureau of Land Management files.)

These figures mean that BLM and its predecessor, the Grazing Service, with lots of cooperation from users, other Federal agencies, and State and local governments, has halted the deterioration of public lands. Those who know the physical and institutional obstacles that had to be overcome know that this is an heroic achievement. Even so it means that at a time of foreseeable stress, the public lands are somewhat like a man recovering from a serious illness.

In recent months ugly storms sweeping over parts of the Great Plains and intermountain basins remind us of the Dust Bowl of the 1930's and in some places have brought hardship and even ruin to hundreds of farmers.

Soil Erosion

This is bad enough, but more signigicantly, an April 22, 1977 article in Science entitled Soil Erosion: The Problem Persists Despite the Billions Spent on It says the recent dust storms give warning that "effective soil conservation of the kind necessary to sustain the agricultural productivity of the Great Plains and other farming regions over the long term simply has not been achieved. . .although nearly $15 billion has been spent on soil conservation since the mid-1930's."

The problem's persistence probably can be attributed, at least in part, to the belief of many farmers that it's more to their advantage to maximize short-term crop yields while the chances are good than to bank on the long-term benefits of conserving the soil.

In reviewing the problem Science notes that the seriousness of soil erosion in the Great Plains was pointed out two years before in the little noticed report prepared for the Senate Committee on Agriculture and Forestry by the Council for Agricultural Sciences and Technology (CAST). CAST attributed much of the erosion to these causes:

1. Changes from wheat-fallow or wheat-sorghum-fallow rotations to continuous planting of wheat by many farmers in in order to take advantage of high wheat prices even though this was eliminating strip cropping for wind erosion control.

2. Plowing with moldboard plows to bury the seed of downy brome grass (a pest that reduces yield when wheat is grown continuously from year-to-year without rotation) to leave surface bare of harvest residues that help hold the soil in place.

3. Extensive leveling of fields (in some cases creating barren patches or erosion prone sand) to permit installation of new wheeled irrigation systems turning in a wide arc from a center pivot.

4. Conversion of rangeland to grain production in response to high prices for grain and low prices for cattle with attendant greater susceptibility of crop land to wind erosion.

While it isn't possible to quantify how much the semi-arid and arid lands contribute to the calculated four billion tons of sediment carried by surface runoff in the waterways of our contiguous 48 states each year, it is estimated that another billion tons of soil is lost annually by wind erosion. By far the greatest part of this loss comes from the arid lands of the West (Science, April 22, 1977). Spread out over dry-land farming and grazing areas it seems almost certain the losses are well above any tolerable level.

An inventory of erosion conditions on 144.5 million acres of the public lands over the past five years indicates that 91 percent of this land is presently sustaining some soil loss.

Erosion on ten percent of these public lands is either critical or severe. Erosion on 35 percent is considered moderate, meaning that they sustain soil loss during heavy storms and have sustained some loss of productivity. A further loss of vegetation could accelerate the destruction of these lands and create downstream flooding and deterioration of water quality.

Over 46 percent of the public lands is considered to be in a slightly eroding condition but no real damage as yet sustained. Management practices over the next few years will determine whether they can be stablilzed or if they will further deteriorate losing productivity at an ever increasing rate. (Erosion figures from Bureau of Land Management files.)

Energy Development

The energy crisis has caused deep apprehension.

President Carter has labeled the crisis urgent and called for conservation of petroleum resources. Forecasts predict that the Nation may be faced with shortages possibly as early as the mid-80's and shortages by the end of the century are almost a certainty. Of course, nobody knows exactly what reserves we have or what might be available in other countries. We do now know when the world will run out of oil and gas, but it is certainly prudent to clamp down on waste and urgent that we develop alternative sources of energy. This is the only safe course for the Nation.

In the West thousands of wells have been sunk in the last 75 years and only a few areas remain where geology seems promising for future discoveries of oil and gas.

The President has called for an increase in the production of coal from the present 665 million tons a year to 1.1 billion tons by 1985. Much of this increase will come from the vast reserves of stripable low-sulfur coal found in the generally semi-arid regions of the northern Great Plains and Rocky Mountains.

The changing energy picture-- decline of production in the numerous oil and gas fields and the increase in the enormous coal areas in the West will mean tremendous social and economic changes. And there is the prospect of much higher environmental costs in degraded air quality unless there are vast investments made in smoke-stack scrubbers (estimated by Time in May 16, 1977 issue nationally at $57 billion by 1985) for mine-mouth electric generation and industrial plants. There is also the prospect that some strip-mined lands cannot be fully restored by land reclamation and revegetation. Power lines, slurry pipelines, roads and railroads directed to the transportation of energy will bring changes in the western landscape unwelcome to many.

Water Demands

It seems that the competing demands for water for irrigation, energy development, domestic and industrial uses, recreation, navigation, and for sustaining natural ecosystems--will outrun the supply. The current drought threat has emphasized this worry.

It is hard to generalize about our water supply. The amount and use in each river basin is governed by weather, soil, geology, vegetation, the number of people, industrial and commercial development, conservation measures, storage facilities, and many other factors. Important as some overall estimation of the amount of available water may be, how much will still be left is more pressing. A recent Conservation Foundation report on the "Water Outlook" (Conservation Foundation letter, April 1977) makes this assessment--"for many rivers, all conflicting water uses by farmers, energy industries, and municipalities may be a drop in a bucket compared to three other claims that loom larger and larger."

1. "The maintenance of 'instream flows' at minimum levels designed to preserve the biologic and recreation values of streams."

2. "Federal reservation. . .(of) enough water in national parks, national forests, and other lands to serve the purposes for which these areas were established."

3. "Considerable water is needed by Indian tribes. . .the Supreme Court has made is clear that Indian water rights are independent of any State water laws. . .Indian claims on water have proliferated only in recent years, and the potential demand is largely unknown. But is may be very large."

Deficiencies of surface water in the semi-arid and arid Western States have spurred greater use of ground water than in the more humid areas. In many places more water is pumped than can be replaced by the limited rainfall--in short "we are mining our water."

Ground water provides about 59 percent of water used in the western United States. Ground water levels have dropped in areas where there is heavy pumping for irrigation, industrial and municipal use. Notable areas where this has happened include Southern California, Central Arizona, and on the high plains of Texas where water tables have been lowered by 200 feet or more. Proper management would allow most, if not all, the aquifiers to recover in a few decades.

Urban Spread

The urban centers of the dry West have sprawled over vast areas. The Los Angeles urbanized area is projected to occupy 2,057 square miles in 1980, San Diego 563, Phoenix 522, San Antonio 343, San Bernardino 315, Denver 303, Sacramento 283, and Salt Lake City 228 square miles. (Pritchard, Jerome P., Dimensions of Metropolitanism, Urban Land Institute Research Monograph No. 14).

The inhabitants of these and other urbanized areas of the West have been in overdrive for many years, at enormous expense in energy, land, and environmental quality.

Many of the urban centers of the West, with their congested freeways and great reliance on oil and gas for utilities and industries face drastic changes in meeting the national energy conservation measures. Our dependence on cars, our habits, long distance commuting, and the smog in many urban areas, will make it difficult to adapt to the national energy program. It seems that cities of the future must become more compact, be serviced by mass transit systems, and depend on less polluting forms of energy.

While these imbalances between man and nature have been developing, there has been a large population shift to the Sunbelt area of the Southwest. In the period 1960-1975 California increased its population by 5,450,000 people. Most of this increase took place in the sunshine part of the State. Texas, which is partly semi-arid, had an increase of 2,654,000; Arizona 918,000; Utah 316,000; Nevada 306,000; and New Mexico 195,000. (Statistical Abstract of the United States, 1976.) If the big freeze of 1977 with its attendant unemployment and enormous heat bills is followed by another one or two severe winters, the migration to the parched Sunbelt may become a flood.

As one might expect with such Sunbelt growth the desert urban centers have also grown tremendously. In the same period the metropolitan areas of Anaheim-Santa Ana-Garden Grove in California grew by 957,000 people, Los Angeles-Long Beach by 887,000, Phoenix by 508,000, San Diego by 485,000. Riverside-San Bernardino-Ontario by 404,000, Sacramento by 257,000, Salt Lake City by 190,000, and Tucson by 168,000. (Statistical Abstract of the United States, 1976.)

California Desert Study

About ten years ago, the Bureau of Land Management here in California decided to try to get ahead, or at least catch up, with some of those forces which spring up from this urban outburst. It organized the California Desert Study whose focus was the 12,000,000 acres of the Southern California Desert. I won't go into detail because the study and California Desert Conservation area has already been discussed here.

However, I would like to highlight two things about the evolution of the California

Desert Conservation area.

1. Despite all the use that had taken place and despite all the use that was foreseen, BLM quickly learned that there was a great deal to learn about the resources of the desert and that next to nothing existed in usable knowledge upon which we could base our management decisions.

2. There has been close and extensive public participation and intergovernmental cooperation from the very beginning. Title VI of the Federal Land Policy and Management Act established the California Desert Conservation Area and it makes continued public participation mandatory.

I think this is important not only in thinking about the deserts of the United States but in thinking about the deserts of the world. Looking ahead to the UN Desertification Conference and the UN Plan of Action to combat desertification we must be very realistic. It is the people who use desert lands who determine their fate. Laws, policies, and programs are really licenses for government to persuade people to use lands in the ways that maintain and sustain them. This is true whether the user is a nomadic Bedouin living the pastoral life of Abraham or a Californian with a Ph.D. in nuclear physics relaxing in the desert with an off-road vehicle.

While the trend of population growth was going on in the sunny climes of the dry land states, certain important new eddies of population distribution began to emerge in the 1970's. These, especially if they continue, must be taken into account in connection with future land use planning and management, in Federal revenue-sharing with State and local governments, in the distribution of Federal community development funds, and in Federal financial assistance to relieve the social and economic impacts of energy and other resource developments in western communities.

Studies of Population Redistribution

A recent report of the Population Reference Bureau entitled Rural Renaissance in America, (Vol. 31, No. 3, October 1976) identifies these new patterns.

- Certain metropolitan areas, especially some of the larger ones -- Los Angeles--Long Beach and Beaumont--Port Arthur--Orange stopped growing altogether.

- All culturally distinct subdivisions of the arid and semi-arid West, i.e. the Rocky Mountains, the Great Basins, the Rio Grande watershed, and the Northern and Southern Great Plains have shown non-metropolitan growth and a reversal of net outmigration.

Population losses have also occurred in the metropolitan areas of Bakersfield, Beaumont-Port Arthur-Orange, Corpus Christi, and Los Angeles-Long Beach.

These trends have caused new stresses and created new issues. The decline of population in older central cities has spread outward, and more than a few suburbs and semi-rural areas are no longer experiencing rising employment, industrial development and construction of new shopping centers, an expanding tax base, and general prosperity. On the other hand, many rural areas that failed to grow (some despite years of Federal economic development assistance) are now growing spontaneously. This reversal has disconcerted top-level planners, government policymakers, and country supervisors. Energy developments in arid lands have caused some of this spontaneous growth.

Congress has recognized the logic of providing assistance to help communities impacted by energy and other resource development with their planning and appropriate legislation is now pending.

Whether these trends are temporary or long-term is not yet clear. Some shifts may be due to the economic recession of the past several years. If so, shifts in the opposite direction should occur as the economy improves, but, as many suspect the taste for rural living may persist so that urban and rural growth will co-exist, and establish a new population distribution pattern.

The resources of the West are limited. This limit on resources may well preclude the fantastic growth in affluence of the individual that characterized the post World War II era. The necessities of the moment will force the West to use greater care in the management of resources and to make drastic technical and institutional adaptations for conserving and developing energy to regulating land utilization, including strip mining, transportation, and the use of range and farm lands.

MULTIPLE-USE AND SUSTAINED-YIELD FOR DESERTS

In talking about our western arid lands it has not been my intention to expose their limitations and some of their misuses, but to set forth the context of a discussion of an alternative which takes those limitations into account.

The deserts and their margins are fragile. I need hardly report to this conference that there are definite and easily breached limits to their uses. There are signs that big projects involving great sums of money that transform environments will no longer pass muster solely on the strength of a development mystique.

Cecil Andrus, the new Secretary of the Interior, has clearly set forth the doctrine that he will support development projects when the environment is protected and will oppose development projects when the environment is not protected. President Carter in his Environmental Message of May 23, 1977 set forth a similar doctrine in relation to coal mining.

In this context and given the several major trends I have outlined, multiple-use and sustained-yield management offer an attractive alternative.

Multilpe-use means combination. Sustained-yield means the preservation of the productivity of the land. Most misuse comes when we try to maximize any one use. Multiple-use combinations and sustained-yield considerations hold down the instinct to maximization.

Multiple-use relies more on the advice of trained professionals and the cooperation of users and less on massive alteration of the environment. Human talent is cheaper than steel and concrete. So multiple-use is less expensive economically, and far safer ecologically than major alterations.

BLM is, for example, de-emphasizing mechanical and chemical means of restoring vegetation, i.e., plowing and spraying and relying more on grazing systems where the grazing cycle is synchronized with the plant's physiological cycle to provide favorable conditions for the growth of selected species. BLM is getting encouraging results in the recovery of desirable vegetation.

Multiple-Use on Public Land

As far as applying multiple-use and sustained-yield to the deserts of the American

West, the Federal Land Policy and Management Act has dedicated 166 million acres to these principles. It also requires a continuing inventory of public land resources, to provide data in a usable format for management decisions.

The Federal Land Policy and Management Act also requires BLM to set up a procedure to plan for land use. BLM has been practicing multiple-use management and sustained-yield production for many years. In fact, the term was used in the documents which created the Bureau of Land Management in 1946. During the past ten years BLM has been building a land use planning system. The importance of the Federal Land Policy and Management Act is its statutory recognition of multiple-use and the planning system and the directive to increase the pace.

The land planning provisions of the Federal Land Policy and Management Act require public participation. This is in recognition of the point I developed earlier "that users determine the fate of lands" and hence must be given opportunity to be involved in plans for their use.

From my remarks today you might think that BLM invented multiple-use and sustained-yield. It did not. These principles evolved from the management of the national forests. BLM has applied them to arid and semi-arid lands. We are proud of BLM land use planning procedure as a mechanism for converting multiple-use and sustained-yield from statutory principles to actual combination of uses on the land. I don't want to go into details any further than to say that the procedures for identifying actual places of complementarity and conflict shift discussions from a theoretical plane, where strongly held principles can clash interminably, to a real world of certain resources located on very specific tracts of land which can be discussed, accommodations developed, and specific decisions made. It is not a system for planners. It is a system for resource specialists with active involvement of the many publics associated with public land management putting together land use combinations on definite areas.

Except on a demonstration resulting from a rigorous land use planning process, the arid and semi-arid public lands of the United States have been dedicated to multiple-use and sustained-yield.

Multiple-Use for Non-Federal Lands

If the current drought demonstrates that any significant acreage of arid and semi-arid land outside Federal ownership cannot sustain maximized economic use over the long term, the success and prospects of multiple-use and sustained-yield indicates a management alternative for them.

For all the long effort to get multiple-use recognized on the public lands, it will have been relatively easy compared to any effort to bring multiple-use to other desert lands. Bringing multiple-use to other arid and semi-arid lands would require an entirely new set of institutional arrangements.

Arid and semi-arid lands not in Federal ownership are in state or individual ownership. If multiple-use is to be attained, the individual's property right to maximize economic gain must be reduced and paid for.

Given the political will to do this, it could be done in several different ways. I will discuss them without any judgment on technical or political feasibility. I might say here that I am entering unmapped land as far as Federal policy is concerned. I am speaking as a professional land manager offering an alternative for academic discussion. I am not proposing or advocating.

Land use could be "de-maximized," to be jargony, where maximized use poses a threat to the long-term desert environment by:

1. Enactment of state land use plans and zoning ordinances halting or severely limiting maximizing use and requiring multiple-use.

2. A return to the Bankhead-Jones program of the New Deal, i.e., Federal purchase of economically and/or environmental marginal land for multiple-use management.

3. Adapting the soil bank concept keyed to environmental fluctuations--not economic fluctuations--with some mechanism for public authority to decide when production could be temporarily maximized.

4. Federal or other purchase of the "maximizing rights" of lands.

All of them presuppose some form of multiple-use to include farming and mineral extraction under limited conditions, domestic livestock grazing, wildlife enhancement, outdoor recreation, preservation of historic and archaeological values.

Management alternatives include:

1. Leaving management to individuals who have sold or had their maximizing rights zoned away.

2. Organizing some form of public authority to purchase maximizing rights, and to manage land for multiple-use including economic use but not maximized economic use.

3. Creating a new Federal agency or extending jurisdiction of and existing agency.

SUMMARY

The need in a post frontier era of land use adjustment through political and administrative processes is a unifying management principle that permits us to treat the desert as the unity it is in nature.

I commend multiple-use and sustained-yield as such a principle.

It is being applied on 166 million arid and semi-arid acres of the public lands. To you students of desert management, I urge you to follow closely the evolution of the California Desert Conservation Area. It may well be the test case on land use adjustment through political and administrative processes in this post frontier era.

As for arid and semi-arid lands which do not lend themselves to long-term maximized economic use although in private ownership, I challenge you as desert land managers to think through the institutional arrangement needed to bring such land under multiple-use and sustained-yield.

STRATEGY FOR THE DEVELOPMENT OF CHILE'S IVth REGION: A TRANSITIONAL ZONE BETWEEN DESERT & SEMI-ARID AREAS

Carlos Correa Sanfuentes

ABSTRACT

The present report attempts to explain in synthetic form the strategy of Regional Development of the IVth Region, Coquimbo, in the context of the National Development, emphasizing the territorial unity of the country. Special transitional situations have been produced, products of the ecological and social-economic conditions in which this country finds itself. This is the specific case of the IVth Region, Coquimbo.

GENERAL BACKGROUND OF THE REPUBLIC OF CHILE

Chile stretches along the west coast of Latin America, from the Peruvian border down to the South Pole. It extends down the continent 4,280 kilometers with an average width of 138 kilometers, resulting in a total territorial area of 756,626 km^2. It has 10,500,000 inhabitants, which is 31.7 persons per square kilometer. Due to the length of the territory, there is a variety of climates. However, in spite of such diversity, the moderating effect of the sea excludes extremes in climatic conditions, and places Chile in the category of countries with a moderate climate. The yearly average temperature in the extreme northern zone (Arica) is 19.5°C. and in the southernmost point (Punta Arenas) 6.6°C.

The Government of Chile is central, and political power in its fullest attributes and functions is exercised by one and the same national authority under an identical juridical system throughout the country. However, the modernization of the Chilean government, which is meant to safeguard this unity, has made it indispensable to evolve a system which permits a progressive administrative and territorial decentralization because of the rather special geographical characteristics of the country. This facilitates in the best possible way the coordination and the participation of the regions* as regards integration, security, socio-

*Region: the territorial administrative units into which the country was divided with their special physical and socio-economic characteristics. The IVth Region was so designated and formally began to function within its new category beginning the 1st of January, 1976.

economic development and national administration. This process, which has meant a profound change, was begun in 1974, with the division of the country into 13 regions, each one of them assuming the responsibility for its own development, and has continued consistently and gradually to the present day. In a broad sense, Chile could be divided into three noticeably different zones:

The Desert in the North (Mining)
The Central Zone (Agriculture and Forestry)
The Southern Zone (Forestry and Animal Husbandry)

Fishing is practiced throughout the entire length of the country, with greater emphasis in certain zones.

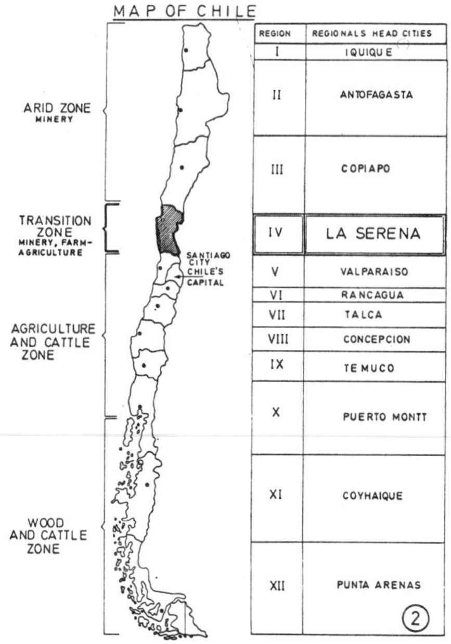

CHARACTERISTICS OF THE NORTH OF CHILE

The arid and semi-arid zones of Chile extend principally from the capital of the country, Santiago de Chile, up to the boundary in the north, and constitute 40% of the surface of its national continental territory.

The hydrological, geographic, climatic and ecological factors of these zones present special characteristics which raise problems seriously compromising zonal development in its physical and socio-economic aspects. These limiting factors have to do principally with the natural resources of water, energy and food, which in turn are interrelated with the adequacy of an economic structure and a socio-cultural development.

In addition, these limiting factors render impossible the best use of other resources (mineral, metallic, non-metallic, local climate favoring intensive cultivation, tourist trade, etc.), which in many zones of the arid region of Chile offer a remarkable potential of a high economic value.

DIAGNOSIS

The IVth Region is located within the territory of the arid and semi-arid zones, and its capital is La Serena, 500 km. north of the national capital (Santiago de Chile). The area of the IVth Region is 39,647 km^2, representing 5.2% of the national total, with a population of approximately 396,000 people, and a density of 10 persons per square kilometer of habitable land. 60% of these people live in urban areas and 40% in the country.

This Region is characterized as a "Zone of Transition" because it is situated between one zone that is agricultural and another that is desert, and hence cannot be unreservedly classified as one particular sector but rather as being simultaneously one of agriculture, fishing and mining. The potential natural resources have been under-exploited due to a lack of appreciation as to their real value and also because of obsolete techniques of exploitation and management of such resources.

The annual rainfall of the Region varies between 50 mm. in the North and 220 mm. in the South, with a wide variability. In addition, the high seasonal temperatures and the dryness of the area, causes accentuated evaporation which in turn creates a semi-arid zone. La Serena has 94 mm. precipitation, annually averaged over a period of 39 years, with some areas having less than 33 mm. and others having more than 175 mm. This perpetual drought is broken once in every 6 or 7 years with relatively heavy downpours. The snow that falls in the cordillera is a permament supply of water running through the principal transverse valleys; 80% of the precipitation falls between the end of May and August, making possible the cultivation of dry crops such as wheat, lentils and comino in the southern part of the Region.

Rivers are fed with the melting snows of spring. Their flow is harnessed by dams and reservoirs, which permits an extended area for cultivation and constitutes guaranteed irrigation throughout the entire year. The IVth Region is formed by 3 hydrographic basins, whose principal rivers the Elqui, Limari and Choapa have given rise to 3 provinces (administrative units into which a region is divided) with those same names. It is here in the basins of these valleys, alongside the streams where an intensive irrigation agriculture is practiced, that close to 2/3 of the population has congregated.

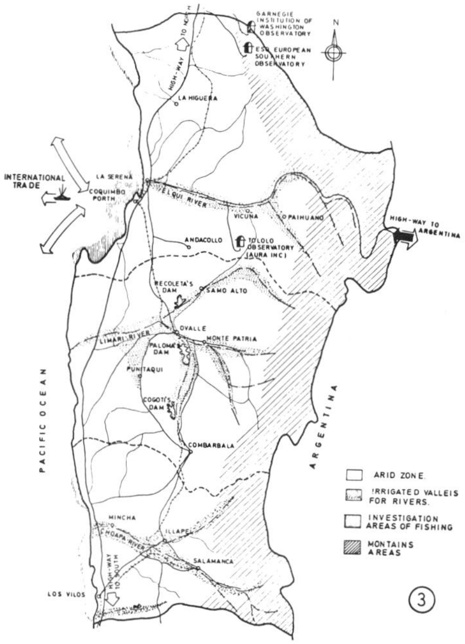

Of a potentially total agricultural area of 2,133,000 ha., 2,030,771 ha. are too dry to be tilled, 22,954 are dry but arable, and of the 104,074 ha. of land that could be irrigated only 50,000 ha. in fact are, and that is only 23% of the land potentially suitable for agriculture. This is due principally to a lack of proper control of water and to a deficient road infrastructure, since the principal highway connecting the Region is the Pan American Highway which follows the coast line and leaves innumerable towns and villages in the interior completely isolated. This situation has become worse with the suspension of railroad service.

The greater part of the territory of the Region is found in these valleys which is an arid zone in the strict sense of the term. One-third of the people of the Region live here and own the land in a system of communes; in fact it is called "The Zone of the Communidades". It is the poorest area of the Region, depending for its subsistence on rain for the crops or on the raising of goats. Indiscriminate sowing, poor management, over-production of goats and the use of the few trees still existing there for fuel, has practically exhausted the topsoil, and has accelerated erosion and has hastened the inroads of the desert in alarming proportions.

As for mining, the region is rich in metallic deposits, especially copper, iron, gold, and manganese, natural resources which can be recovered and which constitute one of its principal possibilities of development. Only iron, found in the northern part of the Region, belongs to the large mines, the other metals being extracted by the intermediate and small mines; it is with this last type of extraction that the highest percentage of mining manpower is to be found. The "pirquinero" (the small miner) is everywhere, up and down the entire Region, and works for the most part in very precarious conditions, with a bare subsistance salary.

The deposits of the National Company of Mines in Andacollo in 1980 and Los Pelambres in Choapa in 1990 incorporated the IVth Region into the Major Mines of copper. For the deposits of Limari and Choapa, a detailed investigation of reserves and of economic feasibility has to be undertaken in order to attract capital, create stable sources of work, and assure the plants, already existing in both provinces, of sufficient raw materials.

The sea holds out considerable hope because it possesses such a variety of replaceable resources (among them - mollusks, abilones, giant oysters, "shoe" clams, shrimp), and fish as for example: jurel, Spanish sardines, and deep-sea fish. Algae too, because of the jobs they give, are important and are found in abundance all along the sea coasts. Both the mollusks and the fish that frequent waters near the shore have been exploited irrationally.

As for sources of energy, these are scarce in the Region and more costly than in the center of the country, reaching as high as 30% more for electricity, and with consequent negative repercussions for the industrial area. In all, the electrical plants in the Region have a potential of 62,735 KW. (74.5% of this energy corresponding to thermo-electric plants). As for oil, coal and natural gas, there are no such resources in the Region, and therefore any power so generated has to come from outside the Region.

REGIONAL ROLE WITHIN THE NATIONAL CONTEXT

The basic objective of the Region is prolonged economic and social development that would permit:

> guaranteeing the entire population of the Region a basic level of subsistance that would mean the eradicating, totally and forever, of extreme poverty; this would be based on a consolidated development: of the mines (copper and iron fundamentally), a more intensified way of farming, a more rational and modernized exploitation of the sea; and on the creation of an agro-industrial complex to process the raw materials of the Region. All this will promote considerable increase of regional exports and also a greater use of the port of Coquimbo (Province of Elqui).

Parallel with the above, is the necessity to consolidate the Region as one functional interrelated system, with all its parts integrated with its center which is the source of regional development. To that effect, it will be necessary to create and amplify the network of interchange (channels of communication, highways into the interior) between the regional center (La Serena) and the provincial heads, and again between the latter and the smaller populated areas.

- detain the advance of the desert through a more adequate conservation of the resources still existing; this to be implemented, among other things, by a more rational control of cattle, reclamation of soils and the conservation of what vegetation that still remains.

An adequate agricultural study will determine which crops are most indicated for these valleys and the Region in general to compete rationally, taking into account our special conditions of water, climate and light intensity.

REGIONAL LINES OF ACTION

The policies of regional development are oriented toward taking advantage of the comparative benefits to be found in locating and developing the regional resources, the proper functioning of a market economy, the placing of non-strategic productive activities into private hands, the stimulation of new non-traditional exports, and the benefiting from the unique advantage that Chile has with its gateway to the Pacific, in relation to the nations along the Atlantic.

On the other hand, it is basic to achieve a regional development that is autonomous and planned, and which takes the concrete form of:

- consolidation of the regional system of decision making.

- strengthening of the bonds between the technical services and the centers of decision making in matters relating to regional development.

- institutionalizing of the participation of groups depending on the Regional Ministerial Secretariates within the system and in the process of planning with SERPLAC as the center. Especially important is the participation of groups for social development dependent on the Ministry of the Interior. It is not logical to maintain a dualism between social planning and economic planning, keeping in mind the great overlapping of assistance programs with those of social development.

The problem of the extreme poverty of the people in the Region has necessitated the adoption of emergency measures; yet it remains absolutely necessary that all these measures stay within the framework of the regional policy of development for a considerable period (1978 - 1981).

The idea is that:

- the total assistance to those affected should lead to the undertaking (on the part of these same people) of actions conducive to the complete solution of extreme poverty.

- the actions just mentioned take place in areas and sectors of activity that hold out greater promise of development even when this might cause persons and families to move from one place or occupation to another.

All this necessitates an adequate organizing of individuals within groups that represent their activities.

AGRICULTURAL SECTOR

Agricultural development will have to be oriented within a reasonable time towards promoting fruits, vegetables, seeds and flowers, oil-bearers such as higuerilla, ajonjoli, fojosa and alquinas, as well as types of essential oils.

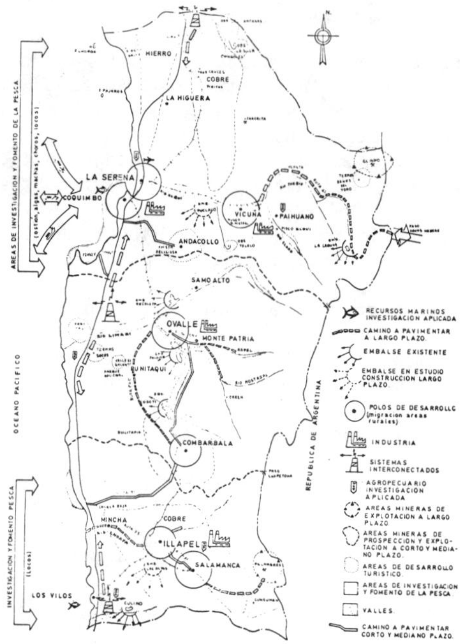

As can be concluded from the preceding pages, the greater part of the agricultural land is dry; in this sense it will necessitate an orientation for conserving existing natural resources, promotion of bush and forage species plantations and protection of water basins.

As regards studies, we shall divide this into two sections:

> <u>Irrigation</u> - Direct and promote studies tending towards a greater crop profitability, searching for new varieties and introducing others chosen from a commercial viewpoint.
>
> <u>Dry Areas</u> - Promote studies aimed at the conservation of the environment and at a more rational use of same. Also intensify the search for new varieties and kinds of seeds and plants, without excluding those already known to do well.

MINING SECTOR

Within a short time the "pirquinero" (small miner) will have to change over from his individualistic style of work to the formation of work teams, organized and versed in the lastest techniques, with the double aim of raising productivity and reducing production costs.

The Mining Sector will shortly inaugurate a change destined to become the most important for the increase of the gross production of the Region and in the investments of the next decade (1980 - 1990). Its direct and indirect help will maintain current and anticipated levels of production in the small gold mines.

Many predictions concur in that the remainder of the present age will see an increase in the international demand for copper. In this context, arguments have been advanced to substantiate hopes for a better world price for this product within the coming years. These hopes, along with the implementation of the Andacollo Project and others of lesser importance (Chacrita, Talcuna, Polimetalicos, etc.) help us foretell a sound future for the mines of the Region.

The help needed by this sector must keep in mind activities that would transfer some of the installations to private enterprise, and would adequately exploit existing and new copper mining districts.

The government agencies should also increase their activities for the search of minerals, and for the formulation of the corresponding blueprints, in order to interest private enterprise in investing in this area.

FISHING

The expansion of activities relating to the extraction of fish, mollusks and algae shall continue, but with greater rationalized control (alternating zones of extraction, systematizing the periods of prohibition, etc.). A system of organization of fishing villages shall be undertaken to satisfy the needs of the villagers, taking advantage in the best way possible of the local marine resources and furthering the development of sea nurseries.

The help needed by this sector will be along the lines of applied study, education, infra-structure, credit and marketing. The success of such assistance will depend in great measure on their effective, integrated and continued delivery at the local level of each village.

INDUSTRY

Within a reasonable period, important plans should be forthcoming related to the exploitation of known regional resources (farm industries, canning factories of fish and mollusks, metal mechanics, etc.), in which national or foreign investors could be interested. It is indispensable to inaugurate studies to detect the comparative advantages present in the Region in order that industrial development has a solid base.

SYSTEMS OF HELP

Policy of the Development of Social Sectors

It is important to take into consideration all the asepcts related to social development which must go hand in hand with progress in many of the basic sectors of the economy.

Among others, public health occupies a special place, and must shortly provide greater attention to the rural area with the creation of conditions for future progress in this zone in the form of programs for the promotion of health improvement, the setting up of health centers in rural areas, etc.

In the sector of education, the trend is to continue improving its quality, through training programs for teachers and studies and evaluations of the programs currently in use. The aim is to establish an equilibrium between the urban and rural educational areas by means of integrated boarding schools.

In the housing sector, it is planned shortly to solve the actual déficit of social housing for the sector of the population unable to accumulate savings.

It is hoped to see within a reasonable period of time the sections of production develop regional technologies based on the resources proper to each zone.

EGYPTIAN EXPERIENCE IN LAND RECLAMATION IN EL-WADI EL-GEDID AREA

Mohamed Aly Ezzat
Undersecretary for Desert Irrigation
Ministry of Irrigation
Cairo, Egypt

ABSTRACT

The purpose of the present paper, is to give the Egyptian experience gained in land reclamation in the EL-WADI EL-GEDID area of the Egyptian Western Desert, and in the exploitation of already proven groundwater resources.

Land reclamation started in 1960, in the two southern depressions, Dakhla and Kharga Oases. 23,000 feddans (9,200 hectares) were reclaimed from 1960 to 1970. (A feddan is approximately equal to one acre.) Since then reclamation work stopped, in order to make an agro-economic study for the reclaimed area, and to give recommendations for future development.

Studies indicated that the present method of irrigation by small basins is in fact a good one as regards the uniformity of application of the water on the field. Net water requirements have been determined for the major crops suitable for the area. Proposed values are based on evaporation pan data and from lysimeter data.

In order to choose the cropping pattern suitable for the development of the area, 20 models of farm development were analysed. The most suitable one for farm settlements is as follows: one third date plam trees, one third winter forage crops (alfalfa and berseem) and the rest broad beans and onions which are cultivated as intercrops in one third of the orchard. In summer, 40% is devoted to forage production (alfalfa and sorghum) while the rest of the available land is cultivated with rice and watermelons.

For future development, it is recommended to reclaim plots of 500 feddans each. Such a plot could be irrigated by a group of 5 wells. The 500-feddan units should be managed and the wells monitored according to crop requirements. A system of drains is necessary everywhere from the start, especially at the leaching stage. Due to the necessity of executing good land preparation at the start, including sufficient leaching and proper planting of trees, the land to be allocated to farmers should remain under the control of the government for at least two years.

A critical economic analysis has been made of a farm unit of 4.5 feddans owned by one family. The financial return is about 6.4% return on investment. The average farm family income is expected to be L.E. 1350/year after 27 years.

THE PROBLEM

The Western Desert in Egypt comprises an area of 750,000 square kilometers and supports a population of about 200,000. The desert includes 90,000 square kilometers of low land, part of which is potentially arable land that could be irrigated if an adequate supply of water were available. That part of the Western Desert which is designated as EL-WADI EL-GEDID area extends diagonally southeastward from Siwa Oasis, north of 29^0 Latitude, through Bahariya, Farafra, Dakhla and Kharga Oases to a point on the southern boundary of Egypt where the Nile crosses the Sudan boundary. The lowland is extremely irregular in shape and consists of a series of five principal oases depressions in the Plateau.

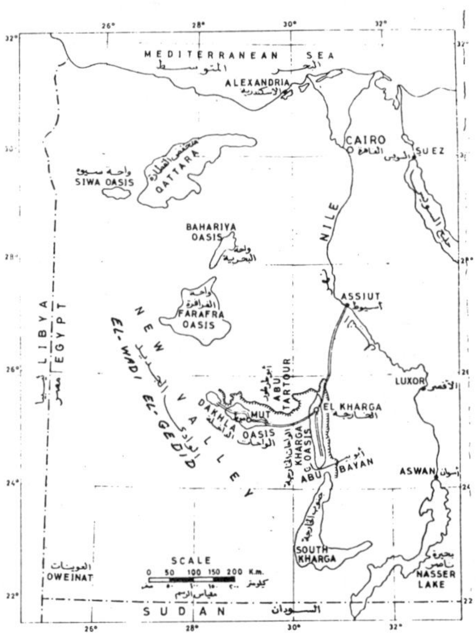

LOCATION MAP

EL-WADI EL-GEDID area, is the eastern portion of the Sahara, the world's largest desert, where huge natural reservoirs of fresh water are lying unexploited. The Western Desert, including EL-WADI EL-GEDID, area is the largest groundwater basin of the Sahara with an area of about 1,800,000 sq.km., a water storage capacity of 6,000 milliard cubic meters with a natural recharge of over 2.3 milliard cubic meters per year and artesian flow in an area of 57,000 sq. kms., including the Qattara depression.

The area of the five depressions is 14.5 million feddans. A reconnaissance soil survey at a scale of 1,100,000 was carried out over 6,250,000 feddans, of which 3,300,000 feddans were classified from "very good" to "good" soils, subdivided into 2,100,000 feddans below 200 meters elevation and 1,200,000 feddans above 200 meters.

A semi-detailed soil survey at a scale of 1:25,000 was carried out mainly in Kharga and Dakhla. Also in the Farafra depression over 600,000 feddans of which

200,000 feddans were further surveyed (detail survey) at a scale of 1:10,000 to identify the project area to be reclaimed and to give recommendations as to how to reclaim the soil.

The two southern depressions were chosen for pilot development. Land reclamation started in 1960. Before 1960, the area under irrigation was 13,000 feddans. At present the land under irrigation in the two depressions is about 41,000 feddans distributed as follows:

Kharga : 5,000 feddans
10,000 feddans with new wells since 1960

Dakhla : 8,000 feddans
13,000 feddans with new wells since 1960

West Mawhoub : 5,000 feddans with new wells (50 kms. West of Dakhla).

Below the arid surface of the Western Desert and of the depressions of EL-WADI EL-GEDID area is a huge reservoir of groundwater which could sustain human settlement and productive agriculture. The water bearing formation is the red Nubian sandstone of lower and middle Cretaceous age. This sandstone which in many places is interbedded with shales, is 600 - 800 meters thick in the Kharga depression, more than 1200 meters thick in Dakhla and other depressions and rests on Pre-Cambrian basement rocks. The French name equivalent to "Nubian Sandstone" is the "Continental Intercalaire." The groundwater reservoir is a confined aquifer overlain by a semi-impermeable stratum. It is likely to be under artesian pressure that will cause the water to rise above the top of the aquifer when the aquifer is penetrated by a well. Whether or not the water rises high enough to flow at ground level, will depend on the ground elevation. Along with the development of groundwater, the area of artesian flowing wells will progressively be transformed into an area of non-flowing wells.

At present, discharge of flowing wells range between 300 cu.m./day for shallow wells (150 m. deep) to 3520 cu.m./day for deep wells (400 - 800 m. deep). Production behaviour of shallow and deep wells from 1956 till 1975 is as follows:

Annual Discharge (Million cu.m.)

	Shallow Wells		Deep Wells		Total Wells	
	1956	1975	1956	1975	1956	1975
Dakhla	90	52	6	145	96	197
Kharga	38	16	13	62	51	78
Total	128	68	19	207	147	275

Thus total production increased from 147 million cu. m./yr. in 1956 to 275 million cu.m./yr., and discharge of the deep wells is nearly constant. However, discharge of the flowing shallow wells is decreasing, due to the effect of the deep wells on the shallow aquifer.

The quality of water developed by wells generally improves with depth. The Nubian sandstone beds yield highly corrosive water creating well construction problems that need careful study. In the Dakhla and Kharga area, the water from the wells is low in total dissolved solids, not exceeding 250 milligrams per litre, low in chlorides, not exceeding 40 milligrams per litre and contains some sulfur in mineral salts and gas (carbon dioxide). The water temperature increases with the depth of the producing beds and ranges from 30^0 to 40^0C. The temperature of the artesian water in the oases, therefore seems to be abnormally high.

The present paper deals with the experience gained through the reclamation of 28,000 feddans in the Dakhla-Kharga area of the Egyptian Western Desert. The paper discusses the proper cropping pattern, which is most profitable. Net crop water requirements for the different crops as well as the gross water requirements are discussed. The present paper deals also with the appropriate method of irrigation, which is suitable to the type of soil of the area and to the climate. Finally the construction cost of newly reclaimed land in a prototype farm is discussed.

NATURAL CHARACTERISTICS

Climate

The climate is very hot. In the Kharga oasis maximum air temperature occurs in June (47.4^{0}C). Minimum air temperature occurs in January (1.0^{0}C). Average daily humidity ranges between 21% in June and 44% in January. Rain is practically nil. Maximum wind speed may reach 10 m/sec during Khamasin south wind storms (in March). Maximum evaporation (piche), may reach 54.3 mm/day in June. Minimum evaporation, 5.5 mm/day occurs in December. Solar radiation in gm. cal, per sq. cm. ranges between 6095 in December to 17918 in July. Percentages of sunshine duration range between 82% in December to 94% in August. Table 1, shows a summary of the climatological data for the Wadi El-Gedid area.

The high temperature, dry air and continuous winds cause a very high free water surface evaporation rate which amount to 5 m. per year.

Table 1: Summary of Climatological Data
Wadi El-Gedid Area

	Siwa	Bahariya	Farafra	Dakhla	Kharga
Temperature ^{0}C.					
Monthly mean max. temp. ^{0}C	29.8	29.5	30.3	31.7	31.4
" " min. "	13.0	13.5	13.3	17.8	15.6
" " average "	21.4	21.6	21.8	24.7	23.6
" absolute max. "	41.5	41.5	41.0	43.4	42.5
" " min. "	5.1	5.7	3.1	7.3	8.5
Average annual precipitation	9.6	3.6	1.8	1.3	0.30
(mm.) relative humidity mean					
monthly (%)					
Max. (December)	58	65	56		40
Min. (June)	30	42	28		21
Evaporation (Piche) mm./day:					
Max. (June)	17	14.0	24.1		37.1
Min. (December)	5.2	5.1	7.2		17.7
Wind speed (m/sec.)					
Max. (March)					10
Min. (December)					1.3

Soil

There are various classes of soil in the Dakhla-Kharga area. They range from class 2 to class 4. Soil classification depends on soil properties, texture, structure, depth of profile, salinity and alkalinity ... etc. The construction of a drainage network is essential to prevent soil salinity and a high evaporation rate requires increased water consumption for irrigation in comparison with the Nile Valley.

Roads

The five oases of the Western Desert, are located in remote areas, far from the Nile Valley. The Kharga oasis, is connected by an asphalt road with Assiut in the Nile Valley. The Bahariya oasis, is connected with Cairo by an asphalt road. Also the Dakhla oasis is connected with the Kharga oasis by an asphalt road.

An air-port exists in the Kharga oasis and landing strips exist in the other four oases.

Population

The total population of EL-WADI EL-GEDID area, amounts to 117,000. They are distributed as follows:

Siwa Oases	15,000
Bahariya Oases	10,000
Farafra Oases	2,000
Dakhla Oases	40,000

Mineral Resources

A large iron deposit exists in the northern part of the Bahariya oasis. The iron deposits are now under exploitation.

At the Abu-Tartour area, about 40 kilometers west of the Kharga oasis, phosphate deposits exist. At present, detailed studies are in progress to design the phosphate plant which is to be put under exploitation by 1980.

CROP WATER REQUIREMENTS

Net Crop Water Requirements

Agricultural development in the EL-WADI EL-GEDID area depends solely on irrigation, since rainfall is practically nil. A good knowledge of the crop water requirements in the area is therefore essential, and all the more necessary as the cost of exploitation of groundwater appears to be high.

The bulk of the investigations concerning crop water requirements in the EL-WADI EL-GEDID area, has been carried out at the Water Requirements Station in the Kharga oasis, and activity in this field is continuing.

The finally proposed values for net crop water requirements, Table 2, are based on evaporation pan data and from lysimeter data.

Efficiency of Irrigation

The gross crop water requirements is the quantity of water to be delivered by the wells in order to supply all the crops with the necessary water. The gross requirements exceed the net requirements and the ratio E of the latter to the former

Table 2: PROPOSED BASIC NET CROP WATER REQUIREMENTS. (mm. per day)

	J	F	M	A	M	J	J	A	S	O	N	D	Total mm.
Wheat, barley	3.5	5.0	1.5								2.5	3.0	463
Broad Bean	3.5	5.0	2.0								3.0	3.0	494
Onion	3.0	4.5	7.0	7.0							3.0	3.0	845
Watermelon					4.5	9.0	12.0	9.5					1060
Sorgum					5.5	11.0	12.0	5.5					1043
Rice					12.0	12.0	12.0	12.0	9.5				1760
Berseem	3.0	4.5	7.5	9.5							5.0	3.5	995
Alfalfa	3.0	4.5	6.5	8.5	10.0	10.0	10.0	9.0	9.0	7.0	5.0	3.5	2635
Citrus	1.5	2.5	4.0	5.0	6.5	8.0	7.0	6.5	6.0	5.0	3.0	2.0	1768
Date palm	2.0	2.5	5.0	6.5	8.5	11.0	10.0	9.0	8.0	5.5	3.5	2.5	2257

is the global irrigation efficiency.

Part of the difference between the gross and net requirements is lost in the conveyance and distribution system and the rest in the field itself. If "d" denotes the efficiency of the distribution system, i.e., the ratio of the quantity of water delivered to the field to the quantity of water delivered by the well, and "f" the farm irrigation efficiency, i.e., the ratio of the quantity of water actually consumed by the crops or net water requirement to the quantity of water delivered to the field, then

$$E = df.$$

The current practice in the EL-WADI EL-GEDID area, is basin irrigation. Farmers are accustomed to preparing small basins which they flood one after the other. The method would be satisfactory if the land preparations were correct.

The water is usually conveyed from the well to the fields, first through a lined canal at a distance ranging from 50 to 2000 m, then through earthern canals for 200-300 m. The water flows into the fields at discharges of about 20-25 l/sec. (70 - 90 cu.m./h).

Distribution losses include infiltration along the irrigation canals, evaporation in the same canals, losses at outlets not correctly closed and unused water, that is to say, flowing directly to the drains.

Field losses are due to uneven application of water to the total area of every plot. Too much water subsequently percolates below the root zone. Excess water accumulated at the bottom of the fields and running off to the drains is also included here. The losses are equal to the difference between the net water requirements, and the actual quantity of water delivered to the field.

A survey was carried out during the winter of 1974 - 1975 in order to measure irrigation efficiency. On 8 wells, permanent records were kept of discharges of the wells and of the areas and crops actually irrigated. By comparing the water actually delivered by the wells with the required water, calculated by applying the rates adopted in Table 2, the global irrigation efficiency (E) can be calculated.

Table 3 gives the values of E obtained from the survey, while Table 4 gives also the value of "f" deduced from "E" and "d". The two tables, clearly show the low efficiency during the months of December and January, due to excess of irrigation at a period of low crop water demand. The high values of (E) in February and March should also be noted. This proves that the farmer, using the traditional small basins, is able to irrigate his field without high water losses when he receives the correct amount of water.

Expected Irrigation Efficiency

It is assumed that the distribution system efficiency, d, will be improved by introducing better water management in the conveyance system. In the present irrigation networks where only the main canal is lined, it will remain at about 0.70. New areas to be cultivated should be equipped with irrigation networks having more lined canals than at present. If only the third order irrigation canal (the canal which delivers the water directly to farm) is unlined, d, should reach 0.85. If all canals are lined, d is expected to be 0.90 and if all the distribution is made of pipes, 0.95.

The present method of irrigation by small basins is in fact a good one as regards the uniformity of application of the water to the field.

Table 3: GLOBAL IRRIGATION EFFICIENCY (E), WINTER 1974/75

	Type of Well	Irrigated Area (feddan)	N	D	J	F	M	Main crops
El-Sherka 1	Pumped	110.25	0.59	0.44	0.53	0.81	0.91	Wheat, berseem
Ginah - 1	"	45.8	0.24	0.16	0.20	0.27	0.30	Orchard, veg.
Ginah - 6	Flowing	95.5	0.51	0.32	0.34	0.48	0.58	Orchard
Bulaq - 3	"	147.5	0.60	0.49	0.65	0.97	0.91	Wheat, berseem, onion
Garmashene - 1	Pumped	69.4	0.71	0.43	0.57	0.59	0.75	Date palm
Garmashene - 6	"	84.6	0.43	0.35	0.39	0.58	0.77	Alfalfa
Bulaq - 4	Flowing	25.6	0.62	0.48	0.57	0.69	0.91	Wheat, berseem, beans
Baris - 10	"	77.7	0.45	0.37	0.25	0.71	0.62	Wheat, alfalfa

Table 4: CALCULATED VALUES OF FARM IRRIGATION EFFICIENCY (f)

	d	Nov	Dec	Jan	Feb	Mar
El-Sherka - 1	0.90	0.66	0.49	0.59	0.90	-
Ginah - 1	0.70	0.34	0.23	0.29	0.30	0.33
Ginah - 6	0.80	0.64	0.40	0.43	0.60	0.72
Bulaq - 3	0.7-0.9	0.67-0.84	0.54-0.70	0.61-0.93	-	-
Garmashene - 1	0.7-0.9	-	0.48-0.61	0.57-0.73	0.66-0.84	-
Garmashene - 6	0.85	0.51	0.41	0.46	0.68	0.90
Bulaq - 4	0.8-0.9	0.69-0.77	0.53-0.60	0.57-0.64	0.77-0.88	-
Baris 10	0.7-0.9	0.50-0.64	0.41-0.53	0.28-0.36	-	0.69-0.89

For basin irrigation as used at present, the farm irrigation efficiency f, is considered to be 0.80 for orchards and 0.70 for other crops. Orchards have to be planted on lighter soils with low capillarity and consequently less risk of salinity rising from the subsoil, which means lower leaching requirements. With a deep root zone as required for trees, only a small amount of water will percolate to the deeper layers. The 20% excess water resulting from a farm application efficiency of 0.80 therefore appears sufficient. On the other hand, for crops with a generally shallower root zone, and for the most part on heavier soils for which leaching water is necessary, f, has been reduced and estimated to be 0.70.

The resulting global efficiency E, is shown in Table 5.

Table 5: EXPECTED GLOBAL EFFICIENCY

Type of Distribution System	d	f		e	
		Tree crops	Other crops	Tree crops	Other crops
1. Main canal lined, 2nd, 3rd order earth	0.70	0.80	0.70	0.56	0.49
2. Main and 2nd order canals lined; 3rd order earth	0.85	0.80	0.70	0.68	0.60
3. All canals lined	0.90	0.80	0.70	0.72	0.63
4. Main and 2nd order; 3rd order earth	0.90	0.80	0.70	0.72	0.63
5. All distribution systems piped	0.95	0.80	0.70	0.76	0.67

On the other hand, it is well known that drip irrigation involves a proper filtration of water. The groundwater of the Nubian Sandstone contains iron compounds which present a serious risk of clogging the orifices. It will therefore be necessary to carry out trials in order to find out to what extent drip irrigation could be used.

CROPPING PATTERN

Selection of a Cropping Pattern

The climate and the soils of EL-WADI EL-GEDID area are favourable for the following crops within the limits and conditions given in Table 6, below.

As EL-WADI EL-GEDID has to be developed step by step, a cropping pattern has to be selected for each increment of the reclaimed area depending on the conditions prevailing at that time. It is obvious that for any new area to be put under cultivation, the cropping pattern should include for the most part, the crops which are considered as giving the more substantial returns without special difficulties regarding the marketing of such crops.

In the EL-WADI EL-GEDID area, the best kind of crop is undoubtedly the date palm tree. For the time being, Egypt imports dates and there is therefore the possibility of extending the cultivation of date palm trees over several thousands of feddans without particular problems of marketing.

Table 6: SUITABLE CROPS IN EL-WADI EL-GEDID

Degree of suitability	Crops	Observations
Very suitable	Date-palm trees	In soils of light and medium texture
	Sorghum or Indian corn	
	Rice	
	Watermelons	
	Perfume plants	
	Berseem	Winter crop only
	Wheat	" " "
	Onions	" " "
	Broad Beans	" " "
	Vegetables	" " "
Suitable	Alfalfa	Low production in summer
	Oil seeds	
	Citrus	In light soils-low yields as compared with those in Nile Valley and Delta
	Gouafa	

The second selected group of crops is constituted by forages in order to develop cattle breeding. Although the latter seems to be not very beneficial in terms of money, Egypt needs more meat and milk products and any local production which entails a reduction of present imports is a gain for the country. Anyway, the coupling of a high return crop such as date palm with forage whose return is less, is a good principle since the production of the area as an average, will be high enough to offer an acceptable repayment capacity for the necessary investments and current expenses. In addition. alfalfa culture is of interest due to its improvement of the soil structure which in the long term is a factor of increasing productivity.

However, cultivation of forage for cattle breeding requires a continuous production all year long to be able to feed the animals in all seasons. Due to low productivity of alfalfa in the summer time, the necessary cattle food in summer should therefore be supplied by sorghum; although it is a high water consumer it grows perfectly well in the area and in any kind of soil.

Date-palm trees and forage being the ossature of the agriculture production life, alternatives have been envisaged for completing the cropping pattern. They are defined by Table 7, to which the following comments have to be added.

Crop Pattern I: This is characterized by full occupation of the land in summer as well as in winter. This intensive cultivation is also reflected in the intercropping between the date-palm trees where it is assumed that one third of land is put under various crops. The table indicates broad-beans in winter and watermelons in summer but by rotation one will find all other crops listed in the

Table 7: ALTERNATIVES OF CROPPING PATTERNS

(For a unit farm of 4.5 feddans.)

Crops	CP I		CP II		CP III		CP IV		CP V	
	W	S	W	S	W	S	W	S	W	S
Date Palm trees	1.5	1.5	1.5	1.5	1.5	1.5	1.5	1.5	1.5	1.5
Intercrops:										
Broad-beans	0.5	-	0.5	-	0.5	-	0.5	-	-	-
Watermelons	-	0.5	-	-	-	-	-	-	-	-
Wheat	-	-	-	-	-	-	-	-	0.75	-
Alfalfa	1	1	1	1	1	1*	1	1*	1.5	1.5
Wheat	1	-	1	-	1	-	1	-	0.5**	-
Onions	0.5	-	0.5	-	0.5	-	0.5	-	-	-
Berseem	0.5	-	0.5	-	0.5	-	0.5	-	1.5	-
Indian Corn	-	1	-	0.5	-	-	-	-	-	-
Watermelons	-	-	-	0.5	-	-	-	-	-	-
Sorghum	-	0.5	-	0.5	-	1	-	1	-	-
Rice	-	0.5	-	0.5	-	0.5	-	0.5	-	-
TOTAL	5	5	5	4.5	5	4.5	5	4	5.7	3

* Irrigation is limited in summer

** One crop of what every three years, wheat and alfalfa being sown together on the same land.

table except alfalfa and rice.

The cropping pattern includes rice which in spite of its high water requirements and its relatively low return is of interest because the leaching effect resulting from the cultivation of this crop is a means of keeping control of the soil salinity.

Crop Pattern II: There are no summer crops between the date-palm trees and the area under Indian corn which is reduced from 1 to 0.5 feddan.

Crop Pattern III: This is the same as CP-II with the difference that alfalfa receives only limited irrigation during the summer. In order to compensate the corresponding reduction of forage production, the area under sorghum is increased from 0.5 to 1 feddan while Indian corn is suppressed.

Crop Pattern IV: This crop pattern compared to CP-III, has no watermelons which reduces the area under cultivation in summer by another half feddan.

Crop Pattern V: There are no summer crops at all. Wheat is cultivated on half the surface between the trees and mixed with alfalfa when the latter is at its early stage. However, Crop Pattern V is not as well equilibrated as the others. A part of the alfalfa harvested in winter has to be kept as hay for summer where there is only the alfalfa summer production for feeding the cattle. The absence of rice, obliges the control of salinity through leaching which means additional expenditures for water without any return.

The five cropping patterns have been compared from the point of view of economics. Table 8 shows for each of them the expected net return on the basis of yields and farm prices. Net returns are obtained by subtracting from the gross returns the production costs which exclude the farmers labour and costs for water. Then for each cropping pattern an amount of L.E. 202.5 is subtracted which represents the living allowances for the farmer and his family and repayment of his debt for land and house. The remaining amount of money is considered as the capacity of repayment for water.

Under this criteria cropping pattern III appears to be the most profitable, since its capacity of repayment for water is estimated at L.E. 5 for 1000 cu.m. of required water. It must be pointed out that net water requirements of the crops are considered here. For relating the capacity of repayment to the water to be actually used, the figures should be multiplied by the global irrigation efficiency.

It is assumed that each farmer will be the owner of a unit farm of 4.5 feddans. The crop pattern selected (crop pattern III), gives the farm family a balance between a high-value but slow developing crop (dates), and the essential annual crops to provide the essential food requirements for the family and to support the livestock. The percentage distribution of land between the crops in a double cropping year allows the annual crops to carry the development costs of the dates while still providing the farmer with an acceptable level of income.

Returns to the Farmer

The farmer will be provided with one cow, one sheep and one donkey. At full development, he will have arrived at a livestock balance of two cows, three calves, ten sheep and one donkey. With a forage production of over 60 tons per year from 1.5 to 2.0 feddans of alfalfa, berseem and sorghum, the livestock maintained by the farmer at full development is considered to be conservative.

The farm family net return from labour builds up from a minimum of L.E. 0.17/man-

Table 8: CAPACITY OF REPAYMENT FOR COST OF WATER (in L.E.)
(For the unit farm of 4.5 feddans.)

Crops	Net Return per Feddan	Net Return				
		CP I	CP II	CP III	CP IV	CP V
Date Palm	143.6	215.4	215.4	215.4	215.4	215.4
Intercrops						
Broad beans	32	16.0	16.0	16.0	16.0	-
Watermelons	66	33.0	-	-	-	-
Wheat	32	-	-	-	-	24.0
Alfalfa	31	31.0	31.0	25.0	25.0	46.5
Wheat	32	32.0	32.0	32.0	32.0	16.0
Onions	67	33.5	33.5	33.5	33.5	-
Berseem	25	12.5	12.5	12.5	12.5	37.5
Indian Corn	17	17.0	8.5	-	-	-
Watermelon	66	-	33.0	33.0	-	-
Sorghum	14	7.0	7.0	14.0	14.0	-
Rice	25	12.5	12.5	12.5	12.5	-
Total		409.9	401.4	393.9	360.9	339.4
Farmer allowance (to be subtracted)		202.5	202.5	202.5	202.5	202.5
Capacity of repayment for cost of water (for a unit farm)		207.4	198.9	191.4	158.4	136.9
Net water requirement (cu.m. per year for the unit farm)		42,799	40,806	38,340	36,460	36,647
Capacity of repayment for cost of water (for 1000 cu.m. of water)		4.85	4.85	5.01	4.34	3.74

day in the first year that land is farmed (project year 4), to L.E. 0.72/man-day in the second year, to about L.E. 4.5/man-day at full development. The average return from labour over the 27 years of the expected production period would be L.E. 3.36/man-day; considerably higher than the average labourers wage of L.E. 0.40/ man-day. The average farm family income would be about L.E. 1350/year.

SETTLEMENT

The opening of new areas to development implies the recruitment of an agricultural labour force from outside the EL-WADI EL-GEDID area. The related human problems should therefore not be overlooked.

Agricultural development should be based on the experience gained over the past 20 years. The opportunity of expansion is made possible by the proposed crop rotation. The proposed unit for ownership is 4.5 feddans for each family. In order to insure the success of operation of such small units, good selection and essentially good training of the labour force is necessary.

Due to the necessity of executing good land preparation at the start, including sufficient leaching and a proper planting of trees, the land to be allocated to farmers should remain under the control of the Government for at least two years, during which intensive training could be given to the workers in order to prepare them to become good farmers. The above operations are part of the capital investment and should be considered as such instead of letting them be carried out by the farmers which apparently reduces the investment cost but leads to an incomplete reclamation. This alternative also allows the farmers to avoid facing the difficulties of the first agricultural seasons when the revenues will remain low.

Another difficulty pertaining to settlement rests with the way of recovering the operating costs and even a part of the investment costs, since the farmers are very unlikely to accept to pay commensurate water charges. In the case of state farms, this problem is avoided as all sales of production will bring in revenues directly to the Government.

DEVELOPMENT

When speaking of development in the EL-WADI EL-GEDID area, it is necessary to introduce the concept of a development unit. To drill a well and depend on its discharge to reclaim and cultivate a certain area in the vicinity of the well is not a good approach to development. The management of a large number of small independent units is not easy and can result in poor water usage and land development. The discharge would drop with time, in the case of flowing wells and the actual cropped area would decrease accordingly.

The grouping of a sufficient number of wells, say at least five, gives the possibility to connect them to a single distribution system serving about two feddans. The addition of a reserve well will help maintain the discharge at the required level. The management work for the area supplied by a group of wells, will be about the same as for a single well. The association of farmers into a cooperative or similar organization is easier if they depend on the same facilities.

The 500 feddan unit, could be organized as a state farm or a settlement scheme of about 100 farmers living and working in good cooperation. The regularity of the water supply and the control of distributed water would also be more easily obtained. Such development by units remains very flexible since it does not require the initial development of a large scale infrastructure.

CONCLUSIONS

The efforts of the last twenty years to develop the EL-WADI EL-GEDID area have already given very valuable results. The small town of New Kharga has been created, the same for the Governorate of EL-WADI EL-GEDID in addition to about twenty new villages. Orchards have been planted which produce fruit and new land has been cultivated. All this can be considered as a real success since it has recovered more land from the desert and provided new life for many farmer families. Furthermore, an invaluable sum of knowledge has been acquired.

From an economic point of view, the situation is not very encouraging, however, if the development process is executed according to the kind of methodology discussed earlier, the project provides an acceptable return on investment. With the present increase in population and the acute shortage of agricultural land and cultivated area, Egypt will have to develop the EL-WADI EL-GEDID area. Other urgent national tasks might have priority but this should not mask the necessity to develop the EL-WADI EL-GEDID area, either in the short or a longer term. It is therefore necessary to be prepared for that day, which means to continue to support agriculture, to extend the irrigated area even at a low rate of progress, to improve farming practices especially water management and to maintain, in the area necessary trained personnel for the implementation of new development.

DESERTIFICATION IN IRAQ & MEASURES FOR DEVELOPMENT

M. Said Kettaneh, Sami R. Daghistani & Laika K. Al-Jiburi
Scientific Research Foundation
Iraq

INTRODUCTION

In view of the limited possibilities of the environmental conditions and natural resources in arid and semi-arid regions of the world and despite the present advance of science and technology, the rapid increase of population is causing greater pressure and negative effects on the natural resources of such regions. In many cases, catastrophies have occurred and further problems will be facing the settlers of these regions if proper measures are not taken. Therefore, in December 1974, the United Nations General Assembly passed Resolution 3337 calling for an international conference on desertification to be held in 1977. Since then, preparations for such a conference have been made. These preparations will definitely clarify the problems properly and it is hoped that measures to control desertification with proper scientific and technological implementation will be found.

Iraq, with its global location and environmental conditions, is considered as arid and semi-arid. Therefore, it is necessary for the country to be involved in all the universal and regional activities aiming to control desertification. In the meantime, national programs should be in accordance with such activities. At the present time, the wise leadership of the 17th July Revolution in Iraq considers desertification control programs to be very important and vital. Therefore, a considerable part of the development programme's budget of the present 5 year plan (1976-1980) has been allocated for such activities.

This report deals with the general conditions of desertification in Iraq, taking into consideration a brief survey of natural resources and socio-economic problems, causes and measures. Accordingly, a general plan for coordinated action will be developed.

LOCATION AND AREA

Iraq is located in the southwestern part of Asia and covers an area of about 438,466 square kilometers (43,400,000 hectares). It extends from 28°59' to 37°20' north latitude and between 39°-48° longitude E. It is bound by Turkey on the north, Iran on the east, Syria and Jordan on the west, and Kuwait, Saudi Arabia and the Arabian Gulf on the south.

Physiographically, the Tauros mountain chain bounds the country on the north, the Zagros mountain chain on the east, and the Arabian plateau and the Arabian Gulf on the west and south. It is divided into the following regions: mountains which represent 6.6% of the total area with an altitude ranging from 1000-3650 m. above MSL (mean sea level), foothills which represent 15.3% at an altitude of 300-1000m. above MSL, upper plains which represent 8.4% at an altitude 100-300 m. above MSL, lower plains which represent 19.5% at an altitude 0-100m. above MSL, and deserts which represent 40.5% of the total country at an altitude of 90-915m. above MSL.

Administratively, Iraq is divided into 18 Muhafadha (Provinces) which in turn are divided into "quadas" (counties) and nahias. The population of Iraq is about 12 million, with a density of 23 people per square kilometer and a growth rate of 2.3 percent. Sixty percent of the total population is classified as urban. It is expected that the population will be doubled by the end of this century.

NATURAL FEATURES CAUSING DESERT FORMATION

Topography and Land Formation

Geologically Iraq was formed during the third geological period which was a period of marine submergence. Accordingly, the topography and land formation belongs to the tertiary geological period. In this period, mountains, plateaus, and river valleys of the continents of Euroasia were formed, and the climatic conditions were established. As a result of tectonic movements, the country has been divided into the following three major structural zones:

- The folded zone which consists of the north and north-eastern mountains and foothills.
- The western desert which is part of the Arabian plateaus.
- The Geosynclinal plain which is a depression located between the folded zone and the Arabian plateau.

Topographically the region's characteristics can be summarized as the following: (Table 1.)

a. Mountainous region: This region is part of the Tauros-Zagros mountain chains which consist of parallel mountain ranges directed from the Northwest to the Southeast. The highest peak reaches 3,700 meters above M.S.L. The Tigris River and its tributaries cut deep valleys across the parallel series of mountains. Many fertile plains are formed between the mountain chains.

b. The Foothills: This region is a transitional formation located between the mountain region and the geosynclinal plains, and it follows similar direction of the mountains.

c. <u>The Upper Plain</u>: It consists of the upper part of the geosynclinal plain and is characterized by undulating topography in the northern part and more or less flat topography in the south. Most of the area consists of many closed watersheds. The Tharthar wadi is the largest one, and it ends in the Tharthar depression.

d. <u>The Lower Plain</u>: This area consists of the lower geosynclinal plain mainly formed by the thick sediment layer carried by the twin rivers, Tigris and Euphrates. It is known as the famous Mesopotamian Plain. This region could be divided into two main subregions called:

- The terrace and flood plain, which represents the upper part of the plain, and
- The Delta-marsh plain, which represents the lower part.

e. <u>The Desert</u>: This region is located west of the Euphrates River. The whole area is a barren high plateau with altitudes reaching 1,000 meters and it is part of the Arabian Peninsula. Most of the area is covered with limestone and limited stretches of sand dunes making a belt of 10-15 kilometers wide, parallel to the Euphrates River. Desert pavement forms extensive areas of the region which is a result of weathering processes of the limestone. Many depressions are formed in different locations throughout the area.

Climate

The climate of Iraq is affected by the following factors:

a. Geographical position: Iraq lies between the latitudes 30° - 37°N and is located in the southern portion of the temperate zone of the northern hemisphere. This position causes a special effect of incoming solar radiation which results in the weather and climatic conditions of Iraq.

The meridional ciculation shows that Iraq is located in the southern portion of the middle latitude cell which lies to the north of the sub-tropical high pressure belt. The polar front, due to seasonal oscillations, may on rare occasions affect northern Iraq during a severe winter.

The entire country is more or less under the effect of the westerly winds at all levels, except the lower levels which are affected by the rugged topography in the northern parts.

b. Physical features: As was mentioned previously, Iraq's physiographic regions affect the climatological conditions differently. The climate of the mountain regions shows more or less the Mediterranian climate pattern during the winter which is rainy and mild, while the summer season is continental. The rest of the regions are continental during most parts of the year.

c. Natural vegetation: Due to the climatic conditions and misuse of land, the natural vegetation is scarce. However, different types of vegetation have been established.

Forests cover only five percent of the total area, mainly in the mountain regions, while grasses, herbs, legumes, and shrubs cover the rest of the physiographic regions. However the effect of the natural vegetation on the climate is more or less limited. The climate of the four seasons and their main features are as follows:

- Winter (December - February): Usually this season is cold and fifty percent of the annual rain occurs. January being the coldest month, the average minimum and the mean temperature from the north to the south varies from 2.5°C to 7°C and 60°C to 12°C respectively.

- Spring (March - May): The transition from winter to spring is very gradual. Thirty-five percent of the rainfall occurs during this season and decreases rapidly in May.

- Summer (June - September): The change to summer is quite fast. The rapid increase of temperature and decrease of relative humidity with attendant drought are the characteristics of this season.

- Autumn (October - November): Temperature decreases and humidity increases rapidly. Convective showers are characteristics of this season. Only fifteen percent of the annual rainfall occurs during these two months.

Table 1. Physiographic Regions of Iraq

Region	Area Km^2	% Area	Elevation Range M.
Mountainous	29,000	6.6	1,000 - 3,650
Foot Hills	67,000	15.3	300 - 1,000
Upper Plains	37,000	8.4	100 - 300
Lower Plains	134,600	30.6	0 - 100
Deserts	167,000	38.1	90 - 920
½ Neutral Zone	3,500	0.8	
Regional Water	900	0.2	
Total	438,000	100.0	0 - 3,650

Soils

Soils of each physiographic region will be discussed briefly as follows:

The Mountain Region: In the mountain regions where the precipitation is in the order of 700 - 1200 mm per annum with distinct plant cover including pine and oak forests and grasses, mountain chemozim and rendzina soils have developed, while in the lower parts limited areas of chestnut soils have developed. Most of the areas in this region are subjected to severe erosion processes. Soils on the slopes are characterized by a shallow profile, while in the valleys they are deeper due to sedimentation processes.

Foot Hills: Where the annual precipitation varies from 350 to 600 mm per annum in the Northeast and from 200 to 350 mm in the Southwest, the steppe chestnut, desert steppe brown, and reddish brown soils are present. The chestnut soils have heavy textures with low humus content in the upper zone. Generally all the soils of the region are subjected to severe erosion. Gradually the soils are changed to semi-desert steppe brown and reddish brown towards the southwest direction. High calcium carbonate content and low humus content are the characteristics of these soils.

Upper Plain: Upper portions where the rainfall ranges from 300 to 400 mm per annum, semi-desert brown and reddish brown soils are formed, while in the southern part where the rainfall ranges from 100 to 300 mm, shallow semi-desert brown and grey brown soils are more dominant. Most of the soils are calcareous with very low humus content and are subjected to severe wind and water erosion.

Lower Mesopotamian Plain: This region is situated within the geosynclinal depression and filled with deep strata of marine and continental deposits which are covered by the quartenary alluvium of the Tigris and Euphrates Rivers. The soils of the region are characterized by a complex combination of alluvial-meadow soils of different textures, salt content, hydromorphism, and desertness. Generally, the soils could be grouped as: (a) alluvial-meadow soils of levees and flat areas in the river basin, (b) depression alluvial-meadow of different basins and meadow marshy soils of lake depressions, (c) strongly saline-alluvial-meadow soils (solonchaks).

Desert: This region is divided into two main sub-regions, namely the Northern and Southern Desert. The Northern Desert is composed mainly of limestone, while the Southern Desert is composed of limestone and sand deposits. In the valleys, the soils are a combination of desert grey-brown soils. The soil profile is shallow, underlain by limestones. The soils of the central part of the desert are shallow, poorly developed, and stony. The upper part of the desert is characterized by a shallow stony soil profile, while in the extreme west the area is characterized by deep desert grey-brown soils. Sand and gravel-sand deposits are the characteristics of the extreme Southern Desert (Dibdiba Plain).

Water Resources

Iraq depends primarily on the Tigris and Euphrates Rivers and their tributaries for its water resources. Sixty percent of the water yield of these twin rivers originates from watersheds located in Turkey, 20 percent from watersheds in Iran, and 20 percent originates from the territories of Iraq. The following table shows areas of the Tigris and Euphrates watersheds located in the above mentioned countries.

Table 2. Watersheds of the Tigris and Euphrates Rivers

RIVER	COUNTRY	WATERSHED TOTAL AREA (Km^2)	SIGNIFICANT WATERSHED AREA YIELDING WATER (Km^2)
Tigris	Turkey	57614	57614
	Syria	834	834
	Iraq	253000	83237
	Iran	160158	130158
	TOTAL	471606	271843
Euphrates	Turkey	125000	125000
	Syria	76000	76000
	Iraq	177000	------
	Saudi Arabia	66000	------
TOTAL		444000	201000
GRAND TOTAL		915606	472843

The Tigris River originates from the southern slopes of the eastern Tauros Mountain Range in Southeastern Turkey, north of the city of Diyarbakir. Three main tributaries (Batman, Botan, Garzan) join the river within the Turkish boundaries, while four more tributaries (Khabour, Greaterzab, Lesser Zab, Adhaim, Diyala) join the river within the Iraqi territories, before the confluence with the Euphrates, the final tributary (Karkha) which drains from Iran joins the Tigris. The average amount of water flowing into the Tigris and tributaries throughout the year is 48.69 Km^3. This amount is subject to increases and decreases according to annual rainfall. The highest flow reached 106 Km^3 in 1969, and the lowest 19 Km^3 in 1930. The Tigris floods normally during winter and spring seasons when heavy rains and snow melting occur. Major flood control and water storing schemes of the Tigris River and tributaries in Iraq are: Dokam Dam and Lesser Zab, Derbedi-Khan on Diyal, and Samara Dam-Thar Thar Reservoir on the Tigris River. More schemes are planned for the near future, such as Mosul and Fatha Dams on the Tigris and Bekhme Dam on Greater Zab.

Originally, the Euphrates is combined of two branches, Firat Su originates from the upper parts of the Anatolian Plateau near the city of Erzrum, while the other branch, Murat Su, originates from the high lands of eastern Anatolia, near the city of Agri. The two branches join north of Keban Dam and form the main Euphrates River, which flows across the Tauros Mountain chains into the southern plains of Turkey. After leaving the Turkish border, it flows through arid lands of Syria for 700 Kms, then enters the arid lands of Iraq and joins the Tigris River near Kurna City, forming the Shatalarab River which pours into the Arabian Gulf. Ninety percent of the water yield of the Euphrates is drained from watersheds in Turkey. The average annual water yield of the Euphrates is 28.7 Km^3, which is subject to increases and decreases according to annual precipitation and snow melting. The highest amount was 63.4 Km^3 in 1969, and the lowest 12.5 Km^3 in 1930. Generally, the Euphrates floods during the spring months. The main flood control and water storage projects of the Euphrates River are: Keban Dam in Turkey, Tabka Dam in Syria, and Ramadi Dam-Habbaniya Reservoir in Iraq. Plans are under preparation for constructing more dams in Turkey (Karakaya) and Iraq (Hadihta).

The Shatalarab River is a result of the conjunction of the Tigris and the Euphrates. After about 100 Kms distance it pours into the Arabian Gulf. The only tributary joining this combined river is the Karun River, whose watershed is entirely located in Iran.

Ground water resources of Iraq are characterized by the physiographic and geological regions. Waters of the mountains, foot hills, and partly of the upper plain are considered the best in quality and quantity. Chemical characteristics of the ground water varies according to geological formation of the regions, which produce salt contents and other dissolved materials. The salt contents of different regions varies from 200 - 10,000 mg/1. There are more than 2,000 deep wells ranging from 20-600 m. deep and several thousand hand-dug wells 2-15 m. deep throughout the country, mainly in the foothills, plains, and deserts. Ground water is extensively used for municipal and animal drinking purposes.

Several thousand natural springs are distributed in Iraq, mainly in the mountains and foothill regions. These are considered to be the source of discharge of ground water of the rivers in the country during the dry seasons. More than half of the spring water is being used for agriculture in the northern parts of Iraq.

Vegetation Cover

Natural vegetation in Iraq is divided into the following (Physio-climatic) regions:

a. Mountainous Region: This region gets sufficient annual rainfall (600-1200 mm) to support forests, good range plants, and safe dry farming. Dominant forest tree species are: *Quercus aegilopidis*, *Q. infectoria*, *Q. libani*, *Pictacia khinjuk*, in certain localities, *Pinus brutia* and *Juniperus oxycedrus*. In valleys and along streams, *Platanus orientalis*, *Faraxinus rotondi Folia* and *Salix safsaf*. Forest limits now start in the elevation of 600 m. and in upper limits reach 1800 m. above MSL. Above the forest upper limit *Astrapalus* spp. are dominant. All forest areas are very rich in range plants as legumes, grasses, and herbs.

b. Foothills and Northern parts of Upper Plain (Jazira): This area is a narrow strip below the mountainous region. Rainfall (300-600 mm) is enough to support shrubs, rich range vegetation, and safe dry farming crops in the upper parts; natural vegetation considered as moist steppe vegetation due to dense population. Most of the area is under cultivation now and most of the original vegetation is lost. At present, remnants of dominant shrubs such as *Cratagns*, *Azarlus*, *Pistacia khnijuk*, *Pyrus syriaea*, and *Prunus* spp. could be seen. Grasses as *Poa bulbosa*, *Hordium fragile*, legumes such as *Medicago* spp. and herbs as *Galium coronatum* are dominant plants.

c. Lower plains (Mesopotamia) and South of Upper Plain: The rainfall (100-150 mm) is not enough to support a vegetation cover except a few very drought and salt resistant species (halophytes) growing in areas not under cultivation such as: *Suaeda baccata*, *Salsola spp*, *Nitraria retusa*, and *Tamarix* spp, or thorny shrubs as: *Prosopis farcta* and *Alhagi maurorum*. In the lower parts the marsh lands are extensive, hygrophilous plants are dominant such as: *Phragmites communis*, and *Typha angustata*.

d. Desert region: In this region rainfall (90-110 mm) is enough to support only very drought resistant vegetation. A number of habitats differing with respect to exposure, micro-climate water regime, and soil texture exist. In general, *Haloxylon salicornicum*, *Artemisia herba-alba*, *Rhanterium epapposum*, and psamophytic and halophytic communities constitute the major cover types of the desert. A few of the most common annual plants are: *Schinpera arabica*, *Mathiola okyceras*, *Trigonella stellata*, *Medicago minima*, and *Onobrychis squarrosa*.

Natural Resources Balance

a. Soil-Water-Plant Relationship: Evidence shows that soil-water-plant relationships were balanced at the beginning of civilization in Mesopotamia. During those periods vegetation cover was dense, impact of man did not exist and natural erosion was the source of soil formation. With the advance of civilization and the increase of human and animal population, the natural balance was disturbed, erosion became severe and waters turbid. At present the studies in the watersheds of the Tigris and Euphrates show that severe water and wind erosion covers 32 percent, moderate erosion 60 percent, and little or no erosion 8 percent, of the whole area of Iraq. Also, studies show that there is a loss of 10 mm soil per year, which equals 1mm of top soil per year from the steep mountainous areas, while this is about 5 M^3/ha/year in the foothills and plains. Grazing, misuse of land, and cultivation of the steep slopes are among the main reasons for the disturbance of the natural balance.

b. Wildlife and Fishery: Evidence shows that wildlife and fish population was in keeping with vegetation and water resources. Also it is known from history that big game animals, such as lions, bears, gazelles, and wolves, existed in most parts of the country. At present, few gazelles, wolves, jackals, rabbits, and other minor game animals can be seen in most parts of the country. In the mountains, a few wild mountain goats exist in rugged unreached places. A fish population exists in inland waters (streams, rivers) marshes, and regional sea water of the Arabian Gulf. Water areas are estimated to be about 1500 Km^2 (150,000 ha). The production per year is estimated to be 10 kg/ha./year. Fifteen percent of the production comes from regional waters, 25 percent from streams and rivers, and 60 percent from marshes and reservoirs. Most of the fish species are local. New species are being introduced and raised in fishery farms.

MAN'S ACTIVITIES CAUSING ACCELERATED DESERT FORMATION

Socio-economics and Land Use

Sixty percent of the population is considered urban. The largest cities are Baghdad (2.5 million), Mosul, Basra, and Kirkuk (a half million each). They hold more than half of the whole population and more than 80 percent of industry. With the present rate of population growth (2.6 percent), the population will be doubled by the end of the century.

The national economy, depends on industry (47 percent), agriculture (17 percent) and others (36 percent), which include construction, power-gas production, transport, trade and general services. Oil extraction and related industries comprise 2/3 of the gross industrial output. Recently, many types of industries have been developed such as: machinery, food, chemical and pharmaceutical. However, the present agriculture and industry do not meet the requirements of the population. Considerable amounts of consumer goods and food have to be imported. This amounts to about 35.8 percent of the gross national product. Communication systems, which include railroads (2,230 km), asphalt roads (5,400 km) and macadamized roads (5,200 km), do not meet the rapid development situation of the country. According to statistics of 1971, the gross national income of the country reached 1,104 million Iraqi Dinars; this is about 3 times that of 1957.

Regardless of the majority of the population being urban, more than 50 percent are engaged directly or indirectly in land use (agriculture, forestry and range). Land ownership is a result of long history and tradition, and accordingly is classified as: state owned (97 percent), Mosque or church owned (2 percent) and privately owned (1 percent). Agriculture is of the extensive type, depending mainly on cereal production (wheat, barley, rice) and animal husbandry including sheep raising (13.4 million head). Cattle raising (1.9 million head) is of a nomadic type and depends on natural grazing which is affected by the poor condition of the grazing area. Estimation of land use by their gross catagories, areas and percentage of the area of the country is shown in the following table.

Soils and Cultivation

Besides the natural condition and socio-economical situation causing desertification, misuse of soil and land for centuries has caused accelerated desertification. Among these problems, soil erosion in the undulating areas, salinity and water logging in the plain region, lack of soil treatment, incorrect plowing, use of unsuitable machinery, and fertilizer use are among the most important ones. Rotation and cultivation of fodder and forage crops is not correct; accordingly animal husbandry depends on natural vegetation and affects land use dangerously.

Table 3. Land Use and Area Estimates in Iraq

Type - Land use	Area ha x 1000		% of total area
Forests	2 000		4.5
Range	27 000		61.3
Water Surface	2 000		4.5
Rocky & Sand	1 000		2.5
Cultivable	12 000		27.2
dry farming		4 000	
irrigable		8 000	
Cultivated	9 000		
irrigated		6 000	
dry farming		3 000	
TOTAL	44 000		100.0

Water Use

Although irrigated agriculture in the Mesopotamian plain was started long ago, the application of modern irrigation activities started in 1911 where the first barrage was established for flood control and other problems related to the water resources utilization on the Euphrates River. The first irrigation structures on the Tigris River was constructed later in 1937. Many other projects followed since then and until 1950 most of the projects failed to satisfy the needs of the country due to improper management and lack of experience. After 1950 new projects were established according to scientific investigations. Among these were the Ramadi Barrage on the Euphrates River, the Sammara Barrage on the Tigris River with the Tharthar flood control complex, the Dukan Dam on the Lesser Zab and the Darbandi-khan Dam on the Diyala River.

Besides that, investigations for the establishment of drainage systems started for the purpose of reclamation of salt affected soils in the Lower Mesopotamian region, such as: The Greater Mussayeb Project, the Dujaila Project, the Anana and Saqlawiyah Experimental Stations. Some leaching experiments were conducted in these projects. Although these experiments were not perfectly designed, the findings allowed the drawing of a number of conclusions related to the reclamation of saline and saline-alkali soils in Iraq.

In most of these projects, the irrigation and drainage networks were either incomplete, or imperfectly designed. As example, most of the irrigation canals and hydraulic structures were designed to satisfy fallow rotation farming without any consideration for future development. Besides that, irrigation practices were primitive and new techniques in irrigation were never applied. It was clear that irrigation efficiency in most of the irrigation projects was very low and losses from the irrigation canals were high in addition to the over-irrigation phenomena, which was very familiar in most of the irrigated areas. Naturally, because of that, water logging and salinization problems started to build up very soon in these projects. So every year, a new irrigable area came out of production and the total bare lands were continuously increased.

As a result of over-irrigation and seepage from the irrigation canals, water logging problems struck a considerable area within the boundaries of the Lower

Mesopotamian Plain (about 10,000 km^2). In the southern areas (regions of lakes and marshes) ground water almost out-crops on to the surface. The accumulation of excessive amounts of soluble salts in the soil is characteristic of arid and semi-arid regions, although not entirely limited to such areas. Throughout history, irrigation in arid zones has often led to soil salinization, thereby creating a problem which has faced man for centuries.

The Lower Mesopotamian Plain is one of the most famous plains in the world due to its historical background. This plain has been intensively cultivated since 6,000 years ago, and one of the oldest civilizations has been established there. This plain comprises an area of about 25 percent of the total arable land of the country. Due to the flat topography and more or less stagnant ground water, natural drainage conditions do not exist in this plain. The misuse of irrigation leads to a gradual rise of the ground water table (average depth 1.5 m below the surface). With the absence of active artificial drainage systems in combination with an arid climate, a severe salinity struck the area since very long ago. At present, about 70 percent of the total area of the plain can be considered as salt affected.

The accumulated salts in the Mesopotamian Plain soils can be attributed mainly to the following sources: irrigation water, ground water, sea water, and transported sediments. The prevailing salinity type is chloride-sulphate, changing to sulphate-chloride in the south of the plain. It is quite common that one can find a mixture of different salinity types in the same area. The main soluble salts present in Iraqi soils are CaC_{12}, MgC_{12}, NaCl, $MgSO_4$, Na_2SO_4, and to some extent $NaHCO_3$. Calcium carbonate is present in most of the Iraqi soils in a high percentage (average 20 - 30 percent and sometimes reaches more than 80 percent in the north). Morphologically, saline soil in this plain can be distinguished either by its dark color due to the presence of some deliquescent salts such as calcium and magnesium chlorides or by its white salt crust on the surface, mainly sodium chloride and sodium sulphate.

As long as surface water plays an important role in salinization processes, it is necessary to have an idea about the quality of the main water resources in Iraq. The following table shows the average results of many years of analysis for the main river waters, Tigris, Euphrates, Diyala, and Shat Al-Arab.

Table 4. Water Quality of Rivers in Iraq

River	Location	ECX (IO^6) Micromhos	SAR*	Salinity Class**	Sodicity Class**	pH
Tigris	Mosul	404	0.3	C_2	S_1	7.2
	Amara	650	-	C_3	S_1	8.5
Euphrates	Alkaim	575	0.9	C_2	S_1	7.2
	Qurna	960	2.2	C_3	S_1	8.6
Diyala	Baquba	738	1.2	C_2	S_1	7.3
	Baghdad	1050	1.8	C_3	S_1	7.9
Shat Al-Arb	Qurna	900	2.0	C_3	S_1	7.5
	Fao	2800	7.0	C_3	S_2	8.5

*SAR = Sodium Absorption Ratio
**According to the USDA Classification

In many places salinity is associated with sodicity, but it is very rare to find a non-saline alkali soil, although the area has been subjected to leaching processes during irrigation for centuries. Many investigators attribute this to the presence of abundant calcium in the soil [in the form of gypsum (1.5 percent) and calcium carbonate (20-30 percent)] and in the irrigation water. In the northern regions, due to undulating topographic features, high annual precipitation and good natural drainage conditions, salt affected areas are negligible, except in small patches located in low lands.

Natural Vegetation Cover Use

Vegetation cover (forest and range) have been freely used for centuries without any planning or control. Forests of the north and northeast, riverine groves everywhere, and perennial vegetation in all physiographic regions have been subject to overgrazing, cutting, destruction, burning and shifting cultivation. There is evidence that the forest natural lower limits (border) have been shrunk by about 100 km. At present, the forest lower limits are at 600 m above sea level. A few hundred years ago it used to be at 500 m and in some places at 400 m above sea level. In the meantime, all of the forest areas were severely grazed. This has caused the decrease in density and quality. Riverine forests have been cleared for agriculture. Severe grazing has caused the same problems for all areas under range plants. In the foot hills, upper plain (Jazira) and the deserts, legumes and other good range plants have been depleted; instead, unpalatable plants survived. Introduction of machinery a couple of decades ago also has caused the cultivation of most parts of the upper Jazira and foot hills which used to be under grazing. This ultimately caused more pressure on the remaining range land. With the increasing population and animal stock, the pressure on vegetation cover will be increased if measures are not being taken into consideration.

The Natural Resources Balance Disturbance

Besides the critical natural condition of the arid and semi-arid regions, because of the misuse of land for centuries with increasing population and animals, the natural resources balance among the renewable resources has been lost. Accelerated erosion (wind and water erosion), salinization, water logging and the depletion of vegetation cover have resulted. Wildlife has been depleted, in many cases lost, due to severe and improper hunting; fish population due to turbidity and pollution has been decreased. As a result of the overall disturbance of the natural resources balance, the present extensive types of agriculture, nomadic animal husbandry and unstable cultivation have been established.

DESERT DEVELOPMENT MEASURES

Climatic and Topographical Measures

Generally, arid and semi-arid regions are the result of global, topographic and climatic conditions. It seems that the human being, with his limited scientific and technological possibilities, cannot progress toward affecting the mentioned conditions positively. However, by increasing vegetation density, establishing shelter belts, exploiting ground water, bringing water from outside regions, terracing, proper cultivation, using suitable plants and crop species, microclimatic areas could be created. The areas of existing agriculture and ranges could be increased. Finally, aridity boundaries could be fixed. For the time being, till these kinds of activities are established, projects of desert and arid area reclamation are increasing. These projects include settlement, ground water exploitation and management, range management, cultivation, shelter belt establishment, and erosion control.

Socio-economics and Land Use

The problems of socio-economics and land use have to be solved in relation to the natural resources balance. The most important measures are:

- Adjusting land use, management and ownership to its capabilities for production. This is aimed to be accomplished through the Agrarian Reform Law and by regulation.
- Adjusting population increase to the available natural resources without affecting the good balance existing among the different elements.
- Increasing education toward understanding the natural resources balance and proper land use. This is more important in critical areas of severe erosion, salinization and aridity.
- Population concentration should be avoided. This has to correspond to the productivity of land.
- Nomadic life and animal husbandry and unstable severe grazing should be avoided and regulated according to land productivity.
- The use of proper machinery and avoidance of shifting cultivation is necessary.
- Industry and communications should meet the needs of agriculture, animal husbandry and forestry activities.
- Possibilities of tourism, wildlife and game and health resorts should be envisaged.

Soil Treatment and Cultivation Measures

Soil treatment for the improvement of texture, structure, fertility, and good soil and water management is very important. This includes measures for erosion control, prevention of salinization and water logging, increasing natural water percolation and drainage properties and finally, increasing fertility.

Choice of proper cultivation methods requires machinery and tools, proper rotation, and selection of crops. Methods of land preparation are among the important measures. Intensive agricultural systems should be used for maximum production. All cultivation measures must coincide with the topography and climate of the regions.

Water Use, Irrigation and Drainage Measures

Water use in arid and semi-arid zones can be considered in many cases as the most limiting factor. With the present water resources and proper use of these resources, more than 13 million hectares could be put under cultivation. Accordingly, maintaining these resources is quite important. This includes minimizing water losses, proper irrigation activities, and preventing water pollution. Irrigation systems and methods should be revised for better efficiency. In dry farming areas, maximum use of ground water without affecting the water balance of the area should be provided, especially in dry years.

Establishment of an efficient drainage network in central and southern Iraq is the only way to reclaim the salt affected soils and prevent salinization processes. Under the prevailing conditions in the Lower Mesopotamian Plain, it is necessary for the drainage network to be built together with the irrigation network and they should be operated together. Maintenance of these systems should be kept at optimum condition. The central drainage system (under construction now) is the only way to get rid of the saline drainage water. This system follows the general pattern of irrigation systems along the Tigris and Euphrates rivers from north of the plain to the south; it finally ends at the Arabian Gulf. Investigations to find the proper method of reclamation which suits different soils plus the possibilities of re-use of drainage water are necessary.

Vegetation, Forest and Range Measures

Increasing vegetation density and quality should be the final aim of all measures. For protection of the existing forest, planting new areas and shelter belt establishments are among the necessary steps. For range management, protection of the existing grazing areas and improving the existing grazing areas are widely needed. However, severe grazing of forest and range area must be avoided. Since rain and water are the limiting factors for improving vegetation cover, all methods of water conservation should be used. Land capability classification must be the basis for defining forest and range areas. Accordingly, protective productive, or mixed purposes should be aimed at.

Natural Resources Balance Measurement

All steps toward the re-establishment of a good natural balance among the elements of the natural resources are necessary. This mainly depends on soil-water-plant relationships which differ according to localities and physiographic regions. In the mountains, especially with steep slopes, protection of forestry and wildlife should be a dominant practice. Land classification according to capabilities should be the key for use and development of all arid and semi-arid regions.

CONCLUSIONS

Iraq is considered one of the arid and semi-arid countries of the world. Although natural features are responsible for the aridity formation, man's activities causing accelerated desertification or aridity in Iraq are unique and of special importance. These are the main reasons:

- Natural conditions causing climatic, land formation, and soil characteristics according to physiographic regions.

- Disturbance of the natural balance. Elements of natural resources have been lost due to misuse and destructive practices.

- Salinization, water logging in the plains, and erosion in the undulating area are among the most important results and need special care.

- Severe grazing all over Iraq has been seen for centuries, accordingly, good plants have been removed, the quality of grazing areas decreased, and vegetation cover depleted. This has to be improved by good management which must be established.

- Socio-economic conditions and land ownership problems are plentful. Planning and coordination of all practices with natural resources conditions are necessary. These include settlements, population increment rates, standard of living, education etc.

- Finally, up-to-date scientific research and technology must be used for the proper development of deserts and the maintenance of all other lands.

REFERENCES

Al-Ani, T. A. et al. (1970): Range Resources of Iraq. 1. Range cover types of western and southern desert. Tech. Rep. No. 16, Inst. Appl. Res. Nat. Resources, Abu-Ghraib. 8 pp. (mimeo).

Clor, M. A. (1975): De-Desertification and Land Ecology in Iraq. Inst. Appl. Res. Nat. Resources Baghdad, Iraq, 15 pp. (mimeo).

Clor, M. A. and Dougramaji, J. (1977): Country Report, Republic of Iraq. U.N. Conference on Desertification, Nairobi - Kenya, 18 pp.

Gangopadhyaya, M. (1974): Agroclimatological features of Iraq, 88 pp.

General Scheme of Water Resources and Land Development in Iraq, Report of Phase 1, Ministry of Irrigation, Republic of Iraq. 1975.

Gillet, J. B. (1948a): The Steppe Vegetation of Iraq, Min. of Agri., Iraq, 7 pp. (mimeo).

__________ (1948b): The Desert Vegetation of Iraq. Min. of Agri., Iraq 5 pp. (mimeo).

__________ (1948c): The pasture and wild fodder plants of Iraq. Min. of Agri. Iraq, 14 pp. (mimeo).

Guest, Evan (1966): Flora of Iraq. Vol. 1: Introduction, Min. of Agri. Republic of Iraq. 213 pp.

Ijaz Hussian, (1975): Range Management in Northern Iraq. Report prepared for the Government of Iraq. Technical report 13. FO: DP/IRQ.

Kaul, R. N. and D. C. P. Thalen (1971): Range Resources of Iraq. A problem analysis. Tech. Rep. No. 200 Inst. Appl. Res. Nat. Resources, Ahu-Ghraib, 28 pp.

Kettaneh, M. S. (1976): Upper watersheds of Tigris and Euphrates rivers, Tech. Bul. No. 103, Inst. Appl. Res. Nat. Resources, Baghdad, Iraq, 23 pp.

Kettaneh, M. S. et al (1976): Evaporation losses form a free water surface within the watersheds of Tigris and Euphrates rivers, Tech. Bul. No. 103, Inst. Appl. Res. Nat. Resources, Baghdad, Iraq, 23 pp. (mimeo).

Kettaneh, M. S. et al (1974): Water Balance, Higher council of Agriculture. Study No. 1-1, Baghdad, Iraq.

Kimitz, G. and A. Hawa (1971): Introduction and method of planning water resources management in Iraq. Tech. Rep. No. 36 Sci. Res. Foundation, Baghdad, Iraq.

Mineral Resources Organization. Republic of Iraq. 1975.

Ministry of Irrigation, D. G. of Reservoirs General scheme of water resources and land development in Iraq. Report on Stage I.

Rainfed and Irrigated Agricultural Projects on Desert Improvement (1974): Study No. 3-6 Min. of Agri. and Agrarian Reform.

Range Management Report (1976): Republic of Iraq.

Seminar on Salinity Problems (1971): Baghdad, Iraq, FAO Rome Paper No. 7.

Al Taie, F. H. (1968): The soils of Iraq. Ph.D. Thesis, State University of Ghent.

TOURISM DEVELOPMENT IN THE ARID AREAS: THE TUNISIAN EXPERIENCE

Azouz Ennifar

ABSTRACT

In studying the feasibility of the touristic development of an arid region such as the South of Tunisia there are many criteria which must be considered in order to determine the profitability and success of any project.

The Tunisian South already offers an interesting product which could appeal to a segment of international travellers especially those travellers seeking an exotic, out-of-the-way destination. This product has to be packaged in such a way as to offer as attractive a destination as possible in the conditions of security and comfort.

There are a certain number of engineering conditions to be resolved: ie access via road and airports, utilities, sanitation, medical facilities and telecommunications must all be provided. One of the major problems remains the question of providing the hotel facilities with water.

It is necessary when implementing these modernizations of the area to also preserve the cultural, social and physical environment. This end can be attained by adopting the architectural styles of the areas, by avoiding pollution, and by establishing a ceiling on the projected hotel capacity.

Finally the development of touristic activity must guarantee a dual purpose:

- the profitability of the envisoned investment,
- offer to the concerned arid area a new economic opportunity.

In the specific case of Tunisia, the development of tourist activity in the arid zones has a double purpose within the framework of the Tunisian national economy:

- to pursue by the opening up of a new region the growth of the tourist industry which constitutes one of the fundamental sectors of the economy,
- to contribute to the development of a region which has few possibilities for the implantation of agricultural or industrial activities.

It is mostly in the double economic and social context that the impact of tourism in the Tunisia South shows its importance. However, on the global scale of the total volume of tourism in Tunisia the Sahara tourist activity will remain marginal but will be economically profitable.

INTRODUCTION

In the market competition between Mediterranean beach resorts, Tunisia offers a particularly attractive product: more chance of sunshine, and a physical and human environment which borders on the exotic. For Tunisia is the Maghreb . . . AFRICA, and therefore a little of the Orient.

As a country of the Maghreb, Tunisia benefits from a privileged location. Fixed essentially on the Eastern coast of North Africa, face to the rising sun, on the eastern side of the Mediterranean basin, extending to the Gulf of the Great Syrthe.

This favorable location, the realization in ten years of a hotel infrastructure of 65,000 beds, and the increase in European tourism have resulted over the last ten years in an ever increasing number of visitors attracted to Tunisia, which in 1975 hit the one million mark.

Certain factors have prompted the Tunisian Government to examine the optimal conditions for future tourist activity. Certain economic considerations are now pushing us to adjust our promotional efforts, efforts which should aim toward two primary objectives: to extend the comforts offered in existing tourist establishments and to improve upon the basic product offered to date, the seaside resort, by bringing in an element of value which would permit Tunisia to make itself stand out vis à vis, the competition.

When you consider the range of possible touristic resources in Tunisia in addition to the "seaside Tunisia", one feature stands out as having sufficient weight to drum up clientele and to present itself as a self-contained product: Southern Tunisia and the Sahara regions.

DEFINITION OF THE PRODUCT

The South and the Tunisian Sahara represent nearly one half of the area of the entire country. In fact, of the 63,000 square miles of Tunisia, almost two-thirds consists of arid zones where the average rainfall is less than 8 inches per annum.

The area opens out over 225 miles from east to west, from the Lybian border to the border of Algeria, and over 300 miles north to south from the latitude of Gafsa to that of Bordj el Kadra.

If you consider human settlements and simultaneously geophysical conditions, two separate and distinct entities are brought to light; the Sahara and the South.

In schematizing, we could imagine a line passing through Nefta-Douz-Tataouine which would establish the link between these two different areas of different content.

The Sahara

The Sahara represents a clear and simple vision, strong in all minds. It belongs to mythology, it conjures up the troubling image of a world from which man is excluded. It is also the image of the absolute: oppressive heat, total dryness, endless reaches; an immense and merciless man-eater.

Only the nomad, who does nothing but cross it and who has paid dearly to figure it out and the military man, whose cry is risky adventure, have been able to acquire a dangerous familiarity with it.

Even if the oil man has been able to partially demysticize the symbol, we know that he has only been able to domesticate it through exceptional technical means.

The Sahara also arouses a highly emotional image. This image is particularly troubling and it is perhaps essential at least in terms of the tourist to preserve the dimension of the exceptional, of the adventure out-of-the-ordinary, and beyond man.

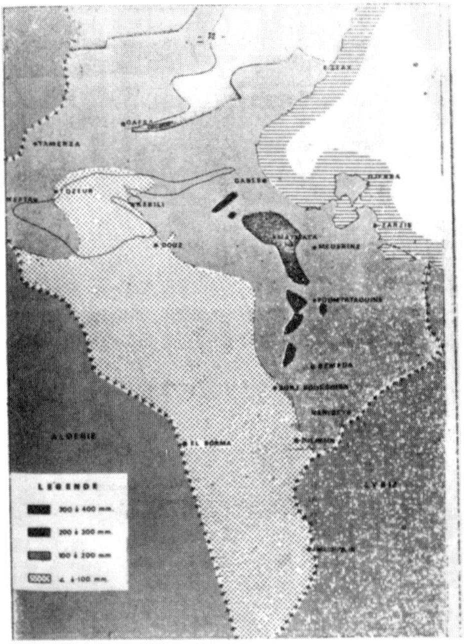

The South

The Tunisian South involves another neighboring mythology but with one fundamental difference. The South is the world of men. It is above all victory of man, whose tenacity has been able to control a ferocious world without complacency.

There is one entire segment of the Tunisian populace which clings to and fights against the desert. It is a whole other adventure, human and tenacious. More than 700,000 people live there is small communities, dispersed throughout the area which bear witness today to the art of adaptation of those who, without any elaborate means, have succeeded in creating a physical environment, however, ruthless, and exploiting it to the fullest to extract whatever can be gotten from it. Here are to be found the two most important oases of Tunisia: Tozeur and Nefta. They go far back into history, geographically they were at the crossroads of passing camel caravans coming back up from the Soudan and were part of the Roman Empire. Here also is the "Nefzaoua" which is the region at the confines of the steppes. These steppes, corroded by crusts of salt of the Chott el Djerid and smothered under sand dunes, represent confines also of human settlement, last civilized posts on the edge of the desert, oases bearing witness to the miracle of water.

DEVELOPMENT OF A TOURISTIC POLICY

Tourism in the Sahara is by its very nature a tourism of movement. Its purpose can be expressed as follows:

- arouse curiosity,
- satisfy on an emotional level,
- offer the best possible conditions for rest.

Taking into account those remarks we have thought it important to organize the tours in such a way as to alternate adventure and rest.

A thorough study of the construction cost of lodging as well as the establishment of operating conditions so that this highly restricted sort of tourism can function constitutes the foundation of the development of tourism in these regions. We have established that the experiences of improvisational management of those "way stations", previously set up in Tunisia, clearly demonstrates the impasses to which amateurism will lead in this domain.

On the other hand, we have tried to outline the conditions for commercialization capable of insuring those lodgings an occupation rate compatible with the saleability of the operations. We have moreover examined the impact which can be expected on the economy of the zones concerned.

Such are, in our opinion, the various aggregates which must be taken into consideration in evaluating the possibilities of the development of profitable tourist activity in an arid zone. We will briefly study these various elements which have been taken into consideration at the time of preparing our touristic development policy.

The Physical Characteristics of the Region

Regions such as the Jerid, the Nefzaoua, the Matmata Massif and the extreme South constitute highly differentiated zones. The climate is continental and arid. The implications of the climate - which constitute the most important physical characteristics from the tourist point of view - will be dealt with in the heading concerning hotel construction.

Precipitation

Precipitation is relatively light and varies from 0 to 12 inches a year, going from south to north. There are about 30 days of rain a year. Moreover, the rain falls are irregular: some years can be very dry with less than 1.5 inches of precipitation, whereas others can be quite wet and the rain will exceed 12 inches.

Temperature

These regions experience extreme high and low temperatures in the course of a day.

Vegetation

Vegetation is varied - certain zones being complete desert, others consist of alternating steppes and oases.

The Infrastructure of Transportation

The existing network of roads. Roads in the South are satisfactory. This network is composed of:

- main roads (600 miles),
- secondary roads (350 miles),
- classified tracks (3,200 miles).

In general it has been established that the construction of new roads poses no great problems. The good soil found there - sand, limestone or chalk - allows for the reduction of layers and therefore, reduces the cost of construction.

Airfields. Tunisia has three international airports: Tunis, Monastir and Djerba. The construction of a fourth international airport has been undertaken at Tozeur which will constitute a direct port of entry. Arriving in Djerba, also permits the operation of Sahara tourism but, as a satellite of beach tourism. Another airport is envisioned for Bordj el Khadra, in the framework of the second stage of development of tourism in the Sahara.

The Infrastructure of Utilities

The supply of drinkable water. There is an essential problem. At the present time the south of Tunisia is supplied with water essentially by its local resources: spring, surface wells and artesian wells.

The development of tourist activities gould pose a problem in certain regions where the water tables are showing signs of running out. We have decided accordingly to supply water to some of those regions directly from the cities of the North and are making sure that hotel units, which will be implanted outside urban areas, will resolve individually their water problems (drilling, etc.). More often, however, we have planned hotel structures based upon the availability of water.

The treatment of water. It is important to point out the fact that water in these areas is generally hard. The amount of salt varies from 1 to 6 grams per liter. The hydrometry title is also very dense: 80 for the drillings in Tozeur. Therefore and in order to maintain the hotel equipment in good running condition, it is always recommended that water softeners be installed.

Water consumption. While the average water consumption per day and per bed in beach resorts is 150 to 175 gallons it is necessary to foresee more modest water consumption in the Sahara hotels. For our part we have adopted the following bases:

- 65 gallons per day per bed for the Sahara and stop-over hotel categories,
- 130 gallons for tourist class hotels.

This brings us to mention another problem of priority: sanitation.

Sanitation, It is very important to avoid disposing of used water in cesspools or in the "oueds" (rivers) without prior processing. Not only would such a situation create a great discomfort (mosquitoes, unpleasant odors, etc.), but it could also generate risks of pollution and epidemics which would be difficult to control.

It is therefore important that the cities, where hotels will be located, have adequate sewer systems. Hotels outside of cities would have to have individual sewage units (compact units using biological oxydation or septic tanks).

Electricity supply. Apart from the main centers, the "Great South" of Tunisia is without electricity. The majority of hotel units that we plan to build must have electricity, in order to offer minimum comfort. Hotels in remote areas would have to have individual generating sets. We estimate that the average consumption in these hotels will be:

- 800 watts per room for the Saharian category,
- 2,000 watts per room for the way stations,
- 2,500 watts per room for the tourist class.

The last two categories of hotels will provide air conditioning.

Telecommunications. The totality of the hotels, in order to be operative, will have to provide national and international telephone service. In most cases, in Tunisia, this condition is already fulfilled or soon will be.

Medical equipment. It is important to provide all regions with adequate medical facilities, in order to insure the foreign visitor against any possible medical difficulties.

In the light of this study of the natural environment and of the various infrastructures, - existing or planned, - which must necessarily precede the development of any tourist activity, we have defined the kind of hotels to be built.

HOTEL PLANNING

The Tunisian Sahara is unpretentious but it could offer a very wide number of variations for tourist tours. We have made it a priority to offer a quality product, especially as far as the hotel infrastructure is concerned.

While planning hotel development we have taken into account three different limitations:

- types of architecture: concept of "hotel/product,"
- criteria of comfort: water and air conditioning,
- conception and use: type of program.

Types of Architecture (the concept "hotel/product")

As opposed to the beach resort stay, the Saharian tour must not be banal. We have already defined its purpose which is the discovery and integration into an "other" world. It is composed of itinerant movement on the terrain and of the relaxation and rest at the hotel.

As a result, the hotel itself will be a component of the touristic product. It is a stopover during the expedition. It must also be an integral part of the nature of the Sahara tour. Therefore, it is of primary importance to carefully plan the architecture of the hotels in accordance with the environment in order to avoid spoiling the aesthetics of the antique dwellings. On the other hand, new style construction will necessarily be taken as an example by the local population. It is not only an aesthetic problem but a cultural problem as well. In Tunisia we are doing our best to adopt the traditional styles for the Sahara hotels. We have even restored a certain number of traditional dwellings and transformed them into lodging facilities.

If the social effects of tourism are not easily mastered, the architectural aspects can be.

Criteria of Comforts (water and air conditioning)

Climate conditions constrain the types of architecture. We have found that traditional architechture adopted several centuries ago is particularly well adapted to the climate of their respective regions. However, it is also a necessity to quarantee foreign travellers the best possible conditions for their stay.

Modern systems of air conditioning must be reconsidered for, very often, maintenance and electric energy supply are uncertain. The system of air conditioning required will depend upon the category of the hotel.

In Tunisia, we have adopted three main types of lodging facilities in the south.

Sahara Relay Stations (relais sahariens)

They consist of small, isolated units whose aim is to provide lodging and food for itinerant groups whose stay will not exceed 24 hours. Their average capacity will be 30 or 60 rooms. Construction cost cutbacks could be achieved by the use whenever possible, of local materials and the usual techniques employed by local artisans. Participation of specialized contractors will be necessary for installation of electricity, plumbing, and other modern equipment. Air conditioning could however, be provided in the local traditional way, such as cooling by evaporation, etc.

Stop-over Hotels (hôtels d'étapes)

They offer the same kind of services as the Saharian relay stations but with better comforts. Their average capacity will be of 60 to 120 rooms and they will be located in small urban areas. The construction techniques used must also take into consideration local qualifications and materials.

Tourist Hotels (hôtels de grand tourisme)

These hotels will also receive long staying clients and must therefore adopt conventional facilities. However, they must also preserve the image of prestige and respect traditional architecture.

We could also mention another type of lodging: desert camping.

Traditional nomad camping would offer a romantic night in the desert, an exceptional night in the great Saharian mythology. It should be one of the strong points of the tour. It should be a real cultural change however, it must be offered in its traditional integrity and in the best possible hygienic conditions. With this kind of lodging, the possibility of a sand storm or a sleepless night is without importance since it is an exceptional situation; it is an integral part of the adventure.

Apart from the lodging facilities the development of tourism in this kind of area requires special transportation: buses, Land Rovers, camels.

MARKETING APPROACH

At present Sahara tourism in Tunisia is essentially a tourism of excursions with clientele recruited during their stay at the seaside hotels. In the future, such activities could be extremely successful provided we could guarantee the conditions of comfort and hygiene -clientele of Saharian tours will be different. Climate conditions will also be important. We expect an important increase of tourist arrivals in the winter and Easter seasons. At present, most of our traffic arrives during the summer season. This could help balance the occupancy rate in the hotels throughout the year. Contrary to the sea, the Sahara must be commercialized with precaution and excellent conditions of stay. It is especially important to insure a highly valued image of prestige in order to insure the profitability of the investments.

We have decided to develop this tourism in two main phases.

The first one is being implemented now. It consists of renovating and improving the lodging facilities and rectifying and enriching the product. Once completed, it could offer, inside our borders, a wide range of tourist products: (oases tours, discovery of the South, fly and drive, the Jerid and the Sahara, and Djerba and the Sahara).

The second is a long term project. This will be the Great International Sahara Tourism of which Tunisia - with its airport at Tozeur - would constitute a point of departure. These will be the Trans-Saharian Expeditions from the north to the south with extensions to the Tassili, the Hoggar, the Niger, the oases of Fezzan and even to the Tibesti, an even more extraordinary world.

The achievement of such a tourist policy - simultaneously qualitative and progressive - would result in giving to Tunisia an ideal operational position: a first step towards coordinating the developments of Sahara tourist resources of the whole northern Sahara Africa.

ECONOMIC IMPLICATIONS

In any study of tourist development the economic analysis is designed to not only verify the viability of the project but - and this could be its main purpose - to define criteria justifying the nature and the extent of the action to be undertaken.

The development of tourist activity in the Tunisian arid zones has a dual purpose in the framework of the Tunisian national economy:

- to pursue, by the opening of a new region, the growth of one of the fundamental sectors of the national economy,
- to contribute to the development of a region which has few possibilities for the implantation of agricultural or industrial activities.

Let us briefly develop the second point. Because of the means required, the initiation of Sahara tourism can only make a modest contribution to the strategy of developing the southern part of the country. It is extremely difficult to verify the hypotheses regarding job creation and contribution to the local economic activities. The best method would be to evaluate monetary flows created by tourism. The aim would be to identify the additional revenues brought about by tourists and to determine who is benefiting from them; in other words to define the support that tourism could offer to the regional social and economic life. This is based on the concept of added value which is calculated from the difference between the turnover (foreign currency income) and the foreign currency costs of this activity (the required imports).

The conclusion we could draw would consist of recommendations. Tourism activity could, under certain conditions, make a substantial contribution to the economic and social enrichment of arid countries of the Third World. However, it is extremely important that the forseeable impact of tourism be taken into consideration when the touristic development plans are formulated. This activity should in no way be allowed to develop haphazardly but should be carefully planned as a part of the country's development effort. This, in order to avoid spoiling the natural social and cultural environment. It is also essential to establish a ceiling on the planned hotel capacity in order to insure that this activity be profitable and to maintain the preponderance and the characteristics of the milieu.

On the global scale, and compared to the total wolume of tourism in Tunisia, the Sahara tourism - which we have voluntarily decided to limit to 10% of our total trafic - will remain marginal but will be economically profitable. On the other hand, it will have a great promotional impact on the touristic sector and will, to a certain degree benefit the economy of an underprivileged region where it would create a new activity.

It is therefore in a context linking the economic to the social that the impact of tourism in the South will show its importance:

- by participating in the development of the economy of a region; in other words by permitting a larger number of people to live off an activity exercised in their own region; thus avoiding the expatriation of some to other parts of the country or abroad with the social consequences involved (splitting up of the family unit, etc.),
- by facilitating the creation and development of an infrastructure of services (airports, roads, water supply, sanitation, electricity, telecommunications, medical facilities, schools, security) that tourism requires,
- finally, by maintaining an activity at the edge of the desert to contribute to the fight against desiccation.

PERSPECTIVES ON THE DEVELOPMENT OF THE ARID AND DESERT ZONES OF MEXICO

Rodolfo Echeverria Zunó
Executive Director, Balsas River Bassin Commission, Mexico

The utilization of the arid zones of Mexico constitutes an important alternative, in the framework of the national objectives to increase production and create employment.

The development of the Mexican arid zones can contribute to the satisfaction of the country's needs for food, to an increase in employment and, above all, to a reduction in the dependence on centrally planned or developed market economy countries.

This dependence, generated by the historical process of a polarized world economic system, represents in fact the most serious danger for Mexico, unless we succeed in our efforts to develop areas of low rainfall that comprise about 50 percent of our territory.

The world economic space is divided into central and peripheral regions, linked by a market system that imposes on the latter, the conditions and supposed economic reality of the former.

This situation has resulted in an unequal development among the countries which participate in the international economy. On one hand, we find the peripheral nations characterized either by backward technology, and low salaries or while possessing advanced technology are unable to absorb, within the modern sector of the economy, the large majority of the active population. On the other hand, the central economic powers are characterized by a great accumulation of capital and the capability of making full use of their manpower in combination with the most advanced modern technology.

The rational utilization of the arid zones cannot be considered either as an isolated aspect of the world economy or defined by changes in the magnitude of Mexican dependence in respect of the dominant economies. The possible links between the centers of power and the models of development to be applied to our arid lands would be identified through the specific technologies adopted, the orientation of the research relative to the productive activities to be undertaken and the effect upon the national economy as a whole of all these activities.

When modern exploitation technologies are imported, the less complex economy of an undeveloped country increases its dependence respect of industrialized nations. In addition, the goods needed to apply the technology are very often not produced in the country but imported from abroad. Finally, the importation of advanced technologies reduces the stimulus for developing appropriate national technologies. This explains why Mexico has used traditional technologies for the optimum utilization of soils and water, which do not demand high investments or highly qualified manpower.

The Mexican experience shows that those native or acclimatized plants that can prosper in arid zones should be cultivated not for exporting but to satisfy domestic demand. The same experience has proved that certain food plants can produce high yields in zones of low rainfall. Genetic research has developed wheat and corn varieties resistant to drought. The utilization of the arid zones of the developing countries, particularly in Mexico, can be frustrated if we try to apply costly imported technologies based on capital goods not produced domestically, thus increasing the economic dependence on foreign nations.

The high costs involved in the transformation of the arid zones from marginal to developed areas demand that this must be done taking into account the national interest.

Any inadequate planning of arid zones' development divorced from the world reality of an economic space divided between islands of power and archipelagos of poverty, can aggravate the disparity already existing among the various economically productive sectors; lead to the disarticulation of the nation's economic system and augment the degree of dependence of the underdeveloped countries.

The problems of developing the arid zones constitutes one more link in a chain of elements and factors that conform to our time's realities, that is, the movement towards a new international economic order more just and humane.

From the above we can conclude that the development of our arid zones must be undertaken according to plans different from those applied to the developed countries.

In the advanced economies, desert exploitation consists essentially in the application of their extensive infrastructure for research and production to determine the appropriate technology for arid regions. The problem is thus reduced to the spatial extension of the productive apparatus when the economic or political value of the desert is high enough to justify the effort and the money involved.

In the Third World, the problem has other features and should be solved differently. The Mexican arid zones, due to their large extent, would continue to support themselves, developing their own resources and benefitting as far as possible from their scarce and irregular rainfall, or from the limited sources of surface and underground waters. At least for the short-run, is not economically feasible or socially necessary to bring water from far away places.

AN APPROACH FROM SAHEL

Michel Keita

Sahel, originally an Arab word, is the term consecrated since ancient times to designate and characterize the group of lands adjacent to the southern part of the Sahara. As one descends southward, these arid and semi-arid lands form a belt around the Sahara.

Due to this particular location the Sahel is at first sight considered a geographical entity and conventional definitions of the Sahel refer to rainfall, vegetation, or soil morphology. It is in this perspective that the Sahel is presented as a geographical interval between the desert, namely the Sahara, in its strict sense and the Sudanese zone, between the isohyets of less than 100 millimeters and those of more than 90 millimeters. In spite of this reference to phenomena as objectively identifiable as the quantity of rain received, the type of vegetation and the nature of the soil - all efforts for a precise geographical delimitation of the Sahel have confronted the following obstacles:

First, due to the transitory character of this "strip of land" which bridges the arid and desert regions of the north with the sub-humid regions of the south;

Secondly, due to the variable character of its expanse and movement created by the progressive passage from aridity to humidity. Also, numerous authors resign themselves to purporting the Sahel as a strip of land of varying width, within which the rainfall is very irregular, characterized by the contrast between humid and dry periods, rainfall being, on the average, less than 600 millimeters in the summer and 400 millimeters in the winter.

Fortunately, however, the conventional geographical approach isn't the only one possible and surpassing it allows for a clearer, more operational description of the Sahel as an ecological, human and economical milieu.

A Unesco document summarizes thus the general consensus expressed on this question of the delimitation of the Sahel; it indicates:

First, the boundary between arid and semi-arid can be roughly determined between the places where rain-fed agriculture is possible even with great variations from one year to the next and those where it is impossible to have a harvest most years;

Secondly, climatic "norms" do not exist in reality and it is more exact to evoke the probability of the existence of specific phenomena, such as the three seasons of consecutive rainfall less than the standard;

Thirdly, within the framework of the examination of constraints to development, it is necessary to recognize a very large diversity of physical, biological and cultural elements in these arid and semi-arid regions. In addition to the climate, the principal factors that come into play include the topography, the social organization and the diversity and availability of mineral resources.

These are indications which permit one to envision the Sahelian strip as an area of possible livelihood because it provides, indeed, in a precarious equilibrium, minimal conditions which allow for a certain kind of agriculture and a certain type of stockbreeding. These indications are operational.

These two operations, as a matter of fact, are the principal food and revenue source of a large proportion of the population in the Sahelian area. Agriculture there is possible a small portion of the year - the hot and rainy season - (May to August) while stock-breeding requires frequent mobility over large distances, due to the alternate dry and rainy seasons. That is to say that the conditions of existence in the Sahelian zone are harsh and especially precarious even during normal weather conditions. To these difficult conditions in a milieu already unfavorable are added the recent unbalancing effects of colonial exploitation - a common fate of all the Sahelian countries.

The family of Sahelian countries encompasses all those States having at least a portion of their territory included within the variable boundaries of climatic conditions prevalent in the Sahel. Countries in which at least a portion of their territory is adjacent to the Sahara, like Mauritania, Mali, Niger, Chad, Sudan, Ehtiopia and Somalia in this respect are considered as being a part of the Sahelian zone. Also included are the countries in which the Sahelian zone extends over vest stretches and submits to the grave effects of this climatic condition: Senegal, Gambia, Upper-Volta and Cape Verde. This family is important and deserves to be well understood because it is on this new basis of modern states that the Sahelian countries conceive and organize the implementation of their economic take-off.

A sub-distinction with no purpose other than localization, could of course be established between the Sahelian countries of East Africa (Sudan, Ethiopia, Somalia) and those of West Africa (all the other countries cited above).

Thus, what identifies, characterizes and unites the Sahelian countries are:

- Their geo-climatic conditions, aforementioned;
- Their basic economic activities (self-sustaining agriculture and stockbreeding);
- Their lifestyles, dominated principally by the complementarity between sedentarity and mobility;
- Their history, marked by the co-existence of sedentary agricultural and shepherd-normadic peoples, then semi-nomadic and semi-sedentary; the precedence of emerging political systems organized, hierarchy-structured and centralized among the agriculturalists, and the proximity between the time of automation and the domination of shepherds over the agriculturalists, marked by recent colonial government and exploitation;

> Their present situation, burdened by the characteristics of under-development in the modern world (domination of capital, export economy deterioration of the terms of trade).

It is in this overall context that the last and, without doubt, one of the gravest droughts recorded in the Sahel occurred. From 1968-1973 it decimated the population and animal herds, annihilated cultures, pushed to the edge of disaster an area where disease and malnutrition were endemic, the average life expectancy being estimated at 37 years.

Thus, drought and development can be closely linked, the effects of the first and its consequences upon the latter.

It is necessary to note that the Sahel, a marginal area ignored by the media, has become particularly well-known since its brutal experience and the dreadful effects of this drought whose consequences continue to be felt.

The consequences of a disaster of this type are violent and instantaneous, they were recently manifest in the Sahel in terms of the loss of human life, the extermination of animal herds and the destruction of harvests.

Less immediate than the effects, the consequences are more insidious and enduring; because of their qualitative nature they dramatically affect even the manner of organizing human entities (states, nations) and sub-entities (regions, human settlements) stricken by the drought and they operate through demographic deterioration as well as that of production. They overthrow value systems; they compel people to make difficult choices. In the Sahel, the struggle against the effects of the recent drought has taken the aspect of an emergency survival operation supported by the international community which appeared more and more as an effort to organize and sustain actions of "rehabilitation, reparation, and development".

In other words, the experience has shown to the Sahelian countries that their sensitivity to disasters of natural origin was the corollary of their state of underdevelopment and in order to remedy the effects of the recent drought, cure its consequences and prevent all risk of repetition of a similar phenomenon, it was necessary that they engage in constructive techniques similar to those of developed nations - competitive economics and systems of production equipped with appropriate techniques.

Thus, the Sahelian approach to the development of arid (deserts) and semi-arid (Sahel) zones appears as a comprehensive approach which takes into consideration the lessons of history, science and technology without overlooking the fact that in every sincere work of development priority must be maintained for the improvement of working conditions and the existence of populations.

In this perspective and in order to reach this overall objective, a certain number of Sahelian countries regrouped into a Inter-Nations Standing Committee for Drought Control in the Sahel (CILSS) - their headquarters is at Ouagadougau in Upper Volta.

These eight countries, going from east to west, are: Chad, Niger, Mali, Upper Volta, Mauritania, Senegal, Gambia and Cape Verde; stretching over nearly 5000 kilometers south of the Sahara. The member countries of the CILSS comprise over 2 million inhabitants.

Administered at the highest level by a <u>Conference of Heads of State</u> "defining the policy of cooperation and development in the Sahel", the Committee is endowed with a Minister's Council, having the responsibility of promoting all actions leading to the realization of its objectives and is assisted by an Executive Secretariat which is its instrument of implementation.

The goals, as established by the convention of Ouagadaougou the 12th of September, 1973 creating a Inter-Nations Standing Committee on Drought Control in the Sahel, are as follows (paragraph II, Article 3):

1. to coordinate all the actions conducted against drought and its consequences at the regional level;

2. to sensitize the International Community to drought problems;

3. to mobilize the necessary resources for the realization of a program defined by the member states in the framework of drought control and for the development of the Sahel;

4. to mobilize the necessary resources from the financing of operations in the framework of sub-regional cooperation;

5. to implement actions of sub-regional interest tending to reinforce the cooperation between member states in their common struggle against drought;

6. to assist the member states and existing organizations in seeking funds for their programs.

The convention that created the CILSS stipulates also (paragraph V, Article 18) "that any African country could be a perspective member of the Inter-Nation Standing Committee of drought control in the Sahel

a. of which the agricultural and pastoral economy is dominated by ecological conditions of Sahelian and Sudano-Sahelian zones

b. which was a victim of drought and recognized as such".

The objectives include:

- to prepare together with national agencies, projects and programs of drought control and development of the Sahel;

- to prepare, in collaboration with national agencies, the pre-investment and investment dossiers with the goal of rehabilitation of drought stricken zones;

- to formulate any propositions judged useful in the framework of Sahelian land-use policy capable of restoring the ecological balance between natural resources and human and animal populations;

- to establish (at medium-range) a coordinated regional plan of mobilization and actions in the event of future droughts;

- to study (at medium-range) the consequences of the drought on animals and animal foodstuff resources in order to establish a regional program for the regeneration of animal herds and grazing lands;

- to define (at middle-range) a policy assuring the permanent supply of water for population and livestock;

- to construct (at long-range) a comprehensive strategy of development which avoids the return of such a calamity. This consists chiefly of:

  - conducting extensive climatic studies including the use of satellites which can help forsee and determine the effect of climatic factors on agricultural and pastural production, and to perfect an improved climate warning system;

  - to implement a program of study and use of grazing lands and water resources in the Sahelian zone;

  - to increase agricultural production through the arrangement of appropriate hydro-agriculture techniques;

  - to control "desertification" (reforestation, protection of the environment, education).

The preceeding clearly shows the will to make the CILSS a concept as well as a working instrument whose program ranges from immediate action to long-range preparation.

The interest triggered by the CILSS and the effect of its message are already perceptible; a clearer tendency to reserve the term, "Sahel", to the group of states which are part of the CILSS and the founding and recent actions of the Sahel Club. This interest and message result, in my opinion, from the fact that CILSS is an original form of response to a given situation. This originality rests on a correct estimation of the situation and of its implications; the struggle against drought calls for a development struggle and the latter demands the building of adequate supports for:

- reinforced regional cooperation

- coordinated and sustained international cooperation.

This approach of CILSS' member countries is a model proposed for underdeveloped countries. It is based on the awareness that a common situation is an element of union and unity is a resource saving factor, especially when these resources are limited. The approach strongly suggests counting on one's own potential and encourages all forms of assistance based on equality and mutual respect.

In terms of development strategy, the Sahelian approach implies from the very beginning the development of dry farming and the reconstitution of livestock planning for the immediate future, the development of irrigated agriculture by planning and development of big river basins, the development of fishing. For longer range, research, study of appropriate technologies, the conversion of people's mental attitudes and change in technology is planned. All this without neglecting to protect the natural environment (reforestation, research for new sources of energy) and to encourage the uplifting of human societies by educational, sanitary and participative action.

These guidelines were taken up again by the Sahel Club; the latter appearing at first as a response of the international community to the creation of CILSS and then as an enlarged framework for organizing the work and coordinating cooperation. A fact sheet on the Sahel Club indicates:

> Following the drought, the Sahelian countries, regrouped in the Inter-Nations Standing Committee for Drought Control in the Sahel (CILSS) studied the methods of assuring relief to the region and its prospects of development. Certain members of the Aid to Development Committee (CAD), as well as multilateral organizations, proposed the creation of the Sahel Club, according to a flexible and informal formula of association to examine methods of implementation and encouragement of long-term development of the region.

To reach its objectives the Club, which combines in the same effort the member countries of CILSS and CAD, created a task force whose first session was held in Ouagadougou (Upper Volta) in June 1976. On this occasion the task force defined its method of functioning which is described in detail in the document of the "Propositions for a Strategy and Program for Drought Control and Development of the Sahel". This document, which you can consult at the Conference Secretariat, is organized into three parts, of which the first is the presentation of large strategic options and programs, the second is the exposition of sectorial programs and the third the illustration of coherences and priorities as well as conclusions.

However, the document doesn't pretend to have exhausted its objectives, and its adoption by successive ministerial meetings of CILSS and by the Club is an encouragement to the Club to complete its work. However, there is already a program for a group of projects, a so-called first generation, for which the total cost of implementation reaches approximately 3 million dollars. To this must be added the cost of developing the big river and lacustine basins.

The documents relative to the program of the so-called "first generation", as well as those concerning specific programs such as the Institute of Sahel and L'Agrymeth, are available at the Conference Secretariat and at the CILSS (P.O. Box 7049 - Ouagadougou - Upper Volta) and OCDE (CAD). All these documents supplement and illustrate this brief introduction to the program adopted by the Sahelian countries for rehabilitation, reparation and development.

I will terminate by addressing those who not directly knowing the Sahel would have desired a more evoking description of human settlements, their situation and their problems. I hope that the following pages will give them consolation. They are extracted from the proceedings of the seminar, "Women and Development in African Environments" (Niamey, November 8-12, 1976). The seminar was jointly organized by the Government of the Republique of Niger and the Training Program for the Environment of Dakar (ENDA). These proceedings were the result of a field investigation conducted by participants of 12 different nationalities to the seminar (29 Nigerians, 24 non-Nigerians) and they shed light particularly on the following observation of the "Propositions for a Strategy and a Program" (3rd part, conclusions, page 130, 3rd paragraph):

The approach should not make us forget that the work of establishing a program is made for the people of the Sahel and above all the rural populations which represent the vast majority and most deprived of the population. It is a fact that these peasants, shepherds, or craftsmen of the Sahel are the most interested in development. They know better than anyone their real needs and after all, they will be the designers of their own future. The majority of sectorial teams, after having formulated their strategic goals, have emphasized how much it will be necessary in the implementation of this strategy and in the designing of projects to obtain active participation of the rural masses. It is equally necessary to have them express their problems, their needs, their constraints, and incite them to really be the artists of their development.

DEFINITIONS, CHARACTERISTICS AND WORLD DISTRIBUTION OF ARID AND SEMI-ARID LANDS

A. Sasson
Division of Ecological Sciences
UNESCO, Paris

INTRODUCTION

For over two decades, the international community has been actively concerned with problems of arid and semi-arid zones. From 1951 to 1962, a world-wide programme of arid zone studies was carried out by UNESCO, to promote and stimulate research in various scientific disciplines bearing upon the problems of arid regions. This programme resulted in the publication of some 28 volumes of Arid Zone Research, covering hydrology, plant ecology, human and animal ecology, climatology, energy resources, etc., and the setting up of some 200 research units in 40 countries. In addition, during the past ten years, UNESCO has been involved in the preparation of several thematic maps presenting up-to-date scientific knowledge on the natural resources of arid areas.

The current UNESCO programmes most directly related to the problems of arid and semi-arid zones are the Man and the Biosphere Programme (MAB) and the International Hydrological Programme (IHP). Two of the 14 major project areas under the MAB Programme are directly concerned with these problems. Project 3 deals with the impact of human activities and land use practices on grazing lands, including those in arid and semi-arid areas, while Project 4 concerns the impact of human activities on the dynamics of arid and semi-arid zone ecosystems, with particular attention to the effects of irrigation.

Recently, widespread droughts and famine in the Sahelian zone of Western Africa drew worldwide attention to the problem of the management of arid and semi-arid areas. Two resolutions put forth by the United Nations in 1974 requested specific action.

Resolution 3337 (XXIX) assigned high priority to developing concerted international action to combat desertification, and requested the Secretary General of the United Nations to convene a UN Conference on Desertification. The United Nations Environment Programme (UNEP) was given the responsibility to execute this resolution and is convening a Conference on Desertification in August/September 1977.

Resolution 1898 (LVII) of the Economic and Social Council requested the Secretary General of the United Nations to convene an ad hoc interagency task force on arid zones, which would work to identify the obstacles not yet overcome by science and technology that are encountered by the developing countries; and the social, economic and institutional obstacles preventing the application of available technology.

An interagency working group which met in October 1974 decided that UNESCO, in view of its long experience in the field of arid zone research, should convene the task force to prepare a report on the obstacles to the development of arid and semi-arid zones. This report was completed in early 1975 and was submitted to an intergovernmental working group of the UN Committee on Science and Technology for Development, in order that the task force could make its own recommendations to the Committee on Science and Technology for Development. The comments of this working group, and those of several organizations which are members of the ad hoc task force, were included in the original report.

The original document was expanded to include a more detailed consideration of the physical aspects of arid and semi-arid zones, irrigation problems; urban, industrial and tourism development, and obstacles to the transfer of technology. A world aridity map was also prepared, together with an accompanying brochure explaining the methodology adopted for the preparation of the map.

CHARACTERISTICS AND WORLD DISTRIBUTION OF ARID AND SEMI-ARID LANDS

The map, while serving as a general background document for the UN Conference on Desertification, also aims to update the climatic map by Meigs (1953) which was based on the knowledge of the arid zones at that time, essentially Thornthwaite's data and index (1941, 1948). Since then, the climatological data have increased in number and in quality, so that the limits and boundaries of the arid and semi-arid climates and zones could be improved and made closer to the field observations.

The UNESCO map shows the geographical distribution of the arid and semi-arid zones of the world, including the desert zones, and also the subhumid ones which were not represented in Meigs' map.

The scale is 1/25 000 000. At this scale some generalizations had to be made, and the detailed and peculiar properties of some regions could not be stressed, although these regions do not cover large surfaces. The geographic frame of 1/25 million was a reduction of the 1/5 000 000 map edited by the National Geographic Society of New York.

The basic source of information was the evaluation of the degrees of aridity on a bioclimatic basis, considering the relative importance of precipitation, and evaporation and transpiration from soil and plants. This was done for 1600 stations by the World Meteorological Organization (WMO) and the Meteorological Institute in Bonn (Federal Republic of Germany, Drs. Flohn and Hemming). Whilst Meigs' map on the world distribution of arid and semi-arid homoclimates was based on the humidity index of Thornthwaite (1948), calculated from the values of evapotranspiration from the mean monthly temperatures, the values used for the UNESCO map are those of the ratio P/Etp, where P represents mean annual precipitation and Etp potential evapotranspiration, calculated according to Penman's formula. This formula takes into account the radiation balance and the aerodynamic aspects to evaluate the evaporating power of the atmosphere. The same ratio has been adopted by Rosetti and Riquier at FAO, to design a map on the desertification risks for northern Africa and the Middle East at a scale of 1/5,000,000, and a 1/25,000,000 map of the world on the same topic. The calculations of the P/Etp ratio were made for all the 1,600 stations. However, the boundary lines between zones of differing degrees of aridity (including the subhumid zone), were drawn by also taking into consideration local topography, soils and vegetation, based on the most recent publications and personal expertise.

Four classes or degrees of aridity have been distinguished and they allow for the identification of the main geographical zones usually recognized. These are the hyperarid, the arid, the semi-arid and the subhumid.

The hyperarid zone corresponds to the true desert areas, for which P/Etp is less than 0.03, with a very low and erratic rainfall occuring in any season. These regions have almost no perennial vegetation, but ephemerophytes could develop during some favourable years. Neither agriculture nor livestock husbandry are possible, except in the oases, although some nomadic grazing is possible.

The arid zone, where P/Etp is between 0.03 and 0.20, supports steppe or shrubland vegetation. It is sparse and depending upon the regions consists of shrubs and bushes, succulent or thorny plants or even leafless plants. Extensive nomadic grazing is the usual form of land use, and no rainfed agriculture is practised. The mean annual rainfall for this zone varies from 100-150mm to 300-350 mm.

The semi-arid zone where P/Etp is between 0.20 and 0.50, and contains steppes and shrublands, some savannas and tropical thickets. These are often good livestock husbandry regions, where rainfed agriculture shows great risks due to the high annual and interannual variations in rainfall. Annual rainfall varies from 300 to 600 and even 800 mm in regions with maximum summer rainfall, and between 400 and 600 mm in areas with winter maximum rainfall such as the Mediterranean and subtropical regions.

The transition to the subhumid zone is difficult to delineate as it is extremely fluctuating. One could indeed consider the whole semi-arid zones as a transition between the dry regions and the humid ones, as, depending on the years, they could be arid or, on the contrary, clearly humid. It really depends on the duration of the dry season beyond seven to eight months.

The subhumid zone, where P/Etp is between 0.50 and 0.75, mainly has an indicative value in showing the boundaries of the semi-arid zone. However, it includes all the regions which are submitted to a progressive aridification, mostly under human influence. Vegetation types vary: tropical savannas, Mediterranean maquis, or thickets, North American chapparal, chernozem steppes of the USSR, etc. Agriculture is the predominant land use pattern.

In order to avoid repetition and redundancies, these four zones are not detailed in the brochure of the map; the presentation is rather of large regions, characterized by a series of common features, such as the European Mediterranean countries, the Near and Middle East region, Central Asia, the Indian subcontinent, Australia, the Saharan, Sahelian and Sudanian regions, North America, and South America. Such a subdivision mainly refers to geographical regions, which are presented with reference to the map; for each, the climate features are shown and illustrated as diagrams, especially for those climates or areas which could not be mapped (due to the scale). The natural vegetation of each region, including the most characteristic species and the most used among them are also given. This information is completed by data on the main soil types and the main patterns of land use.

Within each of the four classes or degrees of aridity, the temperature regimes are defined by mean temperature (in °C) of the coldest (winter) and the warmest (summer) month of the year. These temperature regimes are the same as those distinguished by Meigs, and the consideration of winter and summer temperatures expresses the annual thermal range, which depends on latitude and continentality.

Instead of being indicated by letters and numbers, as in Meigs' map, the temperature regimes in the UNESCO map are shown as colours:

warm winter (20 to 30°C minimum monthly temperature) -- light purple and deep orange;
mild winter (10 to 20°C) -- deep orange to deep yellow;
cool winter (0 to 10°C) -- deep to light yellow;
cold winter (less than 0°C) -- deep green to light green.

These classes are subdivided according to the maximum monthly summer temperature, i.e., very warm summer (> 30°C)
warm summer (20-30°C)
mild summer (10-20°C).

Besides this zonal aspect, the map also shows the length of the drought periods or seasons and the precipitation regime for approximately one thousand representative stations, which are indicated by colored circles.

The number of dry months, i.e. with less than 30 mm rainfall, is indicated by the diameter of the circle, from 1, 2, 3 months for the lowest diameter, to 12 months for the highest diameter; a triangle means less than 1 dry month. It is to be stressed that in those low rainfall climates, the number of dry months does not differ very much from the result which may be obtained by applying the definition of a dry month, used by Gaussen and Bagnouls (1957) and by Walter and Lieth, as $P<2t$.

A blue circle indicates a dominant summer drought and winter precipitation regimes, sometimes shifted towards spring; a green circle also indicates a dominant summer drought, although there is a shorter and less pronounced winter drought; there are two wet seasons, one towards the end of autumn, the other at the beginning of spring.

Red and orange circles indicate a dominant winter drought, with, in the former case, summer precipitation regimes, and in the latter two wet seasons -- one towards the end of spring, the other at the beginning of autumn.

Purple and grey circles indicate transition regimes: the former, with two wet seasons, one in summer, the other in winter; distinct drought periods, in spring and autumn; the latter, irregular regimes, rains either occasional and unpredictable or, in wetter zones, distributed throughout the year without well-defined maxima, or with unpredictable maxima.

It is to be stressed once more that the scale of the map imposed a high degree of generalization. Although the boundaries of the aridity zones were drawn to reflect the value of the P/Etp ratio for every station, there still is some arbitrary decision with regard to these boundaries, especially in the regions where there are information gaps. Phenomena vary in the field continuously and progressively (except in clear-cut situations, e.g., steep slopes or relief, coastal lines, etc.) and it is rather difficult to assign a clear-cut limit to the mapped phenomena. This is a shortcoming of all climatic mapping undertakings. Therefore the subhumid boundary was not indicated.

That is why, whenever possible, information on soils, vegetation, and topography was taken into consideration, e.g., in the western African coast. The map was established according to the available data which usually are means over a large number of years (25 to 30 years for example). But the calculation of means overshadows the actual phenomena, which may be important for the natural vegetation, livestock rearing, or agriculture. Thus, two stations considered in the same rainfall category may have, during the same year, two very different total amounts of rainfall. For the same station, the total annual rainfall, the number of days of rainfall, and of dry months can vary from one year to the next. Such a variation increases according to the degree of aridity and of the length of the dry season. Such a variation could lead in subtropical or mediterranean regions to an interannual variability which gives, one year, a tropical regime to the station (summer rainfall), and another year a mediterranean regime (winter rainfall, i.e., during the short-day season).

So the extreme variability of precipitation, in time and space, does appear as one of the essential characteristics and common features of the arid and semi-arid regions. However, it is worthwhile to recall the main factors of their diversity.

MAIN FACTORS OF DIVERSITY

Climatic Conditions

The diversity of climatic conditions gives rise to very different situations. Beyond a simple classification into arid and semi-arid regions, it is important to consider the duration of the sparse rainfall, and the duration and timing of the dry season. It is obvious, for example, that the problems of irrigation, "dry farming" and livestock movements will be dealt with differently on the northern fringes of the Sahara, where the rain falls in winter, and on the southern fringes, where the rain falls in summer. In the case of rainfed cultivation, dry farming techniques are well suited to zones with winter rainfall, for evaporation is relatively low and part of the water can be stored in the ground and used by the plants the following year. On the other hand, they are not very effective in zones of summer rainfall, because evaporation is so high at this time of year. The climatic originality of a given semi-arid zone thus appears for a large part to be a function of the general characteristics of the adjacent subhumid zone, and that is another important factor of diversity.

Moreover the precipitations registered by rain gauges cannot always be considered reliable indicators; one of the original features of arid and semi-arid coastal or sub-coastal regions is the presence of hidden precipitations (for example dew), resulting from a greater humidity of the atmosphere (Chili, Morocco, Mauritania, Tunisia, etc.).

Finally, thermic conditions vary widely from one arid or semi-arid region to another: the hot deserts lying in the tropics (Sahara and deserts of Arabia) and the deserts of the temperate zone, characterized by cold winters and greater annual thermic ranges because of their continental location, have often been opposed. Generally, as we cross the tropics towards higher latitudes, the annual range of temperatures becomes greater than the diurnal range. In the hot deserts, the annual range is between 15 and 25°C (24, 4°C at Timimoun in the Algerian Sahara). It is markedly higher in the continental deserts of the temperate zone (37, 7°C at Turgai in Central Asia) where the range of extremes is rather exceptional (sometimes over 90°C). As a consequence, the harshness of winter temperature in the steppes of Central Asia creates conditions altogether different from the relatively mild ones prevailing on the fringes of the Mediterranean world, where the constraints on man are not as great.

Likewise the exceptional heat of certain deserts (in absolute value as well as duration) imposes various types of adaptation to animal life and exercises physiological and neurophysiological constraints on man which limit his activity.

Allogenic Rivers

The presence of allogenic rivers, rising in non-arid regions (Colorado in the United States, Amu Daria and Syr Daria in Soviet Central Asia, Indus in Pakistan, Nile in Egypt, the wadis which flow down from the Atlas in North Africa, etc.) may radically modify the development of a region. These rivers have often made it possible to extend the irrigated areas considerably, but in other cases their utilization remains very incomplete (Niger, Logone-Chari, Senegal, etc.) or might be oriented differently.

The nature of the flow of these rivers introduces considerable variation: for example, the flow of the Nile, governed mainly by tropical precipitations, is very different from that of the mountain rivers of higher latitudes, which are subject to phenomena of snow retention, combined in places with karst retention. Further, the conditions for using surface and underground water resources vary according to topography and a number of geological features, thus are contrasted the Sahara and the Arabian peninsula on the one hand, and Central Asia and the Andean countries on the other.

In the particular case of ground-waters, essential to the development of agriculture and the settlement of populations, a distinction must be established between regions where the geological substratum is made up of generally resistant and impermeable crystalline rocks and those where it is made up of sedimentary rocks with varied facies. In ancient crystalline terrains (shields), the importance of groundwater remains local and its reserves limited. Furthermore, it is usually difficult to assess its extension and force precisely. Most interesting then is the ground water with underground flow (inferoflux) found in alluvium produced by the mechanical disintegration of the crystalline substratum.

In sedimentary terrains, on the contrary, the extension of ground water is considerable and may be found at a number of different levels (water tables of the continental intercalary zone and of the terminal continental zone in the Sahara separated by terrains not suited to the storage of water).

In order to exploit these resources, it is essential to distinguish between ground water, which is being constantly replenished, although often irregularly, and fossil ground water (not being actually replinished). The advantages of sedimentary basins over crystalline shields are obvious, and the geological structure of arid and semi-arid regions, thus constitutes a non-negligible factor of differentiation as regards its repercussions on water resources.

Soil Conditions

Soil conditions vary greatly according to the distribution of the parent rocks, climatic conditions and topography. In many instances there is no real soil and the original substratum is often apparent. When a soil does exist, sandy soils seem best suited to the storage of water. In irrigated zones, silty soils may appear in certain respects more propitious than sandy soils, but present severe risks of soil degradation, especially in regions where natural drainage is poor (Nile delta, plains of the lower Euphrates, etc.).

The accumulation of various salts (NaCl, Co_3Na_2, SO_4Na_2, SO_4Ca, etc.) is generally important because of the intensity of evaporation. Concentration of salts is at its worse at the bottom of depressions where, for climatic and topographic reasons, the flow of rivers to the sea is impossible. Haloplytic plants alone are able to develop in these conditions (playas of North America, takyrs of Central Asia, sebkhas of North Africa).

Finally, all soils of the arid and semi-arid zones remain to varying degrees vulnerable to erosion, hence the overriding importance of all operations of soil fixation, conservation and restoration. The urgency of these operations will appear even more clearly when we consider that in certain cases we are dealing with fossil soils. These soils were created under climatic conditions altogether different from those which now prevail, and therefore once destroyed, would not reconstitute.

Continental Localization

The continental situation of many arid and semi-arid zones has a great influence on development possibilities, insofar as modern development depends heavily on contacts with the outside world. Thus, coastal regions have a definite advantage over regions located thousands of miles from the sea. The ease and frequency of relations with technologically advanced countries with a high standard of living, or even with the vital centers of a nation, are usually decisive factors in the development of these zones. This is illustrated by the contrasting conditions of development in the Chad Basin and in Northern Mexico, respectively.

The position occupied by the arid and semi-arid regions in the countries concerned (marginal, important, essential, or even vital) obviously affects the attention governments pay to them, and, in particular, the investments, public or private, available for their development. These investments consequently are rarely attracted by areas where the immediate return on investment is less certain than anywhere else.

Population Density

Population density constitutes an essential element of differentiation. In some semi-arid regions development may be hindered by shortage of labor. In other regions, the poverty of the inhabitants is due to their great numbers relative to the possibilities of the environment in a given technical, economic and social context. Large sized human, as well as animal populations can lead to dangerous overexploitation and cause serious damage to the environment. The maps of livestock and population distribution established in 1976 for the nations of the Sahel, emphasize some important contrasts and enable one to locate the most endangered zones.

However, similar population densities, subjected to comparable natural conditions, will be faced with vastly different development problems, depending on a certain number of factors. The particular cultural setting (North American, South American, Muslim, Australian) makes for a specific perception of the environment and of its problems.

Lastly, as the social and psychological problems of adaptation to new ways of life are far greater than the technical ones, the approach to development will differ according to the part played by nomads, semi-nomads and sedentary peoples, within a given region.

All this brings us to ask the fundamental question for the future of arid and semi-arid regions: when a choice is possible, should surplus population be maintained in its original environment, or on the contrary, relocated in more favourable areas?

Mineral and Power Resources

The availability of mineral or power resources, as well as the conditions of their exploitation, greatly influence the rate of development. The relative isolation of certain sites and the extraction problems play an important part. However, the evolution of prices on the world market may render profitable the mining of once abandoned deposits, and in many of these nations, highly increase the value of mineral fuels or raw materials. Processing possibilities obviously have a direct bearing on economic and social development. The intermediary stages of the processing of industrial products are poorly represented in the arid and semi-arid zones. On the one hand we have regions where mining industries, or basic processing industries (concentration of an ore for example) predominate, on the other, oases with light industries of a high technical level (electronics for example), as in the United States. Finally, the recent development of tourism has definitely privileged certain areas. This development is governed by a number of factors (presence of beaches or archeological sites, ease of communications with the countries which "supply" tourists), which combine to create a great variety of situations.

Consequently, this great diversity renders extremely difficult the elaboration of development models which do not account for regional characteristics. However, in the presence of common constraints, limited mainly to the irregularity of the water resources leading to extremely hazardous utilization conditions, the most severe problems arise not in the arid, but in the semi-arid zones. There are specific solutions for the use of arid zones, based on division of arid areas into specialized land use units. In semi-arid zones there is conflict between various methods of development (agriculture or animal husbandry) and the choice becomes even more difficult with increasing population densities. Moreover, the threat of desertification by human intervention is obvious in these ecosystems; it is characterized by the spread of desert conditions beyond desert margins, or the intensification of desert conditions within arid regions, which usually manifests itself as a trend towards diminished productivity: lowered carrying capacity for stock, diminishing crop yields, a progressive reduction in real incomes or in social well-being, and/or a reduction in the number of people supported.

ECOLOGY AND ENVIRONMENTAL CONCERNS IN ARID LANDS

E. Lendell Cockrum, Professor and Head
Department of Ecology and Evolutionary Biology
Mammalogist, Agricultural Experiment Station
University of Arizona, Tucson, Arizona 85721

The title of this conference, as has been announced, is: "Alternative strategies for desert development and management." The reason for any pressure to develop on the desert is primarily the result of the ever-growing human population. This results in more and more pressure on planet earth for food and other resources required as necessities or luxuries of life. Any strategy based on the single solution of the reduction in the total number of humans in the world is not acceptable. Such a suggestion must be considered to be only a minor part of the total proposed solution.

During the several days of this conference, a number of papers were devoted to attempts at identifying critical constraints on desert development and management. Time and again we have heard direct or implied statements made that only water is needed to make gardens of all of the deserts of the world! These and certain other statements made at this conference make it evident that many persons striving to help improve the quality of life for humans everywhere - and especially in the arid and semi-arid zones - are quite unaware of some of the controlling factors and basic rate parameters controlling the worlds ecosystem.

Unfortunately ecologists do not yet have enough knowledge to provide precise values for rates nor even for total values. However many have long understood and their importance to humans has been repeatedly underscored by many workers in the past several decades. None of the following material is new. It is based on a whole series of studies that have been summarized in papers such as those of Dregne (1970), Talbot (1972), and Sherbrooke and Paylore (1973). Essentially it is a rewrite of material presented by me (Cockrum, 1976) in Proceedings of the West Africa Conference held at the University of Arizona, Tucson, in April, 1976.

As far as the human population (of the world or any sized subpart thereof) is concerned, the statement is often made - and appears to be a basic premise of the United Nations - that all areas should be developed. However only a few meglamanical individuals have ever thought that he or she personally knew what was "best" for everyone! Within the United Nations, the definition of "developed" appears to mean "an improvement in the quality of life." Almost every sapient member of *Homo sapiens* recognizes that (a) improved "quality of life" should be compatible with the real desires of the majority of the local residents and not something imposed by either an external force or an internal minority. Further (b) the desired improvement is not the utopian dream of the individual - but the realistic improvement that the majority of the local population are willing to work and make personal sacrifices to achieve. Finally (c) both external groups as well as local residents must recognize that many of the "improvements" are associated with technology transfers that are inseparable from various social and politcal changes.

As we shall see in our examination of the ecological parameters one of the basic problems now faced by the world is that the total human population is rapidly approaching and locally exceeds the capacity for past and most present of agriculture, socialogy, political and economics systems to produce enough food, fibers and other items to meet the "necessities" and "demand luxuries" of the humans present.

One obvious solution to the problem is to reduce the total demand for goods and services by slowing the rate of increase or actually reducing the total numbers of humans present. As far as the present world political situation is concerned and as can be easily seen among the various member nations of the United Nations, the question of family planning and population control is political dynamite. At present some nations are trying to restrict population growth while others are offering various official encouragements for the production of larger families. Even within the United States (an area where many people and several socio/political pressures favor small families) certain ethnic groups are encouraging their members to have large families.

In resume, it is obvious that the question of human population control is a highly controversial political issue, especially on the international scene so evident in the United Nations. However, when an ecologist seriously evaluates the present and near future of the total earth ecosystems, several major ecological principles become immediately evident. Unfortunately for man, these appear to operate without respect for man-made political boundaries or laws!

Five major ecological principles that must be clearly understood by everyone involved in "development" include:

I. The primary production of ecosystems control total consumer populations.
II. The rate of productivity of an ecosystem is directly related to the annual evapotranspiration in the region.
III. Agricultural areas are specialized ecosystems that are subject to the same constraints as natural ecosystems.
IV. Ecosystems are variable as to their stability.
V. Recovery of a degraded ecosystem is directly influenced by the rate of primary productivity and the completeness of degradation.

THE PRIMARY PRODUCTION OF ECOSYSTEMS CONTROL TOTAL CONSUMER POPULATIONS

Modern technology is making major strides in producing synthetic carbohydrates, and proteins. Further, tissue culture techniques might well be developed to the point that they might be a source of protein for human diets. However, at present none of these appear to be a practical alternative for furnishing food for significant numbers of humans. The technology involved is complex, the costs of components is extremely expensive (in terms of energy and monetary value), and, at least at present, much of the input is in the form of plant or animal produced products - not in organic components. Therefore for the present as well as for the practical forseeable future, food for humans and animals will result almost entirely from primary or secondary productivity of ecosystems. In other words, photosynthesis and not the chemical laboratory will be the source of food!

As is discussed in any elementary textbook in biology or ecology, published anywhere in the world, the system of life known to us on Planet Earth is based upon a complex biochemical process that depends directly, or indirectly, upon the shortage of solar energy by photosynthesis in green (chlorophyll containing) plants. The various species of green plants have evolved numerous morphological and physiological specializations that maximize photosynthetic success under a variety of conditions. However, this photochemical process is actually quite limited as to the climatic physical and chemical conditions in which it can occur.

The whole process must occur in water even if the water is only a few molecules in the cell of a dehydrated desert plant! Since water is usually rare in deserts, it is little wonder that much of this conference has been devoted to discussions of water, irrigation and so on. However parameters of many other factors including temperature, relative humidity, pH, CO_2 nutrient minerals (including nitrates, phosphates and even other elements in "trace" quantities), and light intensity all are involved in the rates of energy storage as a result of photosynthesis. Usually the result is the transformation of water (H_2O) and carbon dioxide (CO_2) into a complex hydrocarbon molecule (eg $_6H_{12}O_6$)) in which molecular oxygen (O_2) is a by-product.

As already indicated, no precise measurements are available as to how much solar energy is "fixed" (as carbohydrates or other forms) during a given year - even in a given field or forest let alone on a world wide basis.

Educated guesses are that on a world wide scale, including deserts and polar regions approximately 1 to 3% of the annual solar energy is converted to stored energy the the "producer organisms" [= green plants!]. In the laboratory, with all known controlling variables (temperature, CO_2 levels, H_2O, nutrients, Rh, light levels) held at ideal levels, the energy fixation rate has been raised several fold - perhaps to 30%. Even a superficial evaluation of the energy and economic costs of these laboratory demonstrations clearly demonstrate the impossibility of developing such efficient systems in nature.

Most ecologists concerned with ecosystem productivity [= energy storage] recognize the concepts of the pyramid of energy (mass, numbers?) wherever the producer organisms [= green plants] from the base with, perhaps, 1000 C of energy for each 100,000 C of solar energy intercepted by the earth's annual orbital path [- 1%!]. Further, as parts of the green plants and their production (seeds, etc) are consumed by herbivores on the average about 90% of the energy is lost (metabolism, growth, etc).

Obviously any significant increase in efficiency of converting primary ecosystem productivity (ie, plant carbohydrates and other products) into animal flesh suitable (and acceptable) as human food results in a profit making system for enterprising humans. The historical procedure of nailing a goose's feet to a board (to restrain energy loss by exercise) while providing unlimited food and water within easy reach may have been an early system of maximizing protein production per unit of carbohydrate input.

Modern systems of cattle fattening operations and chicken "ranches" - now seen so commonly in Western Europe, Japan and the United States - are specialized examples. Most efficient, in terms of kilograms of carbohydrate food per kilogram of animal flesh appears to be some of the chicken "ranches" wherein as little as 4.5 kilograms of plant foods yields 1 kilogram of chicken flesh. However, such systems are successful only when (a) ambient temperatures are strictly controlled, (b) day "length" is increased by providing artificial light during part of the night time, (c) precisely balanced diet - with necessary added vitamin's, minerals, etc. are provided in proper proportions and (d) the chickens are housed in specialized units in specialized buildings that minimize movements and maximize feeding activities.

Under such idealized conditions, as little as 4.2 kilograms of plant food (ie seeds) plus vitamins and mineral supplements has produced one kilogram of chicken meat! Note, however, that high amounts of energy (for special "chicken ranch" constuction, heating, cooling, and lighting, as well as for the special dietary supplements) in addition to the plant produced feed are necessary to make this system work. So far most attempts to utilize "chicken ranches" and similar activities such as cattle fattening plants, in developing nations have been economic disasters! [The few exceptional successes appear to have been associated with isolated mines (or similar activities) where the people present had adequate income to pay for the extra costs of the inputs necessary to operate the "chicken ranches".]

Attempts to establish such chicken "ranches" in some of the West African countries [especially by the U.S. Peace Corps and U.N. volunteer activities] appears to have been almost invariable economic disasters. High costs of energy for environmental controls [and high mortality of necessary temperate zone stock - from Europe or United States when ambient temperatures were not adequately controlled] and high cost of imported dietary supplements almost invariably resulted in economic disaster as I have seen in Mali, Upper Volta and Chad.

Many ecologists have repetitively pointed out that carnivores depend entirely upon ecosystem energy that has been fixed in the producers and transformed (with an average 90% loss) in the level one consumers. The energy loss to the carnivore I level is again 90%.

As already indicated, no one knows how much solar energy is stored in a given year. Further, much of this primary productivity is in the form of cellulose (not carbohydrates or plant fatty acids) that cannot be broken down into released energy by the enzymes present in many level one consumers. Therefore, it is evident that only a small percentage of the actual annual stored solar energy is directly available as an actual (or even potential) food source to humans.

According to an unidentified source in a recent publication (1:b, Important for the future - publication of the United Nations Institute for Training and Research) only 2% of the total primary productivity of the world's ecosystem is currently being used by mankind. One percent is being used directly as vegetation (including grains and other seeds and nuts) as well as for food of the herbivores devoured as protein. The other 1% is the portion of the primary productivity that is used as fuel (for fires) and fibers (cotton, etc.). According to these estimates the current net productivity is 170 x 10^9 longtons of dry biomass per year.

Unfortunately the source of these estimates has not been given. If we assume that these estimates are correct, then it is obvious that obly a minor shift in the percentage of utilization of current annual primary productivity would automatically result in greatly increased quantities of foods available to humans. Such estimates give the basis for "optimists" that predict that by year 2000 A.D. the world will be plagued by surplusses in everything.

According to this estimate, only 2% of the annual primary productivity is utilized by humans. Therefore, it should be possible for the planet Earth to easily support at least 50 times the present human population - without any major dependence on the chemical laboratory! All we must do is find some mechanism of transforming the other 98% primary productivity into human utilizable resources.

Several different approaches to accomplishing this objective have already been undertaken, probably several will be successful. A similar summary paper prepared a decade in the future hopefully would involve significantly different values.

Currently active research projects designed to significantly change the percentages reported above include (a) the utilization of microorganisms (yeasts, bacteria, etc) to convert currently unused (by humans) primary productivity into usable products.

The current processes of yeast reacting with unrefined sorghum (to produce several food stuffs) and the bacterial reactions with human excreta as well as cellulose products resulting from crop wastes as well as "weeds" (yielding methane as a fuel as well as fertilizers!), are two examples of how current primary productivity can be more effectively used to support even greater populations of humans. Still another procedure is to combine rock phosphate with various types of non-food vegetation to produce a "supercompost" that, added to the soil around the roots of various crops, will measurably increase the yields available for human consumption. From a strictly regional view, one can well argue that there is no need for any given place to be self-sufficient in food production. As an extreme example, let us assume that a large, isolated island has no plants, animals, or even drinking water. However, it is well supplied with platinum nuggets. All a human has to do is go out and pick them off the ground. Obviously these can be exchanged, in the present world context, not only for food and water but also for a huge spectrum of things variously classified by humans as necessities or luxuries.

Thus the land supports a large number of humans. Much truth exists in this view. The exact nature of "platinum nuggets" varies from time to time and place to place. In places, the assembly of transistor radios by local people appears to meet the definition. In other places, tourism, oil, or something else appears to serve the purpose. Two major constraints, however, mitigate against this approach to solving the problems of the arid/semiarid regions.

First is the fact that the supply of "platinum nuggets" is extremely limited. Distance from markets makes all but the rarest or richest deposits of minerals economically suitable for exploitation under present conditions. Similar constraints exist for most other "platinum nugget" solutions. Certainly the few that can be identified should be exploited for the region's maximum benefit.

The second constraint is the recognition that, on a global basis, areas suitable for food production are not unlimited. With the ever-increasing human population, total food requirements are escalating rapidly. Land suitable for expanding agricultural production is becoming more and more limited. As a result, pressures for self-sufficiency are ever increasing. This does not mean, however, that the region should not import foodstuffs. In theory at least, the whole region might be best suited to growing an inedible fiber crop. As long as a world market existed for this production at a price that would permit the region to import all of its needs, then self-sufficiency would exist.

A concept that is the basis of many of our thoughts - but which no one has said much about - is that ever increasing human populations demand marked shifts in ecosystems productivity to forms suitable for human food.

THE RATE OF PRODUCTIVITY OF AN ECOSYSTEM IS DIRECTLY RELATED TO THE ACTUAL EVAPOTRANSPIRATION IN THE REGION

In the given region the plants present have evolved to produce the most growth (greatest primary productivity) possible under the local conditions (eg solar radiation, temperature, moisture, soil, elevation, slope, nutrients, trace elements, etc).

Rainfall, temperature, sunlight, soil type and other factors all interact to control productivity. As has been pointed out by McArthur and Connell (1966) the actual evapotranspiration rate correlates quite well with the total above ground productivity in natural terrestrial ecosystems (Figure 1). Transpiration is a measure of the rate at which water moves from the soil through the plant system and is dependent, in part, on the evaporation rate. Evaporation rates are basically controlled by solar radiation energies, including temperature, but are influenced by such secondary factors as relative humidity and winds. Potential evaporation rates are easily determined by measuring water losses due to evaporation from standard-sized containers exposed to air under standardized conditions.

Figure 1. Relation of the above-ground productivity to actual evapotranspiration in natural ecosystems. 1. Creosote desert 2,3. Arctic tundra 4. Tall grass prairie 5. Pine heath 6. Chestnut-oak heath 7. Beech-maple forest 8,9. Pine forest 10. Tropical forest Modified from MacArthur and Connell (1966), from Cockrum, 1976.

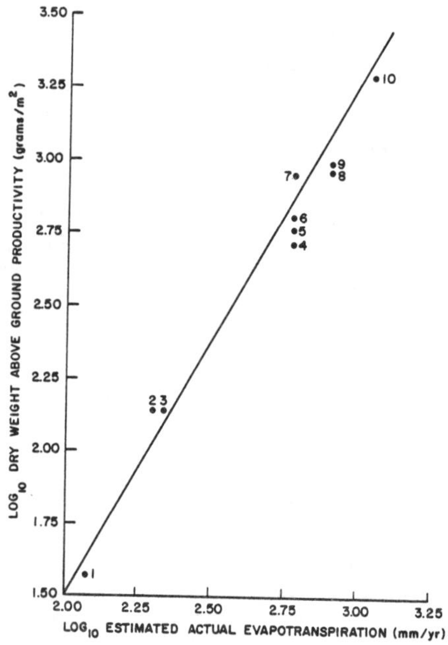

Transpiration rates are generally more difficult to obtain and are influenced not only by the physical parameters of the environment but also by the size, structure, "state of health" and genetic characteristics of the plant. Within the normal limitations of a given plant species, low transpiration values usually result in little or no growth, while high values result in rapid growth. In hot regions, dry winds foster transpiration rates in excess of many domestic crop plants' capacities. Plant "wilting", even if the plant roots are completely surrounded by water, occurs when the water transport capabilities characteristic of the plant are exceeded. Extreme wilt conditions cause plant death.

AGRICULTURAL AREAS ARE SPECIALIZED ECOSYSTEMS THAT ARE SUBJECT TO THE SAME CONSTRAINTS AS NATURAL ECOSYSTEMS

Basically agricultural ecosystems are simply areas wherein, as a result of human activity, species diversity is reduced. Undesired animals are classified by man as "predators" or "pests" and undesired plants as "weeds". Generally there is an increase in the primary productivity and total biomass of the human - preferred species. However, in most situations, the total biomass of the system is reduced - often drastically.

Agricultural ecosystems that have large amounts of energy added from external sources generally produce more human-preferred primary production (=food!). These foreign energy imports include such things as the manufacture, transport and operation of tools (for cultivation, harvesting, preservation, transport to markets, as well as various fertilizers, pesticides and herbicides). Some estimates of the green revolution requirements indicate that as much as 10 C of "foreign" energy (generally oil and other non-renewable resources) is required for each 1 C of harvested food.

Even more extreme are the estimates of expenditures for desert reclamation in Israel. At least some such projects have cost as much as $2500 per hectare. These export out of season crops to luxury markets in Europe.

Obviously such extreme expenditure of energy imported from other ecosystems are only rarely possible and, in terms of food produced for local markets, appear to be usually, economically impractical.

ECOSYSTEMS ARE VARIABLE AS TO THEIR STABILITY

As pointed out by Dr. Petheram of Rhodesia, even the concept of ecosystem stability has not clarified by ecologists. At least two views have been expressed by the past usage of the term: (a) constancy (i.e. no change with time and (b) resistance to perturbations (i.e. ability of system to return from a far different state).

Professor Petheram argues that agricultural ecosystems are potentially more stable than natural ecosystems under similar physical conditions.

RECOVERY OF A DEGRADED ECOSYSTEM IS DIRECTLY INFLUENCED BY THE RATE OF PRIMARY PRODUCTIVITY AND THE COMPLETENESS OF DEGRADATION

Vegetation found in any undisturbed region of the world evolved over many thousands of years. The result is a variety of vegetation types such as deciduous, coniferous and tropical rain forests, and grasslands and savannas, each characterized by a given group of plants.

As used here in this conference during the past two weeks, the destruction of all vegetation, even in a tropical rainforest, would result in an area classified as a desert. The process of vegetation destruction would be called desertification!

Thus the concept that has been expressed several times by several different people. That the planting of a buffer zone of trees to stop the spread of the desert is meaningless. The implied concept that a buffer zone of trees would significantly increase the rain fall of a region also is untrue. Several examples of minor changes in temperature and relative humidity have been documented for small areas associated with such major changes in land use is irrigation or a large city. However, the destruction of most of the native forests and faunas of South Eastern Asia (especially China) in the past 2000 years, in Western Europe in the past 500 years and Eastern North America in the past 100 years appears to have resulted in no proven change in average annual precipitation over any widespread area.

If the previously undisturbed vegetation of an area is destroyed by clearing and plowing, for example, and if the area is then left undisturbed, it will go through a series of generally predictable states (seres) until the original climax vegetation is restored, a process known as ecological succession.

The rate at which an ecological succession moves from bare ground through various subclimax seral stages to climax vegetation depends upon a number of interacting factors. For example, an area in which the vegetation was destroyed by molten lava recovers much more slowly than an adjacent cleared field, because of the time required for the gradual breakdown of lava and the eventual soil formation. Slope, drainage, and temperature as well as other factors influence the recovery rate. Often the single most important factor influencing the rate of ecological succession is rainfall amount and distribution. (Figures 2, 3, 4).

Highly capital intensive restoration processes of ground leveling and contouring as well as addition of fertilizers and water can often shorten the restoration time of ecosystems but rarely result in systems made up of the same species (or preparations thereof) that existed in the original ecosystem before its degradation by human misuse. Statements that imply that such can be accomplished in the next decade or so throughout any significant portions of the bare rock and moving sand deserts of the world are not realistic.

Application of these ecological principle in arid and semi-arid regions lead to the identification of a number of specific ecological statements that must be understood by everyone concerned with regional development projects there. These are:

1. Most semiarid regions are not just temporary extensions of an arid zone.

Aridity has existed throughout historical times and will continue in the foreseeable future. Multiannual periods of successive wet and drought years are normal. Even within a given rainfall zone, local variations in soil, slope and other physical features produce large variations in ecological conditions. As Dregne (1970) pointed out:

> "...The facile belief that all that the arid regions need to solve their problems is water no longer is widely accepted by knowledgeable people. Strong winds and intense solar radiation, with all their accompanying problems, will be there whether the supply of water is small or large. The irrigating of a calcareous soil does not make it into a humid region soil. At best, limitless water would turn the arid regions into numerous oases. Land in much of the arid regions is too shallow, too rocky, too steep, or too salty to be used for anything except the production of a sparse flora and fauna, military ranges, or similar developments."

2. Ecosystems in arid and semiarid regions are extremely fragile.

Destruction of native plant/animal is quick and easy but restoration difficult. Natural restoration rates are estimated to range from more than 200 years to perhaps more than 1000 years. The time required for recovery is directly associated with the degree of ecosystem degradation.

3. Few arid/semi-arid regions are underpopulated either by humans or domestic animals, and parts appear to be overpopulated by both.

Herdsmen, whether nomadic or sedentary, appear never to have existed in equilibrium with their environment. Rather, exploitative over-use has long been recognized by range specialists as being widespread. The tables showing human and animal populations in the Navaho Nation and in Tunisia are illustrative (Tables 1, 2, and 3).

Table 1. Cereal imports and local retail prices in Upper Volta (modified from Becker, 1974. 1971-1973 from UN reports of relief operations). (From Cockrum, 1976).

Year	Imports (Metric Tons)	Percent Commercial
1960/61	NA	NA
1961/62	2,506	100%
1962/63	6,155	100%
1963/64	4,569	100%
1964/65	3,646	100%
1965/66	8,908	69%
1966/67	28,281	30%
1967/68	6,056	26%
1968/69	16,469	10%
1969/70	17,232	0%
1970/71	34,319	0%
1971/72	36,850	0%
1972/73	55,650	0%

Figure 2. This photograph illustrates the extremely slow rate of restoration of natural vegetation in a Southern California section of the Sonoran Desert, where the rainfall is somewhat less than 125 mm (5 inches) per year. The 20-acre area was cleared for a crop in 1927. The photograph was taken in 1974 by Dr. George J. Miller (Imperial Valley College Museum, El Centro, California). From Cockrum (1973)

Figure 3. Photograph of an overgrazed area on Cap Bon, Tunisia. This is on the Mediterranean Coast, in an area of approximately 430 mm rainfall per year (near Kelibia). See Figure 8 and Table 4. (Photograph by Pamela Vaughan.)

Figure 4. Photograph of a 1600-year-old exclosure less than 100 meters from the point where the photograph in Figure 11 was taken. The figs are growing from the vertical walls of a quarry shaft, from which the Romans removed blocks of stone utilized in the construction of buildings and aqueducts throughout northern Tunisia. Grazing animals (now mainly goats) cannot feed further on the vegetation without falling into the shaft. (Phote by Pamela Vaughan.)

Table 2. Comparisons of North African with North American region. See text for discussion.

	Tunisia	Navajo Nation
Area (square kilometers)	164,150	64,750
Rainfall	Varies from 1000 mm (north) to below 50 mm (south). Approximately 1/3 of areas averages 580 mm, 1/3 300 mm and 1/3 100 mm.	Varies from 200 mm (low elevations) to 500 mm (high elevations); 90% between 200 mm and 300 mm; assumed average 250 mm.
Human population		
200 BC	1 million (6.1/sq. km)	unknown
1890	1.7 million (10.4/sq. km)	0.014 million (0.2/sq. km)
1973	5.6 million (34/sq. km)	0.138 million (2.2/sq. km)
Carrying capacity (animal units)		
Calculated	823,875 A.U.	250,000 A.U.
Actual estimate		
1943	?	114,000 A.U.
1960	?	86,000 A.U. (-22%)

Table 3. Total numbers of domestic animals in Tunisia and the Navajo Nation. These estimates, in thousands, are from United Nations, F.A.O. yearbooks (for Tunisia) and from Young 1961 (for Navajo Nation).

Year	Horses	Asses	Mules	Cows	Sheep	Goats	Camels	AU
Tunisia								
1913	37	95	23	217	729	505	112	546
1923	72	123	31	400	1451	777	114	560
1943	96	158	53	525	3239	1954	160	1376
1953	78	160	50	482	3352	1853	202	1366
1959	85	200	54	604	3793	1276	216	1507
1973	100	187	63	680	3200	460	180	1404
Navajo Nation								
1930	75			38	924	280		225
1943								125
1953	41			15	410	69		101
1959	33			22	480	121		110

4. Continued over-use by grazing and especially by browsing, promotes massive ecosystem destruction and desertification, even in subdesert and savanna regions.

Often spectacular examples of the effects of overgrazing can be found in areas where, by various means enclosures have kept grazing from given regions. One extreme case is to be seen on Cape Bon, northern Tunisia, in an area of 430 mm of rainfall per year. (See Figure 4 above).

5. The buildup of animal and human populations during a few wet years leads to an extremely rapid ecological degradation when average or drought conditions follow.

Extreme variability of rainfall is one of the characteristics of arid - semiarid regions. Droughts are generally documented but periods of above average rainfall are generally not recorded and, if continued for a few years, such little pluvial periods are soon assumed to be normal.

Figure 5 of Lake Chad levels in the past 75 years is illustrative. A survey of the literature reveals that some 23 droughts occurred in sub-Saharan West Africa in the past 400 years.

Figure 5. Variations in maximum levels of Lake Chad above the arbitrary assumed zero point equal to the lowest level in the period of record. This is modified from a graph published by UNDP (1972). Note that the period in the late 1920s is considered to be a drought period by some authors. Note also the "little pluvial" decade from 1956 through 1967.

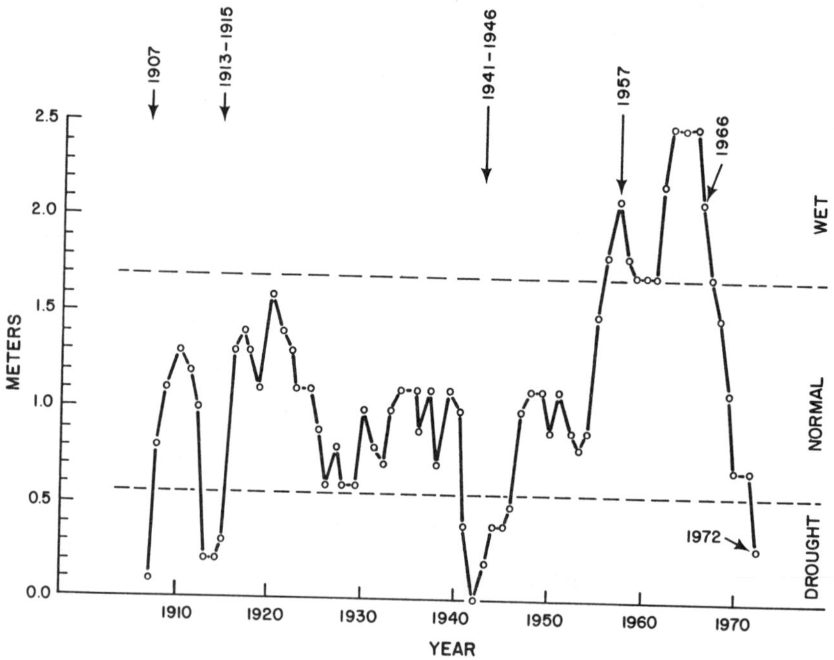

6. <u>Although esthetic considerations are of minimal importance to starving people, ecological balance is of prime importance for long-term survival</u>.

No scheme to increase productivity of foodstuffs for human consumption is likely to succeed unless it takes into account the ecology of the region. For example, developmental schemes such as water development, veterinary activities and predator control that remove environmental limitations on animal numbers but which are not associated with an immediate control of animal use and/or range improvements generally result in range degradation and reduced sustainable carrying capacities (Figures 6 and 7). Most planners agree that some form of animal grazing offers the best potential for a long term sustained yield in areas where water supply is inadequate for crops. However, stocking rates must be planned with the realization that prolonged drought periods lasting perhaps six or seven years will occur.

7. <u>Field crops that depend on available rainfall should never be planted in semiarid areas where adequate rainfall occurs only during exceptional wet years</u>.

Destruction of native vegetation not only reduces available forage for animals but also permits wind and water erosion leading to desertification in average or drought years.

8. <u>Irrigation projects, by changing the primary productivity rates of an area, markedly influence the population structure of a whole series of native consumer species</u>.

Some cannot tolerate the changes and become locally extinct. Others, such as certain rodents, seedeating birds, insects, and aquatic forms, thrive. Some of the arthropods are vectors of human diseases. The destructive effects of crop-eating insects is known and can be counteracted by various insecticides, but the potential damage by rodents and seed-eating birds is often overlooked. Certainly irrigation project designers in the Sahel-Sudan region must take these into consideration and attempt to eliminate habitats for these species. If not, much, if not all, of the increased productivity may well support more rodents and birds - and not more humans.

Even the development of minibasins makes major changes. The example of runoff alongside of paved roads, change in types of plants (and animals) present. Minibasin formation, spread over large areas, make major changes that can make just as many changes in whole system (including <u>pests</u>) as an irrigation project.

ABSTRACT

Five ecological principles and eight statements of ecological relations are identified as being critical to development projects in arid and semiarid regions. Specific examples of each are given.

14

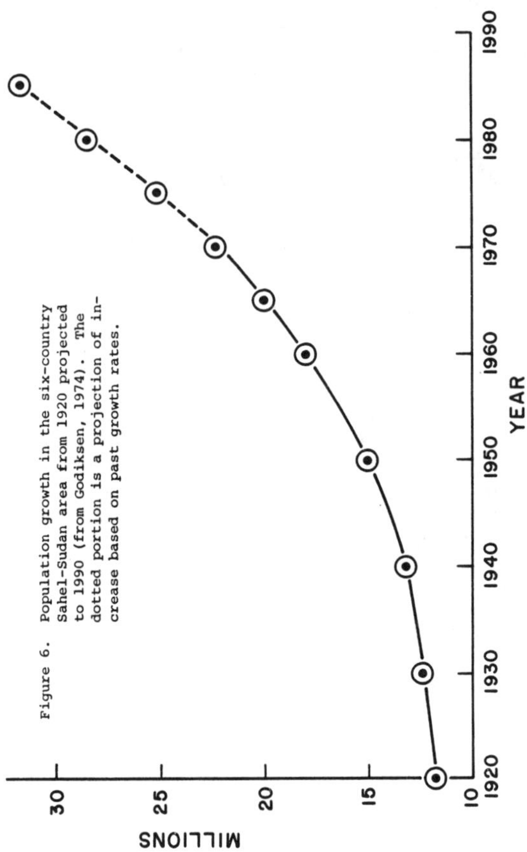

Figure 6. Population growth in the six-country Sahel-Sudan area from 1920 projected to 1990 (from Godiksen, 1974). The dotted portion is a projection of increase based on past growth rates.

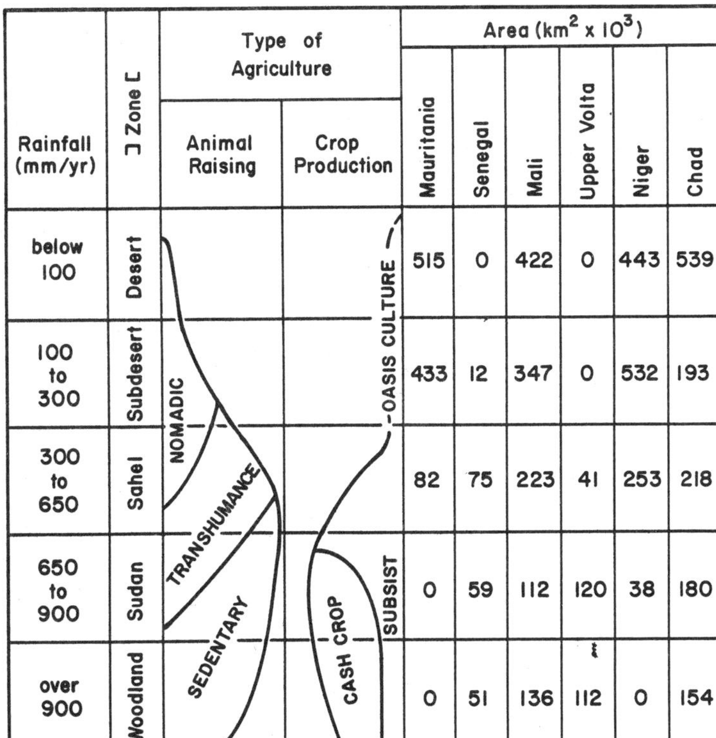

Figure 7. Zonal distribution of agricultural practices and national areas in each zone of the six major West African Sahel-Sudan countries, under average climatic conditions.

REFERENCES

Cockrum, E.L.
1976 An ecological view of the Sahel-Sudan Region of sub-Saharan West Africa, pp. 278-309 in Paylore, P. and R. A. Haney, editors. Proceedings of the West Africa Conference. University of Arizona, Tucson.

Dregne, H.E., ed.
1970 Arid lands in transition. American Association for the Advancement of Science, Washington, D.C., Publication 90. 524 p.

Petheram, R. J.
1976 Ecological Stability - with a bias towards agro-ecosystem management. Rhodesian Science News, 10(11):279-282.

Sherbrooke, W. and P. Paylore.
1973 World desertification: cause and effect. Arid Lands Resource Paper No. 3. Office of Arid Lands Research, University of Arizona, Tucson. pp. 1-168.

Talbot, L.M.
1972 Ecological consequences of rangeland development in Masailand, East Africa. In M.F. Farvar and J.P. Milton, eds., The careless technology, p. 694-711. Natural History Press, Garden City, New York.

THE BREAKDOWN OF ANCIENT DESERT CIVILIZATIONS

McGuire Gibson

INTRODUCTION

The phrase "ancient desert civilizations" conjures up the image of Egypt, Mesopotamia and the Indus, but these great-river-based civilizations are not the only examples of ancient complex cultures that now lie in deserts. Across Northern Africa, in Jordan, Syria, Saudi Arabia, Iran, Afghanistan, Turkmenistan and Peru there are equally impressive ruins that were once supported by a variety of water-supply systems ingenious if not extensive.

It is not within my competence to discuss all ancient desert civilizations. I will, in fact, restrict myself to Mesopotamia with an occasional allusion to Egypt and the Indus. What I have to say about Mesopotamia will, I think, have some relevance to other areas and to the general theme of this conference.

We who work on ancient civilizations are much more accustomed to address ourselves to their origins or maintenance rather than their breakdown. Previous generations of scholars accounted for declines of civilization by recourse to barbarian invasions or deterioration and corruption of the bureaucracy or decay of the ruling class. Rome's fall has even been blamed on the lead in the bathtubs of the elite. Grand schemes such as those of Marx, Spengler, Toynbee and others have encompassed both the rise and the fall of empires through overwhelming forces of history.

Modern anthropological/archaeological work, especially that of V. Gordon Childe, Karl Wittfogel, Robert McC. Adams and an expanding number of younger scholars, has swung from an explicit or implicit evolutionary view through investigations of prime movers such as irrigation to ecologically oriented explanations. In the newer work, civilization is seen as an adaptive mechanism for dealing with environment, as an expression of the interplay of a number of natural and man-made factors. In most of the current work, the emphasis is on origins, but the factors can and must be seen as operating in the long-term maintenance and collapse of civilizations. In fact, for Mesopotamia, the set of constraints that affected ancient civilization can be argued to be in operation today.

FACTORS INVOLVED IN MESOPOTAMIA CIVILIZATION

The factors involved in Mesopotamia were and are land, water, climate, disease, pests, human competition for resources, technology, economics, organization of manpower, and human population dynamics. Each of these elements can be subdivided a number of ways, discussed at great length; I will point out some features that seem at this juncture in our research to be particularly critical.

Land

Under land there are a number of critical points that many in this conference are far more competent to discuss than I. The geological history of Mesopotamia, modern Iraq, has been studied in detail. However, that portion that most directly concerns us, the past ten thousand years when human beings settled in villages and then created the first cities, has been virtually ignored. The hundreds of cores drilled for oil prospecting have, to my knowledge, never been analyzed by a geomorphologist interested in reconstructing the landscape and climatic conditions of the Holocene. Only recently have geomorphologists begun to work with archaeologists on the history of human settlement in the Mesopotamia alluvial plain. I will not go into the complex problem of the relationship of the northern, rain-fed, hills of Assyria (Northern Iraq) to the irrigation-watered alluvium either in terms of natural differences or human, cultural responses. Although the information we have on the alluvial plain is sketchy, it is rich when compared to what we know of the north. The following sections are, then, restricted to the alluvial area where civilization first crystallized (c. 3500 B.C.), went through approximately three millennia of related cultural development, and then took on the coloring of foreign conquerer civilizations.

Natural processes, such as earthquake or gradual changes in landscape resulting from continued techtonic movement or the encroachment of sand must be taken into account. The history of the dunes between the Tigris and Euphrates has not yet been written. Stephen Lintner, the geomorphologist with the Nippur Expedition, has been doing some work on the dunes and has initial indications that most of the material in the dunes is not silt or debris torn loose from the thousands of mounds that mark ancient settlements, (as suggested by Adams and Nissen, 1972) but is composed mainly of grains of a size that argue for an origin in the deserts of Arabia. The dune belts, that now cover a large part of the ancient habitation area of Southern Iraq can be seen as independent of agriculture or lack of it. The dunes would, then, be independent of the supply of water and might have encroached upon agriculture and settlements at various periods in the past as they have been doing over the last two hunded or more years. In Pakistan, near the dry bed of the Ghaggar-Hakra river, where hundreds of Harappan and early-Harappan sites have been located, there are dunes. In this desert, modern nomads live, camping for varying periods near water. Also in this desert, on ancient dunes, there are Harappan materials that must be interpreted as the remains of similar, ancient camps. There was exploitation of marginal environmental niches at the same time as the high civilization in the river valleys (personal communication from Rafique Mughal). Similarly, in Mesopotamia the desert, exploited by herdsmen, existed as an alternative along with swamps, to life in farms and towns. Whether ancient Mesopotamia always had the dune belt cannot be determined as yet. It may have had a smaller belt at times, perhaps closer to the present-day Tigris.

The landscape of Southern Iraq is at present an extremely complex intermixing of man-made and natural features. Almost all relief features are artificial, being the remnants of ancient canals, cities and towns. There are occasional rises that mark ancient river levees, and dunes. The slope of the alluvial plain is very shallow, Baghdad being only 33 meters above sea level. Because of the slightness of gradient and some impermeable layers, drainage is and apparently always has been a problem. The buildup of levees along the Euphrates, and artificial irrigation banks have tended to divide the plain into basins which receive differential amounts of rainwater, flood and irrigation water. Poor drainage and an irregular pattern of water retention have resulted in salinization that is not altogether uniform. It has been proposed that progressive salinization led to the gradual shift of economic and political power from the southern part of the alluvium to the north (Jacobsen and Adams, 1958), but the evidence may indicate a more dynamic, less progressive process, as shall be seen below.

The richness of the soil in the alluvial plain was a marvel to ancient writers. Compared to Greece, Mesopotamia's irrigation agriculture might have seemed extraordinary. But, we have documented indications of the effects of salinity long before Herodotus paid his visit, and the yields indicated, even in early (e.g., 2500-2000 B.C.) texts, are not more than the average gotten by my grandfather from relatively poor rainfed land in Maryland before the introduction of chemical fertilizers. It is my feeling that ancient Mesopotamian soil was probably not much richer than its modern counterpart. In new development schemes, the expected yield has never been enough to justify the installation of field drains to remedy the salinity. Only in the Greater Mussayib project, patterned on Dutch salinity-control schemes, were such drains planned and it is only now, with the increased budgets allowed by the oil boom, that the drains are being installed. Concurrently, enormous drainage canals and new irrigation projects are being cut in the desert where no agriculture has been carried out for four hundred years.

In ancient and modern times, lacking the technology to put in drains, the farmers of Mesopotamia have relied on a system of fallow to slow down the inevitable deterioration of their land through salinization. There seems never to have been a time when the entire aluvium was cultivated, and thus to an ancient or modern developer there might seem to be plenty of room to expand agriculture with a new project, rather than violate fallow but ancient and recent documentation shows that there were violations of the fallow principle. I have argued elsewhere that the violation of fallow led inevitably to the deterioration of the agricultural base through a combination of increased salinization, lowering of yields, consequent indebtedness of farmers, subsequent flight to the cities for work resulting in both an abandonment of the countryside and the creating of an urban population that consumed rather than produced food (Gibson, 1974).

Water

Water in Southern Iraq, past and present, takes two forms. Irrigation from the Tigris and Euphrates and rainwater. The Tigris, coming into the alluvial plain much more directly, with greater velocity, than the Euphrates, cuts into the alluvium and meanders markedly. Its bed is several meters below plain level until it reaches the town of Kut, where today it divides and begins to run above the plain, allowing flow irrigation. In ancient times, the lack of even simple lift devices made the use of a long stretch of the Tigris impossible. Ancient Mesopotamia was for most of its history dependent only on the Euphrates for the bulk of its irrigation. The nature of that river, much longer, slower and more manageable than the Tigris, has been described often (e.g., Adams, 1965; Gibson, 1972; cf. Adams and Nissen, 1972, pp. 44ff.). Originally a braided stream, it was gradually made more and more artificial. I have previously argued that the Euphrates was affected by man's attempts to use and control it and that reactions of the river to human alterations as well as natural forces resulted in oscillations in stream flow from one branch to another (Gibson, 1972, 1973). Shifts in the name, Euphrates, from one branch to another through time, as well as archaeological survey data, were the information on which I based my reconstruction of the river's history. I also linked such changes in the Euphrates to evidence of a large meandering stream-bed visible in the desert (on air photographs), assuming it to have been a major branch of the Euphrates in the time just prior to the development of truly urban settlements. Adams (personal communication), on the basis of new survey data, has decided that this ancient river bed must have been the Tigris in a position some miles to the west of its present bed. Whether this was the Euphrates or Tigris, the abandonment of this streambed, and the consequent shift of the population that had been on it, must be argued to have had a drastic effect on the relationships and organization of the people in the entire Mesopotamia alluvial plain when civilization began.

The early peopling of the plain, dependent on the rivers, was also affected by the Gulf into which the rivers flow. It is an indication of the state of our understanding of the processes involved in early civilization that there is still active debate over the formation of the alluvium. An early formulation by de Morgan (1900), in which the Tigris-Euphrates delta built out gradually to its present form, was challenged by Lees and Falcon (1954) who argued that the bedrock under the alluvium was and is still being compressed downward by techtonic movement and that any siltation is being equalized by that downward movement. For them, the head of the Gulf would have been in approximately its present location for millennia. Recent work has challenged Lees and Falcon, bringing to bear newer research on oscillations in sea level (e.g., Larsen, 1975), but not giving enough weight to techtonics. No one has, as yet, brought into the discussion the effect that irrigation or lack of irrigation would have on the silt load transported to the sea. There is a crying need for concerted geomorphological and hydrological work, done alongside archaeological survey and digging. I have, personally, introduced two geomorphologists to the alluvial plain. The first one found it fascinating and was amazed by the massiveness of change brought on by minor fluctuations in climate, human intervention and sand belt movement in the last eighty years. He found it too complex a set of problems to start to study so late in his life. The second has now worked with us for two seasons and intends to continue. He is specifically involved in research on ancient river channels and changes in landforms, land use, and climatic change.

In outlining the history of the Euphrates, I (Gibson, 1972, 1973) suggested that at any time the Euphrates was pouring most of its water down its western branch, the present-day Hindiya, there would be an expansion of the marshes that even today act as the recipients of surplus water. It seems odd to speak of surplus water in a desert situation, but for ancient Mesopotamia it was a critical problem. Lacking modern capacities to store water in reservoirs, the ancient Mesopotamian had to contend with a river that was at its highest about a month too late for optimum use on crops. The water might come down as a destructive flood. A critical problem was getting rid of the unwanted water. Diversion into less productive areas, down the less economically vital channel and eventually into swamps was the answer. In periods when the major irrigated areas were abandoned, and this seems to have been an occurrence that happened several times, much more water would have flowed down the remaining, western, branch and the swamps would have been even more extensive. The swamps, whether in times of flourishing irrigation or virtual abandonment, were a valuable alternative subsistence area and a refuge for people in trouble with governments, enemies, and the like. The swamps were rich in reeds, always a major building and craft material, in fish, turtles, birds, wild pigs and other animals. Along with cities, farms, and deserts exploitable by herdsmen, the swamps were a valuable alternative subsistence area, available for exploitation in times of stress.

The manner in which water most directly affected man and land in the alluvium was, of course, in the form of irrigation. The technology for irrigation farming, the number and scheduling of waterings, the need for fallowing, the complementary keeping of herd animals as an added source for wealth and as a hedge against the failure of crops (and subsequent switch to nomadism), can be argued to have changed little until the advent of mechanical pumps, modern farm equipment, and hybrid seeds. Flow irrigation in Mesopotamia is relatively simple and can be carried out to a surprising extent by tribal cultivators without government control (Fernea, 1970). I have argued elsewhere (1974) that evidence of government intervention in Mesopotamian agriculture, especially in the construction of new canals, marked the decline of the agricultural base. I saw government attempts, in ancient times, to restore agriculture as a sign that traditional means of production had ceased to function. In somewhat similar manner, it can be said that, contrary to the old notion that the Abbassid Caliphate (A.D. 750-1258) fell because the Mongols destroyed the irrigation system, the Abbassid agricultural base was weakened long before by over-ambitious irrigation schemes. I am not saying that such schemes were the only causes of the decline of the Abbassid state, but that they had a part to play. The major canal systems, which were large enough to sail a fairly large boat through, in their very magnitude created a constant need for cleaning, repair, and regulation because of their greater silt load, velocity and the temptation to over-irrigation, leading to salinization. No bureaucracy could consistently, over several hundred years, carry out the enormous task of maintaining so large a system. This would be especially true if there should occur any economic or social disruptions. Economic disruptions can take the form of either a dip or a rise in the economy. During a dip, a government might not have the money to pay laborers. Conversely, during a boom, labor might be judged more necessary for construction in the cities. Men might also be needed for warfare, the ultimate in social disruption.

Besides irrigation, water for crops is supplied by rainfall. Although rain is not predictable, and varies extremely from place to place and from year to year, it is extremely important for successful crops. This is true today and, in the absence of reservoirs to assured early water, was even more true in ancient times. For ancient Mesopotamia, the variability in availability of water was the most important factor in continued life as a farmer.

Climate

Just as rainfall and the river flow could vary widely, temperature in Mesopotamia, although predictable in its extremes, fluctuates from place to place and from day to day, especially in the growing season. I need not spend much time speaking of the effects of a change in the course of major air currents, the drop or rise in temperature by a few degrees, upon life in the affected areas. It is generally agreed that there has been no appreciable change in climate in the Near East over the past ten thousand years, but what would a change that is expressed as a minor fluctuation around the mean do to an early civilization without recourse to modern means of facing water shortages?

Disease

Disease in humans and in the animals they depend on has not been investigated systematically in terms of its interaction with other factors to bring about or destroy a civilization. We do have, in cuneiform records and in hieroglyphics in Egypt, indications of epidemics. One source, the Atra-hasis epic, directly links plague with overpopulation (Lambert and Millard, 1969). In this work, the gods of Mesopotamia are said to be tired of the noise men make. They send a number of misfortunes, but man always breeds back to a level that annoys the gods again. This particular epic is perhaps of great significance because it was originally composed during the Old Babylonian Period (c. 1635 B.C.) some time after an economic collapse, that will be discussed below, and just before the complete dissolution of the Babylonian kingdom and a slide into an extended Dark Age. Cuneiform records give many indications of disease, but the exact identification of most of them is uncertain. In last year's excavations at Nippur, we discovered many examples of the snail that houses the Schistosomiasis-causing parasite. One such snail shell was embedded in an inscribed clay tablet with the date of a King of Ur (c. 2100 B.C.). This does not mean that Schistosomiasis was present in Mesopotamia that early, but it does indicate the possibility.

Disease in animals, especially the few farm animals that have traditionally made the difference between a reasonable life as a small farmer or tenant and increasing indebtedness leading eventually to abandonment of farming (Gibson, 1974) could be crucial. We have thus far little evidence of die-offs in farm animals, but we must include them as a factor in any reasonable account of change in ancient civilization.

Pests

Pests, in the form of locusts, mosquitoes, and other insects that destroy crops or cause disease, can be shown to have existed both in ancient Mesopotamia and Egypt. They cannot, at present, be shown to have had a serious enough effect that they brought down a civilization, but in combination with a number of concurrent crisis, they could be very effective.

HUMAN FACTORS IN MESOPOTAMIAN CIVILIZATION

The elements to be taken into consideration now are intimately related to those that have gone before, since we are considering civilization as a tool, an adaptation by man to environment. It is more difficult to separate them into categories, since they are all part of that multi-faceted adaptation. A few years ago, several scholars wrote syntheses of ancient civilizations using population pressure, increase in density, and resultant changes in human organization (state formation), technology (irrigation), and economics (trade), as results of competition for resources (warfare). These syntheses, although addressing themselves to the origins of civilization, can be drawn upon to answer the question, why did civilizations collapse.

Competition for Resources

Under the heading competition for resources, which can include internal and international competition for water and labor, there is the factor of economics. Civilization as an adaptation to environment is most basically to be seen as the search for subsistence. Whether one grows something and barters or sells it; whether one takes manufactured goods and trades them for some other items of the same value or at a profit; whether exchange takes place in the form of tribute from one ruler to another, or in the form of a bride transferred from one family to another to cement alliances as well as to gain wealth, economic life is, at its core, the maintenance of life. In an area as poor in resources as Mesopotamia, Egypt or the Indus, with only agricultural and herding produce but little else to trade, it was necessary to develop continually more sophisticated mechanisms for the procurement of wanted goods. Ancient Mesopotamian society was a complex web of relationships. Kings in various times and places could hold unquestioned power, but in others might be mere figureheads. The influence and wealth of members of the royal family and the other ruling elite would vary from reign to reign or even year to year. The power and the composition of the temples and temple personnel would also vary, and the relationship of state to temple was a complicated one. It is probable that in all periods the important families, the personnel of temples, and the personnel of state institutions were related and connected on several levels. Kings made certain that their daughters were appointed high-priestesses of important gods in various parts of their kingdoms. These priestesses, in at least some cases, had important functions in the economy of their cities and the country as a whole. Temples may be seen as serving the function of banks. In them was kept the surplus wealth, prsented by the king as booty from conquests or annual gifts, and by the pious in the form of objects, land, and personnel. Temples may also be seen as welfare institutions, taking in the destitute, orphans, and children of the very poor who gave them as a form of votive offering. Like royal establishments, temples had sometimes hundreds of agricultural and craft personnel. Rich individuals could also keep large estates. There are records on long-range trade which show in some periods a seeming monopoly by the crown in certain ventures, but in other periods the crown shred in trade with individuals, large families, and temples. The known economic records, and there are thousands of them, cannot account for all the thousands of objects of stone, metal and other materials found in archeaological sites. Clearly, there were many networks of trade in items over which the record-keeping large institutions wanted no control. There are surprisingly few records of customs or taxes paid on most kinds of commodities, and although in some periods the crown kept close account of some perishables, including thousands of fish, turtles and birds, for centuries there was no written record of such basic foodstuffs. That there was no cessation in eating such foods can be shown by hundreds of fish, bird, turtle and other animal bones (including wild animals) found in excavations.

Economic crisis, price fluctuations, failed ventures, can be inferred from some records. Such crises, if they occurred at the wrong time, could easily bring about a general collapse of an empire. I would stress, <u>if</u> they occurred at the wrong time. In any civilization, we are dealing with a complex system of inter-reacting parts and a state might easily survive one crisis or many, then collapse if it occurred when some other elements made it impossible to meet what would otherwise be a routine flare-up. We have, in Mesopotamia, a growing body of evidence on a major economic crisis that must coincide with at least one major natural crisis. As mentioned above (p. 6) there seems to have been an economic collapse during the Old Babylonian Period. After the reign of Hammurabi (1792-1750 B.C.), when Babylonian held much of Iraq, part of Syria, and other areas, there was a very swift change of fortunes. His successor Samsuiluna lost most of the empire almost immediately and by about 1737 B.C. a great silence fell upon most of the great cities of Mesopotamia proper. Only the cities along the western branches of the Euphrates continued to produce cuneiform tablets. Excavation at Nippur and other sites confirms that there was a sudden end to several major cities. At Nippur, houses seem to have been abandoned without any sign of violence, with bowls and grind stones left on the floor. The records from the last two or three years at Nippur, just before the silence, show a sharp increase in loans to farmers, then a selling off of farmland and later of houses in the city. House and farm prices dropped sharply. A few rich men ended up with a good deal of property. Then, abandonment. Elizabeth Stone, a student in the anthropology department of the University of Chicago, has just completed her dissertation on this material and she argues that there must have been some major change in the water supply, causing the loss of a crop, resulting in the need for loans, then foreclosures in the next year as, apparently, the water failed again. The amazing thing is that there are, as far as I know, no independent records of the loss of at least one, probably as many as three branche(s) of the river during this time and the abandonment of the main irrigated area. Where did all the people go? Is this event really reflected in the Atra-hasis epic (p. 5)? Can we suggest some related events or processes that preceded the crisis?

Technology and the Organization of Manpower

Above, I mentioned that intervention in irrigation agriculture by centralized government could precipitate processes that weaken the agricultural base. States in the alluvial plain have traditionally drawn their revenues from taxes on agriculture. Pressing for increased taxes or production, without modern technology to offset salinization, must lead to degradation and depopulation. Governments, if led by vigorous leaders surrounded by dedicated, sensible officials, and backed by a good army can survive for a long time on conquered territory, captured commodities, and imported labor. In other situations a reduced or no longer adequate food supply at home can be offset by the fruits of commerce and/or new technology for industrial production. Factories producing cheap goods for export can take the place of farms as long as the lines of supply are not disrupted and as long as the goods are accepted by the trading partner. But ancient Mesopotamia was not a unitary civilization. It was seldom held in both the north and south by one power. Only with the Assyrians, in the first millennium B.C., did one part of Mesopotamia hold sway over both the north and south with any consistency. Even with the Assyrians, there was constant shifting, factional politics within the Assyrian court, and intrigues that acted to break the hold on Babylonian, in the south. For most of its history, Mesopotamia was really an area split not so much into north and south (the Assyrian hills and the Babylonian alluvial plain), but into two competing, basically duplicating trading systems. The alluvium, feeding on the Euphrates and using it for transport of heavy goods such as stone from Syria, wood from Lebanon, and so on, had access to trading areas in Turkey and on the Mediterranean as well as to the Gulf. Assyria, based on the Tigris, had similar links to Turkey, Syria, the Mediterranean, and Iran as well as the Gulf. Each power found it difficult to control all routes supplying the other; both could supply or be supplied with virtually the same sorts of goods from similar sources, and through most of history an unstable truce seems to have held. Occasionally, a king in the alluvium would make some move that would be interpreted as a threat to the north, such as a raid over the Tigris. Sometimes, an Assyrian would hold a point on the upper Euphrates, but even when such moves were being made, trade still may have gone on between them. Textiles made in the south would be sold to Assyria for transport farther north and west. We are only beginning to understand the intricacies of Assyrian/Babylonian relations, and economic texts do not give many direct hints of their historical and political contexts. But, it can be said that the more intricate and far-reaching the trading system, the easier it was to disrupt it. If it were disrupted at the wrong time, at the same time as a major crop failure say, it could cause internal problems, plots, hostile military action and finally, collapse. This would be especially true if, in order to support a vigorous manufacturing economy, people were allowed or encouraged to leave farms and come to the cities to work. A rise in living conditions, even if the wages were small, would help to make the city or cities boom, bring about housing construction, create another labor need, act as an added incentive to leave the countryside. Bring into this situation, in which an empire is living on its manufacture and is importing a good part of its foodstuffs besides other goods, a disruption of the supply chain, and the results could be catastrophic.

Human Population Dynamics

Implicit in some of the sections above is the assumption that over time population tended to increase and that population pressure was a major factor in the move toward urbanization, toward expansion of trade networks, and so on. But, population need not rise to bring pressure on a social system or a natural environment's ability to support it. Population density in restricted areas can have the same effect. I have already mentioned the evidence of a major river course's having gone out of use just as Mesopotamia civilization was beginning. I have already cited disease and pests as factors in civilizational processes. There is no reason to expect that the European Medieval experience of the Plague was not preceded one or more times in the ancient Near East. Atra-hasis would seem to confirm it. Demographers, especially historians dealing with population records of Medieval and post-Medieval Europe, have only relatively recently begun to quantify the diverse records on population fluctuations. The light such quantifications shed on political history, previously written in terms of great men - or small men in great places - is sometimes astounding. We lack such precise and numerous records for the ancient civilizations, but there have been attempts made to estimate ancient population on the basis of site size. Even if accurate estimates of population cannot be reached with this method, at least some idea can be gotten of the types of settlements and some ranking and suggestions of functional relationship can be made. Such ranking, seeming to indicate trading and/or administrative/religious hierarchies (city, town, village, farmstead) allows the archaeologist to add substantially to the model of complexity gathered from written sources. Rankings imply hierarchies and bureaucracy. Distance of a settlement from a major center implies a loosening of authority, the possibility of autonomy and the formation of a base of alternative power within a state. The larger an empire becomes, the more liable it is to be subjected to divisive fractures, forces that work against unity. The farther an empire expands, the more disparate the units it encloses and the more likely it is for the whole to break apart. Attempts to hold the edges will result in weakening the core of the state. Compromises made in one reign, or in one year, although contributing initially to the unity of the state can in later times act as disturbing elements. A loyal soldier rewarded with land or an office in a distant part of the state may tend to favor his sons or kinsmen and wish to be succeeded in his office by his son, to the harm of the state as a whole.

I have, in this section introduced the personal element, because I believe that the interrelation of persons can bring about great change in the life of a civilization.

CONCLUSION

I was asked, in this paper, to discuss the breakdown of ancient desert civilizations. I have spent most of my effort outlining some basic natural and human factors that were as much in play at the beginning and continuation of civilizations in Mesopotamia. I say civilizations rather than civilization, because it is a bit difficult to talk of one Mesopotamian civilization. Likewise, it is impossible to speak of one breakdown. We generally take the end of Mesopotamian civilization as the time when the Achaemenid Persians conquered Babylon (539 B.C.), but this is just a convenient stopping point. We could just as easily speak of the <u>breakdown</u> of a number of earlier or later states. There is a cyclic character to Mesopotamia civilization. States rise, grow in complexity, begin to expand geographically, and at about this point there is evidence of violation of fallow, new irrigation projects, flight of tenant farmers to the city, and unwillingness of city officials to send people back to the country when they are needed as urban labor. Eventually, usually sooner rather than later, there comes a collapse and a concurrent or subsequent defeat by another power. There may be a long pause before recovery, then another area begins to grow and repeat the cycle. Usually, in Southern Mesopotamia, the new power is located in a place that is different from its predecessor. Salinity certainly plays a part in the process, and I think it is central to it, but rather than a progression from south to north as suggested by Jacobsen and Adams (1958) I would see an alternation between the south, center and north of the alluvial plain. Beginning in the Early Dynastic

Period, about 2400 B.C., the centers of power went from Kish, in the north of the plain, to Akkad (location unknown), to Ur (south), to Isin (center), to Larsa (south), to Babylon (north), to the Sealands (south), to Dur Kurigalzu and/or Babylon (far north and north), back to Isin (center), to the Sealands again (south), and finally back to Babylon (north) just before the Persian conquest. Salinity, rather than creeping north gradually, seems to have occurred wherever the greatest irrigation was being done. When it had had its effect on one area, another region, which had not irrigated as strongly or had had a long period to recover from a previous experience of salinization, would come to the fore.

Salinization, however, must not be seen as the only cause for the shifting of power. Changes in streamflow, the oscillation among the eastern branches of the Euphrates and the western branch or branches with related fluctuation of lesser or greater marsh area, must also be appreciated. I find it extremely interesting that in two periods (17th century B.C. and early first millenium B.C.) when we have evidence of abandonment of the main, eastern branches, the power center shifted to the Sealands - a marshy territory that included the head of the Gulf.

Natural and economic factors were and are very much affected by administrative response. Just as today in any country long-term planning is often counteracted by management decisions based on short-term, pressing, political realities, there were grand schemes countered not only by a lack of technology to solve the salinity problem, but also a competition within states between traditional social values and new values embodied in new men.

Other desert areas may share some of the features outlined here, but none will duplicate them exactly. There are, however, definite environmental constraints that have limited civilization in these areas and the great likelihood is that they are still operable. Study of present conditions will illuminate ancient realities and an appreciation of former civilizations and the factors that brought about their rise and fall might help to plan more wisely for the future.

REFERENCES

Adams, Robert McC. (1965), Land Behind Baghdad: A History of Settlement on the Diyala Plains, University of Chicago, Chicago.

Adams, Robert Mc C. and H. J. Nissen (1972), The Uruk Countryside: The Natural Setting of Urban Societies, University of Chicago, Chicago.

Gibson, McGuire (1972), The City and Area of Kish, Field Research Publications, Coconut Grove, Fla.

__________ (1973), Population Shift and the Rise of Mesopotamian Civilization, in C. Renfrew, ed. The Explanation of Culture Change: Models in Prehistory, Duckworth, London., 447-63.

__________ (1974), Violation of Fallow and Engineered Disaster in Mesopotamian Civilization, in T. E. Downing and M. Gibson, Irrigation's Impact on Society, Anthropology Papers of the University of Arizona, No. 25, University of Arizona, Tucson, 7-19.

Larsen, Curtis (1975), The Mesopotamian Delta Region: A Reconsideration of Lees and Falcon, Journal of the American Oriental Society, 95, 43-57.

Lees, G. M. and N. L. Falcon (1954), The Geographical History of the Mesopotamian Plains, Geographical Journal 118, 24-59.

Lambert, W. G. and A. R. Millard (1969), Atra-Hasis, The Babylonian Story of the Flood, The Clarendon Press, Oxford.

de Morgan, J. (1900), Etude geographique sur la Susiane, Memoires de la Delegation en Perse, I, 4-32.

THE KALAHARI DESERT

David I. Field[1]

ABSTRACT

This paper refers to the section dealing with the long-term carrying capacity of desert areas in relation to reliable water supplies and other physical constraints. I have tried to set out the present situation and then indicate the factors affecting the future potential. The Kalahari Desert constitutes about 1,000,000 km^2 of which 40% lies within Botswana. Within Botswana Kalahari sub-desert soils cover 70% of the country. Unlike the Sahara Desert, the Kalahari Desert is in fact a shrubland savanna, although the vegetative cover is part of an extremely vulnerable ecosystem. The basal cover of less than 5% and the loamy sand to pure sand soil with a rainfall generally below 400 mm per year emphasise the delicate situation which exists.

The grazing livestock population in Botswana, as opposed to the Kalahari Desert, consists of approximately 3,000,000 cattle, 1,000,000 goats and 400,000 sheep. In addition there are considerable populations of wildlife especially antelope species. The cattle density varies markedly from one part of the country to another, ranging from as low as 2 ha/mature livestock unit (LSU) in the east to as high as 75 ha/livestock unit in the west. Surface water is extremely limited and although there are a few perennial rivers on the eastern borders, the Kalahari itself relies almost entirely upon ground water obtained from wells or boreholes. Depending upon the year or season of the year, pans also form a rather unreliable source of water. Given the species of grasses and shrubs which exist in the Kalahari and the sandy soil, the major factor affecting the carrying capacity is the rainfall. The potential carrying capacity based upon mean rainfall varies from 8 ha/LSU in the north to 27 ha/LSU in the south west without improved management such as paddocking and the reticulation of water.

There are two aspects of Kalahari development discussed in the paper: first, the potential for expansion into new areas and second, improved management in the

---

[1]Range Ecologist

existing cattle post areas. With regard to further expansion, constraints such as lack of water, absence of communications and services, low availability of fodder and the competition with wildlife are discussed. With regard to improved management in existing areas utilised by domestic stock, the problem of desertification as it occurs in Botswana is described, referring particularly to the localised overgrazing around watering points. A new policy for grazing lands, initiated by the Government of Botswana in 1975, is briefly described in which the major aim is to zone the country into communal, commercial and reserved areas. In the course of implementing the new grazing policy, comprehensive land use plans are being prepared for each district. It is important however to stress the need for a strong educational programme aimed at all sectors of the community. The livestock industry is given high priority in Botswana and provided some of the measures discussed in this paper are fully developed, I see no reason why the Kalahari Desert, as it is called, cannot continue to support a valuable export industry.

INTRODUCTION

This paper refers to the section dealing with the long-term carrying capacity of desert aras in relation to reliable water supplies and other physical constraints. I have tried to set out the present situation and then indicate the factors affecting the future potential.

The Kalahari Desert as defined by Lebrun (1947) and modified by Devred (1975) constitutes 1,000,000 km^2 of which 40% lies within Botswana. Within Botswana Kalahari sub-desert soils cover 70% of the country (see Figure 1.) Unlike the Sahara Desert, it is in fact naturally a shrubland savanna as illustrated in Figure 2. However the vegetative cover constitutes part of an extremely vulnerable ecosystem. The basal cover is less than 5%, the soil is a loamy sand to pure sand and the rainfall is generally below 400 mm per year (see Figure 3) which together emphasise the delicate situation which exists.

The shrubland savanna (Weare et al 1971) consists of an upper canopy of Acacia species such as *A. girffae*, *A. tortilis*, *A. leuderitzii*, *A. mellifera* together with *Terminalia sericea* and smaller shrubs of *Grewia* spp. *Boscia albitrunca* and on stony areas *Combretum* spp. The layer of grasses and herbs is quite rich in different species, all of which are widespread throughout the Kalahari. Further mention of the grass species will be made later in connection with the result of overgrazing.

PRESENT DISTRIBUTION OF GRAZING LIVESTOCK

The grazing livestock population in Botswana, as opposed to the Kalahari Desert, consists of approximately 3,000,000 cattle, 1,000,000 goats and 400,000 sheep. The population of wildlife is by no means well documented but an approximation shows the following main species:

Elephant	15,000	Hartebeest	6,000	Impala	14,000
Zebra	30,000	Springbok	50,000	Kudu	40,000
Wildbeest	30,000	Eland	12,000	Waterbuck	1,000

The present distribution of cattle is of course closely associated with the human population (see Figure 4). An indication of the present distribution by districts is shown in the following Table 1 and the last column shows the variability in stocking rates between districts whose boundaries are shown in Figure 5.

Table 1. Cattle Distribution by Districts

District	Area (km^2)	Cattle ('000)	Stocking rate (ha/L.S.U.)
Chobe	20750	6	432
North East	5400	112	6
Central	145750	1108	16
Kgatleng	7650	174	5
Kweneng	36800	224	20
South East	1800	93	2
Southern	27000	268	12
Kgalagadi	106000	182	73
Ghanzi	115000	192	75
Ngamiland	109850	308	45
	576000	2667	27

(L.S.U. refers to livestock unit, i.e. one mature cattle weighing 500 kg.)

PRESENT SOURCES OF WATER

Surface water is extremely limited and although there are a few perennial rivers on the eastern borders, the Kalahari itself relies almost entirely upon ground-water obtained from wells or boreholes. Depending upon the year or season of the year, pans also form a rather unreliable source of water. Thus boreholes are the major source of water for the cattle and the distribution of boreholes corresponds to that of the human and livestock populations. There are about 5000 known boreholes in Botswana and probably a total of 8000, including those which have not been registered. Although the pans are unreliable, in recent years of above average rainfall, they have provided considerable quantities of water which have lasted for almost six months.

PRESENT CARRYING CAPACITY

Given the species of grasses and shrubs which exist in the Kalahari and the sandy soil, the major factor affecting the carrying capacity is the rainfall (see Figure 6). I have been amazed at the growth potential of the cattle in Botswana even though the fodder available in many areas appears sparse and of low nutritive value. The carrying capacities indicated in Figure 6 (Field 1975) are those suggested when neither the rangeland has been degraded, nor any form of ranch development has taken place. In some areas where the grazing pressure has been excessive, the present carrying capacity is lower than the general situation for that zone. In these areas, grasses such as *Anthephora pubescens*, *Brachiaria* spp. *Digitaria* spp. or *Schmidtia pappophoroides* have been replaced by *Eragrostis pallens*, *Aristida* spp, or *Pogonarthria squarrosa*; and where the pressure of grazing has been extreme the short *Eragrostis* spp have been replaced by *Aristida congesta*, annual species such as *Chloris virgata*, herbaceous weeds and finally only bare ground remains. This succession of grass species is associated with an increase in the proportion of undesirable thorny shrubs such as *Dichrostachys cinerea*.

POTENTIAL UTILIZATION OF PRESENTLY VIRGIN KALAHARI

The potential carrying capacity of the Kalahari Desert is influenced by numerous factors and only some of them can be mentioned in this short paper. Figure 4 indicates the present area of the country used by domestic stock and Figure 7 shows the areas of present land use. It can be seen that 41% of the Kalahari remains unutilized by domesic stock. The question then is: what prevents the further expansion of domestic stock into the Kalahari area?

Water Availability

The further west one moves, the greater the depth of sand over the underlying rock, the less that is known of the geological structure of the area and thus the increasing frequency with which one encounters black boreholes. One project in Ghanzi District obtained only two reasonably yielding boreholes (400 gallons/hr or 30 /min) out of 22 drilled. This points obviously to the considerable cost of such exploration in the unutilized areas.

In many cases even when water has been obtained it contains excessive levels of salts, rendering it useless for livestock let along human consumption.

Communications and Services

The relatively small human population of Botswana, about 800,000 means that there are few roads across the country, and services (clinics, schools, post offices, etc) are limited to the major towns or villages. This is a further constraint on people moving westwards. Farmers need materials, repairs and access to centres of facilities. A specific item of importance here is the marketing facilities: there are a few cattle trek routes with boreholes maintained by the Government, but there is a need for more even with the present distribution of cattle. Any expansion into new areas would again provide difficulties in marketing arrangements.

Pressure on Land Availability

There has been and still is a slow move to the west but it is not expected to increase in rate, since some people are leaving farming for urban or industrial life, and people tend to keep more animals round an existing water source rather than drill a new one.

Availability of Fodder

The areas which are unutilized at present are those which are drier and generally have a lower carrying capacity.

Competition with Wildlife

Although the areas devoted to wildlife are indicated in Figure 7 they are not of course restricted to those areas and in fact, their seasonal movements involve most of the Kalahari area. I should point out that as far as the carrying capacity of the desert is concerned, wildlife could well provide the optimum biomass to utilize the fodder which is available, both grasses and browse. The Wildlife Department is working on this subject and an extensive monitoring project is hoped to be financed with the help of the European Development Fund.

PROBLEMS IN UTILIZING THE RANGELAND

Even if the various physical constraints limiting the expansion of the domestic livestock industry westwards could be overcome, there are certain aspects which could seriously affect the areas both in use now and possible areas to be used in future. The overriding factor is that of management, which is closely associated with that of education.

Desertification

Desertification, defined as the intensification of areas of sparse or absent vegetation, is happening in the Kalahari today and will expand dramatically if measures are not adopted to combat the situation. Desertification in the Kalahari takes the form of localised degraded areas associated with excessive grazing pressure from domestic stock. A water source with 400 - 500 cattle shows little or no signs of deterioration, whilst a water source with 2000 - 3000 cattle can be completely destroyed for 2 or 3 km. and markedly degraded for up to 15 km from the water source. Since 71% of Botswana is classed as tribal land, as opposed to 5% freehold and 24% state land, this vast area is grazed in communal fashion with nobody bearing any responsibility for the condition of the rangeland. Until the people have been educated into the principles of sound range management, this deteriorating situation will continue.

The Government of Botswana has already realised this and commenced an active programme to rectify the situation. The first legislation was an Act of Parliament (Republic of Botswana 1972) which involved the setting up of Conservation Committees in the ten districts and an Agricultural Resources Board to be actively concerned with the agricultural resources of Botswana.

A Grazing Land Policy

In 1975 a Government White Paper (Republic of Botswana 1975) was published setting out the proposed policy to, in effect, combat desertification. The Policy called for the zoning of the country into Communal, Commercial and Reserved areas: the Communal Zone would remain much as at present - an area around villages where crops are grown and cattle for ploughing or milk production can be kept within certain limits. The Commercial Zone would be areas where either individuals or groups of farmers would be given exclusive rights to a specific area and in return they would have to pay a rent. The size of the areas allocated and the number of stock permitted would be detailed in a lease agreement. The Reserved Zone would be earmarked for such uses as mining, wildlife, forestry, future commercial or even future communal.

This Policy, dealing with the subject of land, is of vital importance to all those living in the Kalahari, and the Government of Botswana is proceeding with great care in order to ensure that decisions taken now will benefit the long-term carrying capacity of the Kalahari Desert.

HOPE FOR THE FUTURE

The moderate levels of stocking which have been indicated in Figure 6 could no doubt be doubled through the use of improved ranching methods: husbandry techniques, water reticulation and controlled grazing. The areas which are degraded at present could be rehabilitated, using existing technical knowledge, provided the socio-cultural problems can be overcome. I therefore forsee a bright future for the Kalahari Desert with the gradual expansion of the domestic livestock industry and possibly the wildlife industry into new areas, whilst at the same time restoring the presently degraded areas. I must reiterate the key factor in all this is a strong educational programme aimed at all sectors of the community, utilizing all available facilities of training centres, radio and mass communication media. There is a need to instil a sense of responsibility from school age up to senior government official level, and special training for ranch managers is an essential. The livestock industry is given high priority in Botswana and provided some of the measures discussed in this paper are fully developed, I see no reason why the Kalahari Desert, as it is called, cannot continue to support a valuable export industry.

BOTSWANA

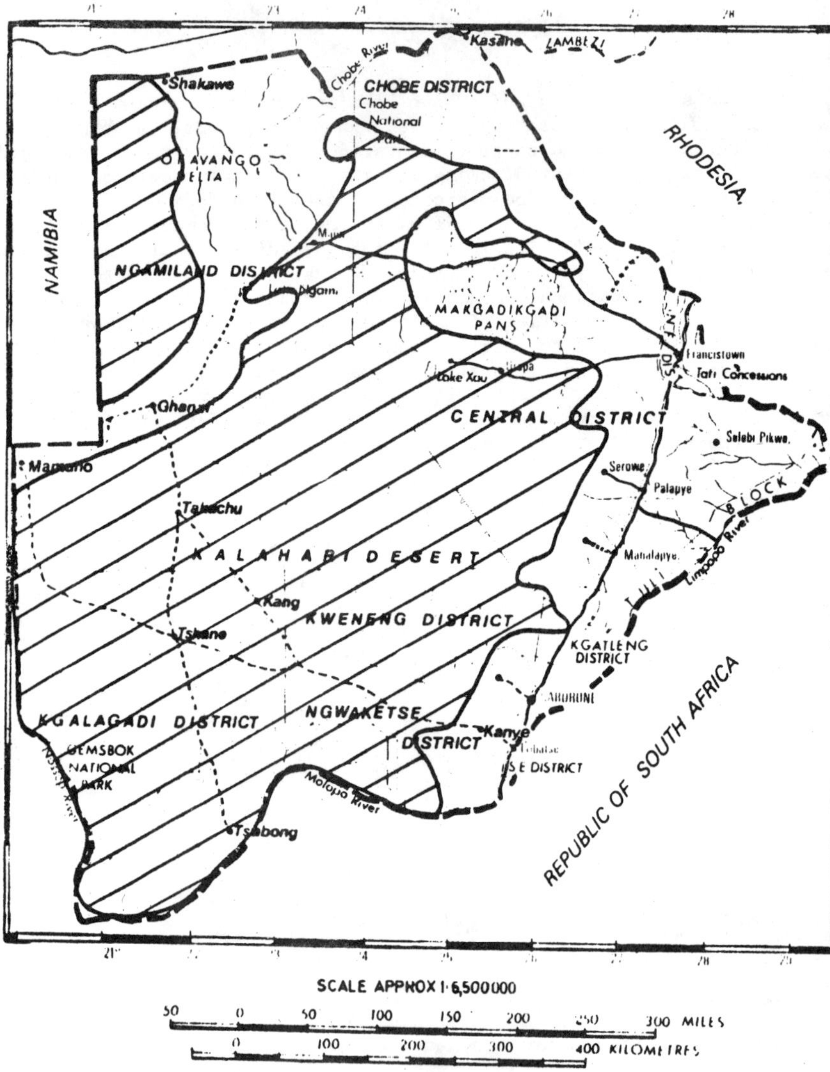

Figure 1. EXTENT OF KALAHARI SUB-DESERT SOILS IN BOTSWANA

Shrubland savanna vegetation

Sand dunes in S.W. Botswana

Figure 2.

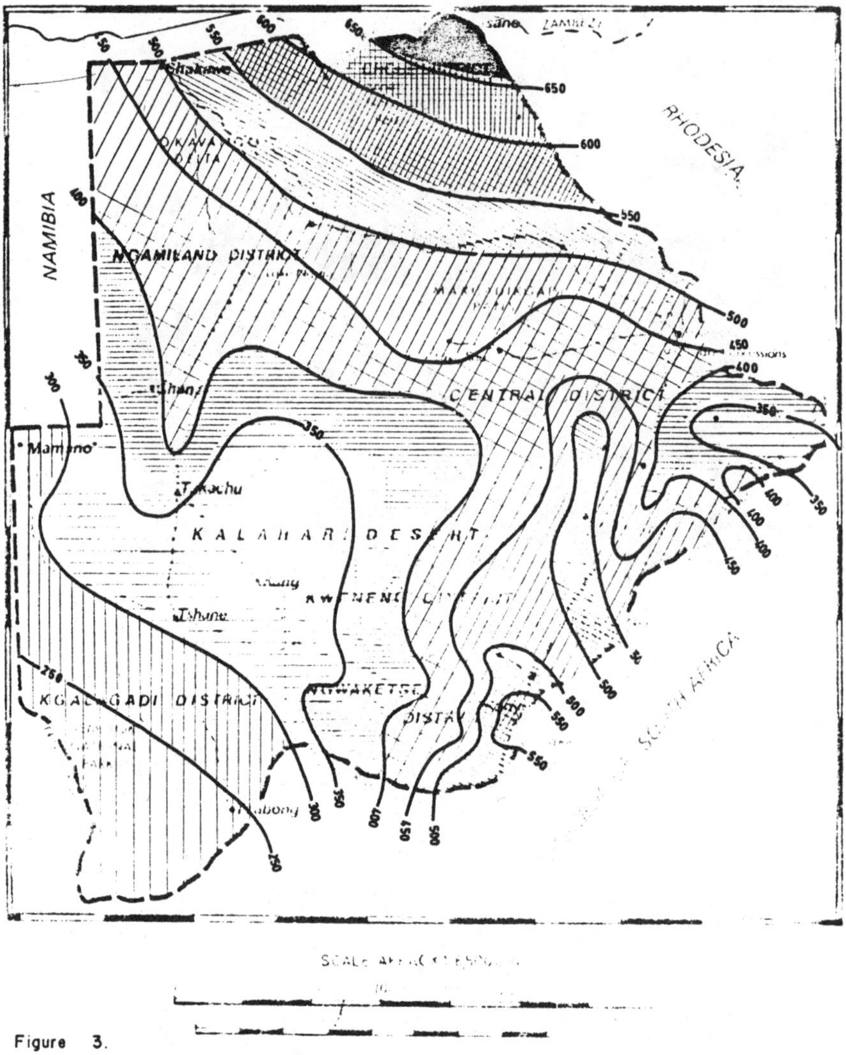

Figure 3.

MEAN ANNUAL RAINFALL

RAINFALL in Millimetres

Above 650	400 - 450
600 - 650	350 - 400
550 - 600	300 - 350
500 - 550	250 - 300
450 - 500	Less than 250

Based upon FAO/UNDP Project 359 Technical Document No 1 by Pike and Hearn 1971

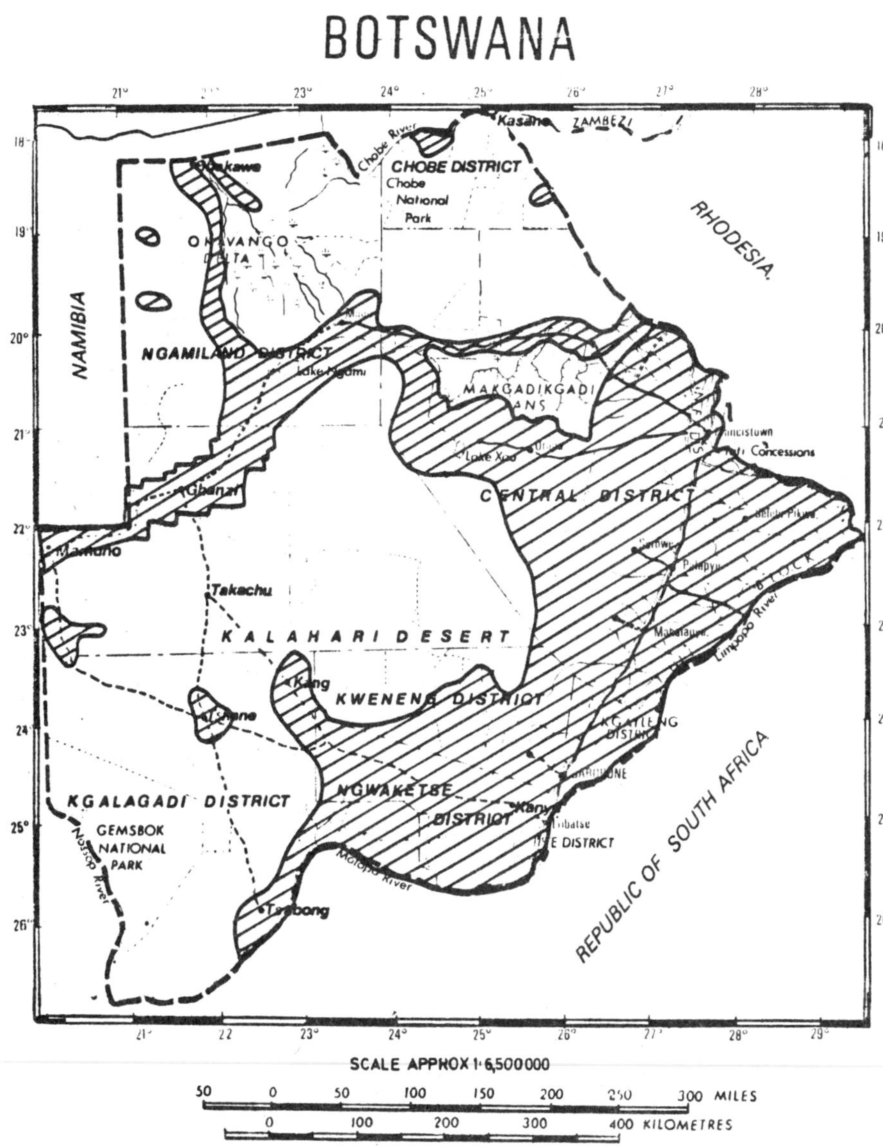

Figure 4. DISTRIBUTION OF CATTLE IN BOTSWANA

BOTSWANA

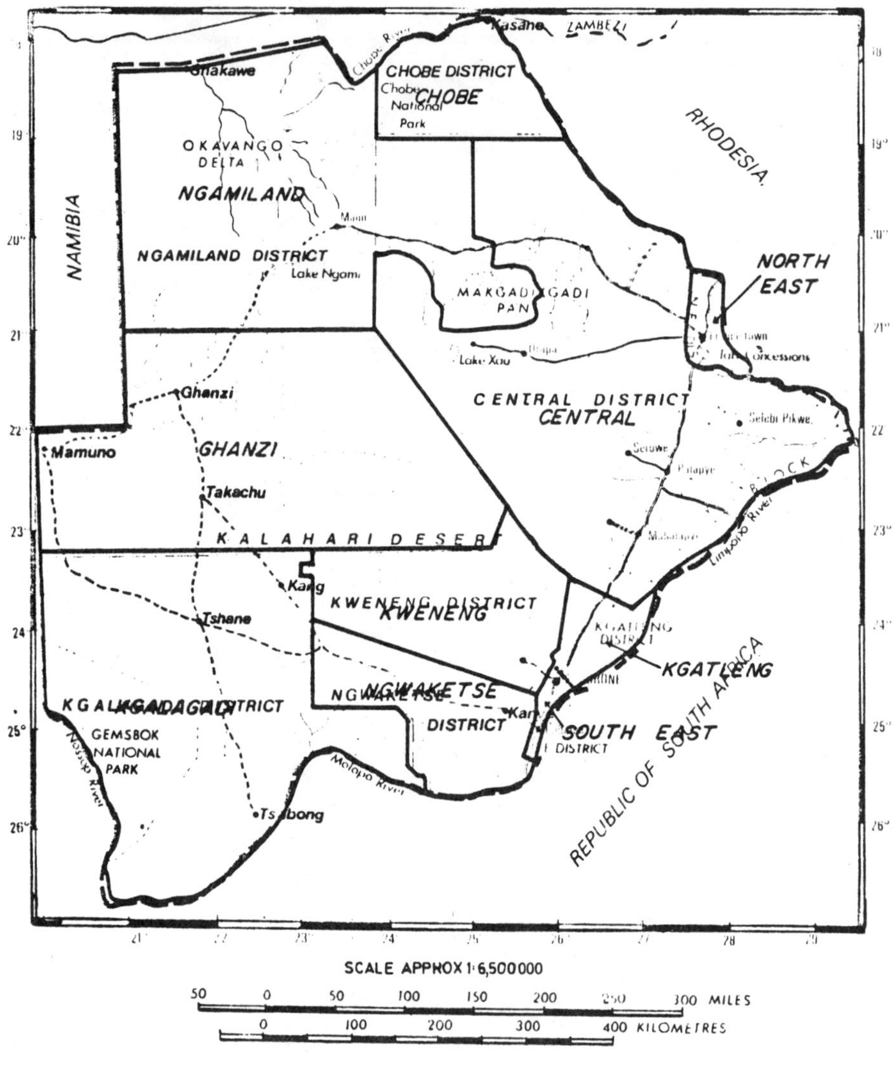

Figure 5. DISTRICTS OF BOTSWANA

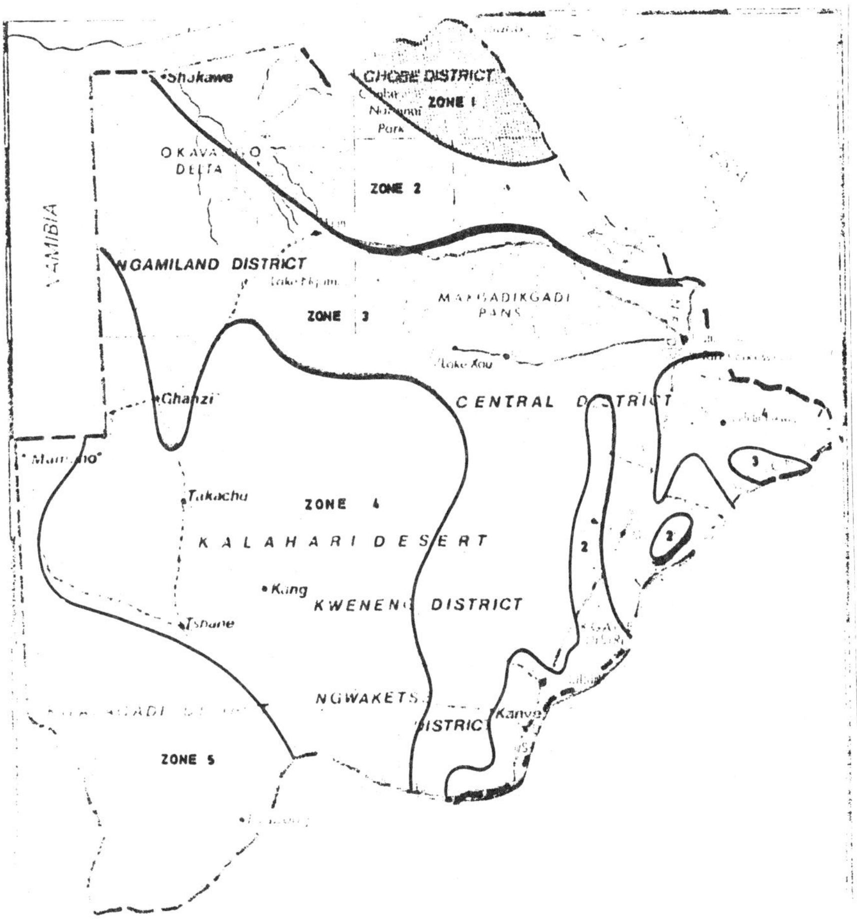

Figure 6.

POTENTIAL CARRYING CAPACITY

	ZONE	RAINFALL	CARRYING CAPACITY
☐	1	ABOVE 600 mm	8 ha/ L.S.U.
☐	2	500 – 600	12 "
☐	3	400 – 500	16 "
☐	4	300 – 400	21 "
☐	5	LESS THAN 300	27 "

By Field. 1975.

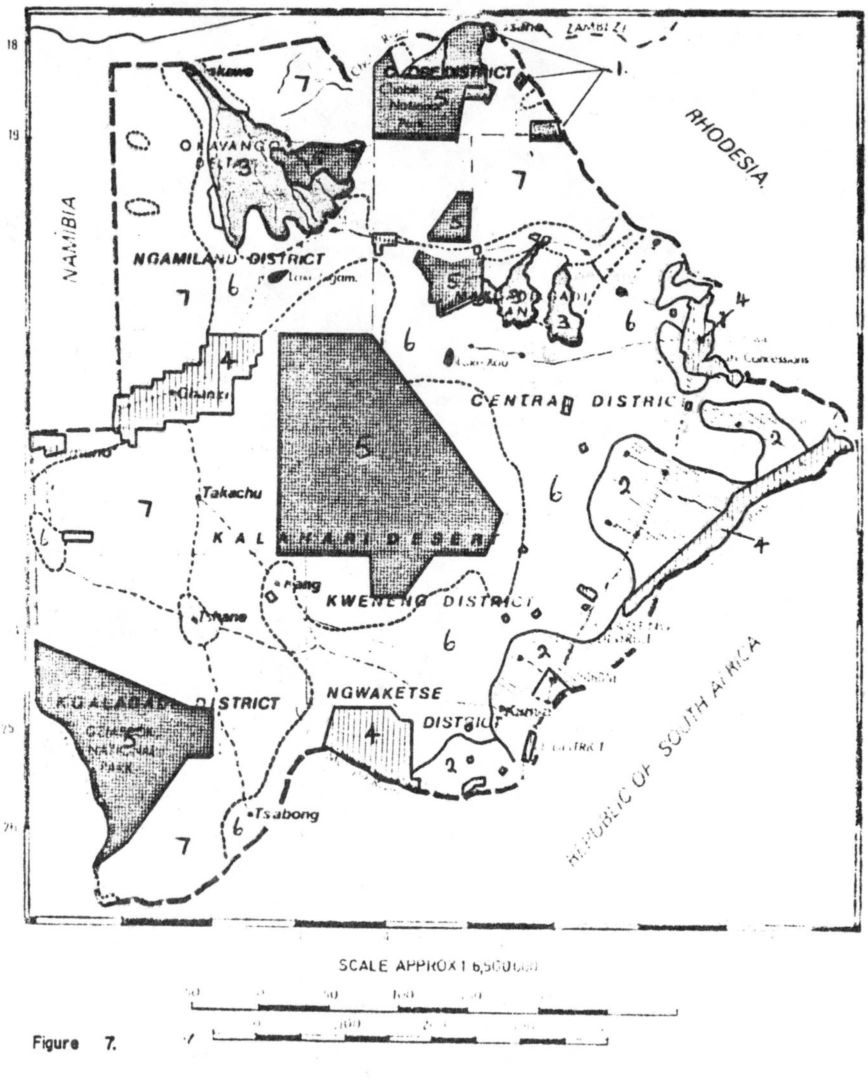

Figure 7.

PRESENT LAND USE

1	FORESTS	5	NATIONAL PARKS AND GAME RESERVES
2	ARABLE	6	CATTLE POSTS
3	OPEN WATER AND SWAMPS	7	UNUTILISED
4	FENCED		

Compiled by Field 1975

REFERENCES

Devred in Africa, The Devastated Continent? by Antoon de Vos, Dr. W. Junk b.v. Publishers The Hague 1975.

Field, D.I. A Handbook of Basic Ecology for Range Management in Botswana. Ministry of Agriculture, Gaborone (1975)

Lebrun, J. La Vegetation de la plaine alluviale au Sud du lac Edouard. Inst. Parcs Nat. Congo Belge, Explor. Parc. Nat. Alb., Miss. Lebrun (1937-38) I. 800pp 1947.

Republic of Botswana. The Agricultural Resources Conservation Act, 1972. Botswana Government Gazette 5 January 1973.

Republic of Boswana. National Policy on Tribal Grazing Land. Government Paper No. 2 of 1975.

Weare, P.R. and Yalala. Provisional Vegetation Map of Botswana. Botswana Notes and Records Volume 3. 1971.

A BRIEF REPORT ON DESERTIFICATION

Processes in Afghanistan[1]

INTRODUCTION

The general problem of arid and semi-arid areas with large populations is essentially one of human ecology. In the absence of adequate scientific knowledge and due to prevailing attitudes, the local population has over-exploited the natural resources resulting in an ecological imbalance.

Where large scale water resources from outside the arid region can't be tapped, the inherent meagre and fluctuating water resources within the region set the ultimate limit of production of plant resources which sustain both human and animal population.

Erratic rainfall results in fluctuating production, and this in turn leads to frequent scarcity conditions. As the population increases the stress becomes greater. Eventually, there is an increasing rate of imbalance between the human and animal population on the one hand and plant water, and land resources on the other. As the human demands persist and intensify, the resources tend to become further depleted. Thus there is set into operation a process of progressive degradation of resources, if continued unchecked, it leads to permanent damage of vegetation through excessive grazing or wood cutting for fuel leading to replacement by bare land, or at least by less useful plant communities, to the loss of surface soil by water or wind erosion or lower soil fertility and ultimately to production of large areas of wasteland or of barren sand dunes which are not only unproductive but may involve considerable public expenditure where they encroach on community services.

As a result of the cumulative effect of all these processes, changes are brought about in both the hydrology of the region and its micro-climate, which tend to create a more arid and less productive environment. Nevertheless such vast areas with abundant natural assets like plenty of sunshine, soil responding to manage-

---

[1]Prepared by the Forestry Department, Ministry of Agriculture, Kabul.

ment around the year, possibility of plant production and a healthy climate have definite potentiality for development.

Inherent in the desert problems is the necessity to recognize that the arid areas are supporting a substantial population and that methods of reclamation or utilization must be applied while the population continues to subsist on land. Arid and semi-arid regions have definite limitations imposed on their potential productivity, and in certain areas the existing population has already exceeded the limits of permanent supporting capacity.

It should be recognized that the physical and biological effects of desert expansion are important to the world community mainly because of their impact on human beings. Desertification is an aspect of the wide spread deterioration of ecosystems under the combined pressures of adverse and fluctuating climate and excessive or ill-advised exploitation. Such exploitation has diminished or destroyed the productive potential of food and fibre producing ecosystems at a time when increased productivity is needed to support growing population in quest of development. Important features in modern society, the struggle for development, population growth and change, and the effort to increase food production interlock in a network of cause and effect. Progress in development, planned population growth and improvement in food production must therefore be dealt with together. The deterioration of production ecosystems is an obvious and serious threat to human progress.

In general the quest for ever greater productivity has intensified exploitation and has carried disturbances by man into less productive and more fragile lands. Over-exploitation gives rise to degradation of vegetation, soil and water, the three elements which serve as the foundation for development in many nations. Therefore it is essential to combat desertification before it reaches disastrous stages.

CLIMATE

The climate in Afghanistan varies; it is under the strong influence of geographical altitude, relief and the Indian monsoons.

All except the extreme Northeast of the country is semi-arid or arid and precipitation decreases with altitude from an average 400 mm. to below 100 mm. in the lower plains. Due to low humidity evaporation is high and further exacerbated by strong winds.

The country is situated between 29^0 and 38^0 north latitude. The climate of Afghanistan is the climate of subtropical deserts and the Mediterranean. The desertic climate is the characteristic of the desert climate of the Marge Desert in the Southwest of the country, and for the deserts of the north plains near the Amon river. It is certain that the Mediterranean climate is characterized by dry and hot summers and cold humid winters.

In the eastern part of Afghanistan the climate is equally under the influence of the Indian monsoons, that come from the Indian Ocean. It should be noted that the influence of the Indian monsoons is more visible in the mountains at an altitude of 2000 meters and more. It is because of this reason that the best forests of conifers (*Cedrus deodora*, *Pinus gererdiana*, *Picea morinda*, *Pinus excelsa* and *Abies webbiana*) are located between 2,000 - 3,000 m of altitude. The high mountains of the east of Afghanistan play a positive role, because it is over these that the Indian monsoons condense water vapor as precipitation. At the same time these mountains are a barrier which do not permit the monsoon to penetrate further to the west in Afghanistan. For the same reason the humid monsoon climate covers only a small part of the eastern territory of the country. The high mountains are equally a barrier to the air currents of Siberia that can't penetrate into the country from the south.

In winter time, in the high mountains, precipitation in the form of snow is very high. That is why winter is the time for water accumulation, which in turn supplies water for the rivers of the country in Spring and Summer. Water from these rivers is used for irrigation to cultivate lands of the regions with dry or semi-desert climate in low altitude areas. And so, as we have seen, climate in Afghanistan depends upon many factors and because of this it is so changeable.

TOPOGRAPHY

The majority of reliefs in Afghanistan are mountainous and disjointed. It is the massive Hindu Kush, that begins from the high plateau of Pamir, from the east to the north of Afghanistan and continues towards the southwest with many divisions, that dominate the country. The mountains of Afghanistan have an average altitude of 3,000 m and most of these peaks are over 4,000 m. Some of the peaks have an altitude of 6,000 m and more (the plateau of Nouristan, Safid Kouh, etc.).

The north of these massive mountains with an altitude of 300-400 m are the northern plains of Afghanistan. In the Southwest of the country with an altitude of 300-400 m is the great desert of Dashti Marge. In Afghanistan we see most of the plains covered by deserts; these deserts form part of the great deserts of Asia, which are situated in Afghanistan, Iran, Pakistan and the USSR.

VEGETATION

When we cross Afghanistan by the principal roads of the country from the north to the south and from the east to the west we get the impression that there is no greenery or natural vegetation, and somehow most of the country if not the whole, seems to be like a desert.

We will see further that the study of ecological conditions in Afghanistan from the point of view of rain is quite difficult. There are desert regions, semi-desert regions, dry regions and humid regions, and because of the insufficient rain in these areas it is difficult for the natural vegetation to grow easily, especially in the presence of human activities, which destroy most of it (for example by over-grazing and deforestation). Therefore, we must underline the fact that natural vegetation growing in the country corresponds to the ecological conditions of Afghanistan and they are very tough. The problem is to find a scientific way to begin recultivation.

Man has always profited from the natural vegetation as a principal mean to survive under the forms of pasture areas, materials for construction and fuel. The process of destruction of the natural vegetation has gone on during 2,000 years, as a function of the progressive growth of population. In other words this process is caused by the over-exploitation of the natural resources and by not undertaking cultural measures to improve it. The problem of destruction of plant resources in Afghanistan has become very serious, because it is the question of destroying the natural environment.

DESERTIFICATION PROCESSES IN AFGHANISTAN

Historically, the predominant form of land use was transhumant. Large sectors of the population moved seasonally from the plains, exploiting each altitudinal niche at the most favourable time of the year.

Though naturally arid, some parts of the desert areas of Afghanistan have supported considerable populations in the past. Four major river systems rise in the major mountain ranges (Pamir, Paropamisus, Kuh-i-Baba, Hindu Kush) that occupy the central and northeastern parts of the country. These rivers - the Oxus, Harirud, Helmand and Kabul rivers - divide the country into four major catchment areas. Nimroz, which is watered by the Helmand and adjoins Iran in the southwest of the country, was known as the granary of southwest Asia in the 9-10th centuries. Historical records attest great expenditure even then, not only on maintenance of the irrigation systems, but on measures to reduce waterlogging and salinity.

For the purpose of the coming United Nations Conference, desertification is defined as the loss of productivity in arid and semi-arid lands, especially where such loss is associated with human activity and might be prevented or reversed. The Helmand valley in the southwest of Afghanistan is at once one of the most arid and potentially one of the most productive areas of the world. It has been heavily exploited in the past and it was here that economically serious desertification was first experienced on Afghan territory. After a long period of decline, redevelopment of the Helmand River Valley was begun in 1948. It still continues, but the same problems were very soon encountered and presently constitute a major planning concern.

This type of desertification is most severe on the Helmand - both in terms of cultivated area affected and economic value lost - but it is not unknown in the rest of the country. Under normal conditions as estimated 3.9 million hectares of land are cultivated annually, of which 2.6 million hectares are irrigated. A number of surveys show that the cultivated land at present constituted half of the total cultivable area of the country while land presently irrigated is 50% of the total irrigable land (though due to lack of irrigation networks or proper water control facilities only 1.4 million hectares receive adequate water throughout the year). Nearly 85% of the irrigated land receives water through canals - from the Helmand and other major rivers (though not the Amon, which mostly flows below the level of the arable land on its Afghan bank) and smaller systems that flow north out of the central mountains and are lost in the plains before they can reach the Oxus. The remaining 15% comes from such sources as springs, <u>karez</u> (qanat), and wells. A significant preparation of all these is the sum of thousands of small village systems in and around the central mountains, many of which suffer from salinisation.

Whereas many studies have been made of irrigated farming especially in the Helmand, and much is now known of the nature of the ecological problems that develop from it, other types of land use have received very little study. Not only is little known of the desertification problems that are associated with them, but in most cases there is not even public or explicit Government consciousness of their existence or significance. Though at first this may seem surprising, it should not in fact be so. Afghanistan's major development effort is only beginning now. With the exception of waterlogging in the Helmand, little has yet happened that might be expected to lead to a national consciousness of desertification.

The unfavorable climatic conditions in the years 1970-1971 reduced agricultural output severely. The net value of agricultural products, which was Afs. 51

billion in 1968, dropped to only Afs. 43.4 billion in 1972, constituting a 15% decline. However, as climatic conditions returned to normal and important measures were taken by the Republican Government, productivity recovered.

Nevertheless, the symptoms of desertification are already there - glaringly so for the natural ecologist. The current Seven Year Plan lays emphasis on agricultural development as the basis for industrial development. Continued population growth (currently estimated at 2.3% but likely to rise), together with extension and intensification of agricultural production will very soon exacerbate the ecological problems which the specialist already points to, and lead to a general decrease in productivity and concomitant lowering of living standards.

The following desertification processes are already evident in varying degrees in Afghanistan.

1. Waterlogging, salinity and alkalinity due to increased irrigation
2. Loss of soil due to extension of dry farming of unsuitable soils
3. Reduction in quality and quantity of vegetation cover due to overgrazing
4. Accumulation and movement of sand on to arable land and rangeland
5. Deforestation

Waterlogging, Salinization, Alkalinization

The Government has for some time given high priority to increasing the land area under irrigation, but relatively little attention has been given to the provision of adequate drainage. The drainage problem has now reached such a level of severity that major efforts are essential to prevent further deterioration of valuable agricutural land. Unfortunately, as already stated, few figures are available. Only on the Helmand, where the heaviest investment has been made, have quantitative studies been carried out. The following figures indicate the severity of the problem:

Land Distribution by Salinity Level (hectares)

Tracts	Total ha.	Very severe ($ECx10^3$-16+)	Severe ($ECx10^3$-0 to 16)	Moderate ($ECX10^3$-4 to 8)
Nad-e-Ali	12,406	1,060	4,000	7,305
Marja	19,520	1,820	9,400	8,300
Shamalan	28,800	4,200	4,400	20,200
Darweshan	20,000	2,160	3,240	14,600

According to these figures, severe or worse salinization exists in from 27% (Darweshan) to 57% (Marja) of the land. It has been established that salinity levels correspondng to an EC value of 9 or 10 may be expected to result in a 50% decline in yield for most crops. This means that from a quarter to a half of this development land is yielding no more than a half of its potential.

Already in 1956 according to one report 35% of the irrigated land on the Helmand was believed to need drainage. It is now estimated that as much as 70,000 hectares would benefit significantly from drainage and leaching. In the upper and lower Helmand areas taken together (in the provinces of Helmand and Nirmorz) there are 518,000 hectares of potentially suitable land of which it is presently estimated that 310,000 hectares could be developed and irrigated. So far less than 100,000 hectares of this is provided with an adequate water supply. The Government considers irrigation systems to be the basic infrastructure of development. In order to exploit to the full the productivity of this vast resource, heavy investment and long term planning are required in order to avoid further desertification and to reclaim the productivity already lost. This is fully recognized in the current Seven Year Plan.

Soil Erosion

Given Afghanistan's climate and topography, soil erosion is a constant danger. Traditional land use patterns have increased the danger. Pastoralism and wood cutting may lead to the removal of the protective cover of vegetation, and thence to soil erosion, but an even greater danger can develop from the extension of dry farming. The current Seven Year Plan aims at increase in productivity per hectare, rather than in cultivated area. In 1975 the total cultivated area was 3,882,000 hectares, which is not significantly different from that of 1968. But the per hectare yield of grain, cotton and sugar beet went up to 15.5%, 10.8% and 61.3% respectively. During the same period the Agricultural Development Bank financed the sale of 1,395 tractors and 1,224 water pumping sets. The current Plan aims at the continuation of these trends. By 1982 the total area under cultivation will be increased by only 1.4% to 3,936,000 hectares. The most important nonirrigated crop is wheat. Out of a total 2,350,000 hectares under wheat in 1975 1,050,000 were unirrigated. According to the Plan this unirrigated wheat area will not change between now and 1982. However, reports suggest that large areas of former rangeland have been taken over for cultivation during the last decade. The increase in the number of tractors, facilitating cultivation, lends credence to these reports. Where this has happened, it is likely that productivity will fall off after the first year or two of cultivation, and the land will be abandoned. Deprived of its protective cover of vegetation and exposed to wind and runoff the soil will start to erode. The danger is greatest on the northern slopes of the Paropamisus where climatic conditions are most attractive. Careful socioeconomic planning and environmental monitoring and called for to avoid this desertification process.

Over-Grazing

Over-grazing is a second example of a desertification problem that has long been known to exist in Afghanistan, and is of great economic significance but has not yet become critical enough to ensure action.

Sixty two percent of Afghanistan or 40 million hectares is currently considered to be rangeland, but there has been no evaluation of their capacity. They are presently estimated to be stocked with a maximum of 30 million animals, of which over 80% are sheep.

Livestock is one of Afghanistan's main economic sectors. Animal products constitute 18% of the Gross National Product. The animal population, which had attained a peak level in the late sixties, was drastically reduced as a result of the unfavorable climatic conditions of 1970 and 1971. There was an estimated loss of 40% in sheep and 6% in cattle. During the last few years, however, the volume of sheep stock has grown substantially on account of favourable climatic conditions. In view of the size of the livestock sector and the extensive possibilities for growth, the current Plan aims at a 49% increase in the net value of animal products by 1982.

In order to make this possible, and to ensure optimum use of the country's rangelands, a detailed survey is planned. Based on the findings of this survey, a long term policy will be formulated for the development of the livestock sector. Special attention will be given to increasing productivity and establishing a rational balance between animal population and range productivity. Experimental centres will also be established, and the cultivation of fodder will be encouraged in areas covered by development programmes.

Conversion of rangelands to dry farming, which continues to increase in response to the demand for foodgrains for the human population, has already been noted as a problem. In sixteen provinces, in the northern, northeastern, northwestern and central parts of the country, it is planned to set aside some 30 million hectares of public pastures for protection and management.

Despite these measures, there is need for much greater effort in range management in order to ensure optimum long term exploitation of the country's rangelands, which are among its most important basic resources. Due to their vastness, an efficient monitoring programme is essential as a basis for management.

Sand

No detailed information is available on Afghanistan's two large areas of sand: one in Helmand province in the south, and the other bordering the Amon in the north. Both areas are grazed; the former encroaches on cultivation in the Helmand Valley. Stabilisation has been attempted on 500 hectares in the north by means of saxaul plantation. Seventy hectares are reported to have been successfully stabilised. Detailed studies are needed on sand movement and associated human use systems in both areas.

Deforestation

In Afghanistan, forests cover a total area of 1,978,000 hectares of which 450,000 hectares are located in Paktia Province, 444,000 hectares in Kunar and 1,093,000 hectares are scattered in other parts of the country. The existing stock of timber in the coniferous forest regions is estimated at nearly 15 million cubic metres, and the volume of wood suitable for fuel is estimated at about 14 million cubic metres.

According to preliminary surveys, it appears that these forests are being exploited too fast. The annual growth of timber is estimated at 126,500 cubic metres, of which the quantity exploitable is estimated at 70,600 cubic metres. However, the actual exploitation of this item is equal to nine times the reasonable exploitation capacity. Similarly, while annual production capacity of firewood is estimated at 60,000 cubic metres the rate of exloitation is 200% more than that. From the above figures it is concluded that an estimated one million cubic metres of wood are cut every year while the annual productivity is 100,000 cubic metres. Moreover, the wood is cut unsystematically and with no provision for reforestation.

Assuming that 100 cubic metres of wood is produced per hectare nearly 10,000 hectares of forest area is presumably being destroyed annually. If this situation continues, one of the most valuable natural resources of the country will be destroyed in as little as 10 to 15 years.

A number of programmes are being introduced to correct this tendency, including surveys, road construction, conservation and reforestation. All the forests have recently been declared the property of the state, and the Institute of Environmental Studies in the Forestry Department of the Ministry of Agriculture is engaged in the preparation of detailed vegetation maps. The Government recognizes that forests constitute one of the country's most valuable natural resources and that exploitation of them must be carefully monitored. It will however take time before a fully effective management programme can be implemented.

Responsibility for dealing with these problems is spread over a broad range of government departments. Many of these departments are in the Ministry of Agriculture, and several of those come under the Department of Forests and Range. In the past, coordination between these departments was not adequate, but it is hoped that with the creation of a National Committee on Desertification in accordance with the proposed Transnational Monitoring Project, coordination problems will be solved.

Man-power problems remain. Though the quantity and quality of trained personnel varies greatly among the relevant departments, in general there is a severe shortage of people trained in ecological subjects. One of the most important results of any Conference on Desertification for Afghanistan could be assistance in the training of personnel in ecological subjects, and provision of foreign experts until such training can take effect.

Apart from the institutional and manpower base for the fight against desertification, attention must be given to the provision of an adequate natural resource inventory. Certain basic information is available. For example

1. Topographic maps: the entire country has been mapped at large, intermediate, and small scales, and aerial photographs are available at a scale of 1:60,000 from 1956-1958.
2. A climatic station network was begun in 1939 with the establishment of a meteorological service and nine stations. Prior to this Kabul was the only place for which any records exist. The service was reinforced in 1956 with the incorporation of Ariana Afghan Airlines. By 1969 thirty three stations had been established, most of them with instrumentation set up in accordance with the technical regulations and guidelines of the World Meteorological Organisation of the United Nations. Since then the number has been further increased, and the present total is 51 synoptic stations, 9 agroclimatic stations, 100 climatological stations, and 130 pluviometric stations. An eventual total of 440 climatological and pluviometric stations are planned. In addition to these, there are seventy or more stations with varying periods of record in Pakistan, the U.S.S.R., and Iran which provide data of direct relevance to Afghanistan. Monthly and annual publications make these data available.
3. Detailed work on soils is available only for major agricultural areas. The only generalized soil map that is available for the whole country is at a very small scale ($1:12.5 \times 10^9$).
4. Vegetation mapping is being carried out by the Institute of the Environment. A land use map of the entire country has been prepared at a scale of 1:50,000, and

vegetation mapping (with an emphasis on forests and rangeland) is already available in different degrees of detail for different parts of the country at scales of form 1:50,000 to 1:500,000.

5. 136 hydrological stations of various types have been set up. Their location and the basic drainage network is mapped, and records are held by the Ministry of Water and Power.

Though gaps exist and improvement is continously being sought, Afghanistan has already developed a natural resource data base for the monitoring programme envisaged in the proposed Transnational Project with India, Iran and Pakistan. At our present stage of economic development we stand to benefit greatly through this project and from other knowledge, expertise, and training facilities that are expected to become available as a result of conferences on desertification.

Coordination is a problem not only on the national and regional levels. In a country like Afghanistan with serious manpower shortages and organizational problems, a major difficulty is presented by the apparent lack of coordination among the large number of offers of assistance and invitations for participation that arrive continuously from various agencies of the United Nations. We should like to take this opportunity therefore to request that much greater care be taken to explain each new international activity explicity in the context of previous activities and in particular of past and present projects of other United Nations agencies that might be considered relevant.

In conclusion, it may be said that although desertification is not new in Afghanistan's experience, it has not yet been widely recognized as a national problem, but it is expected that it will be an increasing problem from now on. Our main difficulty in planning to combat it lies in shortage of trained personnel. Ideally we need high level training abroad for potential decision makers that will enable them to conceptualize and plan. We need assistance in improving domestic training facilities for the medium and lower levels of personnel. Study tours of areas with comparable technical and management problems in other countries would be useful at all levels.

DESERTIFICATION IN THE UNITED STATES[1]

Jack D. Johnson[2]

ABSTRACT

This paper is an attempt to indicate the extent of the desertification problem in the United States. It includes a historical review of the land-use problems in the arid and semi-arid regions of the United States, discusses some of the current problems and emphasizes the need for concerted and sensible land-use planning and management if the U.S. is to curb its degradation of the semiarid and arid western lands and avoid the creation of deserts in those areas where land is being misused. Desertification is not a problem for the developing countries to tackle alone. Indeed it is a problem which the world must confront. The United States shares this problem, can offer some solutions, and will definitely benefit in the exchange of ideas and methodologies which will be presented throughout and following the U.N. Conference on Desertification.

OVERVIEW

The total population (based on 1970 census) of the United States presently exceeds 203 million. The population of the 12 western states exceeds 45 million and is a little over 22 percent of the total population of the U. S. These same 12 states comprise about 40 percent of the total land area of the U.S. and almost all of the arid portions of the nation. (See map) Approximately 23 million people reside in the arid regions of the U.S.

The 12 western states constitute a land area of about 1.4 million square miles. Seventy-two percent of this land, or in excess of one million square miles, is within the region currently or potentially subjected to desertification processes.

---

[1]Presented by E.L. Cockrum

[2]Director, Office of Arid Land Studies, University of Arizona, Tucson, Ariz.

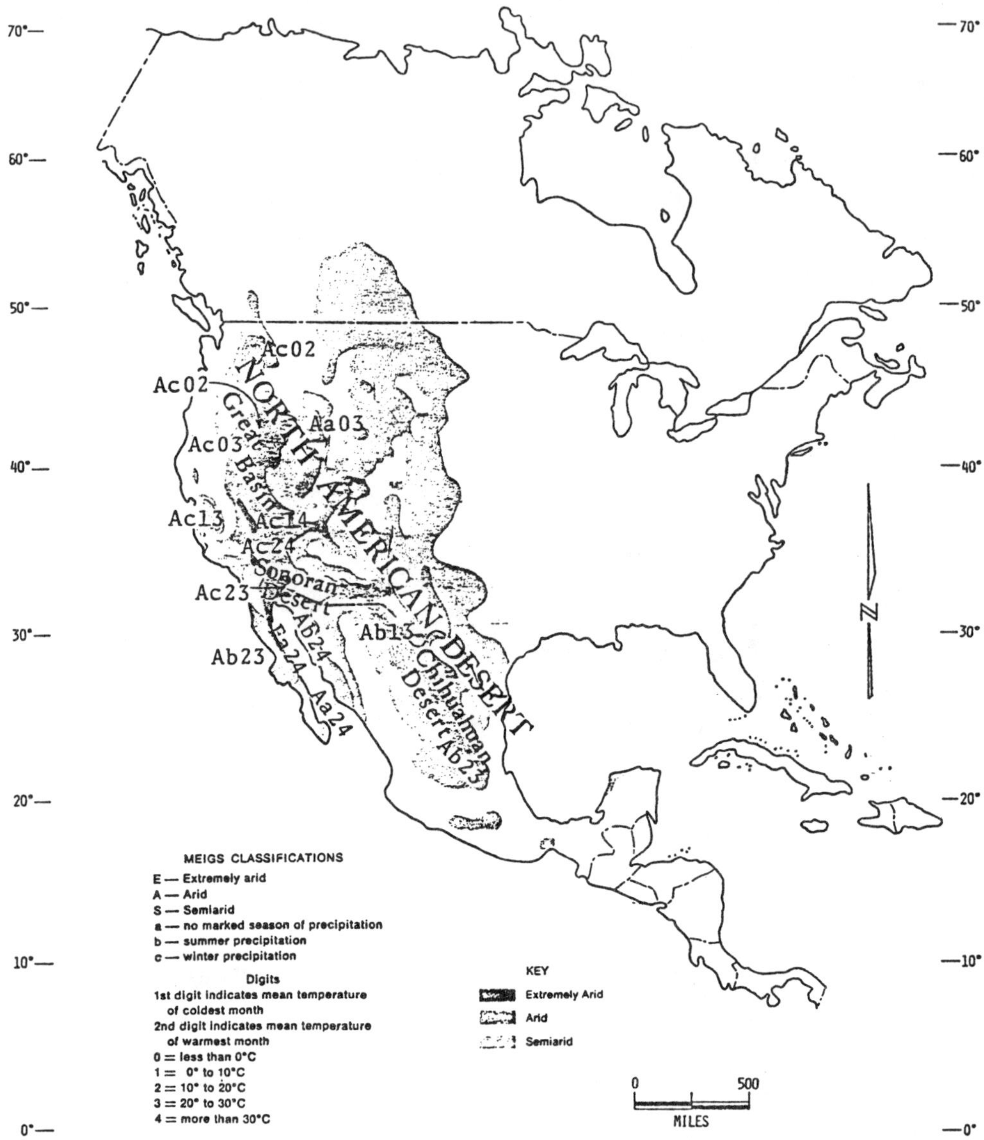

Arid Lands of North America (after Meigs)

Population growth in the West during the last decade almost tripled that of the eastern states. The combination of 1) immigration, 2) migration from Mexico, north-central states and eastern states to the West, and 3) general population growth because of local birth rates is causing an alarming increase in the population of many of the western states. Land developers, chambers of commerce and individual businessmen expecting financial rewards are advertising and exploiting the arid regions.

In the populous eastern states, the population density is high. In New Jersey, for example, there are over 953 people per square mile of land area, and in Washington, D.C. there are more than 12,400. In California, even with its large urban centers, the population density is only about 128, and within the other western arid states the population density is considerably less. For example, Arizona has about 16 and Wyoming about 3.4 people per square mile. This seems to indicate room for growth, but unfortunately growth from the large urban centers is rapidly expanding into the deserts and up into the mountains with little regard for the resources of arable land, water, vegetation and wildlife.

Archaeological sites, ghost towns, abandoned ranches and empty farms that dot the arid West are dramatic examples of man's early failure to obtain a balance between his natural arid environment and his desire for expansion. When the earth was sparsely populated or uninhabited, land was abundant; man could abandon old environmental failures and move onto new lands. Today, the situation is much different; with expanding world population it is impossible to abandon or ignore the arid lands of any country.

Within the U.S., 60 to 70 percent of the arid lands are in federal or other public ownership. Some 130 million acres in arid lands states have been placed under multiple-use, sustained yield management within the national forests, which are administered by the U.S. Forest Service. Recent passage of the Federal Land Policy and Management Act has overcome some of the policy indeterminacy of the U.S. public lands, which are administered by the Bureau of Land Management. The new Act addresses the multiple-use and sustained yield management concept through an interdisciplinary land-use planning system.

One of the attributes of the rugged, arid western United States is that one can drive from the hot desert floor or large urban area into the dense mountain forests, often within a matter of minutes and never more than a few hours.

Irrigated agriculture is an important aspect of arid regions, and some irrigated portions of Arizona and California deserts comprise the most productive land in the U.S. It may not, however, be the most desirable or economically efficient use of water, and the expansion of irrigation farming farther into the desert requires a critical review. Much of the arid and semiarid areas is used as grazing land, and some of the arid regions are rich in mineral deposits, particularly in copper and coal. (Over 50 percent of U.S. copper is mined in Arizona.) Energy potential (solar, coal, uranium, oil and gas, and hydroelectric) is being exploited for use in large urban centers, particularly for Phoenix, Las Vegas and Los Angeles.

Recreation is rapidly becoming an important economic use of arid regions in the U.S. Surveys in desert recreational areas of Southern California during 1968 indicated that there were 570,000 people on one-day field trips; 1.1 million overnight trips; and about $30.9 million expended by these recreation users. Projections based on 1968 statistics of 4.9 million visitor days are that Southern California deserts will see 7.8 million visitor days in 1980 and 13.2 million by the year 2000.

One of the recreational uses which has recently received much publicity (mostly negative) is the use of motorcycles, "dune buggies" and four-wheel drive vehicles in remote desert areas. Because of the climatic and biotic conditions, physical scars on the landscape remain for centuries before nature can hide them. (For example, some of the tracks left by wagon trains crossing the arid regions of the western United States in the 1840s are visible today.) The lack of moisture slows natural decay processes and allows discarded waste products and containers to remain almost unaltered for long periods of time. (For example, a discarded paper tissue on the desert floor may remain as a visible symbol of human thoughtlessness for many years.) Pits from mining operations could possibly serve as recreational facilities or as sites for other industrial development, but usually they simply stand as a reminder that man can dig a big deep hole in the ground.

Desertification, as a definitive process, has received only recent attention in the United States. But if we accept the concern regarding misuse of arid and semiarid lands, desertification concerns began sometime in the 1800s. In 1947 Dr. Walter P. Cottam authored a University of Utah Bulletin (Vol. 37 No. 11) entitled "Is Utah Sahara Bound?" Had he been able to predict today's use of the term "desertification," he might have used some other title such as "The Extent of Desertification in Utah." While economic data have not been researched for presentation herein, desertification does have serious economic impact. Arid land recreation and agricultural land are producers of a significant national income, and any reduction of that income through the desertification process has had and will have national consequences.

A recent disastrous drought of a few months' duration on the Papago Indian Reservation required emergency funds and services from several federal agencies and brought crippling losses of cattle and crops. The earlier experiences of the "dust bowl" in Oklahoma, the land reclamation programs, and the assistance to Indian reservations have totaled billions of dollars - and both the benefits of these expenditures and the burden of these costs are borne nationally. It should be emphasized that the U.S. experiences with desertification, reported herein, should be relevant to other arid and semiarid areas of the world, with the possible exception of some of the recreational uses.

HISTORY OF LAND USE IN ARID REGIONS OF THE UNITED STATES

There is evidence that man has lived in the arid and semiarid regions of the U.S. for as long as 11,000 to 25,000 years, and ecological adaptations to the arid environments indicate a zone of extreme aridity in the Great Plains between about 5500 and 2000 B.C. Evidence from caves in west central New Mexico indicates that the period from perhaps 4000 B.C. to a few centuries before the beginning of the Christian era was a time of life based on cultivation. Established agriculture, however, did not replace the ancient food gathering practices until about 300 B.C. It was with the growth of cultivation and the improvement of agricultural techniques that the American Indian became capable of mastering many of the arid and semiarid zones of the continent.

Until the middle of the 19th century the vast arid/semiarid region lying between the Rocky Mountains and the ranges bordering the Pacific Ocean was virtually unknown to the anglo. A century earlier Spanish missionaries had penetrated a small part of southern Arizona, introducting cattle to the scattering of Catholic missions they had founded there. They also made occasional trips across the intervening desert to the chain of similar missions along the California coast. These and other travelers, such as trappers and traders, did not write detailed descriptions of their travels; maps were not published prior to the 1840s.

The discovery of gold, which was in 1848 in California, brought tens of thousands of people across the desert region en route to the gold fields, but the arid area still lay virtually empty and unused until the latter part of the 1860s. From the early 1800s through the late 1860s grazing animals were brought into the Southwest by Spanish and Mexican missionaries and cattlemen. In the late 1860s there occurred an influx of settlers, chiefly comprised of cattlemen from the semiarid regions farther east. California presents some interesting statistics regarding sheep population which had increased to 1.9 million in 1860, 2.8 million in 1870 and 4.4 million by 1890. At that time, conditions in the arid West were far from stable and the nation was still recovering from the Civil War. Partially settled western territories were poorly administered, and the military and police forces were inadequate.

The following half century was the era of the open range. Herders laid claim to the widely separated watering places and thus held practical if not legal control over the vast waterless grazing lands. Land tenure was the major problem in the United States. The first attempt at control was the development of a General Land Office as an agency of the U.S. federal government to administer and apportion public lands of the U.S. Under the Homestead Act of 1862 any citizen could attain title to 160 acres, but this Act was principally created for the purpose of developing the agricultural land of the East and the Midwest. One hundred sixty acres in the eastern and midwestern United States was adequate for maintenance of moderate to high standards of living, but land holdings of such small size were totally inadequate for enterprises in a semiarid or arid pastoral region. In most of the arid regions 160 acres would support no more than two head of cattle.

The Desert Land Act of 1877 provided that title might be obtained to a full section (640 acres), but that Act had only limited effect since it required that part of the land be irrigated. Most of the land in the arid West was not irrigable. In 1879 a federal commission headed by John Wesley Powell turned in its report on the governmental policies concerning the public lands of the arid West. The report stated that..."the Homestead and Preemption laws are not suited for securing settlement of more than an insignificant portion of the country."

Powell's report went on to suggest: 1) that land should be scientifically appraised and classified before settlement and that each category should be handled under laws specifically applicable to it; 2) that land must be disposed of in quantities sufficient to the establishment of working enterprises, and that the price of such a grant must be kept low...at least four square miles (2,560 acres) was proposed as a minimum; 3) that farm and ranch residences be grouped to permit a form of social life not possible on isolated ranchsteads; 4) that surveys should deviate from rectangular systems when necessary to divide irrigation waters properly; and 5) because of the great expense required to develop irrigation, irrigation schemes should be undertaken under the auspices of the federal government.

Unfortunately, no action was taken by the Congress of the United States on Powell's suggestions, which proved to be far ahead of their time.

During the 1890s and on into the early 1900s, particularly under the impetus provided by President Theodore Roosevelt, conservation measures were beginning to evolve. This was a rather stormy period in the grazing land areas where cattlemen fought farmers, sheepmen fought farmers, and cattlemen fought sheepmen. Pitched battles were waged over the right to build fences and section off land. In general the land was heavily overgrazed but the cattlemen were perceived as popular heroes with the "dirt farmer" and sheepmen getting the worst of most battles - both physically and legally.

The Taylor Grazing Act of 1934 was the first major congressional action aimed specifically at the management of grazing lands of the arid and semiarid West. The Taylor Act was stimulated by the result of a 1932 survey indicating that rangeland productivity had declined by 50 percent and that grass removal had exposed 80 percent of the range to erosion. The Act established grazing districts and created the Grazing Service. Under federal control the number of animals was regulated to approximate a balance between range use and the forage-producing capacity of the land. It was fortunate that the Act was passed and implemented during the depth of the drought and the depression of the 1930s, for during that period the number of animals was already sharply reduced. The government therefore had to face principally the problem of preventing an increase in the animal population rather than having to bring about a drastic reduction. For administrative purposes a system of range inspection and evaluation was inaugurated to check on different stocking and management programs. Although relative stability has characterized the western desert ranges since the late 1930s, many areas are still poorly managed, and attempts at scientific grazing control are often frustrated by the complications of land tenure, multiple use of lands and the ease with which legal actions against agencies can be brought into the courts.

Many settlers migrating from the eastern United States brought their more humid region technology into the arid regions and attempted rain-fed farming. This usually accomplished nothing except to lay the land open to erosion by wind and rain, and in all cases failed to provide dependable crop production.

The development of irrigation in the arid portions dates back to the Hohokam and Pueblo Indians, but it was not until the 1840s that the anglos began irrigated agriculture. Among the first to establish successful irrigation farming were the Mormons in the Salt Lake City area. One of the reasons for their early success was the inseparability in the Mormon philosophy of church and state. Thus, it was relatively easy for them to establish a communal irrigation system and thereby allocate the land and water rights and obligations.

It was during the reclamation era of the 1930s and 1940s that the arid regions saw the harnessing of most of the major streams originating in the mountainous West. This resulted in multipurpose dams and the extension of irrigation to vast areas of the desert and semidesert.

Other land uses in the arid Southwest included Indian reservations, large urban centers, electrical energy production, mining and industrial development, and recreation.

RANGE RESOURCE UTILIZATION

As a result of the severe overgrazing which occurred during the late 1800s, the U.S. government has attempted to develop range control on its publicly owned land. Private, state-owned and Indian lands are not under federal grazing control, but through the Agricultural Extension Service associated with the Land Grant colleges and universities many improved range management techniques have been and are being adopted on non-federal lands.

Both the U.S. Bureau of Land Management and the U.S. Forest Service lands are grazed, in keeping with the multiple-use management concept. Multiple use may include recreation, timber production, mining, water resurce development, transportation and power production. While multiple-use policy has its obvious advantages, it also leads to complications for the range manager. Range scientists are as yet unable to predict accurately the synergistic effects of multiple use on the carrying capacity of the land.

While severe overgrazing is generally prevented on most public lands, the result of multiple use is that the associated impact on range vegetation and soil requires continual assessment and re-evaluation of carrying capacity. Land tenure also complicates the problem in the U.S., as grazing may occur in a geographical area which involves land in which various parts are owned by the U.S. Forest Service, the U.S. Bureau of Land Management, Indian reservations, the states or private parties. The Indian reservations are generally severely overgrazed, and control of animal numbers is through tribal councils or cattlemen's associations. Indian cattlemen are reluctant to reduce herd size because to many the existence of a large herd is a source of savings, community power, influence and prestige.

Most range scientists agree that many of the public and private lands are being degraded, and that controls must be stepped up. There are many well-documented historical cases in which overgrazing caused the invasion of low woody plants of little nutritive value, the acceleration of erosion processes and depletion of groundwater.

Western cattle ranches today average over 20,000 acres with a carrying capacity of about one grazing cow per 100 acres. The economic stability brought to the desert ranch is largely a result of the transportation system, the development of water well drilling and pumping equipment, and the application of modern scientific methods in range management, breeding, feeding, veterinary medicine and marketing.

IRRIGATED AGRICULTURE

Irrigation has literally transformed many parts of the deserts of the arid Southwest into agriculture production centers. Irrigation has, in the U.S., contributed to the desertification process via its consumptive use of scarce water supplies, salinization of soils, and, occasionally, the retirement of these agricultural lands to become barren wastelands.

The water problem is most critical in areas in which groundwater is being consumed faster then the rate of recharge to the groundwater system. Several streams and rivers that once were perennial are now ephemeral. This problem is particularly acute in those parts of the semiarid Southwest where groundwater is the only source of water. In the Santa Cruz Valley in southern Arizona, for example, water withdrawal for the urban center of Tucson, for irrigated agriculture and for mining copper is so far in excess of natural replenishment that the groundwater level has declined 200 feet or more in some areas.

One of the problems facing the irrigation farmer, who depends upon a declining groundwater supply, is that he must not only incur steadily increasing costs associated with the energy required to pump from ever-greater depths, but the rapidly increasing cost of energy itself. This places farmers in a position of exponentially rising costs and has caused many of them to abandon their lands. This abandonment of formerly productive agricultural lands leaves vulnerable top soil, which is often saline and does not naturally revegetate with low rainfall. As a result, the barren land serves primarily as a producer of windblown dust and sand. So severe is this problem that many motorists on the highway between Tucson and Phoenix have perished in auto accidents caused by blinding duststorms. This highway now has special sensor signs posted to warn travelers when duststorms are anticipated.

The salinization of soils is, of course, a well-known process associated with irrigation which may contribute to desertification. The general solution to the problem within the U.S. is simply to apply sufficient water to leach out the salts through a subsurface or well drainage system. While the solution clearly solves the problem locally, it results in other complications. It creates a downstream leachate that is higher in dissolved salts than the upstream waters, and, as our friends in Mexico are keenly aware, when the return flows are cycled through several uses, the end user receives water which may be too laden with salt to grow many crops. The Colorado River below Yuma, Arizona is just such an example, where the U.S. is now in the process of building a 100 million-gallon-a-day desalting plant.

Land abandoned because of salinity problems presents special difficulties in reclamation or revegetation with native plants. One of the greatest challenges of the future, one that numerous civilizations have failed to meet, is dealing with the effects of long-term arid lands irrigation.

MINING

While not particularly extensive in areal terms, mining can create a virtual desert locally. Mine reclamation is a field now receiving considerable attention. The effects of mining are different, depending upon the method and the ore body. This paper will discuss only two of the more obvious mining activities that have stimulated considerable public awareness.

Copper mining is extremely vital to the economy of Arizona, and it constitutes the major industry of the state. Most of the copper is mined in open pit operations that may be several square miles in top area and about 800 feet deep. There are several of these huge holes in the Arizona landscape, and the removed overburden plus evaporation ponds, holding processing wastes, create relatively large unproductive areas around the mine.

Architects have looked at the possibility of locating housing developments on the overburden, but for now they lie as unused waste, a tribute to man's ability to move a lot of dirt. This may be called desertification, but in general it is a local phenomenon and does not spread beyond the mine operation. Its visual impact is felt by residents or visitors for miles in all directions. The depletion of water resources required for milling may involve nearby or distant aquifers.

Coal mining in Arizona, Utah, Colorado, New Mexico, Montana and Wyoming has the potential to lay waste to large areas. To date, strip-mining reclamation has had only sporadic success. The Four Corners area in New Mexico has such highly alkaline overburden that revegetation has failed, but in the Black Mesa area of Arizona and in parts of Montana and Wyoming strip-mine revegetation has been more successful. Huge deposits of coal and oil shale exist in the West, and exploitation seems inevitable. Many U.S. citizens, non-governmental agencies and governmental organizations are concerning themselves with the environmental impact of this exploitation, especially that which is related to the shale deposits that have both huge energy potential and huge land degradation potential. Recklessly approached, shale mining has the potential to desertify large expanses of land.

URBANIZATION

Urban development may be one of the best uses of arid and semiarid lands, if proper planning and common sense accompany such use. While wasteful and improper water use in these cities may ultimately lead to ghost towns or concrete deserts, the most publicized problems are those caused by thoughtless land developers who carve up dry desert areas with roads in an attempt to sell these arid plots of waterless land to gullible easterners and midwesterners. A flight over New Mexico, Arizona and Nevada will reveal hundreds of these unpaved roads simply cross-hatching the land with no housing at all. On the ground one sees that many of the "streets" have become gullies which carry away the thin topsoils that typify the arid areas.

Another problem associated with urbanization is the demand for water and land. These demands often force agricultural lands to be retired, either to housing, as in the Los Angeles and Phoenix areas, or to simply unused lands so that the groundwater resource can be used by the city dwellers. Part of the once agriculturally rich Avra Valley in southern Arizona is in the process of becoming a wasteland and potential source of windblown sand and dust, because Tucson needs the groundwater resource for its increasing population. Population growth with the attendant demands for scarce resources is the basis for most arid lands problems.

Many of our western arid and semiarid areas were originally developed for irrigated agricultural usage. Recent population increases (largely attributable to the desirability of the climate for living), reductions in groundwater levels and quality, and increased costs of energy for high-technology agriculture that typifies these areas suggest that U.S. arid lands are on the threshold of a revolution in usage patterns.

RECREATION

The plentiful sunshine of the arid regions, the general affluence of the average U.S. worker, plus more and more spare time have created an interesting and bothersome land degradation problem. As previously mentioned, motorcycles, "dune buggies," four-wheel drive off-road vehicles and campers have invaded the desert areas, and they leave their marks.

Although widespread, this problem is particularly acute in the Mojave and California deserts, on federal lands administered by the Bureau of Land Management. Eleven million Southern Californians view of this unreserved public domain as "their own backyard." Use of the area by off-road vehicles are unregulated until 1972. During one 12-month period there were 151 organized competitive racing events for off-road vehicles involving more than 67,000 participants and 189,000 spectators.

The most extraordinary of the "scramble" races was run each Thanksgiving weekend over a 160-mile course between Barstow, California and Las Vegas, Nevada. Three thousand motorcyclists spread across the desert, traveling at breakneck speeds as they compact the soil, devastate vegetation, stir up clouds of dust and terrorize wildlife. The long-term results of such activities have been denudation of significant patches of the landscape. The Bureau of Land Management has subsequently canceled the races, but the damage remains.

PHREATOPHYTE AND OTHER VEGETATION CONTROLS

The demand for water resources in arid areas is always greater than the supply, causing agricultural, mining and municipal planners to be constantly devising new schemes for diverting larger portions of the limited water resources through the human portion of the desert ecosystem. Recently great concern has been expressed about phreatophytes. Rivers and washes, the arteries of our watersheds, are frequently bordered by stands of tall trees in significant riparian communities. Many of the trees are heavy water users, phreatophytes, and their elimination has been advocated in some areas. Phreatophyte control programs frequently include channeling water courses.

It has also been noted that vegetation could be stripped from several million acres of Arizona forest and desert (largely chaparral vegetation) to increase water runoff. Such schemes to remove large tracts of vegetation are possibly ill-advised because of the long-range degradation and desertification potential. In the short term, increasing water runoff through vegetation removal may mean less water infiltrating to groundwater, and sediment may accumulate more rapidly in water catchments and reservoirs.

SUMMARY

While desertification is a relatively new term in the U.S. scientific vocabulary, land and resource misuse and degradation are not. Desertification is indeed a serious problem, with overgrazing, salinization of soils, depletion of water resources, land subsidence, population pressures, abandonment of lands, some mining operations, and destruction of vegetation or desert pavement through careless recreation, and other exploitations of the desert resources all contributing factors.

To ignore the impact of desertification on the U.S. is a mistake. While very high-level U.S. technology (and capital expenditures) has tremendous capacity to produce, it has equally tremendous capacity to destroy. If resource planners and managers, private investors and government regulatory agencies do not establish and maintain awareness and vigilance of land degradation in the Southwest, desertification could become a problem of increasing national impact. We have the technical knowledge to control and in some instances to reverse the land degradation, but as in many countries the pressures from a too rapidly increasing population coupled with the social problems of implementing known technology do not permit a careless optimism.

It is for these reasons that U.S. government agencies have undertaken a variety of activities aimed at control and possible reversal of arid lands degradation. For instance, in 1968 the Bureau of Land Management launched a California desert study. Its purpose was to inventory current and prospective land uses in the first step toward introduction of integrated management. The study focused public attention on management needs of the fragile arid lands environment creating a gradual increase in management awareness. As part of the Federal Land Policy and Management Act, the Bureau of Land Management has been directed to effect an interim management plan pending development of a comprehensive multiple-use, sustained yield plan for the Southern California desert conservation area. The interim plan will emphasize orderly public use and authorizes a patrol to ensure public safety and to enforce orderly use of the lands. The comprehensive plan will be developed through the Bureau of Land Management Land Use Planning System. A permanent federal appropriation of $40 million has been authorized to support the undertaking.

BIBLIOGRAPHY

Armillas, Pedro 1961 Land use in pre-Columbian America. In L.D. Stamp, ed., A history of land use in arid regions. UNESCO, Paris. Arid Zone Research 17: 255-276.

U.S. coverage concentrates on the Great Plains, the Colorado Plateau and the Sonoran Desert, followed by Meso-America, Peru, the Andean Plateaus and the Chilean Desert. Particular emphasis is on the development of cultivation and irrigation agriculture. Bibliography.

Ayres, J. E. 1971 Man, the desert farmer. In Hydrology and water resources in Arizona and the Southwest, Vol. 1. American Water Resources Association, Arizona Section/Arizona Academy of Science, Hydrology Section, proceedings of the 1971 meetings, Tempe, Arizona, 373-379.

The pre-Columbian Hohokam Indians occupied the major river drainages of central Arizona and have been the subject of much intense archaeological research. Evidence indicates that the Hohokam began using river water for crop irrigation about 300 B.C. and modified and improved their irrigation systems over time until the maximum extent of these systems was achieved about 900 A.D. Two types of water control seem to have been utilized: 1) the direct exploitation of rivers through the use of irrigation canals, and 2) indirect use through controlled runoff within microdrainages at higher elevations before it reached the rivers. At first, probably only those parcels of land with optimal soils and drainage were used, but apparently population increases fostered by agriculture itself, combined with increasing social and political complexity, necessitated more and more exploitation of marginal lands. Eventually, soil problems increased, imposing severe limitations on agriculture. These involved salt and alkali accumulation due to inadequate drainage, soil density and waterlogging. Additionally, the extension of cropping required the clearing of natural vegetation, which resulted in increased erosion and decreased available native food resources for periods when crops failed. The culture vanished completely about 1450 A.D., probably mainly because of their manner of river exploitation for irrigation. More recent archaeological studies are concentrating not only on river use but also on river abuse.

Bochert, J. R. 1971 The dust bowl in the 1970s. Association of American Geographers, annals 61 (1): 1-22. GA 72C-0326.

Major droughts in the grassland region of the central U.S. have occurred rhythmically during the period of instrumental record. The time for the next one may be near. Early droughts disastrously reduced farm income through loss of crops and livestock. Since the 1930s they have accelerated contemporary basic changes in Americans within this region: fewer, bigger and more fragmented farms; public controls and subsidies; consolidation of urban business and services; and greater management. The next major drought will again accelerate long-term agricultural trends, but a sharp increase in urban federal assistance is likely, and revolutionary changes in settlement could be initiated.

Brown, G. W., Jr., ed. 1968-1974 Desert Biology. Two volumes. Academic Press, New York, London. 1236 p.

The first volume deals with the biological aspects of the world's deserts. It focuses on the forces causing desert and arid regions, the evolution of desert plants, water economy, desert ecology and limnology, poisons, desert animals, and the adaptation and behavior of plants, reptiles, amphibia, birds, fish and mammals (including man). The second volume discusses hydrogeology of arid regions, desert soils, physical and vegetational aspects of the Sahara Desert, desert algae, lichens and fungi. A new approach to water relations of desert plants, desert anthropods and fishes, and man in arid lands (particularly the Pima Indians of the Sonoran Desert and the Mada in Salih Valley in the northwestern corner of Saudi Arabia).

Clyma, W./Young, R.A. 1968 Environmental effects of irrigation in the central valley of Arizona. American Society of Civil Engineers, national meeting on environmental engineering, Chattanooga, Tennessee, May 13-17, 1968. Preprint. 28 p. SWRA W70-07053.

Irrigation has modified the environment of the central valley of Arizona for past 2500 years, beginning with the irrigation systems of the Hohokam Indians. Modification of the physical environment has included changes in climate, groundwater, surface water and vegetation. The social environment has been changed from a rural economy to an urban industrial economic system. Present and future problems related to irrigation in the area are discussed. Some of the problems are water demand, land subsidence, salinity, groundwater management and allocation of Colorado River water. Solutions to some of these problems are suggested.

Cooke, R.V./Reeves, R.W. 1976 Arroyos and environmental change in the American Southwest. Oxford Research Studies in Geography, Clarendon Press, Oxford, England. 213 p.

Cottam, W.P./Evans, F.R. 1945 A comparative study of the vegetation of grazed and ungrazed canyons of the Wasatch Range, Utah. Ecology 26 (2): 171-181.

Two canyons in the Wasatch Mountains near Salt Lake City were compared: Red Butte Canyon had been protected from grazing for 40 years, while Emigration Canyon had been grazed heavily since settlement in 1847. Evidence points to the complete substitution of the original grass type in Emigration Canyon to one with unpalatable shrubs and *Bromus tectorum*, while Red Butte Canyon has maintained a rich plant cover including ten native grasses not found in Emigration Canyon. Sheet and gully erosion are prominent in Emigration Canyon.

Denevan, W.M. 1967 Livestock numbers in 19th century New Mexico and problems of gullying in the Southwest. Association of American Geographers, annals 57(4): 691-703. GA 69A-990.

In the 1880s intensive accelerated erosion began producing large gullies throughout the southwestern U.S. This modern arroyo cutting was originally attributed to deterioration of the protective vegetation cover because of below-normal rainfall and overgrazing by excessive numbers of livestock (4 million sheep in New Mexico in 1880). However, recent studies have stressed the greater importance of increased high intensity rainfall. Additional perspective is provided by an examination of livestock numbers in the upper Rio Grande region of New Mexico during the 19th century, particularly during the Mexican period when the ranges were heavily stocked with sheep (possibly 3 million head in the 1820s), but with little or no gullying. The incomplete record of livestock numbers in relation to climate and gullying backs up the climatic argument but also gives some new support to the older view that overgrazing was a major contributive factor causing severe modern gullying.

Dregne, H.E., ed. 1970 Arid lands in transition. American Association for the Advancement of Science, Washington, D.C. Pub. 90. 524 p.

Invited papers given at a 1969 conference in Tucson, Arizona, jointly sponsored by AAAS and UNESCO on "Arid Lands in a Changing World." (A summary appeared in Nature and Resources 5(3): 7-12.)

Humphrey, R.R. 1958 An analysis of causes. Botanical Review 24(4): 193-252. (Reprinted as Arizona Agricultural Experiment Station Bulletin 299.)

Extensive portions of the desert grassland of southern Arizona, New Mexico and southwestern Texas have been invaded by woody species. Mesquite, creosote bush, cacti of the genus _Opuntia_, burroweed and snakeweed are among the principal invaders. The principal factors commonly believed to have caused this change are change of climate, grazing by domestic livestock, plant competition, rodents and fire. Of these various factors, change of climate seems to have had the least effect. Fires that were formerly frequent and widespread were the chief agency restricting shrub invasion. Since fires have been controlled, the introduction of domestic livestock, plant competition and rodents have been effective agents favoring woody plants at the expense of grasses.

Law, J.P., Jr./Witherow, J.L. eds. 1970 Water quality management problems in arid regions. U.S. Federal Water Quality Administration, Water Pollution Control Reserarch Series 13030. 105 p. Available NTIS as PB-198 125. SWRA W71-06111.

A selection of papers presented at the AAAS committee on arid lands conference, Arid Lands in a Changing World, held in Tucson, Arizona, June 1969. Included are papers covering nitrate removal from agricultural wastewater, the effects of salinity standards on irrigated agriculture in the Colorado River Basin, problems of pollution of irrigation water in arid regions, salinity control in return flow from irrigated areas, natural pollution in arid land waters, distillation of wastewaters, animal waste runoff, water quality requirements, re-use of wastewater effluents, and water quality control problems in inland sinks.

Logan, Richard F. 1961 Post-Columbian developments in the arid regions of the United States of America. _In_ L.D. Stamp, ed., A history of land use in arid regions. UNESCO, Paris. Arid Zone Research 17: 277-297.

A companion review to Armillas' on pre-Columbian America (q.v.), this paper describes the physical setting, including landforms, climate and vegetation. It offers a brief history of the development of a national land use policy: grazing industry, dry farming, irrigation agriculture, recreation and urbanization.

Love, R.M. 1970 The rangelands of the western U.S. Scientific American 222(2): 88-96. GA 71C-1155. SWRA W70-04907.

Cattle grazing is the primary use of the rangelands of the western United States. The western ranges fall into three distinct categories: northern Rocky Mountains, inter-mountain and Southwest. True grassland is rare in the Southwest. Instead, deserts and semideserts predominate. Vegetation management is the major problem in improving the rangelands. Brush conversion programs have shown substantial increases in water yield in the treated areas. This is due to the fact that an acre of brush uses six more acre-inches of water than does herbaceous vegetation. With proper management of brush in the watersheds of the West, the land can provide more meat and lumber, suffer less from erosion and flood, and be more usable for recreation. The rapidly growing need for more recreational land is causing conflicts with other land uses. The advantages of brush control and the use of controlled fires, transhumance, canopy architecture, leaf-area index and exotic grasses are other topics discussed.

McGinnies, W.G./Goldman, B.J./Paylore, Patricia 1968 Deserts of the world, an appraisal of research into their physical and biological environments. University of Arizona Press, Tucson. 788 p.

A book intended for those seriously interested in planning, managing and executing research or development efforts in the arid world. It contains evaluative comments on the work accomplished in weather and climate, geomorphology and surface hydrology, surface materials, vegetation, fauna and desert coastal zones. Extensive bibliographies for each section.

Morris, E.H. 1948 Modern overgrazing by livestock as the direct cause of ruin of southwestern agriculture, with a note by H.H. Chapman. Journal of Forestry 46(12): 929-931.

Reprint of a portion (p.5-8) of the author's 1939 Carnegie Institution of Washington publication, Archaeological Studies of the La Plata District, southwestern Colorado and northwestern New Mexico, dealing with conditions in the pueblo area before the advent of white man and the grazing of sheep and cattle. Overgrazing disturbed the equilibrium of a natural balance of forces such as rainfall, erosion and vegetation.

Paylore, Patricia, ed. 1976 Desertification: A world bibliography. University of Arizona, Tucson, Office of Arid Lands Studies. 644 p.

Complied for the International Geographic Union's 23rd Congress, Moscow, 1976. Consists of 1,750 citations, most with abstracts. Produced from the OALS computerized Arid Lands Information System. Divided into regions covering the Sahara-Sahel, East Africa, South Africa, the Middle East, the USSR, Pakistan, India, China, Australia, and South and North America, with regional introductions by world experts. Maps.

Paylore, Patricia, ed. 1976 The Sonoroan Desert, a retrospective bibliography. Historical perspective, by W.G. McGinnies. University of Arizona, Tucson, Office of Arid Lands Studies, Arid Lands Abstracts 8. 777 refs.

Citations selected from the computerized Arid Lands Information System (ALIS) cover such diverse topics as environment, fauna, geomorphology, geography, surface materials, vegetation, desert grassland, succulents, water, weather and climate. Full abstracts for each with computerized keyword and author indexes.

Paylore, Patricia/Haney, Richard A., Jr., eds. 1976. Desertification: Process, problems, perspectives. Papers presented during a 14-week seminar series, November 1975-April 1976. University of Arizona, Tucson. 125 p.

The topic of this series was addressed by several generalists speaking on its ecology, trends, causes; followed by several case studies to illustrate the theses advanced generally. These dealt with areas in Australia, eastern Oregon and southwestern Arizona and northwestern Mexico.

Powell, J. W. 1962 Report on the lands of the arid reigon of the United States, with a more detailed account of the lands of Utah. Ed. by Wallace Stegner. Belknap Press of Harvard University Press, Cambridge, Massachusetts. 202 p.

A first-hand account of the arid region of the U.S. written in 1878. Subjects covered include physical characteristics of the region, the land system needed, rainfall of western U.S., water supply, questions relating to irrigable lands of Salt Lake drainage system, valley of Sevier River and that portion of Utah drained by Colorado River and tributaries. Author considers character of lands and engineering problems involved in their redemption and suggestions for the legislative action necessary.

Sears, P.B. 1959 Deserts on the march. 3rd ed., rev. University of Oklahoma Press, Norman. 178 p.

This popular account of man's misuse of the land and failure to understand its ecology centers around the historical development of land use in the United States. The cutting of forests, farming of cotton, plowing of prairies, polluting of rivers, overgrazing and general disregard for the welfare of the land are shown to be related to floods, mudslides, duststorms, erosion and natural calamities. The author argues that only through a broad ecological understanding of the environment and man's relationship to it coupled with moral commitment in the employment of technical resources and in our way of living will we be able to change our historic precedents.

Steward, George 1936 History of range use. In U.S. Senate Document 199, The western range, p. 119-133. Washington, D.C.

U.S. Bureau of Land Management 1968 The California desert. A recreation study of the desert public domain lands of California. Preliminary review U.S. National Park Service, Bureau of Land Management, Washington, D.C.

Woodbury, R.B. 1963 Indian adaptation of arid environments. In C. Hodge and P.C. Duisberg, eds., Aridity and Man. American Association for the Advancement of Science, Pu. 74: 55-85.

Briefly traces man's changing roles in the western U.S. Discusses food gathers, big game hunters, village farming life, villages and water sources, water control techniques, decline of village farming life, introduction of adaptable Athabascans, shift to wagework, collapse of Great Basin culture, rise and fall of the Sioux, adjustment to reservation life, Papago struggles in the desert, Indian adjustments and prospects. Climatic change in the 13th century was probably responsible for early decline of village farming life. Grazing abuses by Indians and whites have resulted in deteriorated rangelands. The plight of the Indian in the last century or so is not the result of environmental causes such as aridity; the causes are social. It can be seen as part of the worldwide problem of the non-literate, non-industrial world confronted by the technology and social and ethical systems of the western world. Indian techniques and their philosophy of working with the environment rather than against it may be indispensible ingredients for the successful utilization of arid lands.

200 YEARS OF DESERT CHANGES-THE CALIFORNIA STORY

Herbert W. Greydanus
Chief, Division of Planning
California Department of Water Resources

Deserts cover 25 percent of California as shown on the first illustration. An additional 15 percent is classified as semi-arid. More than one-half of the land irrigated in the State is within those two climatic classifications. This provides an opportunity to review the economic development of a desert area. It has been accomplished with the support of a highly developed urban and industrial economy in the eastern portion of the nation. It is certainly not a typical example of development of desert lands, but nonetheless may provide some knowledge that will be of use for other desert areas.

How have we adjusted to the environment and how have we modified our environment? For the most part we have attempted to modify instead of adjust. The development has required major efforts in water management.

Early explorers have described parts of California as useless and unworthy of claiming. Others have called it a "land of milk and honey". The truth is somewhere in between.

I want to first briefly describe general physical characteristics of the State and then describe its development with emphasis on water management, use, and future needs.

PHYSICAL CHARACTERISTICS

The second illustration shows how California would fit in similar latitudes on other continents. In topography and climate it is very similar to the central third of Chile. Los Angeles is about the same latitude north as Santiago is south. In Africa, Cape Town is at a comparable latitude with Los Angeles, in Australia it is Perth and in the Northern Hemisphere there is comparability with the latitudes of Casablanca and Shanghai.

California is about 1100 kilometres long and about 370 kilometres wide. Its area is about 410,000 square kilometres. The Pacific coastline extends for 1600 kilometres north from Mexico. Three areas are below sea level: The Sacramento-San Joaquin Delta - about 7 metres; the Imperial Valley - about 70 metres: and Death Valley - 86 metres. Fourteen mountain peaks rise more than 4,200 metres above sea level.

ILLUSTRATION I

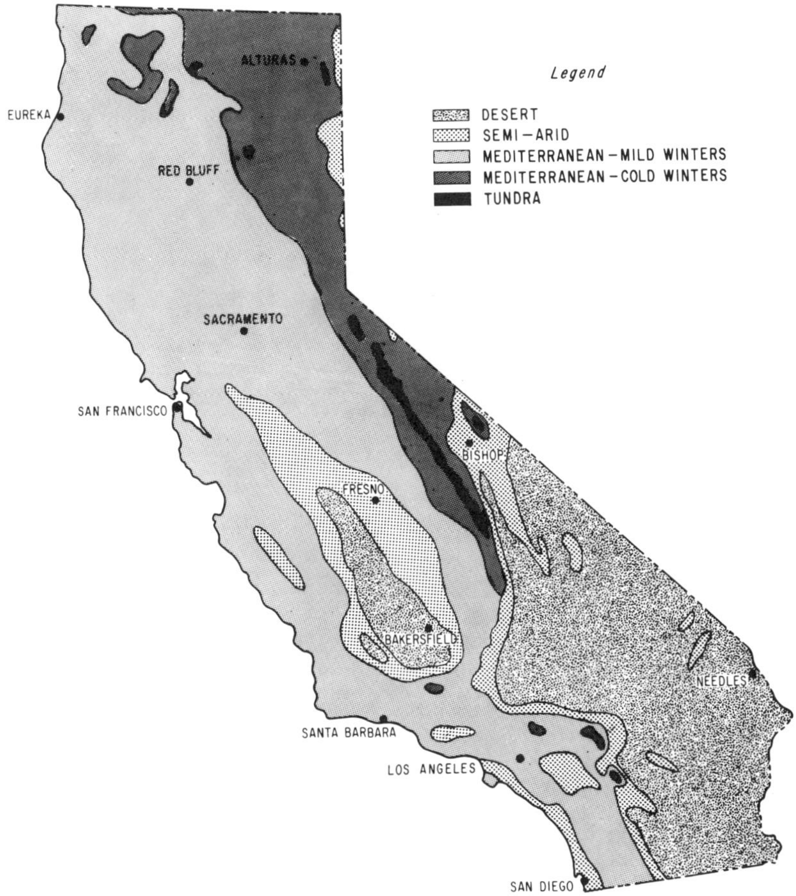

CLIMATIC REGIONS OF CALIFORNIA

Mt. Lassen in the north is an active volcano and fumaroles are active at about 70 metres below sea level in the Imperial Valley.

Perhaps the most outstanding feature of California is the Central Valley, shown on the third illustration. It is bounded by the Cascade Mountains on the north, the Sierra Nevada on the east, the Tehachapi Mountains on the south, and the Coast Range on the west. The single drainage outlet to the ocean for the 160,000 square kilometres basin is through the San Francisco Bay system. This valley is about 800 kilometres long and 200 kilometres wide. The climate and soil provide some of the most productive agriculture in the United States and even in the world.

ILLUSTRATION 2

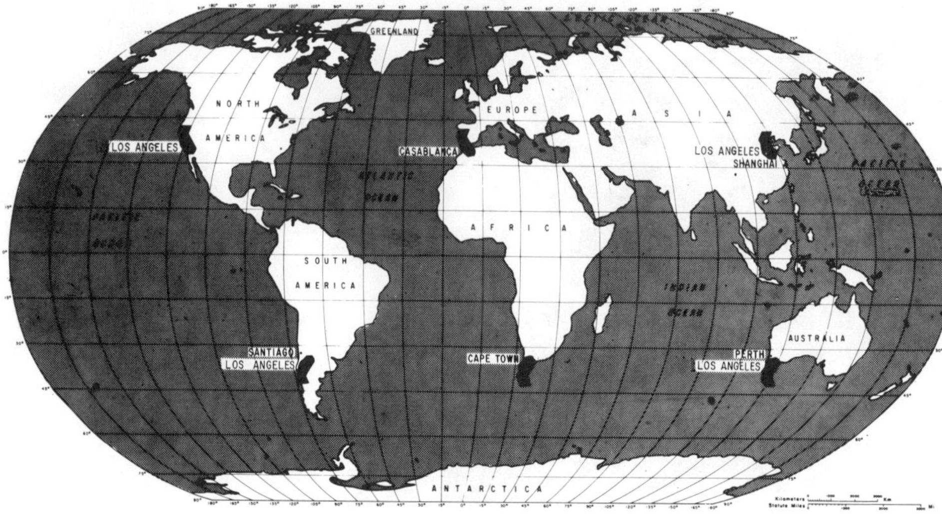

CALIFORNIA IN RELATION TO OTHER CONTINENTS

CALIFORNIA'S CLIMATE

The climate of California is also a story of contrasts and extremes. The northwest coastal area is one of rainfall in the order of 2500 millimetres annually while much of the southwest desert area only has about 50 millimetres on the average. Most of the precipitation throughout the State occurs during the November-April period and the variation from year to year is greatest in the low rainfall area in the south. The May-October period is typically dry and hot. Many days in the Central Valley exceed 38 degrees Celsius and 42 degrees Celsius are typically recorded in the Imperial Valley. The highest temperature of 57 Celsius was recorded in Death Valley. There are five active glaciers on Mount Shasta and a number in the southern Sierra Nevada.

California's climate is strongly influenced by the Pacific Ocean. Winter storms generally develop in the Gulf of Alaska and move southeastward across the State. Occasionally warmer mid-Pacific storms move eastward across the Coast Range and Sierra Nevada with intense precipitation and much flooding. Southern California infrequently experiences tropical storms moving northeastward that cause intense cloudbursts and flash flooding. Much of the precipitation in the Sierra Nevada occurs in the winter as snow and provides extended spring and early summer streamflow as it melts.

EXPLORATION AND SETTLEMENT

European explorers landed in California early in the 16th century. At that time, California was populated by an estimated 133,000 native Indians, about one-eighth of the natives in the area of the United States,

ILLUSTRATION 3

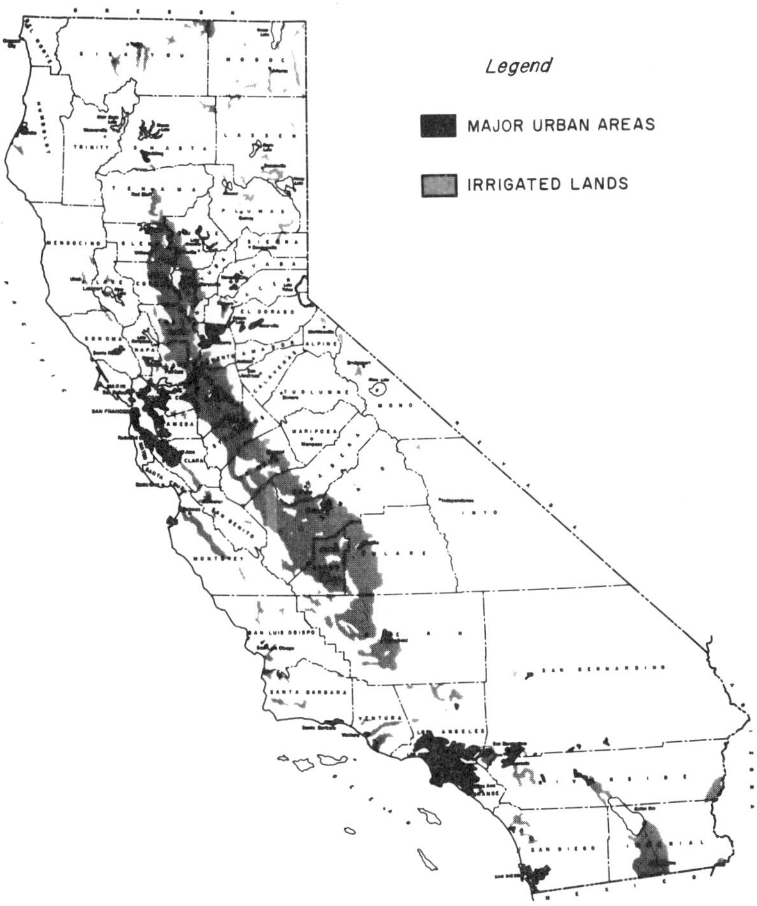

IRRIGATED AND URBAN AREAS IN CALIFORNIA

before the coming of the Europeans. Although there were several exploratory expeditions beginning in 1539 by Spanish, English, and Russian groups, colonization and development did not take a strong hold until the efforts of the Spanish Franciscan Fathers beginning about 200 years ago in 1769.

The Franciscan Fathers established 21 missions along the coastal area from San Diego to San Rafael, a short distance north of San Francisco. While the primary goal of the Franciscans was the spread of Christianity, their efforts of necessity had to involve land and water development in order to be self-sufficient. As many as 40,000 hectares were involved at each mission. At one period about 30,000 native Indians were converted by the fathers and attracted to the vicinity of the missions.

To provide food for the concentration of people when the natives were gathered at the missions, it was necessary to develop water for irrigation and to raise livestock, The oldest masonry dam in California was constructed on the San Diego River in the late 1770s for the San Diego Mission and irrigation in Mission Valley. Close to 150,000 head of cattle, sheep, and goats were raised by the missions around the end of the 18th century.

Between 1777 and 1797 the Spanish Government founded three communities called pueblos at San Jose, Los Angeles, and near Santa Cruz. Only the first two survived and developed as cities. Rights to water for irrigation for the pueblos were granted by the Spanish Government and these have been recognized by the courts as the oldest rights in California. As recently as three years ago, these "pueblo rights" were a factor in a court decision in a lawsuit over ground water.

The next significant change began in 1833, about 12 years after Mexico gained independence from Spain. The Mexican government secularized the missions and allocated their lands to private owners. In 12 years the government granted over 600 ranchos covering most of the best agricultural land in valleys and adjacent hills to private owners. These grants ranged from about 1,700 hectares to 40,000 hectares. Cattle and horses dominated the economic life of the "Spanish Californias" during this period. Hides and tallow provided the basis for outside commerce. There was little use made of the meat.

The first easterners (eastern United States) to come to California were mainly fur traders. Many parties arrived in the 1820s and '30s and eastern settlers soon followed.

The political control by Mexico was not particularly strong during this period. Settlers revolted against the government at Sonoma, north of San Francisco, in June 1846 and in July United States naval forces seized Monterey and announced annexation to the United States. This was part of broader action in the Mexican War and occupation of California was completed in August 1846.

The most influential event on California's growth was the wide publicity given to the discovery of gold in 1848 in the Sierra Nevada east of Sacramento. News of the event brought approximately 80,000 easterners and immigrants from around the world in 1848-49, most of them to Northern California. Impact on Southern California communities was slight. It was the North which later led the drive toward Statehood, which was granted in 1850.

Many of the gold seekers quickly recognized that there was great wealth in commerce and agriculture to serve the needs of the miners. Others came to California specifically for such purposes. The population of San Francisco grew to 56,000 in 1860, and Sacramento was a major jump-off town for the gold fields. Many towns of several thousand people quickly sprang up in the Sierra Nevada.

Southern California, including Los Angeles, remained "cow country" until the 1880s. About that time railroad service to the south was completed. The excellent desert climate with adequate surface and ground water to meet needs at that time led to a "land boom" coupled with sub-divisions for new towns and much real estate speculation. One promoter claimed "The purity of the air in Los Angeles is remarkable. Vegetation dries up before it dies and hardly ever seems to decay. Meat suspended in the sun dries up, but never rots. The air, when inhaled gives to the individual a stimulous and vital force which only an atmosphere so pure can ever communicate". For those who have experienced Los Angeles smog these words are hard to comprehend. Yet even now the climate of Southern California attracts industry and people.

The gold rush created an active market for meat and many parties brought herds of cattle and sheep. This stock was generally much better than the local stock on the ranchos. The cattle census grew from about one-quarter million in 1850 to about three million in 1962. In 1862-1864 a severe two-year drought occurred. The cattle and sheep livestock were devastated. Many ranchos could not any longer continue to operate. Many were forced into bankruptcy and an era came to an end.

The sheep ranching came back quite rapidly and by 1874 there were about seven million head. This number, however, was too great for the land to sustain, particularly since much of it was being fenced for cultivated farming. By 1940 the number of sheep had dropped to about 2-1/2 million.

DEVELOPMENT OF WATER SUPPLIES

Early development of irrigation and municipal water was largely dependent upon diversion of unregulated surface streams and springs. Since much of the Sierra Nevada is normally covered with heavy snowpack, the late spring snowmelt generally sustained water runoff for adequate irrigation. Infrequent dry years and the low flows on the streams were simply endured. As more and more demands were placed on the uncontrolled runoff it became necessary to construct water storage reservoirs. There was a gradual increase in the amount of storage until the 1930s when California experienced a six-year subnormal precipitation period. At about that time the total reservoir storage capacity in California was approximately 10,000 cubic hectometres. Today we have a capability to store about 50,000 cubic hectometres.

Another major event in water utilization was the development of turbine pumps for wells. Much of the arable land of the State is underlain by extensive quantities of ground water. Early utilization of ground water resources was largely limited to locations with artesian water pressures or by extraction through shallow wells with windmills. The turbine pumps with electric motors, however, made possible rapid expansion of irrigation and in many areas demands on the ground water basins have been or still are in excess of the average annual recharge. Initial use of ground water made possible the development of economic bases and that provided the ability to finance more expensive surface water development and importation from distance areas. However, California now has reached the point where careful consideration must be given to controlling further ground water development to avoid future demands in excess of the State's ability to develop surface water. It will be very difficult to face the economic cost and social disruption of farmers and communities which may run out of water when the ground water supplies are exhausted and we are unable to meet their supplemental needs.

Most of California's precipitation falls in the northern one-third of the State, but the water requirements and the irrigable areas are largely in the southern two-thirds. In addition, as already noted, the precipitation occurs in a few winter months and we have an extensive, almost rainless, dry summer season. The need to redistribute water both geographically and seasonally was recognized in early plans by State Engineer William H. Hall, about 100 years ago. His concept was to develop storage for surplus water in the north and construct aqueducts for conveyance into the southern San Joaquin Valley.

The water redistribution plans were more fully developed over the years and led in 1933 to voter approval of a project and authorization to issue revenue bonds. However, the economic conditions of the worldwide depression prevented sale of the bonds. The State then turned to the Federal Government for assistance. A project was undertaken by the U. S. Bureau of Reclamation. The results are the Central Valley Project with storage in Shasta Reservoir on the Sacramento River, Folsom Lake on the American River, and Millerton Lake on the San Joaquin River. The Friant-Kern Canal carries water southward from Millerton Lake and the Delta-Mendota Canal delivers replacement water to the San Joaquin River for areas formerly served by water diverted at Millerton Dam. The aqueducts are shown on the fourth illustration. The Bureau of Reclamation also participated in the construction of the California Aqueduct and San Luis Reservoir for service in the San Joaquin Valley as part of the Central Valley Project. When construction of dams for two additional reservoirs is completed, the Bureau of Reclamation will have capacity to eventually provide about 11,300 cubic hectometres annually to the Central Valley. A number of other U. S. Bureau of Reclamation projects serve water to areas in California and major storage reservoirs on the Colorado River provide water for the Palo Verde, Coachella, and Imperial Valleys and the urban area from Ventura County to the Mexican Border.

During World War II and immediately following there was rapid expansion of population and irrigated land, primarily using ground water. It became apparent that additional water storage and distribution facilities would be needed. The State constructed the State Water Project to provide eventually 5,200 cubic hectometres annually to the San Francisco Bay area, the San Joaquin Valley, the central coastal area, and Southern California. This project, which was initiated about 1960, has its principal storage in Oroville Reservoir, the San Luis Reservoir in the San Joaquin Valley, and several smaller terminal reservoirs along the California Aqueduct and its branch aqueducts. Water is pumped from the Sacramento-San Joaquin Delta to most of the service areas including a single lift over the Tehachapi mountains for Southern California of 587 metres.

CALIFORNIA TODAY

At this time there are about 3.5 million hectares under irrigation and 22 million people in California. The growth of irrigation and population over the last 200 years is shown on the fifth illustration. Over the last ten years 90,000 hectares of irrigated land have been converted to urban uses but the expansion of irrigated land has exceeded this reduction by about 240,000 hectares. The normal present water use is 46,000 cubic hectometres annually of which 33,000 cubic hectometres is net depletion.

California has an average surface water runoff of nearly 95,000 cubic hectometres. Of this supply, the net use for urban and irrigation purposes is around 35 percent. Another 28 percent is committed to environmental purposes for water quality in the Sacramento-San Joaquin Delta and for wild and scenic rivers in the north coastal region. Two percent flows to the State of Nevada leaving about 35 percent presently uncommitted to meet future needs. Over half of this amount runs off in relatively unaccessible coastal streams or as high floodflows throughout developed regions of the State. The remaining developable water, about 13,600 cubic hectometres, is principally in the Sacramento Valley.

ILLUSTRATION 5

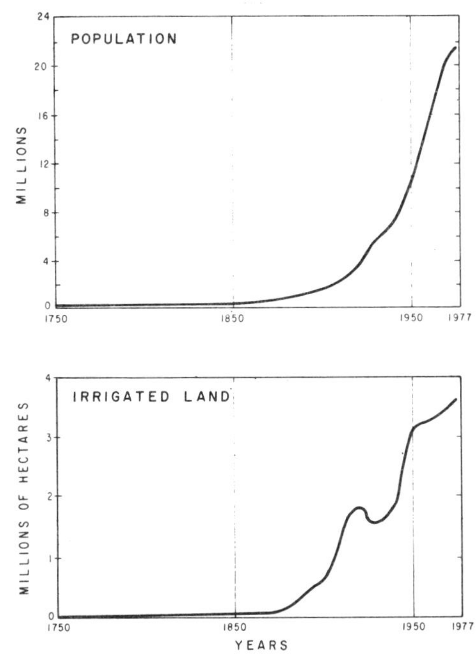

POPULATION AND IRRIGATED LAND IN CALIFORNIA

THE FUTURE

We have looked briefly at the history of California's development and its present situation. Before reviewing the benefits and detriments of the results, we need to look at some indications of the future and at two other influences.

What about the future? The outlook for urban growth is dependent on many factors and a few will probably dominate. The national attitude about family size and birth control will be a big factor since there is free movement of people between states and a current base population of 22 million in California will have a large number of children. The economic opportunities for industry and commerce will be important. Much of the growth in the 1940-65 era was related to aircraft and aerospace industries. The climate of the State and the geographic position for trade across the Pacific Ocean will continue to be important factors. We do not expect to again experience the recent rapid growth rates of 430,000 annually, including about 300,000 net immigration for several years. We can expect, however, that we may have a 40-50 percent increase by the turn of the century and about 75-80 percent increase in the next 50 years.

The future of irrigated agriculture is perhaps even more speculative. At this time there are about 3.5 million hectares under irrigation. There are approximately 5 million hectares of irrigable land which are not yet developed but much of this land would require very costly irrigation service. The most likely areas for economic development are in the southern portion of the Sacramento Valley. The best land and climate are in the San Joaquin Valley but the costs of providing water will be high. While the land in the Sacramento Valley is not quite as good, the costs of water will be less (but still not inexpensive). It is very likely that irrigation will continue to expand by initial use of ground water unless limitations are placed on its development. The cost of supplemental supplies will in large part be dependent on whether costs of future storage and conveyance are melded into past costs, such as with the Central Valley Project and the State Water Project, or whether separate, self-repayment rates are established. Another important factor is whether there are any subsidies. Water from the federal Central Valley Project for irrigation does not need to recover interest and is priced in relation to an area's ability to pay. The State Water Project, on the other hand, requires full repayment of allocated costs with interest. In the San Joaquin Valley current irrigation water costs are about $50 per hectare for the Central Valley Project and range from $85 to $170 per hectare for the State Water Project service areas.

At this time California's principal water shortages are in the San Joaquin Valley. Users are overdrafting the ground water at a rate of about 1,900 cubic hectometres per year. There is also need to develop a like amount to fully serve water already under contract in the State Water Project. Expansion of irrigated areas would require additional water but costs will likely limit the amount of expansion. We should plan to continue to reduce per capita use of water in urban areas and to reuse water. The principal urban areas should be able to meet their needs for many decades with presently developed and contracted supplies with reasonable conservation and reuse goals. Municipal needs in other areas throughout the State can be met from local resources.

In California, the basic responsibility for providing water rests at the local level rather than with the State or Federal Government. State law provides for formation of various types of water organizations, most of which include authority to bond themselves for financing the construction and operation of water facilities. The state and federal projects provide wholesale water to the local agencies. This water is in turn distributed and managed at the local level. There are approximately 4,000 agencies which provide water in California. These include almost every town and municipality. Major cities such as those in the San Francisco Bay Area and in Southern California have undertaken extensive programs to import water from distant watersheds. These are shown on the slide on the screen. Of the total investment in water development, approximately 70 percent is with local financing, 20 by federal projects, and 10 by the State of California.

WATER RIGHTS

The ownership of the water resources is vested in the State Government and rights are granted to individuals and local agencies for beneficial uses. Our water rights laws have, as in most places, developed over time. Much of our early rights stem from English law. Rights to surface water include riparian rights by adjacent land owners to natural flow--that is unregulated water. The largest portion of surface water rights, however, are appropriative rights granted by the State. These rights are granted for beneficial uses and enjoy priorities by the date of filing. As long as water is beneficially used the right to its use can continue.

Ground water law is very limited and has been largely developed through litigation in the courts. All overlying property owners have a right to share in ground water under the land. As long as the water extracted from the underground is beneficially used, there is not any state control on the amount of use. As briefly mentioned earlier, the establishment of pueblos by the Mexican Government, including the rights to water, have been recognized by the courts. Because we are close to full utilization of our readily available water resources, renewed attention is now being given to review of the entire water rights situation. The Governor has recently appointed a special study committee to review this matter and make recommendation for changes.

THE PRESENT DROUGHT

The Desert Conference is taking place at a unique time in our water use history. We are now experiencing the most severe drought of the last 200 years. It is a demonstration that we have new lessons to learn at every stage of development.

Tree ring and other research back to earlier times indicate that there have been periodic droughts of five to ten years. We experienced significantly below normal precipitation for six seasons about 50 years ago. The current drought, however, with two years of very low rainfall is the worst combination for which we have quantitative information.

The present drought actually began in April 1975 with the end of the winter season. We did not, however, realize the situation until about ten months later when it was clear that the precipitation in 1976-77 was going to be very low. It turned out to be the third worst year of record. With the exception of curtailments of water deliveries from the State Water Project and to some extent from the federal Central Valley Project, most areas of California continued to use water during 1976 at near normal levels. The winter of 1976-77 has proved to be the lowest precipitation of record. This situation, following a very dry year, has resulted in even smaller amounts of runoff due to the extreme dryness of the watersheds. In addition, the requirements for water in the irrigated areas are even greater due to the need to offset the lack of soil moisture provided by direct rainfall. In many areas water supplies are being rationed and we can anticipate almost daily reports of new water shortage situations. Much of the shortage will be offset by the additional development of ground water, but as the ground water levels fall many wells will need to be deepened.

The experiences in this drought are providing significant new insights in water management. We have planned our projects to provide enough water during the repetition of historic drought periods. In Northern California we have used the six-year period of about 50 years ago. While we have admitted our inability to operate according to the plans because we cannot see into the future, this current drought has clearly brought this matter to a head. We should have curtailed water deliveries in 1975 to provide additional carry-over water for 1976 and 1977. Perhaps we should be taking even greater deficiencies this year to provide for carry-over water for the next few years. It is essential that we develop new operating procedures for local, state, and federal projects. These procedures must be based on the amount of storage at the beginning of each winter and the anticipated needs the following year. Our plans for water deliveries will need to be adjusted throughout the winter season as additional information on our precipitation and water supply becomes available. It will be necessary to accept more frequent reductions in water supply, but these should be smaller in amount than we must now endure. For the long range we need to develop our capability and skill in making long-range weather forecasts. If we can predict with some assurance the amount of precipitation in future winters, we will be able to greatly improve our water management.

HAS IT BEEN DONE RIGHT?

° 3.6 million irrigated hectares producing 9 percent of the national farm products with a value of $8 billion annually.

° 22 million people living and working on 1.1 million hectares producing $168 billion of output annually.

° 1,300 large reservoirs and thousands of kilometres of aqueducts distributing water throughout the land.

° 3.3 million kilos of gold mined from the land and billions of tons of sand, gravel, and debris turned over with much of it washed into streams and channels.

° 4,000 kilometres of levees and 2,000 kilometres of improved channels constructed to control flooding.

° Tens of thousands of native tule elk reduced to 700 living on 6 preserves.

° Loss of extensive reaches of streams once available for anadromous fish, creation of some greatly improved fishery streams and rivers and many lakes with excellent fishing.

For these and other reasons too numerous to list, how should we be graded? The answer will depend on one's viewpoint and objectives. Our own assessment is partially demonstrated by a strong movement in the last ten years to improve our natural environment. When economic development meets the material needs of most of the population, then environmental needs become a strong influence.

Water quality agencies have been strengthened and air resources agencies have been formed. Coastal preservation and regulation activities are under way. Currently, concerns are mounting to speed up the environmental evaluations, reviews, and permits to create more favorable opportunities for industries and employment.

We want to share our experiences, both favorable and unfavorable, in the development of the California desert. We are likewise interested in learning from the experiences of others. By so doing we can mutually benefit.

June 1977

CLIMATE CHANGE AND DROUGHT IN DESERT AND ARID ZONES

William J. Quirk[1]

ABSTRACT

Examination of the climatic history of desert and arid zones shows that the rainfall in these regions has been extremely variable on all space and time scales. Dry and moist periods lasting years, decades, centuries and millennia have occurred over both small districts and continental sized regions. While on the long term these swings in climate are probably random and beyond the present scope of man to manage, man can in the short term at least ameliorate these changes by proper land management. The large variability of rainfall in desert and arid regions can be compensated for in part by storing world grain surpluses stored against times of need.

INTRODUCTION

There has been much discussion about whether there is a global cooling trend (Bryson, 1974) or a global warming trend (Damon and Kunen, 1976). However, the important fact is not in what direction climate is changing, but that it does change and that man may be influencing that change. This paper will summarize some lessons that can be drawn from recent climatic history, and discuss some possible influences of man on the weather. It will stress that we should be prepared for the unexpected in weather. It will also show that proper land management may be able to ameliorate the large variability in desert rainfall.

CLIMATE HISTORY

The major lesson to be learned from climate history is that weather is extremely variable. While there are long term changes which occur over millenia, it is the

---

[1]NASA Institute for Space Studies, Goddard Space Flight Center, New York, New York 10025

changes with time scale of months to decades that will most affect our strategies of development. That periodic occurrences of drought are normal and must be planned for has already been stressed by several speakers at this conference.

One of the most detailed climate histories of arid lands was recently written by Sharon E. Nicholson of the University of Wisconsin. Her thesis (published in 1976) is entitled A Climatic Chronology for Africa: Synthesis of Geological, Historical, and Meteorological Information and Data. It describes in broad terms the major African climatic episodes of the past 20,000 years. The work shows that there are both long term trends lasting thousands of years and short-term climatic episodes on the order of decades.

Figure 1 shows a summary of climatic conditions during the last ice age. The dry desert areas are shifted very far south and are much more extensive than they are today. Figure 2 shows that in general, Africa was considerably wetter 5,000 years ago than it is today. In a very long term sense, the globe as a whole has been returning to the cooler, dryer conditions of the last ice age. However, during the last 5,000 years, there have been many periods of hotter and wetter weather, as well as colder and dryer weather. Figure 3 shows oscillations in lake levels typical of central Africa. The most certain thing that can be said about the weather is that it is variable.

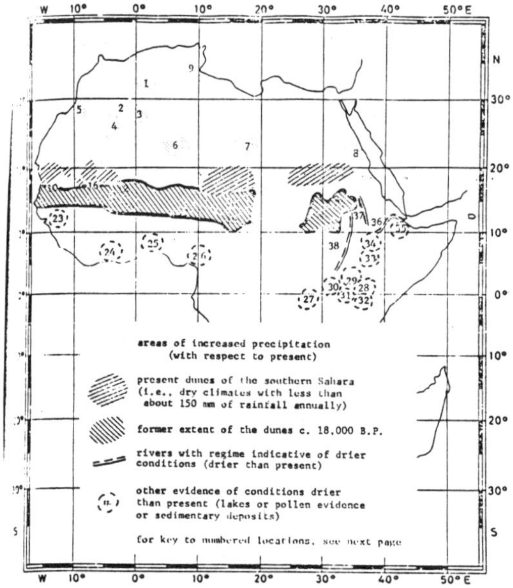

Fig. 1 Summary Map of Climatic Conditions c. 20,000 - 15,000 B.P.
(partially adapted from Rognon, 1976A)

Other lessons from the history of climate have already been discussed at this conference. Dr. Greydanus has stated that tree ring studies show that droughts lasting 5 - 10 years in California have been very common during the last 200 years. Dr. McGuire Gibson has stressed that while the climate in Mesopotamia today is similar to what it was 6,000 years ago, climate has varied so much in that region that changes in weather patterns could have contributed to the breakdown of several civilizations.

The purpose of this paper is not to predict doom. It is just as likely that the next 10 years will bring very good weather as very poor weather. The message of

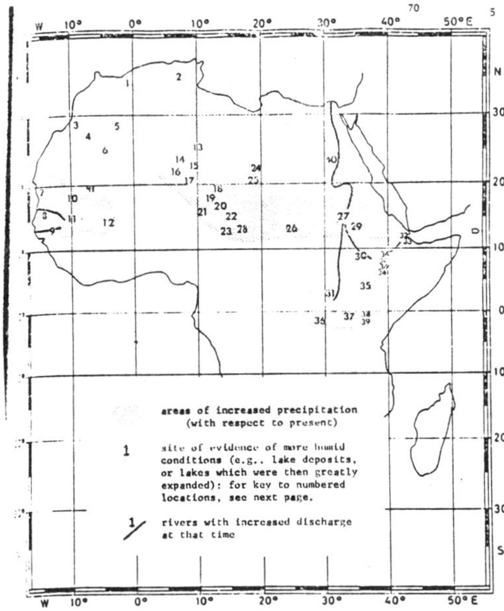

Fig. 2 Summary Map of Climatic Conditions c. 6,500 - 4,500 B.P.

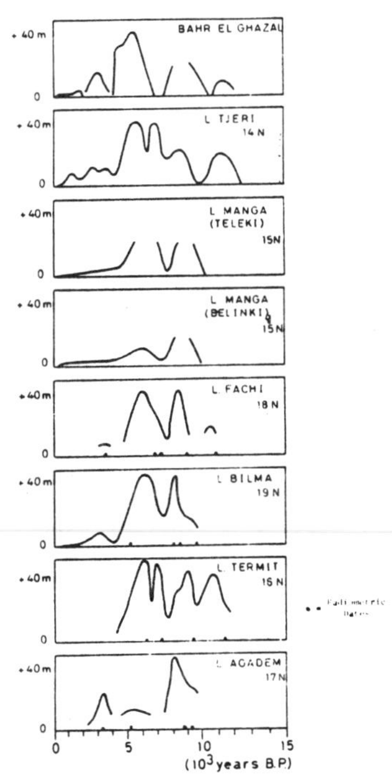

Fig. 3 Lake-Level Variations in Niger and Chad during Late-Pleistocene and Holocene (from Servant, 1973)

this paper is that weather is always unusual, and that plans for desert development should take into account the fact that periods of drought and wet weather are normal.

MAN'S INFLUENCE

It is widely feared that man is inadvertently altering the delicate balance of the present climate (SMIC, 1971). This balance is especially precarious in the semi-arid zones bordering the major deserts. Recently, severe drought occurred in the Sahel at the southern margin of the Sahara when moist monsoon air failed to penetrate sufficiently far northward from the tropical Atlantic. Drought has been a recurrent problem in the provinces of India and Pakistan, which border on the Central Asiatic Deserts. In North America the highly productive lands of the Western Great Plains lying just east of the Great Western Desert have suffered extended periods of drought, including the "dust-bowl" drought of the 1930's and the widespread drought of the 1950's. These regions have in common that (1) large changes repeatedly take place in plant cover and ground hydrology, (2) maximum rainfall occurs during the summer monsoon season and is primarily convective in character, and (3) advection of heat and moisture from outside the regions is too weak to overcome the local effects of albedo and ground moisture changes.

In his Royal Meteorological Society Symons Memorial Lecture for 1974, Charney (1975) showed that the radiative heat loss caused by the high albedo of a desert contributes significantly to the sinking and drying of the air aloft and therefore to the reduction of precipitation. This dependence of precipitation on albedo led him to propose a biogeophysical feedback mechanism linking vegetation, albedo and precipitation as a partial explanation for recurrent drought in desert border areas. In later research Charney, Stone, and Quirk (1976) found that the lack of evaporation caused by large changes in plant cover can be just as great an influence in prolonging a dry period.

If the soil is light, dry and sandy, as it often is in these areas, a decrease of vegetation will lead to an increase of albedo, a reduction of precipitation, and therefore a further decrease in vegetation or at least a perpetuation of the initial decrease. This effect is further enhanced by the reduction in evapotranspiration resulting from the decrease in vegetation. Thus, the vegetative cover at the border of a desert may be regarded as unstable or metastable. If unstable, an initial departure from equilibrium will amplify; if metastable, the departure will tend to persist. The perturbation will remain until external counteracting climatological forces become large enough to overcome the local feedback effect and return the system toward the old equilibrium. These external forces may act to prevent yet another possibility, namely, that if they did not exist or were sufficiently weak, the positive feedback could lead eventually to a new statistical equilibrium, i.e., to a new climate. Such a system would be intrasitive in the sense defined by Lorenz (1968), having the property that different initial states lead to different climates. But if the external forces are sufficiently strong, the new statistical configuration will not persist indefinitely, but will exist merely as a long-term perturbation of a basically stable statistical equilibrium.

Charney et al. have presented evidence only for the reality of the geophysical, not the biological, component of the system. They have shown that the restoring climatological forces, consisting of the global circulations and their random perturbations, are too weak for the limited period of the integration to overcome or even mask the local effects of the albedo changes in the regions in question. Hence, whatever the ultimate behavior of the biogeophysical system as a whole, the feedback mechanism will exist and will have the effect of at least prolonging a given fluctuation, whether positive or negative, i.e., of prolonging both dry and rainy periods. Since drought is prolonged dryness, the postulated feedback mechanism can be expected to contribute to drought.

Control numerical integrations were carried out for the month of July with the general circulation model of the Goddard Institute for Space Studies. These were repeated with surface albedos modified in certain selected regions: the Sahel, the Western Great Plains and the semi-arid Rajputana Desert of southeastern Pakistan and northwestern India adjoining the very dry Thar or Indian Desert. For comparison purposes, a third integration was performed in which the albedos were altered simultaneously in Central Africa, the Mississippi Valley and Bangladesh. These regions were selected as being rainier and more centrally imbedded in their respective monsoon regimes. Some early results for the Sahel integrations have already been presented by Charney, Stone and Quirk (1975, 1976).

Albedos can be changed by natural as well as anthropogenic factors. They have not distinguished between the two, but if the entire system were being considered, in its biological and ecological as well as its geophysical aspects, a sharp distinction would have to be made, since anthropogenic changes can be very persistent whereas natural ecosystems tend to recover more readily from natural perturbations.

Man-induced changes may very well have operated in the semi-arid zones treated in this article. In the Western Great Plains, for example, erosion, agricultural tillage and crop harvesting have at times reduced the darker humus layer to expose the brighter lower horizons of the soil (Buckman and Brady, 1969). In the Sahel, there is strong evidence that overgrazing has caused surface albedos to increase over large areas (Otterman, Walter and Schmuge, 1976), and, as reported by Hora (1952), "the Rajputana Desert is largely a man-made desert...(formed) by the work of man in cutting down and burning forests...(and by) the deterioration of soils."

The last reference is taken from an article by Bryson and Baerreis (1967) who suggest that the dense pall of dust frequently found over northwestern India is a significant factor in reflecting solar radiation and reducing the radiative heating of the air. Since destruction of vegetation destabilizes the soil and thus permits the wind to raise more dust, the possibility exists that reflection by dust may be another feedback mechanism for decreasing rainfall. However, since dust also absorbs solar radiation and traps infrared radiation from the ground, it is not known yet whether it produces a net increase or decrease of radiative heating. Some early estimates suggest that the dust does not greatly change the planetary albedo but that it absorbs solar radiation in the mid-atmosphere and thereby reduces the heating of the ground (J. Hansen, personal communication, 1976). This would tend to stabilize the lower part of the atmosphere and inhibit thermal convection.

Since accurate measurements of actual albedo changes on the earth have not yet been made, Charney et al. only dealt with idealized numerical experiments with simplified albedo distributions and albedo changes. To the extent that the albedo-induced perturbations are linear, their magnitudes can later be scaled for observed changes in albedo as these become available. The albedos for all permanent desert areas in the Northern Hemisphere were fixed at .35, and the albedos of the marginal and comparison areas were assigned the values .35 or .14; the albedos of all other land surfaces save permanent ice cover were taken to be .14; the ice albedo was assigned the value of .70, and the ocean surface albedo was taken to be .07.

The locations of the permanent desert and marginal areas were assigned on the basis of relative regional brightness obtained from minimal planetary albedos (Raschke et al., 1973). To insure computational significance the Western Great Plains and Rajputana areas were made somewhat larger than they are in reality. Figure 4 shows the albedo distributions assumed for the experiments.

Charney et al. showed by numerical simulation (see Table) with the GISS General Circulation Model that appreciable changes in albedo and evaporation rate significantly affect convective cloud and precipitation on the borders of and within certain monsoonal regions of large scale (> 400 km in their smallest dimension) during the summer rainy season. The changes were pronounced in the semi-arid zones adjoining the Sahara, the deserts of Central Asia and the Great Western Desert of North America; they were also pronounced in Central Africa and, to a lesser extent, in the Mississippi Valley.

Table. Precipitation (mm/day)

	OBSERVED	EXCESSIVE EVAPORATION		NEGLIGIBLE EVAPORATION	
		Albedo		Albedo	
		14%	35%	14%	35%
SAHEL	1.2	7.4	4.0	4.0	2.7
RAJPUTANA	2.7	4.9	2.3	2.1	2.4
WESTERN GREAT PLAINS	1.9	3.7	2.2	0.8	0.4
CENTRAL AFRICA	5.1	5.0	1.9	3.1	
BANGLADESH	7.9	8.0	8.0	4.6	
MISSISSIPPI VALLEY	2.3	4.4	3.3	1.2	

There are two difficulties in applying our results to real situations. First, the changes in albedo were necessarily hypothetical in the absence of realiable measurements of actual albedos and albedo changes. The albedo and hydrological changes in the experiments are about twice as high as the changes that actually occurred. Second, although the numerical model compares favorably with others of like complexity, it was found to be too inaccurate in its simulation of the climates in some of the test areas to justify more than purely qualitative conclusions concerning real situations. As better general circulation models are developed, the experiments described will be repeated. But the only way that biogeophysical feedback hypotheses of the kind offered here can be confirmed or disproved is by study of actual observations. If droughts over large regions are to some extent self-sustaining, then satellite monitoring of the albedo, the plant cover, and the hydrology in drought-prone regions, together with numerical modeling will enable us eventually to predict drought. Another possibility is that by proper farming and grazing practices we can minimize rather than maximize the random changes in rain amounts that affect arid regions. The prevention of rapid changes in plant cover would prevent large changes in evaporation and albedo and thus would ameliorate the strength and duration of droughts. However, it should be remembered that changes in albedo and hydrology can probably not change rainfall by more than 10% to 25%, whereas the natural variability of rainfall on semiarid regions can result in changes of 100%. Proper land conservation might ameliorate a drought, but could not prevent one.

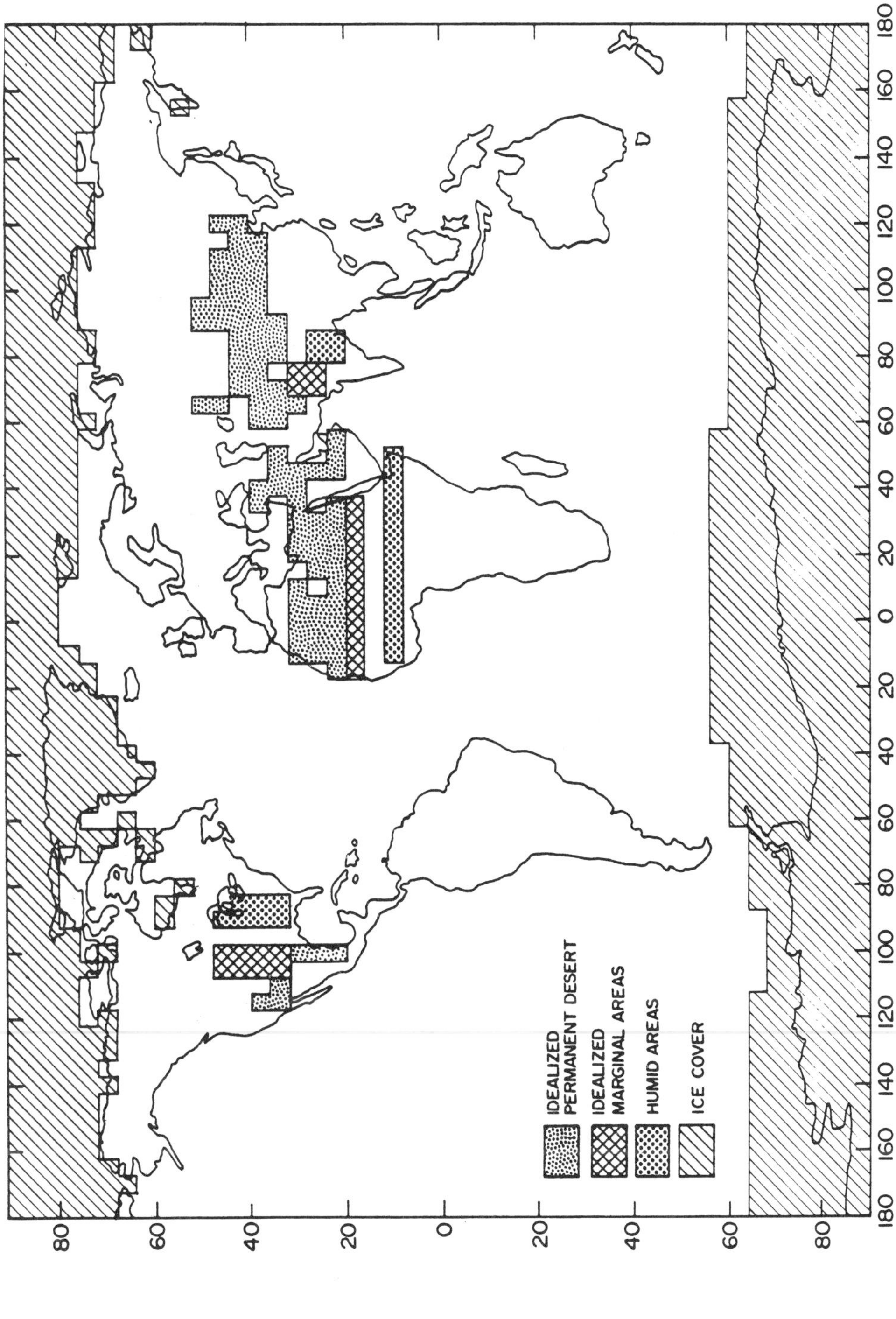
IDEALIZED PERMANENT DESERT
IDEALIZED MARGINAL AREAS
HUMID AREAS
ICE COVER
80
60
40
20
0
20
40
60
80
180
160
140
120
100
80
60
40
20
0
20
40
60
80
100
120
140
160
180

REFERENCES

Bryson, R.A. (1974), World Climate and World Food System III; The Lessons of Climatic History, Institute for Environmental Studies, Report No. 27, Madison, Wisconsin.

Bryson, R.A. and D.A. Baerreis. (1967), Possibilities of major climatic modification and their implications: Northwest India, a case for study, Bull. Amer. Meteor. Soc., 48, 136-142.

Buckman, O. and N.C. Brady. (1969), The Nature and Properties of Soils, Macmillan, New York.

Charney, J.G. (1974), Dynamics of deserts and drought in the Sahel, Quart. Jour. Roy. Meteor. Soc., 101, 193-202.

__________. P.H. Stone and W.J. Quirk (1975), Drought in the Sahara: A biogeophysical feedback mechanism, Science, 187, 434-435.

__________. (1976), Drought in the Sahara: Insufficient biogeophysical feedback? Reply, Science, 191, 100-102.

Damon, P.E. and S.M. Kunen. (1976), Global Cooling, Science, 447-453.

Hora, S.L. (1952), Hora's Satpura Hypothesis: an aspect of Indian biogeography, Current Science, Bangalore, 19, 364-370.

Nicholson, S.E. (1976), A Climatic Chronology for Africa: Synthesis of Geological, Historical, and Meteorological Information and Data, Thesis, University of Maryland.

Otterman, J. (1974) Science, 186, p. 531.

__________, L.S. Walter, and T.J. Schmugge. (1976), Observations from the ERTS imagery of overgrazing and cultivation impact on the Earth's surface, Space Research XVI, Berlin, Akademie-Verlag, 15-21.

Ripley, E.A. (1976), Drought in the Sahara: Insufficient Biogeophysical Feedback? Science, 191, 100.

Schneider, S. (1976), The Genesis Strategy - Climate and Global Survival, New York, Plenum Press, 419 pp.

SMIC Report. (1971), Inadvertent Climate Modification, M.I.T. Press, Cambridge, 308 pp.

Wendorf, F., R. Schild, R. Said, C.V. Hayness, A. Gaitier, M. Kobusiewicz. (1976), The prehistory of the Egyptian Sahara, Science, 193, 103-113.

REMOTE SENSING AND MONITORING OF DESERTIFICATION IN ARID LANDS

N. G. Kharin[1]

DEFINITION AND TERMINOLOGY

i) Desertification is a process of converting an arid or semi-arid area into one with less life. This process (physical and anthropogenic as well) leads to the destruction of the desert ecosystem; natural productivity and economical potential of these areas decline. The process may occur as a spatial extension of desert conditions or may be confined to comparatively limited areas. In this latter case intensification of desert conditions is caused by local phenomena.

According to W. Meckelein (1976) desertification is characterized by the following dynamic processes:

1.) Climatically: Increasing aridity (diminishing of water supply caused by macro- and micro-climatic or edaphic reasons).
2.) Hydrologically: Runoff becoming more irregular, feeding of the groundwater only episodically.
3.) Morphodynamic processes: Intensification of geomorphological processes (for example, salt transport accumulation, accelerated (soil) erosion by water (gullies) and wind, sand accumulation, reactivation of stable dunes.
4.) Soil dynamics: Desiccation of soils and accumulation of salt (salinization).
5.) Vegetation dynamics: Decline of vegetation.

Desertification is as a rule connected with the non-rational utilization of natural resources of arid lands (overcultivation, deforestation, overgrazing of pastures, urbanisation, construction of roads, changes of population etc.).

[1]Head of Remote Sensing and Photo-Interpretation Laboratory, Desert Institute Turkmenistan Academy of Sciences USSR.

Some desert scientists use the term "desertisation" (e.g., Le Houerou, 1968). It means extension of typical desert landscapes and landforms to the areas bordering the arid zones. The word "desertification" is being used by many scientists and by international agencies, so, in our opinion, it may be recommended as the standard scientific term.

ii) Under the term "monitoring" one understands the system of study, control and management of natural resources.

That means that we must find out all about desertification, control this process and take all measures to prevent natural disasters.

iii) A desertification map is a map that shows by symbols and colors, physical features of desert lands and desertification hazards. The map "World Deserts and Desertification", compiled by the Center National de la Recherche Scientifique (Paris, 1976) should be mentioned here as an example of a desertification map. The map represents the four groups of desertification hazards and human control:

a. Natural vulnerability of landscape to desertification processes (formation of dunes and shifting sand, superficial concentration of stones or desert pavement, enlargement of rock outcrops, enlargement of bare areas by sealing, enlargement of salinated or alkalinized areas without vegetation, and areas affected by water erosion).
b. Human pressure (areas of relatively high density of population, areas of relatively high density of livestock).
c. Human action for amelioration.
d. Desertification hazards (high, medium, slight).

iv) Remote sensing studies natural objects from great distances without physical contact with them. Natural objects are studied by means of remote sensors (aerial photographic cameras, thermal systems, radar equipment etc.). Remote sensors are installed on airborne vehicles (aircrafts, helicopters, space platforms), that permit study of the terrain from different flying heights. The next step (after obtaining photographic images) is studying the natural objects by photo interpretation.

TECHNICAL SYSTEMS TO STUDY DESERTIFICATION

It is known that photographic images are based on registration of electromagnetic waves reflected (or emmited) by the ground surface.

Figure 1 shows the electromagnetic spectrum and the bands of registration by mean of different remote sensors (C. E. Olsen, 1971). The photographic system is the most widely used type of remote sensor. These systems are used in inventory of natural resources, for compilation of thematic maps and maps of nature conservation.

Various types of aerial photographic cameras are now used for aerial photography, some of them are shown in Table 1. In the desert zone of the U.S.S.R. cameras with a focal length of 100 mm and 140 mm are mainly used. In flat desert regions with monotonous relief short focal length cameras should also be utilized.

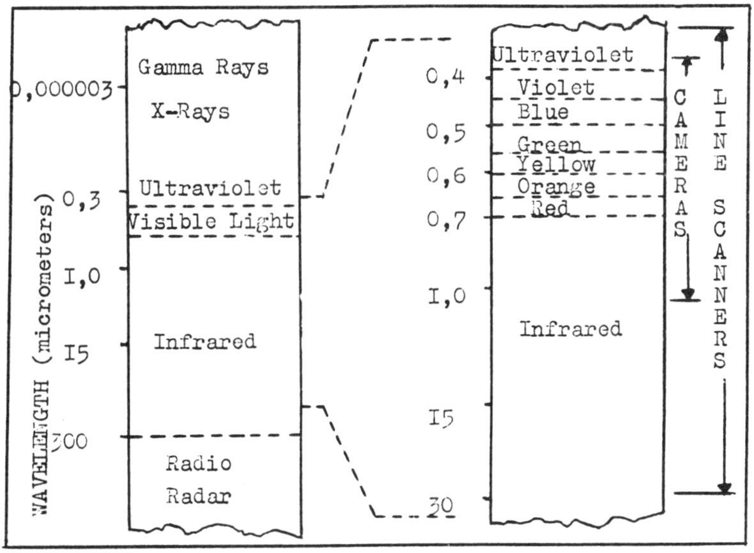

Fig. I. The electromagnetic spectrum and operating regions of remote sensors.

Panchromatic film is the most widely used film type for aerial photography, as it is sensitive to the visible spectrum. Color multi-emulsion film is also widely used. They are divided into two types: true-color films and false-color films (spectrozonal). Soviet spectrozonal film SN-6 represents the second type, this film is used by the Desert Institute, Turkmenistan Academy of Sciences. In the U.S.A. and other countries the films of this type are known as camouflage detection films or color infrared films.

Photographic emulsions have comparatively wide spectral sensitivity. In connection with that, spectral contrasts of natural objects in narrow bands can't be registered by this method of photography. This shortcoming can be avoided by multizonal photography based on utilization of photographic emulsions and a set of filters that cut off narrow spectral bands. Our investigations have determined the optimal bands of multizonal photography in Turkmenistan. They are the following: in the sandy desert 400-525 nm and 600-675 nm; in foothill country 400-500 nm and 620-680 nm; in man-made landscapes 600-675 nm and 750-850 nm (Figure 2). The same results can be obtained by multispectral photography, based on line scanning.

Other methods of aerial surveys should be mentioned here, e.g., radar and thermal sensing. These methods can be used in environmental monitoring to solve special topics in desertification study.

For the most rational application of sensing methods mentioned above, one must take into consideration the physical conditions of the regions to be investigated. Arid lands are characterized by scattered vegetation cover, clear sky during the most part of year, dusty atmosphere, low contrasts between vegetation and soil, non-stable snow cover etc. Annual fluctuation in seasonal events (e.g., plant phenology) is also a characteristic feature of arid lands.

TECHNICAL CHARACTERISTICS OF AERIAL CAMERAS

Country	Camera	Format size	Support	Lens	Focal length	Angle of view
U.S.S.R.	AFA-TE 55	18 x 18cm	film	Rodina-26	55mm	133°
	AFA-TE 70	18 x 18cm	film	Russar-29	70mm	122°
	AFA-TE 100	18 x 18cm	film	Russar-33	100mm	104°
	AFA-TE 200	18 x 18cm	film	Russar-Plazmat	200mm	65°
U.S.A.,	T.11	9 x 9in.	film	Metrogon	6in.	90°
Fairchild	T.12	9 x 9in.	film	Metrogon	6in.	90°
Italy,	Santoni IV	18 x 20cm	film	Aergon	135mm,	90°
Officine-					165mm	80°
-Gallileo	Santoni VI	23 x 23cm	film	Aergon	153mm	90°
Italy,	AFL 92	13 x 18cm	plates	Tessar-Zeiss	200mm	55°
O.M.I.	Foma Nistri 53/B	18 x 18cm	film	Aldebaran	210mm	65°
	Foma Nistri 53/B	23 x 23cm	film	Rigel	152mm	90°
France	Poivilliers	18 x 18cm	plates	Aquilor	125mm	90°
S.O.M.	Poivilliers	18 x 18cm	plates	Orthor	210,300mm	60°,45°
	Poivilliers	18 x 18cm	film	Aquilor	125mm	90°
Switzerland	RC. 7a	14 x 14cm	plates	Aviotar	170mm	60°
Wild	RC. 7a	14 x 14cm	plates	Aiogon	100mm	90°
	RC. 8	18 x 18cm	film	Aviotar	210mm	60°
	RC. 9	23 x 23cm	film	Super-Aviogon	88mm	120°
England,	Eagle IX	9 x 9in.	film	Ross	6in.	90°
Williamson	Mark	9 x 9in.	film	Ross	12in.	55°

Table 1 (T. E. Avery, 1969) shows the effect of day and weather conditions with use of a photographic camera, thermal camera and radar system. It should be mentioned here that modern technical systems permit selection of a suitable remote sensor if it is necessary to study the terrain under any weather conditions. Such natural disasters as forest fires, floods, earthquakes, mud flows etc. should be studied under all weather conditions. Thermal and radar sensing are methods which are very useful to solve this problem.

The aircraft and helicopter are the main airborne vehicles that are used nowadays. During the last years, space sensing also became widely used and its role in environmental monitoring increases from year to year. Photographic experiments from satellites, space ships and orbital stations conducted in the U.S.S.R. and

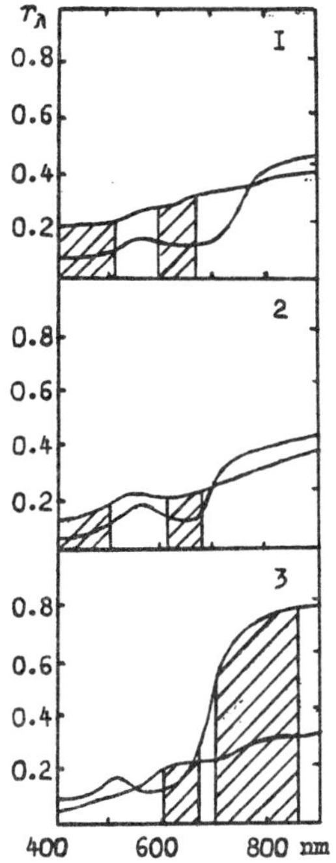

Fig. 2. Spectral bands of miltispectral photography in the deserts of Turkmenistan. I-sandy desert, 2-foothill country, 3-cultural landscape.

U.S.A. have shown the great advantages of this method. By this method, photographic images can be registered on photographic film or can be recorded by electronic systems and then transmitted to the Earth.

Space photos have one important advantage - a synoptic view of big areas. For example, the American satellite "Landsat" circles the Earth every 103 minutes or roughly 14 times a day. The same area is photographed every 18 days. "Landsat" produces the images in 4 spectral bands: green (0.5 - 0.6 micrometers), red (0.6 - 0.7 micrometers) and near infrared (0.7 - 0.8 micrometers and 0.8 - 1.1 micrometers). The Soviet satellite M-18 uses the same spectral bands to obtain images of the Earth. Figure 4 shows the territory surveyed by this satellite. Such satellites allow the study of some natural phenomena of global and regional scale, e.g., the desertification process.

The season of aerial photography is also very important. For compilation of thematic maps in the desert zone, aerial photography should be carried out in Spring. During this time of year maximum optical contrasts take place. But for desertification monitoring, aerial photography should be performed in summer and autumn. Sometimes valuable information can be obtained by comparison of two photographic missions (Spring and Autumn).

Monitoring desertification. The system of environmental monitoring should be designed to study and control, global, regional and local processes in soil, atmosphere and water. Regional phenomena of desertification can be studied on

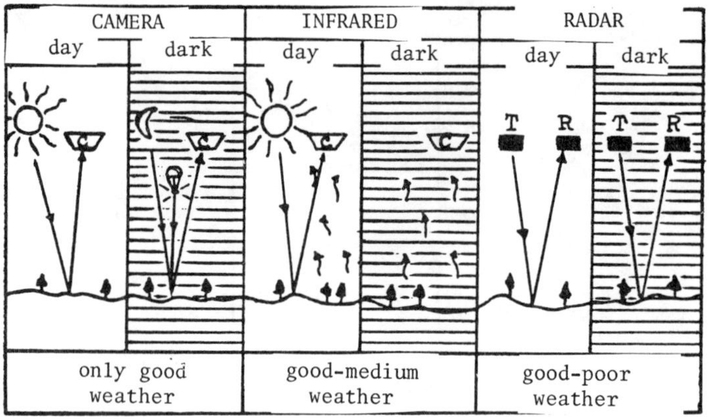

Fig. 3. The effect of time and weather on the utilization of some remote sensors.

space photos, local features should be studied on aerial photos. The ground control is one part of this system.

For development of a rational monitoring system, the location of test areas (sample plots) is of great importance. The location of test areas should take into consideration the application of space and aerial photos. The location of test areas must provide the rapid collection of information about all changes in environment for practical purposes. For study, the desertification test areas should be distributed over the whole territory of arid lands, with the purpose to control the desertification at all phases. This topic is discussed in publications by L. Sayn-Wittegenstein and A. Aldrich (1976), J. Thie (1974) and other authors. The location of test areas (sample plots) is based on the common principles of statistics. In this case, one will ask questions and receive well-grounded answers, e.g.; how to estimate with the predicted accuracy the changes in environment, what is the accuracy of obtained results, how much money and time is necessary to obtain data with the predicted accuracy etc. Now we can only give the general idea of this system because environmental monitoring is in the first stage of development.

Let's consider the location of test areas on the desert territory of Turkmenistan (Figure 5). By random sampling, sample plots are selected by use of special tables of random numbers published in statistical manuals. This method of sampling allows avoidance of personal errors, that the results can be treated statistically.

The second method (random stratefied sampling) differs from the first one, because samples are taken for each strata by the method of random sampling. For our example it means that samples should be taken for each physical region of Turkmenistan. This method is very useful for desertification studies inside physical regions if space photos are used.

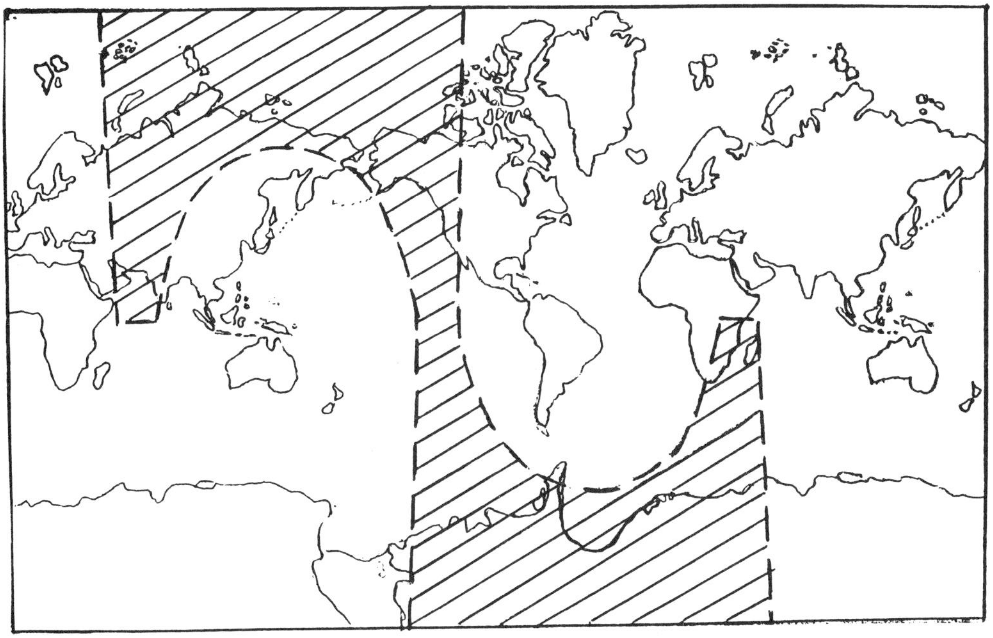

Fig. 4. The territory observed by the Soviet satellite M - I8.

Systematic sampling as shown in Figure 5 provides for the location of samples (test areas) according to the proposed system. It is known from the theory of statistics that this method of sampling avoids taking a great number of plots with low representative value. For example, this method is useful to study the degree of sand stabilization by plant cover. The rational method of sampling avoids taking the unnecessary plots and saves time and money.

The third method can give good results by estimation of degree of desertification on vast territories using space photos. Location of photo plots while measuring optical density of film will provide the best result by systematic sampling.

The methods of sampling considered above are based on ground and photo plots. Some test areas (sample plots) should be taken for continuous environmental monitoring for typical landscapes. Some photo plots (fixed on space photos) may be also considered as continuous plots for periodic density measurements. Sampling locations should be in close correlation with the location of biosphere stations (reservations). The sampling system on the territories of biosphere stations can be analogous to the system considered above.

Lastly, one should take into consideration that desertification monitoring is based on multistage sampling, as illustrated in Figure 6. It consists of the three "stages": space sensing, aerial photography and ground truth. Space photos must cover the whole territory under consideration, (at least for the first period of study). Aerial photography is a method for selective study. On the areas covered by aerial photography, ground truth determination is carried out with the sampling system considered above.

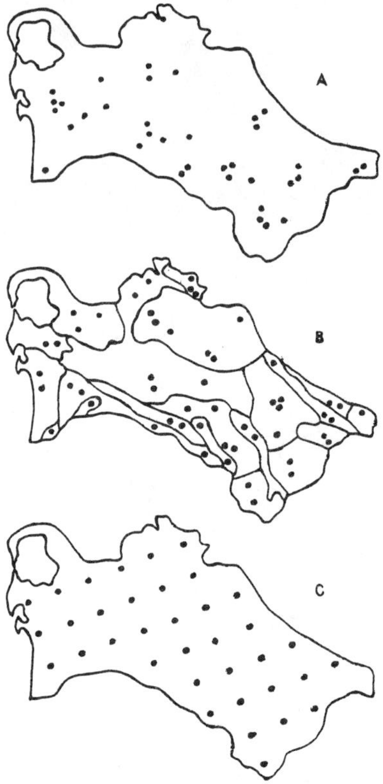

Fig. 5. Location of test areas for monitoring the desertification in Turkmenistan. A - random sampling, B - random stratified sampling, 3 - systematic sampling.

It is difficult now to propose the specifications on scans, seasons and spectral bands for remote sensing because the task of monitoring is very complicated and physical conditions of arid zones vary. So nowadays we can give the general principles of utilization of remote sensing methods. The sketch of specifications on remote sensing is given in Table 2, taking into account the general tasks of monitoring. The four main tasks of environmental monitoring are given as follows: monitoring desertification, natural disaster monitoring, nature conservation and inventory of natural resources. The Table gives information abut the possibility of application of remote sensing methods for practical purposes of environmental monitoring.

The first attempts have been made for application of remote sensing to desertification studies. The publications by N. G. Kharin (1976), M. Mainguet and L. Canon (1976), J. Otterman (1974), B. V. Vinogradov (1976), etc, should be mentioned. J. Otterman studied climatic desertification on "Landsat" photos. He emphasized that natural vegetation cover is pictured on space photos by another tone as compared with deteriorated pasture lands with destroyed vegetation. In his opinion,

Table 2. Types of remote sensing and its use for environmental monitoring in arid lands

N	Tasks of monitoring	Platform	Types and scales of remote sensing: Photographic systems: panchrom	infra	color	false color	multizonal	Non-photographic systems: thermal	radar	Visual reconnaissance	Season of remote sensing (month)	Period between aerial missions
I	2	3	4	5	6	7	8	9	10	11	12	13
	Monitoring desertification											
I.	Climatic desertification	S				I: 1000000	I: 1000000				VII, VIII	5 years
2.	Man-made desertification	S		I: 1000000		I: 1000000	I: 1000000				VII, VIII	I year
		A		I: 50000		I: 50000	I: 50000				VII, VIII	I year
3.	Soil salinization	S		I: 1000000	I: 1000000	I: 1000000					IX, X	5 years
		A			I: 50000	I: 50000		I: 50000				
4.	Water erosion	S		I: 1000000		I: 1000000	I: 1000000				IV, V, VI	I year
		A	I: 50000	I: 50000		I: 50000			I: 100000		IV, V, VI	I year
	Natural desaster monitoring											
5.	Mud flows	S	I: 1000000				I: 1000000				in case of need	
		A	I: 25000			I: 25000						in case of need
		H	oblique photography									in case of need
6.	Floods	S		I: 1000000			I: 1000000					in case of need
		A	I: 25000	I: 25000					I: 100000			in case of need
		H	oblique photography							in case of need		in case of need
7.	Earthquakes	S	I: 1000000				I: 1000000					in case of need
		A	I: 10000					I: 50000	I: 100000			in case of need
		H	oblique photography							in case of need		in case of need

Table 2. Continued

N	Tasks of monitoring	Plat-form	Types and scales of remote sensing							Visual recon-naissance	Season of remote sensing (month)	Period between aerial missions
			Photographic systems					Non-photographic systems				
			pan-chrom	infra	color	false color	multi-zonal	thermal	radar			
I	2	3	4	5	6	7	8	9	10	11	12	13
	<u>Nature conservation</u>											
8.	Study of reserves and biosphere stations	S					I: 1000000				III, IV	5 years
		A	I: 25000			I: 25000					III, IV	5 years
9.	Wildlife study	H	I:1000-I:2000								in case of need	I years
10.	Sea water monitoring	S			I: 1000000	I: 1000000	I: 1000000				I-XII	I year
		A			I: 50000	I: 50000	I: 1000000				I-XII	I year
11.	River water monitoring	S			I: 1000000	I: 1000000	I: 1000000				VIII, IX	I year
		A			I: 50000	I: 50000					VIII, IX	I year
12.	Monitoring the atmosphere pollution by industry	S					I: 1000000				I-XII	I year
		A	I: 50000		I: 50000	I: 50000	I: 50000				I-XII	I year
	<u>Inventory of natural resources</u>											
13.	Forest lands	S				I: 500000	I: 500000				III, IV	10 years
		A	I:10000-I:25000				I: 50000				III, IV	10 years
14.	Pasture lands	S				I: 500000	I: 500000				III, IV	10 years
		A	I: 25000				I: 50000				III, IV	10 years
15.	Soil and landuse survey	S				I: 500000	I: 500000				III, IV	5 years
		A	I: 25000	I: 25000							III, IV	5 years

<u>Notice</u>: The types of platforms are indicated by the letters: S - space platform, A - aircraft, H - helicopter

destruction of vegetation causes formation of vast "thermal depressions" with cooler surfaces. The rate of precipitation decreases here and climatic desertification begins. After some time, vegetation cover may be regenerated, and the rate of precipitation will increase. But due to overgrazing the process of desertification begins again. So the desertification here has a cyclic character. M. Mainguet and L. Canon compiled a map of sand flows occurring nowadays in North Africa. For map compilation they used space photos taken from satellite NOAA - 3.

The authors pointed out that these sand flows coincide with the location of contemporary forms of sand dunes. That indicates a slight change of wind regime during the whole arid cycle in the Sahara. In the other paper, M. Mainguet (1976) gives the proofs for dislocations of the southern limits of the Sahara in connection with climatic desertification. This information was also obtained from space photos.

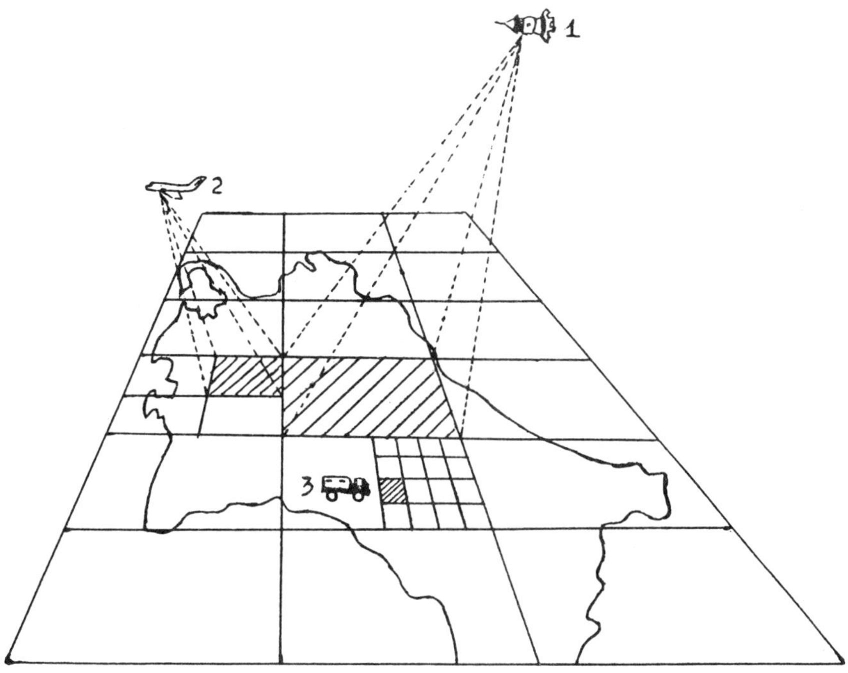

Fig. 6. Multistage desertification monitoring. I-space sensing, 2-aerial photography; 3-ground control.

For desertification monitoring by mean of remote sensing, we must take into account spectral reflectance of landscape elements. Destruction of vegetation, formation of salt efflorescenses on the soil surface and other changes in landscape appearance cause a change of spectral reflectance. These changes can be particular to single landscape elements or can include integral spectral reflectance of land parcels. If these changes go over the contrast threshold, and the resolution is sufficient, it may be affirmed that the spectral reflectance changes will be registered by remote sensing techniques.

It was stated (N. G. Kharin, 1973), that the difference in optical density of the two objects on a negative image can be identified by the human eye if the spectral reflectance ratio of these objects is greater than 1.5. In practice, this figure may vary according to the transparency of the atmosphere, characteristics of the technical system, angles of observation etc. Spectral reflectance curves of some sand surfaces and plants of Turkmenistan are given in Figure 7. One can see that differences in spectral reflectance of barkhan sand and sand over-grown with *Carex physodes* is rather sizable. That means these two objects will be identified on aerial photos. Still greater will be the difference between integral spectral reflectance of sands with vegetation (sand + plants + shadow places) and bare shifting sand dunes.

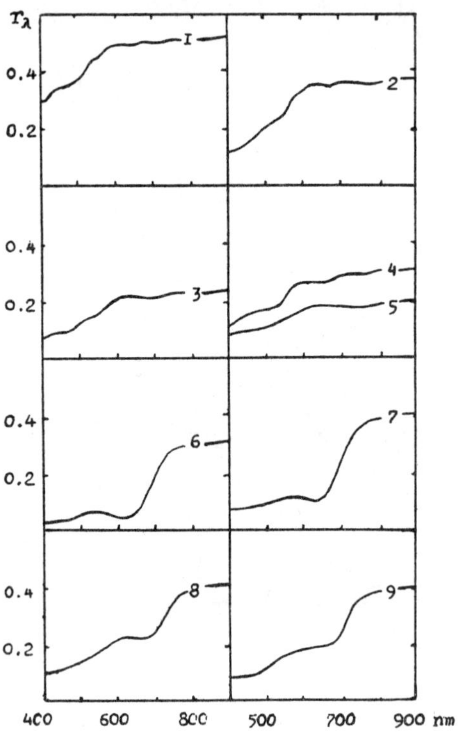

Fig. 7. Spectral reflectance of sands and desert plants. I-whitish quartz sand; 2-eolian sand from the surface of a barkhan; 3-gray alluvial sand; 4-sand surface over-grown by Carex physodes; 5-the same area after rain; 6-Haloxylon aphyllum; 7-Haloxylon persicum; 8-Salsola richteri; 9-Ephedra strobilacea.

Let's consider the case study of desertification monitoring in Turkmenistan. Nowadays the desertification process in this area has the following trends:

i) Climatic desertification (drying up of vegetation during the years with minimum precipitation, appearance of new shifting sand areas as a result of the action of physical factors):

ii) Man-made desertification (desertification near desert wells, population centers, roads, industrial enterprises and other places, soil salinization, water logging):

iii) Desertification appearing as a result of disasters (enormous floods, mud flows, fires, disaster propagation of wild animals etc.).

Man-made desertification is the most active process that is often combined with climatic desertification. We studied climatic desertification in Western Turkmenistan using space photos taken from the spaceship Soyuz -4. Bare sand dunes devoid of vegetation are pictured by a light tone.

Man-made desertification was studied in Southeast Turkmenistan. A number of light diffused spots were identified on space photos. By close observation, it was found that these spots were desertification spots around the wells in the desert. These spots, identified on the light image tone, have diffused outlines. Shifting sand dunes are confined to the wells and they have appeared as a result of enormous concentration of sheep flocks. The transition from plant covered to shifting sands is gradual, so the desertification spots on space photos have dim outlines.

The desertification spots were counted in two regions: the Tedzhen-Murgab interfluviary and South Karakum. 66 spots were registered in the first region and 94 spots in the second. Statistical data of the desertification spots are given in Table 3 (n-number of the identified desertification spots; d-mean spot diameter, km; 6-standard deviation).

Table 3. Statistical Data of Desertification Spots

Region	n	$\bar{d}$	σ	Distribution on spot diameter						
				> I km	I - 2	2 - 3	3 - 4	4 - 5	5 - 6	< 6 km
Tedzhen- -Murgab interfluvial	66	2.2I	± 0.83	6	32	I8	5	I	3	I
Southeast Karakum	94	2.I6	± I.26	I	55	28	6	2		2

It should be mentioned here that overgrazing destroys the vegetation cover and the image tone is in close correlation with it. There is an illustration of this point on the territory of Afghanistan bordering the southeast regions of Turkmenistan. These areas are pictured on space photos by a light tone, which is distinctly differentiated from the territory of the U.S.S.R. Vast areas of fire with destroyed vegetation pictured black are also identified on the territory of Afghanistan bordering the U.S.S.R. It is obvious that fires occurred in the summer of 1975 before the sapce photography.

As to the desertification caused by disasters, such areas are identified only in the Kopetdag mountains. Steep mountain ranges and alluvial fans have traces of mud flows. Other desertification of this type is not identified on space photos. It is obvious these processes have only limited local character, and can be identified on aerial photos.

4. Application of thematic maps and maps of nature conservation. Thematic and nature conservation maps can also be used to study desertification processes. A great number of thematic maps are prepared in the desert regions of the U.S.S.R. All these maps satisfy needs of the national economy or give scientific information on desert environments. Maps of natural conditions include such types as landscape, geomorphologic, soil, vegetation maps, etc. Maps of natural resources include forest pasture, phenological and other types of maps. All maps are prepared on the basis of remote sensing.

Each of these maps or set of maps can be used for planning nature conservation measures. Thematic maps contain valuable information on natural conditions or objects to be mapped. For example, forest maps give comprehensive information about area, productivity, density and volumes of forest crops. On pasture maps, one can obtain information of pasture vegetation, its productivity, degree of pasture degradation etc. All these data can be used in planning nature conservation measures.

But thematic maps have some shortcomings, that should be mentioned here. These shortcomings cause some difficulties in planning nature conservation measures. They are the following:

i) thematic maps are prepared periodically with an interval of 5-15 years. If they are used some years later, the compilation of their contents do not reflect the real situation.
ii) thematic maps are prepared according to the requirements of separate branches of science or of the national economy. That means they contain only specific generalized information on the desert environment.
iii) lack of information for development of nature conservation measures inevitably causes the tendancy to extract from thematic maps information that is not really available on these maps.

Other types of maps that are compiled nowadays should be mentioned here. These are maps on nature conservation. There are two main types of nature conservation maps:

1) Complex multipurpose maps (usually small scale maps), which give information on all aspects of nature conservation. Legends of these maps are very complicated, the maps usually cover vast territories. New methodological problems arise from the compilation of these maps. Such problems should be solved as standard scales, common principles of graphical representation of information, legends, etc. It is absolutely clear that such maps must be compiled quickly and simultaneously on a greater territory, they must be revised periodically to register all the changes in the natural conditions. A desertification map is also a complex map of nature conservation.
2) One-purpose maps (usually large scale maps). These maps are compiled on the basis of field surveys; large scale aerial photos are usually used. Other maps that should be mentioned include soil amelioration maps, insect damage maps, pasture improvement maps, flood control maps, phyto-amelioration maps etc.

Systems of desertification monitoring will be improved in the future. Remote sensing makes it possible to register all the changes in landscape appearance. If scientists and practical workers will obtain this information in proper time, that will be the best promise of success in desertification control.

BIBLIOGRAPHY

(1) Avery, T. E. "Interpretation of Aerial Photographs" Minneapolis, 1968.

(2) Kharin, N. G. "Special Reflectance Characteristics of the U.S.S.R. Main Tree Species". Proc. Symposium IUFRO, Freiburg, 1973.

(3) __________ "Use of Sensing to Study Environmental Conditions and the Desertification Process in Arid Lands". Pre-Congress Symposium K26, Ashkhabad, U.S.S.R., July 20-26, 1976, Kensington (Australia), 1976.

(4) "Landsat Index Atlas of the Developing Countries of the World". The World Bank, Washington, 1976.

(5) Mainguet, M. Teleexploration et Etude Geodynamique des Variations de la Limite Meridional du Sahara. Pre-Congress Symposium L26, Ashkhabad, U.S.S.R., July 20-26, 1976. Kensington (Australia), 1976.

(6) Mainguet, M., L. Canon. Vents et Paleovents du Sharar. Tentative d'Aproche Paleodimatique. Revue de Geography Physis et de Geology Dynamic. Paris, 1976, 2 Vol. XVIII, Tasc. 2-3.

(7) Olson, C. E. "Collection and Processing of Multispectral Imagery." Application of remote sensors in forestry. Freiburg, 1971.

(8) Otterman, J. "Bating High-Albedo Sols by Overgrazing: A Hypothesized Desertification Mechanism. Science, 1974, 84, N8.

(9) Sayn-Wittgenstein, L., A. H. Aldred. "Environmental Monitoring: The Role of Remote Sensing". Forest Management Institute, Ottawa, 1976.

(10) Thie, J. C. Wachman. "Remote Sensing for Environmental Monitoring and Impact Assessent". Proc. Int. Symposium on Remote Sensing, Banff, 1974.

(11) Vinogradov, V. V. "Establishment of Critical Zones of Desertification from Aero and Space Survey Data". Pre-Congress Symposium K26, Ashkhabad, U.S.S.R., July 20-26, 1976. Kensington (Australia) 1976.

THE CHANGE IN DESERT ENVIRONMENTS UNDER HUMAN IMPACT

A. Babayev[1]

Man's influence on the environment has a century old history. Each epoch or period of the development of human society, however, is characterized by a certain degree of influence on the environment. Together with the development of human society its influence on nature has been ever increasing, and in our age of scientific and technological revolution, it has grown to an unprecedented degree.

Man's economic activity is mostly felt in the lands capable of development at a minimum cost. Thus his influence on different territories varies both by intensity and form.

In the USSR, more than 300 million hectares are covered by arid lands, mostly by deserts (more than 210 million ha.) and semi-deserts. Since ancient times the USSR's desert lands have been utilized in man's economic activities, whose ever increasing influence has become more apparent, particularly during the past 30 years.

Hot climate and water deficit in the desert zone have strictly limited the extent and forms of man's influence on nature: human activity in the regions supplied with water is more intensive in comparison to remote and almost inaccessible ones. Agricultural reclamation is one of the most ancient forms of man's influence on desert nature, though water deficit in deserts has always been an obstacle to its development. Only in the Soviet period is land reclamation going on at a very fast pace due to new large hydrotechnic structures. After implementation of a number of large projects, however, still only a small portion of desert territories is being used for agricultural purposes. Thus, of the Middle Asian republics' total area (123.8 million ha.), only 6 percent or 7.5 million hectares are utilized for agriculture, more than 5 million hectares being under irrigation (I. P. Gerasimov, 1967). From 8 to 10 million ha. of so-called ancient-irrigated lands were also used for agricultural purposes at different historical times. In Turkmenistan only 870,000 ha. (17 percent) of the 5 million ha. fit for irrigated

---

[1]Head of Desert Environment Laboratory, Desert Institute Turkmenistan Academy of Sciences, USSR.

agriculture, are being used at present. Though man's influence on desert environments has not been very appreciable, now relief forms have come into being due to his century-old activity: agricultural oases and quite recently industrial ones. (Figure 1)

Range lands occupy most of the desert area. They cover 170 million ha. within Middle Asia and Kazakhstan. Range lands in Turkmenistan occupy 95 percent of all agricultural lands, in Uzbekistan and Southern Kazakhstan - 84 and 89 percent respectively. At present 30 million sheep and goats are grazing upon these vast territories, (N. T. Nechayeva, N. N. Pelt, 1963).

It should be mentioned, however, that desert range lands are still under-used: 4 million sheep, for example, are grazing upon the Turkmenian range lands at present, but scientists show that rational grazing would increase this figure up to 6 million. Considerable grazing areas are still in want of water (Nikolayev B. N., 1972).

Grazing lands of the southern part of Central Karakum and almost all the area of Zaunguz Karakum are underused due to water deficiency and their remoteness from large settlements. Grazing capacity of Zaunguz Karakum, for example, is utilized by only 40 percent. (M. Razhapov, 1974).

Thus man's influence on desert landscapes is far from its maximum intensity, and rational management of all desert resources may become highly effective.

Does it mean, however, that there is no problem of the desert environment? We think it is not the desertification processes that are rather intensive at present; human pressure results in degradation of desert ecosystems. The desertification is caused by industrial development, population growth and intensification of desert natural resource utilization.

One of the most important factors of man's influence on the desert environment in Turkmenistan is the construction of the Karakum Canal, 837 km long, at present, and 1400 km long when completed. The erection of the Canal has considerably contributed to the extension of irrigated lands in its zone. (Figure 2) Before its construction, about 186,000 hectares were under irrigation and this area was increased to 500,000 hectares by 1972 (M. K. Grave, L. M. Grave 1975).

The Karakum Canal's water has essentially changed the ecological situation in the zone of its influence. According to preliminary data, in the zone of the Canal's first turn, lakes due to filtration have already formed along a strip of 305 km. The formation of these lakes gives rise not only to plant inundation but to sand salinization as well. It is these circumstances that have caused the disappearance of many phytocoenoses and their replacement by hydrophilous types of vegetation having no forage value.

Simultaneously, the area of saline soils has widened in the oases, as a result of excessive water use and inadequate improvement. This is vividly illustrated by the table given below.

Great attention is paid to the problems of rational use while developing the desert territories of the USSR. The use of natural resources is accompanied by various measures for their conservation. The application of these measures has prevented, to a great extent, the distribution of undesirable phenomena on large areas. Thus in Turkmenistan, the sand-drift work is carried out yearly on an area of about 40 thousand ha. At present, the area of afforested sands is about 300 thousand ha. Five thousand ha of forest and orchard plantings have been made, and much work is done on the creation of tree shelterbelts (A. O. Ovezlieyev, M. I.

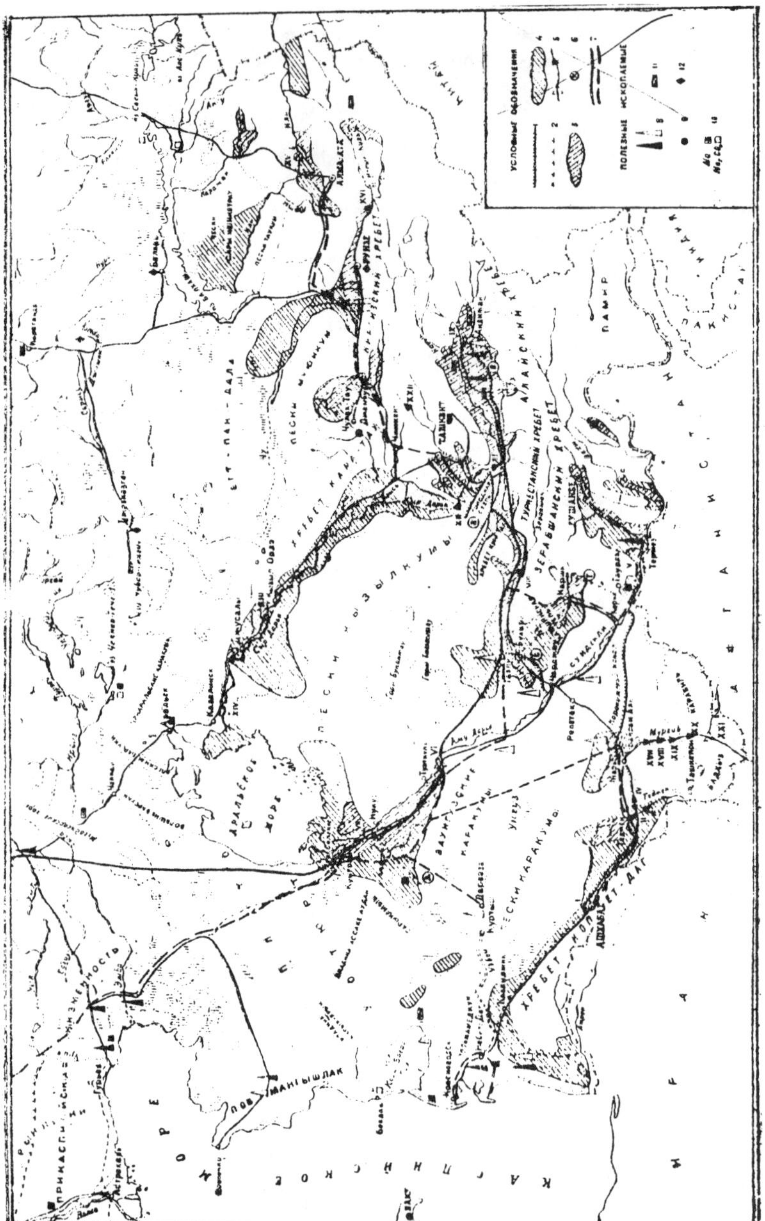

Fig. 1. Map of the development of the desert and semi-desert areas in the USSR - (compiled by M.P. Petrov).

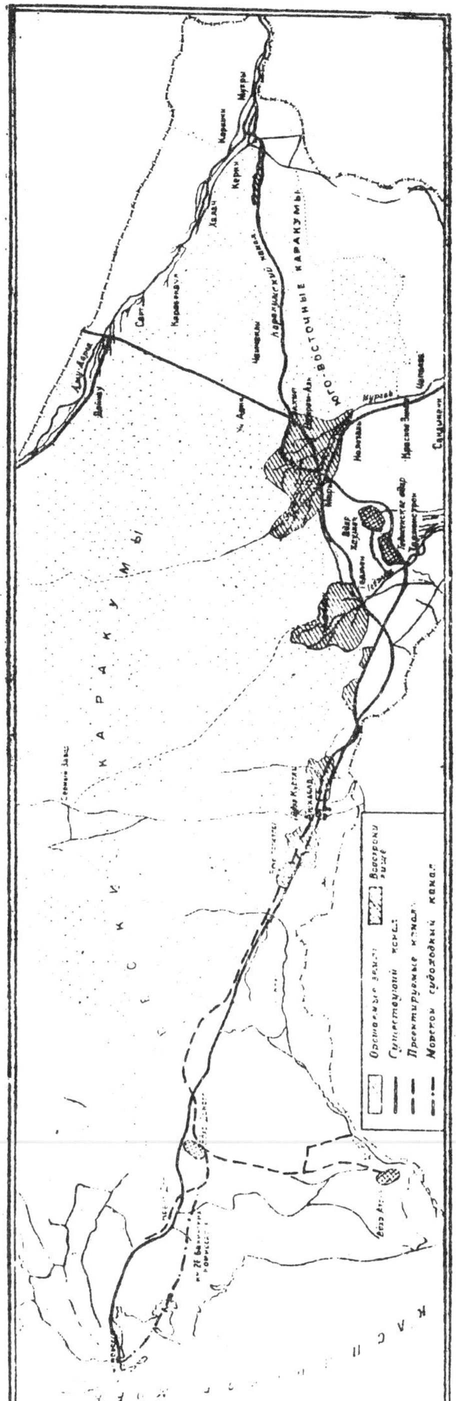

Fig. 2. Route of the Karakum Canal.

Saline Soil Area of the Murgab Oasis Region
(according to K. Redjepbayev and A. S. Ovsyannikov, 1974)

Degree of Soil Salinization	Years			
	1952 - 1954		1963 - 1968	
	area, ha	%	area, ha	%
Unsaline & wealkly saline soils	80304	72	51968	47
Medium saline & severely saline soils	18502	17	41313	37
Solochaks (white alkali soils)	12194	11	17719	16

Frolov, 1970). Work is being carried out for radical improvement of deteriorated pastures and establishment of artificial perennial pastures. Undersowing of wild-growing, high-yielding grass, semi-shrubs and shrubs is used for this purpose. In Western Turkmenia, where large areas of takyrs are spread, experiments are in progress on their plant-growing capacity. These experiments have already given certain positive results. The work on sand reclamation adjacent to oases deserves attention. The area of these sands equals about 2 million ha. Experiments on forage crops growing with application of sprinkler irrigation on these sands have given good results. The following yields have been received from every hectare: 90 metric tons of sorghum, 50 metric tons of corn, and 50 metric tons of alfalfa (A. G. Babayev, Z. G. Freikin, 1976).

Organization of state nature reservations has especially great importance in the system of nature conservation. Such nature reserves as the Ghasan-Kuli ornithologic reserve, the Repetec sand-desert, and the Badkhyz reserve have been organized and successfully operated in the deserts of Turkmenistan. Other reserves and parks are in the process of being organized. The optimal variants of man's influence on the natural environment of the desert are studied in these reserves.

At present, the problem of estimating the degree and scope of the man's influence on the natural environment becomes of primary importance. What is the meaning of the notion "the degree of influence"? Under the degree of man's influence on nature, one should understand the qualitative and quantitative changes in time and space of the landscape elements taken separately, or the landscape taken as a whole. These have different economic importance for man for certain periods of time. Landscapes with different degrees of man's imfluence on them can be singled out depending on this factor. Historical studies of various landscapes and comparative geographical approaches show that within the desert territory one can single out sections with the following degrees of man's influence: 1) the minimum degree, 2) the positive degree, 3) the near-negative degree, 4) the negative degree, 5) the maximum degree.

1. The minimum degree represents such conditions of a landscape or its components when no one of the elements is subjected to any change under the influence of human factors. However, man's influence is observed, although very slightly. These landscapes can be called "natural" in a full sense. Highways, separate

seasonally-used wells or sardobas (Persian: sard-cold and ba-water), and sheep paths can occur at places. Semi-vegetationless sands occurring on the ridge tops are formed under the influence of natural processes. The dimensions of the plants and their density approach the natural state. The availability of dried and broken shrubs and the desert moss covering of the surface can serve as the index of the minimum use of the territory. As a rule, these regions are distributed in the remote territories with bad road accessability, poorly supplied with water, and which are situated far from inhabited settlements. They are used as seasonal pastures.

2. The positive degree characterizes those landscape conditions whose components are being subjected to considerable and regular influence by man and animals, but are in a better than natural state. The investigations carried out prove that limited grazing makes for better development of pasture crops in the desert (L.S. Ghayevskaya, E. S. Krasnopolin, 1956; N. T. Nechayeva, 1954). The main provision of man's positive influence on the desert is regularity and periodicity of the territory use, which stipulate self-restoration of the natural resources. Here, limited grazing treatment and cutting of shrubs and trees are implemented. At separate stretches, the restoring of pastures by planting and sowing of shrubs is carried out. In this connection, separate components (relief and vegetation) are changed to a certain extent (ploughing-up is fulfilled, and more productive types of crop are planted or sowed). At present, necessary nature-preventive measures are carried out on these territories. At various stretches (within the zone of the Karakum Canal influence), favourable conditions have occurred for growth and development of desert plants owing to the modification of the separate elements of the landscape (ground water table rise).

3. The near-negative degree of influence refers to landscape conditions where the vegetative covering undergoes considerable changes. The types most valuable for fuel and pasture (Haloxylon ammodendron and H. persicum) are being annihilated, and new human modifications as Calligonum sp., Salsola sp., etc. appear instead. (Figure 3) Due to repeated pasturage and unsystematic felling, the taxonomic indices of the shrub vegetation are changed. In spite of the normal density, the shrub habitat is negligible. The grass covering is depressed. At places, moving barkhan sands are formed.

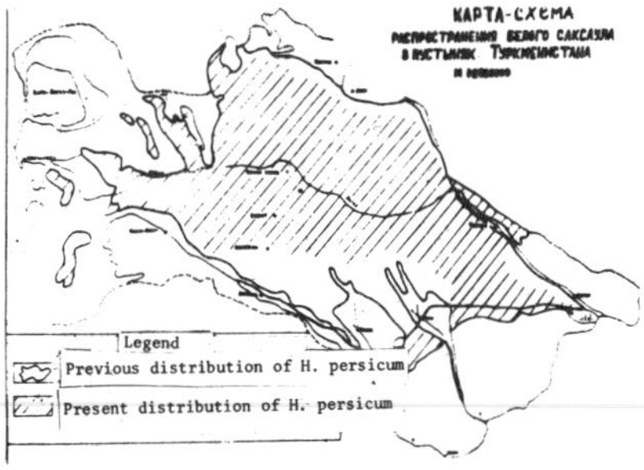

Fig. 3. Distribution Map of Haloxylon persicum in the deserts of Turkmenistan. (Compiled by G.S. Kalenov)

4. The negative degree represents landscape conditions where the main components of relief and vegetation are severly changed. Such stretches are mainly distributed near wells, where the cutting of shrubs is practiced and a large number of sheep is concentrated. No work on pasture restoration is fulfilled here, therefore, favourable conditions are created for the formation of the barkhan sands

with a thin Aristida karelini covering. In some cases, these stretches get overgrown with jurzarlic (Peganum harmala), which is not suitable as sheep fodder. The radius around the well of such man-made desert areas reaches 1.5 - 3.0 km and more. Thus, about 7-9 sq. km of area, representing the negative degree of the man's influence on the desert, appears around each well. Desert areas with negative degrees of man's influence are also spread around large oases, towns, working settlements and enterprises.

5. The maximum degree is a condition of the landscape where all the main components of the natural landscape have been severly changed and qualitatively new cultural landscapes have been formed. The natural vegetation cover is almost completely annihilated and replaced by productive cultural types of crops. These territories are chiefly situated in the river valleys. Nowadays a considerable part of the oases area is ploughed or built on. Besides, in recent years the so-called "industrial oases" have been created in the desert, their landscape being radically changed by man.

The investigation of the landscapes described above, with varying degrees of influence within the studied area, and the determination of their genesis and scale of distribution will help greatly when carrying out preventative measures. It should be mentioned, however, that the above-described types can be distributed rather irregularly. Selective analysis of the landscapes shows that the minimum and near-negative degrees are the most wide-spread types in the desert zone at present. In our opinion this is the qualitative aspect of estimating the degree of the man's influence upon the desert environments.

The problems of working out the criteria for estimating the degree of the man's economic influence on natural landscapes are most burning and complicated. It is only natural that these criteria are connected with the forms of man's influence on the landscapes and can change depending on availability and diversity of natural resources in the given region or geographical zones. Particularly while studying the degree of man's influence on the landscapes in South-Eastern Asia, the extent of land ploughing serves as the index of change (L. I. Kurakova, A. M. Rabchikov, 1971). In the forest zone, it is the woodiness of the territory where separate forest stands have been annihilated up to the present (F. N. Nilkov, 1973).

In the desert, the influence on nature has a local character, in spite of the age-long economic activity of man. Shown in quantitative data, these changes would be expressed in negligible figures, and as a result the desert might appear to be almost untouched. However, this does not correspond to the facts. To determine the real dimensions of man's influence upon the desert zone's natural environment, it is necessary to reveal the indicators which clearly show the dimensions and the degree of influence and to receive quantitative information not only on land conditions, but by use of remote sensing as well. Any human economic activity affects to some extent the vegetation cover, which sharply responds to all the changes made by man, at the same time changing the landscape's physiognomical element, which is rather important for deciphering aerial photographs. The analysis of numerous aerial photographs of various desert landscapes showed that there were different impacts on the natural vegetation cover: undisturbed cover, fellings, tree plantings, excessive grazing, fires, and plant floodings.

To establish the degree of change of the desert landscape, we use the ploughed area as the main indicator for oases, and violation of the vegetation cover structure for the desert. The coefficients of change for the territory may be used for the quantitative expression of the degree of human influence on the natural environment according to the following formula:

$$R_k = \frac{S}{S_1}$$

where S - the area of the changed (violated) stretches, S_1 - the area of almost untouched (weak-changed) stretches. If the coefficient exceeds 0.5, the area is considered to be changed.

Aerial photographic interpretation based on utilization of photo interpretation keys gives great possibilities for mapping the territory according to the degree of landscape change. Photographic keys may be chosen and compiled both out of previous and new photographic missions. The previous missions characterize the landscape make-up at a certain past time and the new ones - their modern state.

As a concrete example, we studied the stages of changes of the desert natural environment under the influence of irrigation structure. Such investigations, (based on the comparative observation data in time and space), obviously show the changes that occurred in the past and are occurring now, and help to determine the tendency and direction of changes, and to forecast the course of their further development.

The key area was chosen in a sandy desert of the Southeastern Karakum in the first turn of the Karakum Canal zone for the study of natural environment change under the influence of the irrigation structure. In 1956, the area considered was covered by aerial photography on a large scale. The region was studied in detail by means of landscape profiles, experimental areas, etc. In 1976, i.e., 11 years after the first missions, the same area was covered by a second aerial photographic mission. The comparison of the aerial photographs, and the use of literary and field data allowed us to single out the stages of Karakum Canal's influence upon the natural environment of the sandy desert.

1. The landscape's natural state before the Karakum Canal construction. Anchored ridged-cellular sands are largely distributed over the area. The cells' depth is 7-10 m, the width - 30-50 m. The plant association is <u>Haloxylon persicum</u> + <u>Salsola richteri</u> + <u>Calligonum sp.</u> types. There are vegetationless sands on the top of ridges at some places. Ground water lies deep. Its depth varies from 5-10 m to 20-40 m depending on the topography. The sands are usually not saline.

2. The landscape state during the Karakum Canal construction. During the cutting of the canal bed, sand landscapes were considerably changed along the Canal track. At a distance of 500 - 1000 and more metres, the ridged-cellular relief has been levelled by the excavated sandy substratum from the canal pioneer trench, as well as by pulp withdrawal. The vegetative covering of a greater area has been utterly destroyed. As a result of the Canal construction, flying sands - barkhans with a height of 1-2 m formed along its trench.

3. The landscape state 10 years after Canal construction. During this time natural plant regrowth began within the flying-sand stretch. But the vegetation is still weakly developed, the pioneer-aristida (<u>Aristida karelinii</u>) occurs. To prevent the banks from filling up, a planting of Russian thistle (<u>Salsola paletskiana</u>) has been made at some places along the Canal embankments. At certain stretches where the sand movement is strong, mechanical protection has been established. A narrow strip of semi-vegetationless sands follows the strip of barkhan sands and presents an intermediate strip between the moving and the anchored sands. The vegetation density is 50-100 plants per ha. there. After this strip, the belt of filtration lakes begins. The lakes are formed both at the expense of the canal water filtration and pulp and water withdrawal in separate years. This water fills depressions, and as a result of this some soil and vegetative cover changes occured and are still taking place. On the stretches

adjacent to the lakes, natural psammophytes are replaced by the hydrophytes of the tugai type (Tamarix sp, Phragmites communis, Karelinia caspica, etc.).

The ground water lies at a depth of 1 m. Sandy desert soils turn into meadows. The filtration lakes are located mainly on the right bank, which is conditioned by the local relief. The filtration lakes strip reaches 2 km. in width on the right bank, while on the left side they are spread within a comparatively narrow stretch (about 300 m). Some of these filtration lakes dry up in summer, and alkali soils (solonchaks) form in their place.

4. The sand landscape state 20 years after the Canal construction. At present, further natural sand revegetation is taking place, and the strip of barkhan sands is turning into anchored ones. The zone of the filtration lakes distribution is considerably extended. We adduce the comparative data (1965 and 1976) of changes occurred at the desert stretch (Figure 4).

The area of the lakes and adjacent stretches covered with hydrophytes and replacing the previous psammophytes, already constituted 14% of the total area of the stretch studied by 1965. By 1976, this area increased by more than twice and was 32% of the total area. However, the area of the lakes' water surface remained almost unchanged during this period and formed 3.5% of the stretch's total area. A comparative analysis showed that significant work was carried out on canal widening. So, if the area of the Canal, located within the given stretch made 4.5% in 1965, in 1976 it was 7.5% of the total stretch area.

Thus the analysis of a certain desert stretch, conducted in time and space, shows considerable territorial dimensions and rates of change of the desert environment under the influence of human economic activity.

On the basis of the elaborated degree classification of man's influence upon the environment, and aerial photography, experimental large-scale mapping of separate key areas has been executed. Fragments of maps according to the degree of human influence on various desert landscapes are given as examples (Figures 5 and 6).

The study of human influence on the desert's natural environment requires not only revealing the past and present changes, but defining the tendency of the landscape development on the whole and the development of its components as well, i.e., elaborating the question of man's influence on separate territories' natural conditions. It is important to use the methods of indicative aerial photography, permitting the definition of the character and direction of the processes occurring according to landscape and vegetative indicators, and to evaluate the intensity of their development in time and space. The vegetation, being one of the physiognomical landscape elements, quickly responds to insignificant changes, made by man in the given landscape and is reliably deciphered by aerial photographs. Thus, forecasting may be realized on the basis of the analysis of the state of vegetative cover, and is closely connected with stipulation or conservation of natural processes.

The regional landscape map is compiled on the basis of the intra-landscape extrapolation method, which includes the selection of areas with analogous aerophotography, and subsequent transference of the detailed photo-interpretation results of the key stretch onto these areas. It may serve as the basis for elaboration of a forecasting map, in which the areas with different degrees of man's influence are singled out. Proceeding from this, a plan of economic measures for prevention of undesirable phenomena and processes may be worked out.

In our eyes the view of the deserts is changing. Accordingly, methods for its study and economic usage should be altered. Thus, in the interrelationship

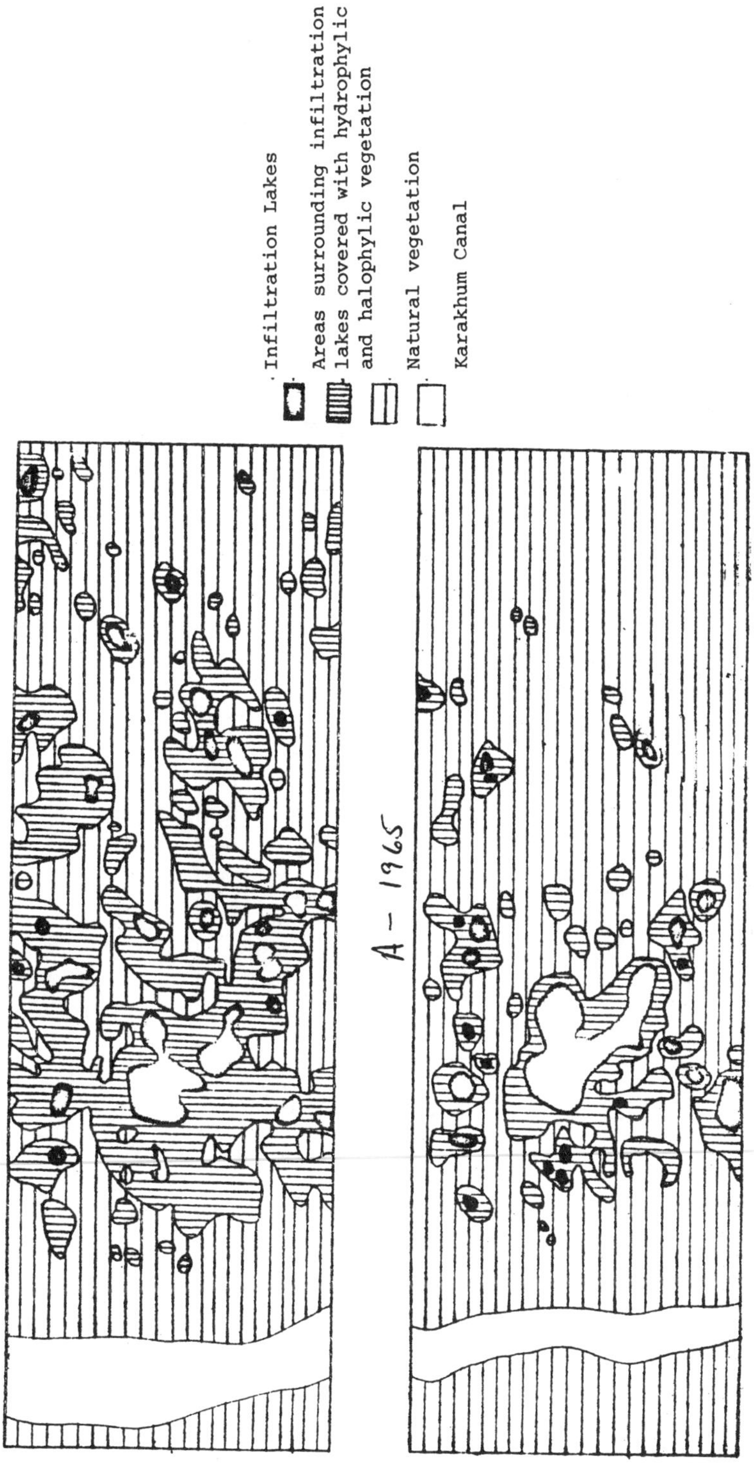

Fig. 4. Comparative Maps showing change in natural conditions in the Karakum Desert.
A - 1965 B - 1976

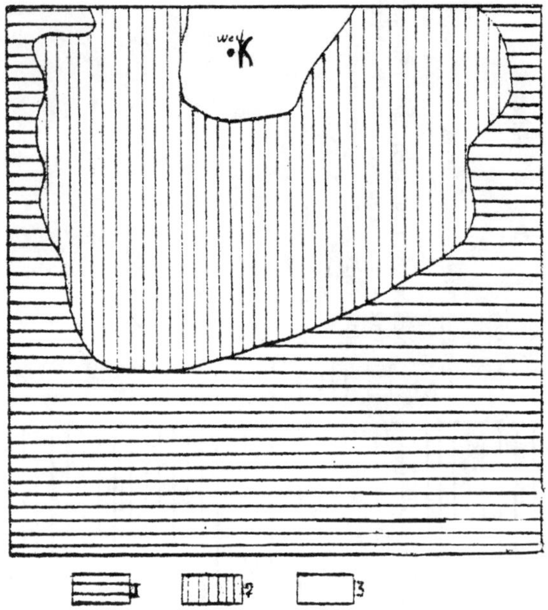

Fig. 5. Photo-interpretation sketch of the degree of human influence upon a desert landscape. (Point K is a well).

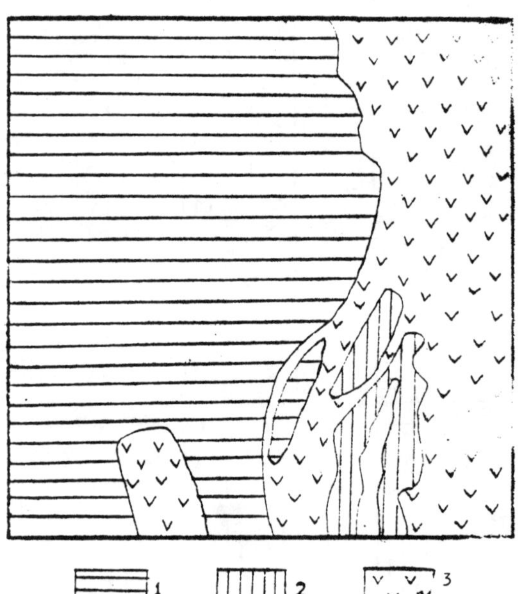

Tillage Waste Land Natural Turgai Vegetation

Fig. 6. Photo-interpretation sketch of the maximum degree of human influence upon flood-plain landscapes.

between the desert and man great changes take place. A new approach appears, based on the principles of reciprocity. The old unilateral form of relations "the desert for the man" should be abolished or at least changed radically. A modern slogan: "the desert for man, and man for the desert" should be the main feature of their relations.

REFERENCES

(1) Babeyev, A. G. The problems of integrated development of the deserts and their solution in the USSR, "The Problems of Desert Development", No. 3-4, 1976.

(2) _________, Z. G. Freikin, Transformation of the deserts in the USSR, Moscow, "Zanie", 1976.

(3) _________, N. F. Ovezliev, S. P. Ratkovsky, Land reclamation of the arid territories of the USSR. "Problems of Desert Development", No. 4, 1974.

(4) Graevskaya, L. S., E. S. Krasnopolin, Vegetative covering changing of the sheep pastures of the Middle Asian clay deserts and piedmont deserts under the influence of pasturage, "Botanical Journal", Issue 7, Volume 41, 1956.

(5) Gerasimov, I. P., The main problems of the nature reclamation in the Middle Asia, "The Problems of the Desert Development", No. 6, 1967.

(6) Grave, M. K., L. M. Grave, The canal in the desert, "Nature", No. 2, 1975.

(7) Kurakova, L. I., A. M. Rabchikov, Anthropogenic landscapes of South-Eastern Asia. In collection: "Natural Resources and Cultural Landscapes of the Mainlands", Moscow. 1971.

(8) Milkov, F. N., Man and landscapes. Essays on the Anthropogenic Landscape Study, Moscow, 1973.

(9) Nechaeva, N. T. Pasturage influence on the Karakum pastures. Collection: "The USSR Deserts and Their Development", Vol. II, Moscow-Leningrad, 1954.

(10) _________, N. N. Pelt, Forage base of sheep breeding in the desert zone. In the book: "Natural Conditions, Animal Husbandry and Forage Base of the Deserts", Ashkhabad, 1963.

(11) Nivolaev, V. N., Natural forage resources of Turkmenistan, Ashkhabad, 1972.

(12) Radjapov, M., The scheme of organization of the USSR grazing territory, No. 4, 1974.

(13) Redjepbayev, K., A. S. Ovsyannikov, The Karakum Canal influence on soil salinization of the Murgab oasis. "The Problems of the Desert Development", No. 5, 1974.

HUMAN ECOLOGY IN SAHEL AND SAHARA

Frank L. Lambrecht, DTMH
Department of Family and Community Medicine
University of Arizona College of Medicine

Abstract

Historical evidence shows that after the désertification of the Sahara, that changed a region of savannah woodland inhabited by a dense and varied animal population into a vast, infertile sand desert, the area became unsuitable for permanent human occupation but was not entirely abandoned.

A network of trade-routes developed after the introduction of the camel in response to an exchange of desert products, mainly salt, for gold and ivory from the Atlantic coastal countries.

Human activities and occupation of suitable Sahara and Sahel zones further developed after the introduction of domesticated animals. Raising of cattle in those dry areas was made possible only by means of seasonal migrations to the south in search of fodder and water for the herds during the months of the dry season. The seasonal herding of nomadic cattle on agriculture land became acceptable through a symbiotic system whereby the cattle was allowed to browse the stubble vegetation of fallow fields thereby fertilizing the field with livestock droppings. Exchange of dairy products against vegetable food further strengthened the symbiosis. The presence of the pastoralists was also of economic importance to market places along their migration itinerary. The socioeconomy-ecology of the vast savannah-sahel-sahara zones has been based upon these relationships for hundreds of years.

Plans for agricultural development of the Sahelian zones are now threatening not only the traditional relationship but the very social structure of nomadic pastoralism. Socioeconomic changes and modification in the environment will affect seasonal migration. Increased demand for meat for the coastal population may change the attitude of traditional cattle raising, at present strictly conditioned to tribal social structure.

A possible solution to a situation which is becoming increasingly unfavorable to nomadic pastoralists, is their sedentarisation in an agro-pastoralist economy, but is doubtful that this solution will be acceptable to all.

The system of nomadic pastoralism is facing its most severe challenge -- its survival is very precarious at best in spite of the remarkable adaptability shown by the nomadic people.

HUMAN ECOLOGY IN SAHEL AND SAHARA

Frank L. Lambrecht, DTMH
Department of Family and Community Medicine
University of Arizona College of Medicine

Arid zones are characterized by parched land, hot days and cold nights, infertile soils and a limited and sporadic rainfall of less than 250 mm. in a good year -- at first approach entirely unsuitable for human occupation. Yet great civilizations developed in arid areas crossed by great rivers such as the Tigris and Euphrates, the Indus and the Nile. It may have been a freak coincidence of history and opportunity but the challenge was met and overcome.

Once the techniques of plant and animal domestication had been well established and the right species selectively outbred from their wild ancestral strains, populations, perhaps under increased pressure of numbers, expanded, discovered and migrated to more fertile lands.

Not all left -- the arid lands were never really abandoned. Those who stayed found means of food production sufficient to sustain their small numbers -- other adopted a life-style that matched adversity.

Across the arid top of the African continent, agriculturists clung to the fertile strip along the Nile, planting and harvesting and waiting each year for the river floods to refertilize their fields. The herders took to nomadic life diligently applying their knowledge on desert ecology to keep and build up their livestock, while the professional traders with access to the Taghaza salt deposits became the first business tycoons, after the introduction of the camel changed the Sahara crossing from an uncertain venture to a regular commercial enterprise.

Camel Ecology

Twenty-five centuries ago, the Greek historian Herodotus spoke of the Garamantes, a people living at the edge of the great desert, who raided settlements by means of four-horse chariots. Long taken as a superb, imaginative tall story, the discovery of numerous rock paintings of horse-drawn chariots in various sites throughout the Sahara have given full credit to Herodotus' tale.

The Garamantes may have travelled as far south as the Niger bend. Herodotus' account relates to the fifth century B.C., but other evidence regarding the introduction of the horse into Egypt suggest that the desert chariots were in use as long ago as 1000 B.C.

Rock drawings indicate routes of travel along two main axes -- one from Tripoli at the coast to the soutwest through Fezzan and Tassali to the top of the Niger bend; the second from the Moroccan coast in a soutwest curve parallel to the continental shore, then turning east to the Niger to what was to become the major trade center of Timbuktu.

It is doubtful, and perhaps it was tried, that horse transport could have gone further south because of the presence of tsetse flies, and the trypanosome-disease they carry, to which the horse is very susceptible. Along the edge of the desert, however, the horse became a symbol of social status, adding strength and swiftness to the armies of the savannah empires.

The real Sahara travel started with the introduction of the camel. Like the horse, the camel came from the East, first to Egypt, then to North Africa, and finally to the Sahara. In contrast to the horse, the camel was an unlovely animal, ungainly in appearance, stubborn and irritable, but still fragile under careless treatment. In the Sahara the camel soon proved its value. It profoundly influenced human occupation, promoted old chariot routes to major trade networks of economic exchange and cultural intercourse, thereby changing desert ecology.

While the horse needed daily watering and ample supply of food, the camel could forego water over long periods. A camel could carry a rider or a full load of 500 pounds during extended desert journeys.

The camel's fine qualities, so well adapted to desert condition, provided a degree of human occupation of the Sahara hitherto unrealized. The animal became the most priced human possession. Wealth of desert dwelling was reckoned in the number and quality of their camels. The importance of his herd of camels was first on the list of a noble or a prince, rather than his holdings in gold or other wealth.

Together with the pastoral Berbers, the Arabs began to dominate the oases and to control the expanding caravan traffic. By the eleventh century the northern desert, as far west as the Atlantic shore, was dominated by Bedouin people, originally from central Arabia. The Tuareg Berbers, located primarily in and around desert uplands as Tassili, Ahaggar and Air, remained largely independent of Arab influence.

Raiding during those days, and till recently, was a way of life, practiced with equal vigour on sedentary villagers, passing caravans, or herds of rival groups. Livestock raids revitalized herds decimated by disease or drought. They helped maintain the ranges, thus preventing overgrazing, and preserve limited water supplies.

In normal times, a peaceful seasonal migration took place when the nomads moved each year during the hot, dry months -- the Arabs of the northern desert pasturing in the Atlas highlands, the Tuaregs southwards in the valley of the Niger.

Tuaregs, and other nomads, developed economic relationships with established desert stations or villages at the edge of the desert areas. In Tuareg country, associations are with small farms in oases tended by negroid people, known as Haratin, raising wheat, barley and millet, together with fig trees, date palms and grapevines. Some of these centers developed into important trading stations and became international markets.

Cattle Ecology

The first domesticated cattle in Africa originated with *Bos africanus* whose graceful lyre-shaped horns adorned the walls of royal Egyptian tombs, and Neolithic cave paintings in the Sahara. Cross-breeding with introduced cattle from the Middle East produced the large variety of types now encountered in the pastoralist area of Africa.

From north Africa, the first tribes of cattlemen spread southwards from the White Nile and Bahr-al-Ghazal region of southern Sudan, through Uganda and western Kenya into Tanzania, following the high rims of the Rift Valley, establishing the ancient kingdoms of Rwanda, Burundi and Baganda, and the powerful tribes of the Masai, Luo, Karamojong and many others.

In West Africa, Fulani pastoralists have spread their cultural, political and economic influence across the northern savannahs from Senegal to Lake Chad. They also helped the spread of the Islam religion. During the 19th century, the Fulani _jihads_ or holy wars, led to the conquest of the Hausa states which converted the Fulani from a subject to a governing people.

Pastoralists have gravitated towards the open dry areas of Africa inhospitable to cultivated plants, and out of reach of the tsetse flies found in the wooded savannahs and forest areas.

The choice of arid lands in protection of livestock against tsetse flies and trypanosomiasis, however, has required compensatory nomadism of seasonal migration in search of water and pasture during the dry season.

The long years of _jihad_ warfare forced large groups of Fulani to abandon their nomadic camps for walled villages later resulting in "settled or town Fulani," their herds tended by hired Hausa.

The great rinderpest epidemic of the late 19th century had a shattering impact upon the social structure of the nomadic Fulani many of whom lost most or all of their cattle forcing them to adopt agriculture as the only means of survival. In northern Nigeria, only one out of five Fulani is a true nomadic pastoralist, the rest having permanently lost their herd and turned to farming for a livelihood, hoping for the day when nomadic life may be resumed.

To the Fulani, cattle are the very essence of life. Cattle are currency, food and a medium of exchange. They are an investment and a means of capital accumulation. They bind social and family relationships. Payment of cattle legitimized a man's marriage and assurred his authority over his children. Sons are their father's labor force but their labor is later handsomely repaid through inheritance.

Livestock, therefore, is basic to African pastoralist social system, serving many purposes of food supply, standard of value and prestige, the seal of social relations and tribal structure.

Fulani Economics

Pastoralism in Africa has been the only way of life for untold generations but even the proudest Masai, Turkana or Fulani have to maintain relations with the disliked world of the farmer.

The cattleman, his herds supporting a half-dozen souls to the square mile had, in many areas, replaced the hunter requiring 5 to 10 square miles for foraging. But the cattleman could scarcely compete with the far greater productivity of the farmer, except in areas of poor soils and low rainfall.

The seasonal scarcity of surface water and of dwindling pastures during the long dry season have forced the Fulani of the Sahel zones to migrate seasonally southwards and renew annual interdependent relationship whereby Fulani herders are invited to graze their animals on Hausa fallow farmland in exchange of the fertilizing by manure provided by livestock droppings.

Among pastoralists, meat from cattle is rarely consumed. Livestock is precious and rarely slaughtered except for old or diseased animals. Direct food derived from herds are various dairy products and, in some tribes, blood. In the course of Fulani peregrinations in Hausa country, dairy products are exchanged for vegetable foods thereby strengthening symbiotic ties. Markets develop along migration routes displaying goods geared to the needs and fashions of the Fulani people, from fancy luxury ornaments to the traditional, decorated calabashes used by the Fulani women for collecting, handling and marketing dairy products.

And when an animal dies from old age or disease, the hide is used for many purposes such as sleeping mats, clothing, utensils and weapons. Skins also serve as a covering for temporary dwellings. Ancillary industries have made places such as Kano internationally known markets of leatherwork and cottage products related to nomadism.

To the pastoralist, a day cannot end satisfactorily but for a man watching his cattle come home. The purpose of his life is to taste that ultimate moment of well-being at each sunset. The most serious business is in or around the cattle pen. It is here that the owner will be buried, wrapped in the hide of his sacrificed favorite ox. Here marriages are discussed, justice pronounced, strategies developed or gossip exchanged.

Access to the cattle enclosure is mostly denied to women. This does not exclude women to retain ownership of stock. Fulani women often manage possession of their own small herd through proper management of inherited cattle. However, the utilization of dairy products by women is traditionally recognized, their preparation and eventual marketing.

Can Traditional Nomadism Survive?

In spite of its great vulnerability, desert nomads have weathered great challenges and adversities through a marvelous mechanism of adaptation, a profound knowledge of the environment and a good symbiotic relationship with adjacent populations.

Heavy grazing due to the build-up of herds in good years for which there is insufficient pasture in less favorable years has always been a problem. The provision of a number of waterholes to bring relief to water shortage has had the reverse effect of livestock starvation in those areas when more cattle was brought in to take advantage of the water supply, resulting in the complete degradation of the pastures in those areas from overgrazing.

Also, available evidence indicates that the nomad population of the Sahel will increase by two-thirds by the end of the century from about 39 million in 1974. Moreover, pressure on marginal lands for the production of more agricultural food may reduce the areas of pastoralism and raises the problem of the sedentarisation to the nomads. It has been predicted that with the greater demand for meat by the ever increasing populations of the coastal areas, and the better organization of marketing and transportation of meat, the pastoralists may look at cattle-herding in a more business-like manner by raising livestock for cast profit, thereby provoke a change in the social structure of their tribe.

Agricultural development in southern Sahel may result in curbing nomadic cattle herding and may have the effect of "fixing" the herders, eventually developing into a agro-pastoral economy with only restricted movement between favorable pastures. This pattern is developing in certain groups such as the Rimaibe, former serves of the Fulani, and the

Iklan, formerly serving the Tuareg. The development of dual economy will be most active and spreading at the margin of desert-savannah areas.

Acceptance of "fixed" pastoralists may vary from one region to the other. For instance, the Dogon in the sandy plains of Bandiagara may well benefit from a semi-permanent Fulani occupation as the cultivator needs the manure, the dairy products, and the care of his own animals offered by the pastoralists. In contrast, further to the south, the cultivators of Seno, farming on heavier soil and preferring shea butter to animal butter, may not wish to see a permanent Fulani occupation of their land.

A type of economic association and mixed culture has developed in regions such as Macina where Fulani extend to the Fimaibe, their former servants, the traditional loan of milk-cows in exchange for rice. Commercial agriculture, however, is generally unfavorable to traditional exchange. It is a sad irony that areas where groundnuts are grown for cash-crops, farmers have no surplus of millet to exchange for milk, so that their diet become improverished.

Nomads have given evidence of remarkable adaptability and a mixed agro-pastoralist economy may seem a possible solution to relieve the pressure and conflicts now building up in marginal Sahel zones as a result of population increase of agriculturists and their encroachment of northern territories which will extend even more with renewed efforts of search for more food production.

It is to be expected that not all nomadic pastoralists will be willing to trade their free living style for that of a sedentary existence in a village.

References

Dickson, B. T. (1956). Challenge of Arid Lands. Scientific Monthly, 82:67-74.

Glantz, M. H. (1976). Grazing in the Sahel. National Center for Atmospheric Research.

July, R. W. (1975). Precolonial Africa. Charles Scribner's Sons, New York.

Ware, H. (1975). The Sahelian Drought: Some Thoughts on the Future. ST/SSO/33, Australian National University.

THE SUITABILITY OF DESERT CLIMATE FOR THE TREATMENT OF CERTAIN DISEASES

B. C. Dando[1]

ABSTRACT

Prevalence of certain diseases is related to climatic conditions; the Kalahari Desert's hot, semi-arid environment can be beneficial or detrimental to a person's health depending on physiological acclimatization processes. The influence of environmental temperature on the thermoregulatory processes is the major climatic factor under consideration. Adverse consequences of a hot, dry environment (heat hyperpyrexia, heat syncope, heat exhaustion) can be escaped by balancing heat accumulated within the body with heat loss. The chief physical means of heat dissipation is evaporation to counter heat gain from solar and ground radiation, air temperature, activity, and the body's metabolic processes and cellular activity. Thermal eccrine sweating under central autonomic control in the hypothalamus together with salt and water requirements necessary to maintain internal balance is involved in some medical conditions. Certain diseases may be ameliorated in the hot and dry desert climate whilst others are aggravated under such conditions. Exploitation of the desert's resources, including tourism, is now underway. It is important therefore that there is a clear understanding of how the human body may react under these conditions of stress. The Kalahari's a typical "desert" environment seen in its abundance of xerophytic vegetation and intense temperatures due to abnormal anticyclonic development offers hope for the increasing health conditions faced in many technically advanced cultures.

THE KALAHARI DESERT

The climate of any place on earth is controlled by various components such as vegetation cover, solar and cosmic radiation, wind, rainfall, general atmospheric circulation and its local effects, and temperature. The prevalance and absence of certain diseases are related to climatic conditions; in the desert, physiological factors are important influences on overall health conditions.

---

[1]Epidemiologist, World Health Org., Gaborone, Botswana

The Kalahari Desert is vast including large parts of Central and Southwestern Botswana (two-thirds of the country) as well as extending into the eastern parts of South West Africa (Namibia) and northern South Africa. The area is around 500,000 square kilometres in area. However, this undisturbed habitat cannot be considered a desert in the fullest sense of the term - it is burgeoning with zerophytic vegetation of grasses, trees, succulents, and flowers. Acacia species are especially prosperous in the Kalahari.

In most desert environments, the denudation of existing vegetal surface or its noticeable absence result in increased environmental temperature and wind factors. In the Kalahari, most of the xerophytic species found are thorny and relatively unpalatable for grazing or browsing, thus they survive readily in this semi-arid environment. This makes the wind factor less pronounced and detrimental. Northern, northwestern and northeastern winds blow daily, rising rapidly to small gale blowings or whirlwinds which bring large quantities of dust. Due to the existing zerophytic vegetation the dust factor is more reduced in the Kalahari than in most desert environments; these remain disagreeable but are not dangerous for health considerations.

MAP OF BOTSWANA SHOWING RELEVANT FEATURES

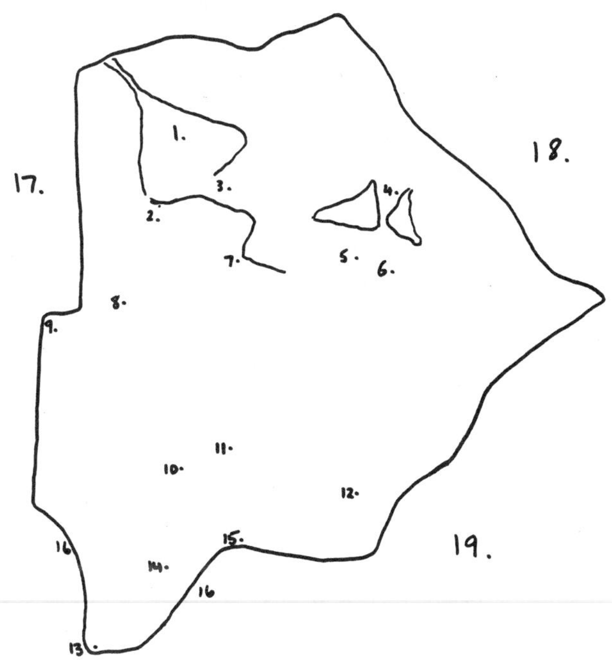

1	Okavango Delta	7	Rakops	13	Bokspits
2	Sehitwa	8	Ghanzi	14	Tsabong
3	Maun	9	Charles Hill	15	Werda
4	Makgadikgadi Pans	10	Hukuntsi	16	Nossop-Molopo
5	Orapa	11	Kang	17	South West Africa
6	Letlhakane	12	Kanye	18	Rhodesia
				19	South Africa

Rainfall is erratic occuring mostly during the summer months of November to April. The average amount of rain varies from 450 mm (a lot for a "desert") in the North to 150 mm in the Southwest; years of drought occur periodically. Evapo-transpiration of the rain water in the top two metres of sand prevents the recharging of ground water. Therefore, water accumulation or surface water is limited to a series of pans - expanses of low ground in which rain water is retained for brief periods. These pans are scattered throughout the Kalahari; some are several kilometers in length and are below the general level of the surrounding desert. Perennial water sources in the form of streams or rivers are virtually non--existent; there is the Okavango and Makgadikgadi which dominates the North and Central areas and the Nossop-Molopo which covers much of the southern region. These two main drainage systems are generally seasonal and of little impact on the general desert climate.

Besides being a dry habitat, the solar radiation is very intense due to the atmospheric circulation. The weather in Botswana is being controlled by sub-tropical high pressure belts. The lack of uniformity in the earth's surface and the influence of travelling disturbances upset the general wind pattern. The pattern revolves around two factors: air from heated surfaces expands and rises being replaced by colder denser air and, air over cold surfaces contracts and forms downward currents. The interplay of these and other factors, i.e. the rotation of the earth, results with the weather in Botswana being controlled by subtropical high pressure belts. Some of the highest climatic temperatures in the world are produced since these belts concentrate their centres of highest pressures over the oceans, leaving land area surfaces too strongly heated by the sun to allow normal anti-cyclonic development (Anderson, undated).

TEMPERATURE AS A MAJOR CLIMATIC CONDITION

One needs an understanding of thermoregulatory processes to evaluate the environmental stress high temperatures exert. All homeotherms, warm-blooded animals, attempt through physical and physiological means to maintain constant body temperature: body temperature is usually lower in the early hours of the morning and highest between 5 - 8 p.m. with the temperature of older people being lower than that of young adults. Daily variation in body temperature is about 1°F. The maintenance of normal body temperature depends upon an equilibrium between heat gain and heat loss.

Heat Accumulation

Metabolic processes create certain inevitable heat gains within the body; further, heat gains accrue from an increase in metabolism, exercise, and absorption of heat from external sources. Primary means of heat accumulation occurs by several means:

(1) Heat flows from a warm mass into a cooler one. Man in a hot, semi-arid environment continually absorbs heat from solar and ground radiation as well as from the intense temperature of the surrounding air. Temperatures in the Kalahari vary considerably with the seasons reaching 120° Fahrenheit in the summer months of October to April and reaching 32° Fahrenheit or even somewhat below this in the winter nights of July.

Daytime temperatures in the winter months of June to September reach 70°-80° Fahrenheit but may remain much colder for a few days at a time. The desert environment therefore exposes man to periodic rather than a constant heat load. High air temperature and intense solar radiation exposes man to a severe heat load for a few hours in the middle of the day.

(2) One's bodyweight and nutritional status together with the environmental temperature determines the amount of heat produced through activity.

(3) Lastly, maintenance of heat is important. Much heat is produced in maintaining the body's essential metabolic processes and cellular activity (including body temperature, circulation, respiration, tissue maintenance, etc.)

Heat Dissipation

Heat dissipation must occur to counter heat production. There are several ways to lose heat:

(1) Daily heat losses occur from the evaporation of insensible perspiration from the surface of the skin, the exchange of gases via the lungs, and the excretion processes of urine and faeces. In hot weather, evaporation is a very important means of heat loss. It is dependent upon the moisture in or on the skin, the nature of the surroundings, and the temperature and humidity of inspired and expire air.

Skin temperature is about 5° Fahrenheit less than that of the internal system. Since in the Kalahari the external temperature greatly exceeds normal body temperature, heat accumulation occurs whose loss can only be accomplished through the production of sweat and its evaporation from the skin (Samson Wright 1953).

Sweat: There are two types of sweat glands, apocrine and eccrine. Apocrine glands are confined to the axillae, nipples, genital areas and the perineum; they play a relatively small role in the maintenance of heat equilibrium.

Eccrine glands are distributed throughout the skin being most numerous on the head, palms and hands and soles of feet (Gray 1950). The sweat produced is a true secretion consisting essentially of a dilute solution of sodium chloride with a little urea. The thermal sweat produced by the eccrine glands falls under the control of the central autonomic nervous system with the heat regulating centre in the hypothalamus. Sweating begins almost instantly all over the body when the skin temperature rises above 94°F. The amount of heat stress determines sweat production.

The efficiency of the above heat regulating mechanism increases with continued exposure to heat stress. When acclimatized, sweating occurs readily with smaller amounts of sodium chloride being excreted which increases the total amount of sweat produced. Those incapable of acclimatization remain especially prone to heat stress.

The fluid produced in sweat reaches the external skin surface: if conditions are suitable, i.e. low humidity, evaporation takes place into the surrounding air. Evaporation requires expenditure of heat thus providing cooling of the body.

(2) Radiation is a process of heat loss due to differences in temperatures of the body and surrounding environment. Since heat flows from a warm mass into a cooler one, the process occurs mainly at night and plays a minor role in dissipating heat during times of crucial heat loads in hot, arid desert climate,

(3) Convection involves air movement patterns surrounding the body. The amount of heat loss depends upon the velocity of air movement, the surface area of the body, the skin temperature, and the temperature of moving air.

Generally speaking, in hot, semi-arid desert environments like the Kalahari, a man is more efficient in producing heat than in losing heat.

HUMAN PHYSIOLOGY AND DESERT CLIMATES

Environmental temperature is the major climatic condition under consideration when discussing human physiology in desert habitats.

In the Kalahari, excessive demands are made on one's thermeoregulatory system. There is an excess of heat intake from the surrounding environment coupled with difficulty in dispersal of this heat due to the high prevailing temperatures. If an equilibrium is not satisfactorily reached then certain disorders may develop, some quite dramatically.

Heat Syncope: (heat collapse, exercise induced heat exhaustion) is a fairly common disorder in unacclimatized persons which may occur in the first few days of exposure to heat. There is an excessive amount of blood deviated to the skin leading to hypotension and cerebral anoxia (lack of oxygen to the brain). The body's response to this is that the person faints, the body is forced to adopt a horizontal position giving a chance for blood to again flow to the brain in sufficient quantities. Rest results in complete recovery. The condition is less likely to occur as acclimatization improves.

It should be realised that the volume of blood in the body is insufficient to fill all capillary beds at all times. In exercise, blood is directed in increasing amounts to the muscles; after meals blood is similarly directed to the intestines; when heat is intense then blood is directed to the skin. However, the brain must at all times receive adequate quantities of blood or else function can be damaged permanently.

Acclimatization is a gradual physiological process in which the body adapts itself to changing environmental conditions by, for example, more efficient heat regulation from more efficient sweating by commencing the process at a lower level of body temperature; further, the sodium chloride content of the sweat falls with resulting better preservation of the electrolyte content of the extracellular fluids. Personal care in avoiding excess heat through appropriate clothing, avoidance of undue exposure to direct sunlight, etc. can assist. Taking of adequate quantities of water and common salt are also very important (various authors, 1972).

Heat Exhaustion is another example of excessive demand on the thermoregulatory system. This can occur from water depletion as happens with persons stranded in the desert where death can occur if dehydration is prolonged. It can occur from salt depletion when sweating is excessive and affects all types of people; it may be an accompaniment of gastroenteritis, malaria and other diseases: heat cramps can also occur in this disorder. Replacement of salt and water leads to rapid recovery. A third cause is from anhydrosis, in which sweating is deficient and occurs from the face and extremities, but not from the trunk; this may follow on severe prickly heat or sunburn: body temperature rises and can be followed by collapse and unconsciousness.

The normal cellular function of the body depends on a finely adjusted relationship between water and electrolytes such as sodium, potassium and others. The complex enzyme systems on which cellular metabolism depends are soon disrupted if this careful balance is upset. Thus the oxygen takeup by cells may be antagonized and oxygen is essential for the life of the cells. Heart rhythm may be affected and can lead to heart failure and eventually to death. Diseases such as gastroenteritis, where large amounts of essential electrolytes can quickly be lost to the body, are important contributory causes.

Other very important disorders of the body occur when there is failure of the thermoregulatory system. The cause of the failure is obscure but the result is that sweating ceases with consequent rise in the body temperature to dangerous levels; dehydration need not be present. The conditions are known as heat hyperpyrexia (excess temperature) and heatstroke.

Hyperpyrexia is failure of thermoregulation associated with conditions of intense heat. Once sweating ceases the body's temperature rises until it reaches 106°F - this arbitrary temperature confirms the condition of heat hyperpyexia. The exposure to intense heat usually has to be prolonged and need not occur in direct sunlight. It is aggravated by too much clothing, poor air movement or enclosure in a confined space. It occurs more commonly in unacclimatized persons but may attack either sex, any age group, and any racial group.

The onset may be sudden and dramatic; the skin feels dry and flushed and sweating is absent. This is a medical emergency and requires prompt reduction in the temperature by any means, the easiest being to bathe the body in cold water and fan it - cooling can be rapid to a body temperature of around 102° F when the additional cooling must be stopped or a state of shock can be induced.

Heat hyperpyrexia may be complicated by coexistent water - electrolyte inbalance and/or by shock - a condition in which there is collapse of the circulatory system, with fast heart rate and low blood pressure; treatment of shock depends on the condition causing it (Conybeare, 1975).

Malaria: One disease that can mimic hyperpyexia is malaria due to Plasmodium Falciparum where high fever and mental confusion similarly occur - a blood slide examined under a microscope will reveal the parasites if the cause of the condition is indeed due to malaria (Adams and MacGraith, 1966).

Cardiovascular Disease: Some cardiovascular conditions such as hypertension, heart failure, may benefit from the desert environment where salt is lost in appreciable daily quantities. Salt in the body attracts water so that an isotonic level is reached; if salt is in excess in the body, as happens when kidney excretion of salt decreases, the concomitant excess water (which shows up as oedema) may induce or aggravate heart failure; its removal from the body is an important part of the treatment of such heart diseases.

Lung Diseases: Lungs are the part of the body where air is taken in for the purpose of supplying oxygen for metabolic functions and cellular life; body waste products, notably carbon dioxide, are passed out from the lungs with expired air. This gaseous interchange take place in the alveolar airsacs whose total area may be that of a tennis court.

At the same time as gases are absorbed or excreted, water is added to the air if the humidity is low, as in the desert. This is a further important part of the heat regulating mechanism.

Lung diseases can be improved in desert climates. Lungs secrete small amounts of mucous which in the normal person is unnoticed. In disease of the lungs, mucuous is usually secreted in increasing amounts so that sputum is produced; this is then coughed up and expectorated. If the amount produced is excessive, the condition of chronic brochitis exists and is a very common disease in industrailized countries where the cause may be due to atmospheric pollution: infection is usually superimposed. The disease is persistent or recurrent and can last for years. Men are more susceptible than women in a ratio of 5:1; cold temperatures favour its development.

Warm, dry, unpolluted desert air can produce a spectacular improvement in this common disease.

A somewhat similar disease is Bronchiecdasis in which the Bronchi of the lungs are permanently dilated and damaged. Cough and sputum again occur and the condition can similarly be helped by the desert climate.

Some forms of asthma may be alleviated in the desert environment. Asthma is characterized by recurrent attacks of dyspnoea and wheezing due to generalized narrowing of the peripheral air ways. Causal factors are usually multiple with extrinsic allergy playing a major role; emotional stress is an important factor with mental tension and frustration being able to trigger off attacks; children can be affected in this way by conflict or tension at home. Asthmatics find a warm dry climate most suitable; they should adopt a general fitness programme with regular exercise, plenty of fresh air and correction of obesity - all of which is available in the desert environment.

Obesity: is excessive weight gain due to a calorie intake exceeding physiological requirements. The causes of this condition are several; amongst the causes are:

(1) psychiatric conditions in which anxieties and subconscious desires may in part be relieved by the pleasure of eating, and

(2) environmental influences, as seen in the adiposity of cooks and hotel owners, children who adopt the overeating habits of parents, etc. Removal to the dry peaceful desert setting where dieting can be carried out together with the reduction of anxieties and environmental influences can be of considerable benefit to such patients.

Skin Diseases may benefit from the desert environment. Thus Eczema, a disease characterised by superficial inflammation of the outer skin layer (Epidermis) together with the vascular papillae supplying it is due to allergic reactions of external or internal origin; it will be aided by the warm dry air and cautious acquisition of a mild suntan.

Dermatitis, a somewhat similar disease, is caused by physical and primary irritant damage of the skin surface. It may similarly benefit from the desert setting. (The terms Exczema and Dermatitis are often used inter-changeably.)

Suntan benefits nearly everyone but must be very carefully acquired in the desert where solar radiation is so intense. Suntan medications assist in this a great deal and are particularly important in fair and ginger complexioned persons. As we have seen above, sunburn can be a precursor of anhydrotic heat exhaustion.

Albino people suffer particularly from the intense solar radiation found in deserts: chronic sunburn is a real problem for them. They are usually found with clothing covering all of the body with only the face and hands exposed. They further suffer with their eyes, as do fair, blue eyed persons, from the severe glare of the desert. This can be markedly alleviated by a good pair of polaroid glasses which are an essential item for many non-indigenous persons in the desert.

Prolonged exposure to intense solar radiation over a period of years, particularly in fairskinned races increases the risk of cancer of the skin. Any skin lesion which fails to heal in a reasonable time should be investigated to make sure that cancer is not the cause. Diagnosis can usually be confirmed early and treatment is then satisfactory.

Trachoma: a communicable kerato-conjunctivitis of the eyes, is found more commonly in dry dusty conditions; flies assist in its transmission though not being essential. Communicability is fortunately low and it is usually contracted after a prolonged stay in desert conditions.

There are some diseases aggravated by dry, dusty conditions amongst which are hay-fever (allergic rhinitis), vasomoter rhinitis - in both of which there is increased nasal secretions and oedema of the nasal lining. The cause can be the pollens of the desert flora and some people may find it unpleasant to continue to live in such an environment even though the case usually occurs seasonally. Asthma associated with allergic causes may be aggravated in the desert.

GENERAL

The population of the Desert is sparse; certain tribal groups such as Masarwa (Bushmen) are well adapted to life in the Desert environment (Encyclopedia Britannica, 1974).

The total population of the Kalahari is estimated at around 100,000 persons of whom some 82,000 are Bantu speaking Africans of Tswana, Kgalagadi and Herero tribes, some 18,000 are Bushmen and a few hundred are of European descent.

The Bantu speaking people live in villages of from 20 to a 1,000 persons in size; cattle are the basis of their economy; these villages are mainly situated near the periphery of the Desert.

The main European settlements are the cattle ranches in the Ghanzi area in the South West; some work as Traders or for the state.

Bushmen are people about five feet in height with light complexions. Many now work as cattle ranch employees; the majority of the remainder follow their traditional pattern of hunting and food gathering; they move around in family bands of up to 16 or so families; each group covers a recognised area of maybe 400 square miles. They know more than 30 species of edible plants of which maybe 8-10 are staples in their respective seasons. Camp sites are moved every 3-4 weeks as local food resources are utilised (Winchester Gould 1968).

Communications are limited. A series of roads are found on the edges of the Desert and there are some across it such as the Kanye - Chanzi road (600 kms): this road consists for the most part of a thick sandy spoor which is suitable for four wheel drive vehicles and larger two wheel drive vehicles: cars cannot pass on this road. Other 'tracks' lead to other parts of the desert and these are usually only suitable for four wheel drive vehicles. The first motor crossing from East to West was made in 1925.

There are several airstrips throughout the desert: these are unsealed gravel strips suitable for small planes of 4-5 seat size.

Radio telephone facilities exist at several centres and these usually communicate with Gaborone (the capital) daily.

Map of Botswana Showing Relevant Features

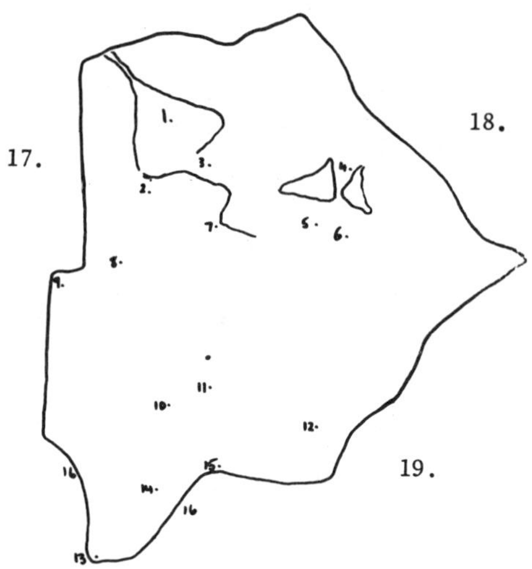

1 Okavango Delta
2 Sehitwa
3 Maun
4 Makgadikgadi Pans
5 Orapa
6 Letlhakane
7 Rakops
8 Ghanzi
9 Charles Hill
10 Hukuntsi
11 Kang
12 Kanye
13 Bokspits
14 Tsabong
15 Werda
16 Nossop-Molopo
17 South West Africa
18 Rhodesia
19 South Africa

Postal services are uncertain and slow.

Health facilities are found in and around the Desert as a series of Health Centres, clinics and health posts; several of these are of recent construction and others are to be built under the current National Development Plan. A health centre is a unit intermediary between a clinic and a hospital; it has beds for inpatients and has out patient services; it serves as a base of preventive health staff who travel to all populated areas; the one at Ghanzi has a Medical Officer. A clinic is a smaller unit and has beds for maternity work; staffed by a Registered Nurse(s) it gives both curative and preventive services. A health post is the lowest level of health care and is a small unit; it has no resident trained staff other than a Family Welfare Educator who is a health motivator: these units are visited at regular intervals by clinic, health centre staff and others.

Health Units at present are comprised of one small mine hospital at Orapa diamond mine, health centres at Ghanzi, Rakops and Tsabong, clinics at Charles Hill, Sehitwa, Letlhakane, Kang, Hukuntsi, Bokspits and Werda, plus several health posts.

Water for human consumption comes from a series of wells and boreholes: for domestic animals additionally from pans of rain water which are seasonal. Some of the boreholes provide fairly saline water, some being too saline for human consumption.

Food: Meat is available locally as are certain vegetables and cereals. Fruits will grow if irrigated, e.g. Citrus, Pawpaw, Bananas, etc. Other foods are brought into the area from the larger centres in the periphery.

There are a whole range of desert root crops, berries, etc. which are utilised by Bushmen throughout the year but these are not usually cultivated in the usual sense of the word. Certainly the Bushmen thrive on such a diet mixed as it is with meat, domestic and wild, and wild birds.

Wild Life exists in large numbers and is protected in five game reserves of which the Central Kalahari Game Reserve covers a large section of the Central desert. Certain species of animals prefer the desert environment and are normally found there - Eland, Gemsbok, Hartebeest, and others.

People coming to the desert need not fear the distances and quietness - airstrips make access to towns and cities relatively easy and distances are soon brought into perspective - specialist care can be arranged at the major centres in this country or neighbouring countries. Shortwave radio keeps one in touch with the world as a whole and even though not advised, one can keep in touch with the stock markets by this means.

CONCLUSION

Man has lived in the Kalahari desert for centuries and certain groups such as Bushmen have adapted extremely well to the environment which may at first sight seem hostile.

The present trend to exploit the desert resources including the large tourist potential, necessitates a clear understanding of human physiology at times of environmental stress plus a knowledge of disease processes improved or aggravated by the dry, hot and dusty climate of the Kalahari Desert.

REFERENCES

Anderson, R.J.F. (undated), Climatic Factors in Botswana, Botswana Notes and Records, Volume 2, 75-78.

Wright, Samson (1953), Applied Physiology, 9th Edition.

Gray (1950), Gray's Anatomy, Longmans, London

Author unstated (1972), Health Hazards of the Human Environment, World Health Organisation, Geneva

Conybeare (1975), Effects of Heat, Conybeare, Edinburgh, London and New York

Adams, A.R.D., Maegraith, B.G. (1966), Malaria hyperpyrexia, Clinical Tropical Diseases, Blackwell, Oxford.

Author unstated, (1974), Kalahari (Desert), Encyclopaedia Britannica, U.S.A.

Winchester-Gould, G.A. (1968), The Bushmen, The Guide to Botswana, Winchester Press, Botswana.

HIGHWAY IMPACTS ON THE DESERT

Robert O. Watkins
District Director, District 09
California Department of Transportation

The California Department of Transportation (CALTRANS) and its predecessor agencies have been responsible for State highways across California's desert since 1911. District 09 is one of 11 regional districts covering the State. It is almost exclusively a "desert" district. The southern part of the District is in the "high desert" environment of the northern Mojave Desert, and the northern part is in the "Great Basin" desert. During the development of the American West, the Great Basin was referred to as the Great American Desert, stretching from the Rocky Mountains to the Sierra Nevada.

While CALTRANS has been building highways in the desert for a long time, it -- like most of the State and Federal agencies which are involved with public works projects -- has been using a formal process of studying, predicting, and monitoring environmental impacts of its projects only during the current decade. In the desert particularly, this process has been hampered by a lack of information about the desert environment, and a limited interest in developing such information.

Nevertheless, CALTRANS has identified some major areas of environmental impact associated with desert roadbuilding and, more significantly, has identified some methods for avoiding or mitigating those impacts. Perhaps the most important impact is soil and vegetation disturbance. Next is the disruption of ecological communities and wildlife populations. Third is the destruction of archaeological sites which were left by the original Native American inhabitants of the desert. Today I'd like to discuss why we believe each of these impact areas is important -- why we <u>should</u> be interested in developing information about them -- and <u>what</u> impacts and mitigation we've identified through our environmental studies.

ATTITUDES

First, let me touch on the traditional attitude toward the desert that we've had to work with, and describe some of the changes that have occurred in that attitude. I mentioned that the Great Basin was called the "Great American Desert". It, and the Mojave Desert, were viewed as obstacles to progress from the time the first European-Americans started westward until quite recently. The railroads first, and then highways, were built to span what was usually considered a useless wasteland. Later some values were recognized in the wasteland -- it contained some valuable natural resources such as minerals and building materials. Even later, during and after World War II, great areas of the desert were devoted to military ranges where the armed forces could conduct exercises and explode ordinance without damaging anything that was considered valuable. At the same time, some areas were "reclaimed" for farming, where water could be found. Most recently, the desert has been discovered by the residents of the metropolitan regions which cluster along the California coastline. Much of the desert which isn't used for military ranges or for farming is used for "off-road" recreation. The trails of motorcycles, 4-wheel drive vehicles, and "dune buggies" can be seen almost everywhere.

Throughout the period of time I've briefly summarized, from the early 1800's almost to today, it's been considered a wasteland, where anything anybody wanted to do, from bombing to biking, was acceptable. No one perceived that there was anything at all out there that could be hurt. Then, in just a few years starting about 1970, the attitude changed. Today, I suspect more people perceive the desert as a valuable, and very sensitive, environmental resource which may prove to be a salvation for the future. Since so much of the desert is not committed to farmland or cities, it provides open space for recreation. Since it has relatively clean air and abundant sunshine, it may provide solar energy to power our metropolitan regions as petroleum resources become depleted. Over the long term, it may provide living and agricultural room for the future as well.

During the current decade scientific and academic interest in the desert has increased considerably. CALTRANS' efforts have helped stimulate this interest. For example, our work with desert tortoises in District 09 focused attention on these animals, to the point where now the tortoise is the State Reptile, the Bureau of Land Management has established a tortoise refuge, and a multistate desert tortoise recovery team has been formed.

SOIL AND VEGETATION DISTURBANCE

With this in mind, let's look at the first environmental impact area CALTRANS is concerned with, soil and vegetation disturbance. In the Mojave and Great Basin, many of the soils are loose, sandy and unconsolidated. Where undisturbed, the soil surface is usually comprised of "desert pavement". Over the years the finer soil particles have blown away leaving heavier particles which are relatively stable even in strong winds. These heavy particles protect the loose underlying soil from wind erosion. Additionally, a plant community will further protect the soil. Perennial brushy plants will be scattered throughout, rather evenly spaced according to the normal rainfall that occurs. As seasonal rainfall and temperatures permit, the intervening areas of ground between the perennial brush plants will support a growth of annual plants which provide a food source for animals and a floral display that can be spectacular for the tourist on the highway.

When this natural soil surface and vegetative cover are disturbed, as they must be to construct a highway, the loose and unprotected subsurface soils are exposed to wind erosion. At times they may also be subject to water erosion, though normally this is not a significant factor. This erosion will continue until the finer soils are blown away from the surface and "desert pavement" is formed again; this may take several years in some instances, if the soil is of a uniformly fine texture. The plant cover must also be reestablished, and that may take an even longer time to accomplish if the highway construction project does not include any provisions to hasten the process.

As an example of the time it takes soil and vegetation to recover, on Range B of the China Lake Naval Weapons Center the signs of a single tank exercise conducted during World War II are still visible. Even the crushed parts of individual shrubs can still be seen, attached to the living remainder of the plant after nearly 35 years.

One problem with disturbing the naturally stable soil surface and vegetation is that footholds for invasive exotic plants can be created. One such plant, Russian thistle (Salsola kali), is a major problem in the California desert. Russian Thistle is one of the first to invade disturbed soils, is prolific, and, when mature, its dry stems break off from the root in the wind and roll like a ball, spreading seed across the landscape. It chokes out other plants and does not provide any habitat value for wildlife. In the western United States it's commonly called "tumbleweed". Each year a significant amount is spent, by CALTRANS and others, to clear tumbleweeds from fences and ditches.

Examples

The results of damaging the stable condition are, first, that large quantities of dust are introduced into the atmosphere. Let's look at a specific situation. Near the community of Boron in eastern Kern County, Californa freeway construction in 1971 involved stripping a large area of the desert to build the roadway and to obtain earth to build the highway fills. The stripping covered a larger area than we had expected, because the highway contractor elected to obtain his fill material by "land leveling". He made shallow excavations covering large areas of land, rather than digging a deep excavation which exposed a minimum surface to the wind.

As a consequence, the community of Boron experienced dense, blinding clouds of dust for several months. According to newspaper accounts, traffic on roads was slowed or stopped outright as the dust limited visibility. Accidents were attributed to the dust couds. Something had to be done.

Mitigation Techniques

CALTRANS controlled much of the dust at Boron by plowing straw into some of the raw surfaces which had been exposed. Also, seeds of native plants, and some non-natives which we felt would grow under the droughty conditions, were planted in the strawed areas. The theory -- which proved out rather well -- was that the straw would mechanically stabilize the soil in much the same way "desert pavement" does. The seeds would accelerate reestablishment of the ground cover.

We found that while the straw seemed to work, the seeds were much less successful. We believe very little germination of the seeds we planted actually occurred, although eventually some seeded shrubs became established. What did happen initially was that seeds in the straw sprouted, and helped stabilize the soil.

At another location near Boron, we tried something else that did seem to speed the natural process of revegetation. A borrow site of about 20 acres (8 hectares) was required to build a bridge over a railroad. The contractor was required to scrape off and stockpile about the top 6" (15 centimeters) of the natural soil on the site, together with all the vegetation that existed. The mixture of soil and vegetation was stockpiled to one side of the pit, and the necessary earth was excavated. Once this operation was complete, the contractor spread the stockpiled soil back over the pit. The pit was then seeded with a mixture of shrubs, annual plants, and grasses, and watered once. Again, we found that few of the seeds we planted germinated -- but the dormant seeds which were buried in the soil we'd scraped from the top of the pit sprouted and greatly speeded up revegetation. Today the pit, while it's obviously been disturbed, is stable and appears to be in about the same stage of recovery as adjacent fields which were farmed and then abandoned many years before our construction project.

We think stockpiling the topsoil and using it to cover new slopes or cuts buys several years in the revegetation process. We also think using straw, spread at a rate of 4 tons per acre (9 tonnes per hectare) and incorporated into the soil, helps greatly to stabilize soil surfaces from wind erosion. Today we're combining these two methods on several projects, and we feel we can successfully mitigate the impact of clearing. We're still trying to plant seeds on new soil surfaces, but we're pessimistic that this will be successful at a cost which we can afford.

We have also, on an experimental basis, planted shrubs which have been started under nursery conditions. Five of our six test plots have been Great Basin locations on completed construction projects where natural revegetation had been slow and conditions were adverse. At these locations we have planted started shrubs into hand-dug holes, watered the plants once or, in one case, 3 times, and then given the plants no care whatsoever. We have survival rates of 70% or better on all but two plots. This greatly exceeds our expectations, and this technique looks promising for spot locations where few enough plants are needed to make it economically feasible. Since a great deal of hand labor is involved, we don't anticipate we'll use this method to cover very large areas of land.

Since every bit of soil we disturb is a bit of soil we have to stabilize, our engineers have also learned to take special care in designing projects to limit soil disturbance to a minimum. We also include requirements in the contracts which prohibit any construction activity outside the specified limits of construction. We're concerned enough that even very small areas a few square yards in size will be protected.

One more concern related to this impact is a maintenance practice which has been standard throughout the state. Along the roadside air currents generated by passing vehicles erode the soil away from the edge of the pavement. This creates a shallow drop-off which can be hazardous for vehicles which happen to swerve off the pavement for any reason. The traditional remedy has been for our maintenance forces to grade up dirt from the roadside to fill in the "drop-off". However the impact of this traditional practice is to perpetuate a disturbed soil surface, prevent perennial shrubs from getting a foothold, and to cause a long-term dust problem. We have now adopted a policy that we will not grade up dirt to fill this drop-off. Instead, we will import gravel or other coarse material to fill the drop-off. The course material will not be as subject to wind erosion, our frequency on maintenance will be reduced, and the soil surface can stabilize.

IMPACTS ON WILDLIFE

Impacts on wildlife are the second major area of highway-related impacts we've studied. Because the desert is dry and food resources are limited, animals that live on the desert tend to use very large areas. Individual animals may depend on range areas many acres or even square miles in extent. Highways may remove acres of land from the ranges of such animals, or may bisect travel corridors which animals use to travel from one part of their range to another. In some instances important water sources may be located on one side of a roadway, and equally important food or shelter resources on the other side.

These impacts are different from the impacts on soil and vegetation in one respect. It is not possible to develop a standard methodology to be used on most projects to mitigate impacts on wildlife. Animals habitat requirements are very specialized, and specific mitigation must be developed for each case. However I can mention a pair of examples that might indicate the direction mitigation measures can take.

The Desert bighorn sheep (*Ovis canadensis nelsoni*) is a shy, wary animal whose population is very limited. A major Interstate highway project in San Bernardino County -- outside District 09 -- passed through important sheep habitat areas. As part of the construction of the Interstate highway, special fences and an undercrossing structure were built to permit the sheep to cross between their various habitat areas. New water sources were also developed. After several years, indications are that the sheep do in fact use the undercrossing structure, and the other design features seem to have mitigated potential impacts.

In another location in Inyo County -- this time in District 09 -- a county highway passes between a group of sheep's winter range and their summer range. The summer range is important because in the particular range of desert mountains where these sheep live, a single spring is the only source of water that wildlife can use the entire summer. In this case, even though we believe the sheep will cross the road after it's reconstructed, we are recommending that a supplemental source of water be developed on the "winter range" side of the road, just to try to tip the odds in favor of the sheep.

The desert tortoise (Gopherus aggazzizi) is the California State Reptile. It's an interesting, slow-moving animal which seems to be particularly vulnerable to highway traffic. Individual tortoises have somewhat restricted territories. They spend much of their time in burrows under the ground. They hibernate during the cold winter months, and estivate during hot, dry summer months. They are out on the surface a few hours a day during the Spring and Fall -- although when we have dry years, it's quite possible that a tortoise will not come out at all! We recognize now that highway construction may directly affect tortoises. It's not possible for a slow-moving animal, which may be hibernating or estivating, to run away from an approaching earthmoving machine. By the same token, during construction it's not possible for the operator of that machine to see that a tortoise is about to be scooped up in the load. Even after the road is completed, tortoises crossing the road are regularly smashed by passing vehicles, because the tortoise can't get out of danger.

The tortoise problem has been difficult to solve. What we're doing is to try to find individual animals which are in areas where construction will occur, and move them to locations where they will be safer. When we started this program, we immediately found that it was extremely difficult to locate a replacement habitat for tortoises. We needed a place to move them where we would not overpopulate the range and exceed its tortoise carrying capacity. We also needed a place where the habitat requirements for tortoises were fulfilled -- and when we started no one knew what these requirements were. Even now we're not certain we can save all the tortoises that might be endangered by our projects, but at the least we can say that we have developed some partial solutions that will help maintain the State Reptile's numbers in the California desert.

IMPACTS ON ARCHAEOLOGICAL VALUES

One common factor in avoiding damage to soils, vegetation, or wildlife is the need to identify potential impact areas in advance, so we can consider environmental impacts during the engineering design of projects. The same need exists for early identification and consideration of archaeological sites.

Man has live on the California desert for centuries, and possibly many millenia. The evidence he left is very difficult to see, because it consists mostly of stone tools which resemble natural rocks to the casual eye, but once we begin looking for the evidence it is clear, unmistakable, and widespread. We realize now that most of the roads we've built in the past probably damaged or destroyed archaeological sites throughout our District. Now, we will try to protect the remaining sites from damage and, where we can't, we'll learn what the sites contain and we'll preserve that knowledge in a form future generations can use.

Why is it important? First, because the descendants of the original Native American inhabitants of the desert have a right to their culture. In many cases these traces have cultural or religious significance to present-day Indians. Second, for those of us whose ancestors have recently come to this continent, the archaeological record is intellectually interesting and may, in the future, show us how to live with more limited resources than we now enjoy. Pre-Columbian Americans lived under environmental conditions which present-day Americans would find hostile and unaccommodating. Perhaps by studying the archaeological record, we will learn to do with less. Our ancestors, our history tells us, learned much from the Indian at first hand. With our historical perspective, perhaps we can now see that there's much more to learn still.

Mitigation

For nearly every proposed construction project, CALTRANS now makes a field survey to locate any archaeological sites which may be affected. If any are found -- and they frequently are -- their significance is evaluated by professional archaeologists. The project engineers, environmental planners, and archaeologists work together to determine what impacts the project may have on each site that's found. Wherever possible, the construction is planned to do the minimum possible damage. If some damage is unavoidable, the data are recovered by archaeologists and the work is documented so it's available to future researchers.

On one project we made some trade-offs between increasing the potential for erosion and reducing the potential for damaging a significant archaeological site. We compromised by building a steeper slope than we would have normally designed -- this required much more extensive erosion-control than we initially proposed. The benefit was that we were able virtually to eliminate any encroachment on the archaeological site. It is still there, available for research at any time. If we had not identified the site and made the design compromises we did, the site would have been significantly impacted and, as another consideration, our project costs would have been higher.

We in District 09 are also working closely with the Indian community to assure that their cultural traditions are respected throughout this process. In the future we will ask that an Indian advisor will be present whenever we are making formal archaeological studies involving excavations or artifact collections. Artifacts that we collect will be returned to the Indian community whenever possible, once the studies which may be required to document the archaeologist's findings have been completed. Special care will be taken to disrupt burial sites as little as possible, since in our area, at least, the Indians are especially concerned that their ancestors' graves not be disturbed.

A SUGGESTED APPROACH

What, then, has our environmental study process shown us about the impacts of highway construction on the desert? In the first place, we recognized that we could not approach the desert as a useless wasteland, where nothing we might do would affect our lives. Rather, we recognize that the desert is an important resource which must not be destroyed for our short-term benefit.

Second, we realized as soon as we started our studies that very little was known about the desert. We had to do basic research, which is continuing, to identify the environmental characteristics of the desert and the particular environmental factors we might affect.

Third, we found that the standard highway construction practices which were appropriate in non-desert areas might not be at all effective in the desert. Standard construction techniques resulted in soil and vegetation disturbance, impacts on wildlife, and destruction of archaeological values.

Fourth, we found that where we identify potential impacts early in the development of a project, designs can be changed or modified to reduce impacts. Special contract provisions to protect environmentally sensitive areas or elements can be written and enforced.

Fifth, we found that effective mitigation measures such as revegetation can be developed, by using experimental research methods which take the specialized nature of desert resources into account.

We have developed a few simple policies to guide our planning, design, construction and maintenance of State highways in the desert areas of District 09. I'd like to suggest that these policies are equally appropriate for all kinds of projects in desert areas. Let me list them for you:

For Soil Stability And Vegetation

- Plan and design projects to disturb natural soil surfaces and vegetation the minimum feasible amount. Consider both construction areas and sources of off-site materials such as earth borrow.
- Where natural soil surfaces and vegetative cover cannot feasibly be preserved, include mitigation measures in the project to stabilize soil and permit rapid revegetation with native plant species.
- Design revegetation to require minimum maintenance, no supplemental water, and avoid the use of exotic (i.e., non-native) plants. Where exotics are used to achieve short-term objectives, be certain that they will not persist or become invasive.

For Wildlife

- Have a good basic understanding of the existing wildlife ecology specific to the project area before undertaking detailed project planning or design. Know habitat requirements and species interrelationships.

- Design projects to the greatest practical extent to accommodate the existing ecological relationships. Do not assume that wildlife can be relocated to new habitat, or will migrate elsewhere, or can adapt to significantly altered ecological conditions.

- Where ecological requirements will be disrupted, include feasible mitigation measures in the project. Be sure the mitigation measures address specific problems identified during the environmental study of the project.

For Archaeological Resources

- Identify archaeological resources in the project area early in the environmental study process.

- Plan and design projects to avoid damaging these resources wherever feasible.

- Where archaeological resources cannot feasibly be avoided, include mitigation measures in the project which will either protect the resources in place, where appropriate, or which will recover the data so that it is available for future study.

- Where mitigation requires removing any archaeological resources from their original location, involve both the Native American and the scientific communities in the work. Consider that the cultural heritage associated with the resources belongs to the Indians -- design the work so that this heritage is respected.

In the end, despite the best advance study and planning, sometimes the mitigation measures don't get incorporated into the project design, don't get implemented once construction is underway, or are unwittingly destroyed during subsequent maintenance of the finished job. It is imperative that the people who actually design, construct, and maintain projects are involved with the development of mitigation measures. They must understand why a particular measure is necessary, and they must know what specific objective the measure is supposed to achieve. They must accept the responsibility to carry out the necessary work, and on occasion can make valuable contributions toward the success of your efforts to reduce damages. Several times we in District 09 have developed a mitigation measure which looked good in the office, but which couldn't be implemented for one good reason or another by the field forces on the job. In such cases, if your field personnel understand the problem you're trying to avoid, they can work out an alternative solution which can be implemented.

A final policy, applicable to any situation where mitigation measures are undertaken, is to monitor their success after the construction is complete. We are all tempted, once a highway is built and we move on to other things, to assume that what we did will work. If we never check to see if this assumption is valid, we may end up repeating the same costly mistake over and over. If we do our job correctly, each trial of a mitigation technique will be an improvement over the previous one, and we can develop cost-effective methods which really do reduce the impact our projects will have on the desert.

TRANSPORTATION FACILITY CONSTRUCTION PROBLEMS RELATED TO DESERT ENVIRONMENTS

Carl E. Forbes, Joseph B. Hannon, and Robert S. Watanobe

California Department of Transportation

INTRODUCTION

The construction of transportation facilities in arid desert-type environments presents multiple problems to the engineer. In the past, those areas of California subject to such conditions were little habitated by man and relatively undeveloped by industry. There was, therefore, very limited need to construct major facilities within an environment so hostile to construction methods and materials developed for the more temperate regions of the state. However, as population centers expand and needs are intensified to develop the resources of the desert regions, the challenge to transportation engineers to develop technology that can cope with construction and maintenance problems in the desert must be met. This task has two main facets: to build and to protect. Construction in the desert environment must be sound in every respect or the heavy investment in time and funding will be dissipated by the ceaseless onslaught of the harsh elements.

This paper seeks to define the various problems that may be anticipated in the desert environment. They relate to the construction and maintenance experience of the California Department of Transportation (Caltrans). Guidelines and recommendations are offered toward solution or mitigation of specific problems. Among these are the effects of severe temperatures, coupled with low humidity and wind on personnel, equipment, and materials. Also examined are problems related to water supply; wind erosion and the transport and disposition of airborne soil; the protection of facilities from sudden and unpredictable storms and their aftermath; and the adaptation of native materials as components in construction.

CONSTRUCTION EXPERIENCE

A section of Interstate 8 in the southeast corner of California, just west of Yuma, Arizona, has provided considerable experience in the construction and maintenance of transportation facilities in a typical barren arid desert environment [1 & 2]. This particular area, known as the "Sand Hills", is 6 to 8 miles (10 to 13 Kilometers) in breadth and consists of blow sand dunes extending from south of the Mexican Border, a distance some 45 miles (72.4 Kilometers) northward. The larger dunes exceed 100 feet (30 meters) in height and move only a few feet per year, while smaller dunes several feet

Figure 1. Remnant of Early Wood Plank Road

in height may move 50 feet per hour (15 meters/hour) in high wind. Winds of 50 to 60 MPH (80 to 97 KPH) are common during the months of March and April. The sand particles in this region are highly mobile, since they are virtually the same size (0.1 to 0.2 mm), are noncohesive, and lack protective cover or shelter. Considerable sand drift will occur with winds of less than 10 MPH (16 KPH), which are a daily occurrence in this area.

The Sand Hills presented a barrier to travel up until the time of major highway construction in 1960. The first attempt to provide a transportation route through the Sand Hills was made in 1902 when it was discovered that laying a patch of Arrow Weed (a stiff, reedy type of vegetation native to southwestern United States) across the dunes, would facilitate crossing by both horse-drawn and horse-less carriage traffic. However, the weight of the vehicles tended to break through the brush and the wheels once more bogged down in the sand.

A second attempt to provide a route through the dunes was made in 1912, when a simplified wood plank road was built. This was replaced in 1916 by an improved wood plank road constructed by the California Highway Commission. The new facility consisted of railroad ties placed side by side and secured with longitudinal iron bars. This road required periodic cleaning by teams and scrapers to keep the road passable. Figure 1 shows the present-day remains of this facility, which was abandoned in 1925 when a 20-foot (6-meter) wide asphalt concrete road was constructed which served traffic up through 1960.

Figure 2. Vibratory compactor in foreground and a dozer in background spreading base material as working table for haul vehicles.

During the Interstate construction in 1960, the contractor experienced complex problems related principally to the instability of materials employed. For example, he had difficulty preparing subgrade to the required tolerances since the tires of the motor grader used to trim the grade rutted the subgrade, despite efforts to achieve compaction. Steel wheel and rubber-tired rollers bogged down in the sand during attempts to prepare subgrade. A satisfactory subgrade was eventually obtained using a vibratory compactor shown in Figure 2 as a finishing machine.

To prevent contamination of the base material with blow sand, hauling equipment was barred from the prepared subgrade. Thus, it was necessary for the trucks to back in and end dump the base material to form a working table for subsequent backing, dumping, and spreading operations (Refer to Figure 2). The contractor also had to restrict his subgrade preparation to areas that could be covered by aggregate base within a day's time. Any subgrade left open to the elements would rapidly dryout and erode, or be covered by drifting sand. Close coordination was, therefore, necessary between the base laying and subgrading operations at all times.

When the contractor attempted to move rubber-tired scrapers into borrow pit areas they were virtually unable to move in dry sand. Saturating the area with water proved of little avail in solving the problem. Although various measures such as lowering scraper tire pressure and filling tires with water or water solutions aided movement, assistance was necessary from other vehicles. The borrow problem was finally solved by using rubber-tired scrapers with power drives on both front and rear axles. It proved fortunate that the All American Canal paralleled the project providing water for the construction requirements, a rare happenstance in desert construction. The application of water permitted the described vehicles and other vehicles, equipped with low-pressure oversized tires, to operate in the sand. However, only track-laying type equipment could negotiate unstable dry sand.

This project has not been as successful as initially expected in reducing the blow sand problem. Even though portions of the facility are constructed on embankment above the tops of adjacent dunes, blow sand continues to deposit on the roadway surface. High velocity winds and the action of passing vehicles tend to keep the pavement clean but low to moderate winds deposit considerable sand requiring plowing at least once a week.

The gravel blanket placed at the time of construction on the embankment slopes has been very effective in preventing wind erosion of the fill material. The blow sand problem is defined in more detail in a subsequent section of this paper.

In most desert environments adequate water supplies for construction purposes are generally difficult to obtain. Either deep wells have to be drilled or water has to be trucked or piped for long distances. It is, therefore, necessary to program construction operations to reduce water requirements to a minimum. Certain structural items should be precast or manufactured for in-field assembly to eliminate on-site concrete mixing water requirements.

Caltrans recently issued a Standard Special Provision which requires embankments to be constructed without water. This special provision has application to projects where water is scarce and the required densities can be obtained without water, or excessive compactive effort to meet the specifications.

PAVEMENT PERFORMANCE

Asphalts used in desert pavement construction harden at an accelerated rate in comparison to those placed in other environments. In some desert areas this rate can be as much as six times faster than normal. Even though asphalt pavements reach an early state of critical hardness, good subgrades generally exist with minimal rainfall conditions and, therefore, traffic load deflections are not detrimental to pavement integrity and service life.

When soft asphalts have been used in the past to lengthen the time rate of age hardening, they have created problems during construction. Considerable delays were sometimes incurred during placement and compaction due to high pavement temperatures and setting difficulties. Bleeding developed in some pavements, requiring use of open-graded surfacing to improve skid resistance. Even though the intent of using a softer asphalt is to improve service performance by maintaining pavement flexibility, there are certain tradeoffs to consider.

The present state of the art in California involves using midrange grade of asphalt to satisfy all conditions and designing for as much asphalt as possible to extend the service life without sacrificing stability.

Stripping of asphalt from aggregates used in asphalt concrete pavement construction leads to ravelling and early pavement failures. Desert aggregates are frequently coated with undesirable carbonate. The carbonate is deposited by hot weather rainfall percolating down through layers overlying the aggregate formations. The aggregates dry out rapidly following such storms, resulting in carbonate build-up with time. Aggregate sources so affected should be rejected for use in asphalt pavement construction, if more suitable alternative sources are available.

Where other sources are not available, or where economics prevent long hauls of materials, aggregates with stripping-prone characteristics can be utilized with special treatment or care to ensure service performance. Lime slurries, for example, have been used with some success in preventing stripping. Asphalt additives may provide another solution to the problem; however, they should be tailored to a particular mix. Harder asphalts are also used when stripping aggregates are encountered.

Desert environments are damaging to cement-treated bases. This material should not be placed on the sulfate soils typical of many desert areas. The combination of low humidity, wind, and severe temperature changes can also produce transverse cracking in new unprotected cement-treated base, similar to the shrinkage cracks associated with concrete pavement. The volumetric changes that occur in cement-treated base under these conditions can result in reflection cracking up through the new asphalt concrete pavement.

Concrete protective precautions are extremely important in hot, arid climates. Basically, protecting fresh concrete for pavement and structures means keeping the concrete cool and preventing loss of moisture as long as possible. There are many practices and techniques that can be used to promote long-term satisfactory service. Many references on the subject are available and a thorough research is advised.

REACTIVE AGGREGATES

Chemically deleterious substances may occur in coarse or fine aggregates as coatings on individual particles or as encrustations cementing them together; or as distinct layers in the aggregate deposits. These contaminations are prevalent in arid or semiarid desert regions.

Certain aggregates undergo chemical changes which may prove detrimental to concrete or other cemented materials. Reactions may occur between the aggregate and the constituents of cement or soluable materials may go into solution and impede the normal hydration of cement.

Reaction between certain aggregates and the alkalies in cement can be associa- with expansion, cracking, and deterioration of concrete. Small amounts of opal, rhyolites, and certain other rocks and minerals can cause excessive expansion and rapid deterioration. Opal is the most reactive constituent, but acidic and volcanic rocks are more numerous. Opaline silica may occur as a minor constituent in many rock types or may form coatings or encrustations on sands and gravels.

Rocks and minerals that are known to react with cement alkalies are: volcanics of medium to high silica content; silicate glasses; opaline and chalcedonic rocks; some phyllites, tridymite, and certain zeolites [3]. Aggregates that are suspected of reactive tendencies, should be avoided or used only with low alkali cement.

Soils containing sulfates or other corrosive substances can be very damaging to concrete structures that come in contact with them. This problem can be averted by overexcavating and backfilling with an impervious barrier, or protecting the concrete with a waterproof membrane. Internal damage can be avoided by the careful selection and proportioning of components and proper practices during placement. Such precautions as special cements, extra embedment to reinforcing steel, and additional cement have proven effective.

BLOW SAND PROBLEM

The problem of blow sand in California has been particularly acute in areas of the Coachella Valley, in Riverside County. Since Interstate Route 10 and other important arterials traverse this region, the impact of the sand problem on transportation has generated much study to seek solutions.

Wind-blown sand will accumulate on the down-wind side of solid vertical obstructions such as a bridge structure. A special railing was developed to eliminate this problem by its self-cleansing action. It also allows easy removal of other foreign objects from bridge decks.

The sandblasting action of wind-driven deposits are detrimental to highway signs in most desert areas. Some signs require replacement after one severe wind storm or as often as once every two years.

The phenomenon of blow sand is the natural physical interaction of sand and wind, plus the natural and human elements which directly or indirectly affect such interaction. Kerr and Nigra [4] define the problem of sand migration and discuss various methods of control relative to oil field operations in desert or semiarid areas.

Blow sand migration and damage associated with abrasion and drift accumulations within transportation corridors has a direct relationship to the rate at which sand movement takes place. The minimum velocity required to initiate movement of the most erodible soil particles (those about 0.004 inch [0.1 mm] in diameter) is about 10 MPH (16 KPH) at 1 foot above the soil surface. These particles have a size and weight relationship which is most conducive to initiation of movement. For dune sand having a mixture of different sized particles, the minimum wind velocity at 1 foot above the surface required to initiate soil movement is usually assumed to be about 13 MPH (21 KPH).

Disturbance of desert soils during construction activities and alteration of natural drainage patterns will aggravate the blow sand problem, if not properly controlled. Once the natural crust is broken, the previously stabilized material is free to move under wind action. An indirect contribution to the problem has been man's lowering of the natural water table in his attempt to develop desert areas. This has adversely affected the growth of natural vegetation, an important element in the control of sand migration.

The effectiveness of any type of direct protection measure is clearly dependent upon its ability to serve as a barrier to sand movement and to facilitate the impoundment of sand. Direct control consists of either stabilization of loose material; redirecting migration around the facility; or permanently stopping migration or impounding material along its present course, before it reaches the facility. General methods for achieving control consist of transporting, planting, paneling, fencing, surface treating, diking, and paving and wind baffling. All these methods provide a certain degree of control but should be evaluated based on the particular situation and the follow-up maintenance requirements.

Planting has been used extensively and found most effective in the Coachella Valley[5]. Planting may include ground cover or large shrubs and trees. Ground cover serves only to stabilize loose sand areas. Long-term effectiveness generally requires that no further appreciable sand-transport from upwind be permitted.

Trees and shrubs have often been planted to intercept oncoming sand. Tamarisk trees have proven effective but are relatively expensive to establish and maintain. Watering is required to achieve sufficient growth to provide an effective barrier within a reasonable period of time, which in turn requires watering throughout the trees' mature life. The availability and cost of water should be evaluated if planting is considered as a direct method of control. Also, if barriers for sand migration are not provided up wind to halt movement in severe blow sand areas, a fulltime maintenance problem develops; that of hauling away sand that has deposited along the tree row barrier. The problem is then converted from one of property damage to one of maintenance.

Fencing can be an effective means of impounding blow sand provided there is sufficient area to do so. Fencing is approximately one-third as expensive as planting-type barriers, both in terms of initial installation as well as long-term maintenance. Fencing is effective immediately upon installation, whereas planting-type barriers generally require about 3 years to provide an effective screen. Fencing can also be adjusted and modified, with supplemental fencing to gain additional height as described in the following section.

Surface stabilization of loose material can be achieved through applications of various bituminous agents, or lime and cement. Watering, paving and the placement of gravel cover afford varying degrees of control. The suitability of a particular treatment will vary with the type of material to be stabilized and the environment involved. Other factors to be considered are previous experience with the treatment, test results, and cost effectiveness.

IMPOUNDMENT BARRIERS

Effective impounding barriers for blow sand have a profile of equilibrium similar to that depicted below by two right triangles which show the effective areas of accumulation.

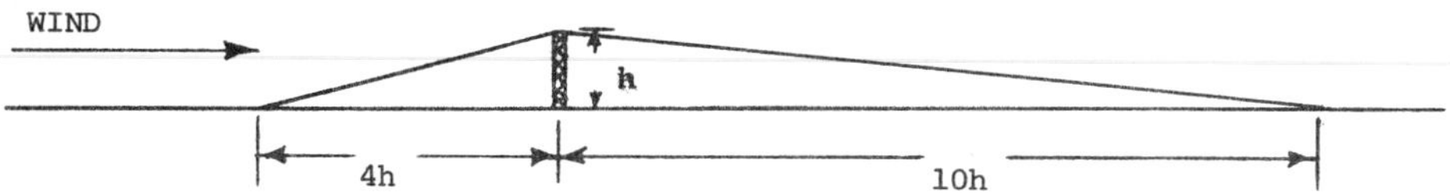

The area of upwind accumulation is given by:

$$A = 1/2 \times h \times 4h$$

$$A = 2h^2$$

Assuming the height of the barrier, h, and adding a dimension of depth (normal to the direction of sand movement), gives an impounding capacity of $2dh^2$ normal to the direction of sand movement. Similarly, the downwind impounding capacity of $5dh^2$ can be determined. The total impoundment capacity, C, therefore, equals $7dh^2$.

Barriers can be classified as either impassable or passable: impassable being those intended to impound only on their upwind side; passable, those intended to impound to their completely stabilized state. Impassable barriers have roughly one-fourth the impounding capacity of a passable barrier of the same height, but offer the advantage of complete protection from blowing sand if of adequate height and located immediately upwind of the area to be protected.

Passable barriers such as snow fence material offer far greater impounding capacity, but have the disadvantage of not being completely effective; that is, some of the oncoming sand escapes impoundment and continues to be transported on downwind.

Being able to determine the impounding capacity of a barrier, and knowing the rate of sand-transport to which it will be subject, enables the determination of the time required for a barrier of any given height and type to become impounded to capacity. This can be determined from the following equation:

$$Y = \frac{C}{S}$$

Where Y is the number of years to reach capacity, C is the barrier capacity and S is the rate of sand-transport to which the barrier is subjected.

A 5 foot (1.5 meter) high passable barrier at a location subject to 10 CY/ft/yr (25.1 m^3/m/yr) sand transport would reach full impound capacity in less than a year's time. However, a second 5 foot (1.5 meter) barrier can be placed atop the first when its total accumulation nears stabilization. The new profile of equilibrium becomes an enlargement of the first -- based upon a 10 foot (3 meter) barrier height as shown below;

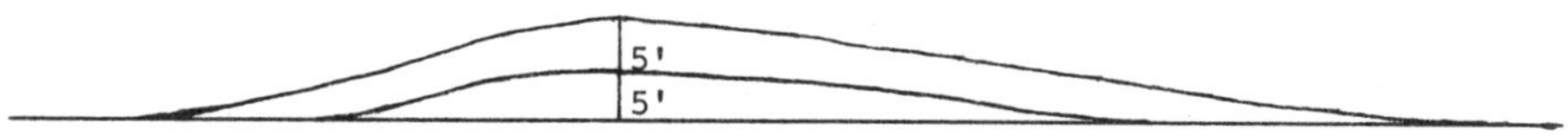

Now the total impounding capacity is 25 CY/ft (62.7 m^3/m); four times that of the 5 foot (1.5 meter) height, and sufficient for more than 2½ seasons. Yet another 5 foot (1.5 meter) increment of height added when the accumulation reaches the 10 foot (3 meters) height provides a total capacity of 58.5 CY/ft (146.7 m^3/m); nine times that of the original 5 foot (1.5 meter) single barrier height. Note that the frequency with which increments must be added also decreases accordingly.

Table 1 indicates impounding capacities of passable barriers of various heights, and also the number of years required for them to reach capacity under several different rates of sand-transport.

The planning of multiple-barrier systems is based upon the product of the two primary factors -- rate of sand-transport and design period -- which is the total impounding capacity required. The upwind, primary impounding barrier is planned to provide 90% efficiency. Its height also establishes the land area, or distance in the upwind direction that the accumulation will ultimately require [5].

TABLE 1

YEARS REQUIRED FOR PASSABLE BARRIERS TO REACH CAPACITY

Barrier Height (Feet)	Impounding Capacity (CY/Ft)	Rates of Sand-transport (CY/Ft/Yr)			
		2	5	10	20
3	2	1	0.4	0.2	0.1
5	6.5	3.2	1.3	0.6	0.3
8	17	8.5	3.4	1.7	0.8
10	26	13	5.2	2.6	1.3
15	59	29	12	5.9	3.0
20	104	52	21	10	5.2
25	163	81	33	16	8.1
30	234	117	47	23	12
35	319		64	32	16
40	416		83	42	21
45	527		105	53	26
50	650			65	32
60	936			94	47
75	1460			146	73
100	2600				130

NOTE:

1 foot = 0.3048 Meters
1 CY/Ft = 2.508 Cubic Meters/Meter
From Coachella Valley Adhoc Committee Report [5]

DRAINAGE DESIGN

The design of adequate drainage facilities presents a special problem in desert construction. In these arid areas where lack of water is a major concern, the ravages from flash floods can easily be underestimated. The typical high velocity of flow and attendant heavy bed load of solid material has caused extensive damage to many desert bridges and roadway embankments. The problem of estimating peak runoff values is often hampered by the lack of reliable rainfall data since maximum flows are generally developed during thunderstorms of short duration.

California's experience in the desert regions of San Bernardino County suggests that drainage design should lean towards the conservative side. Culverts should be oversized and designed to pass the estimated 100-year discharge without static head. Culverts should also be placed so that the invert approximates the grade of the channel it serves to prevent deposition of bed load in the culvert. Box culverts should be designed with single openings since interior walls tend to trap debris and precipitate plugging.

On some of California's highways with low traffic volumes, it was once common practice to set the profile to provide "dips" or depressions that would allow controlled flooding across roadways during peak flows. This was an economical method of avoiding major washouts in the event that storm runoff greatly exceeded amounts that could be handled by conventional drainage facilities. As traffic volumes have increased, most of these overflow "dips" have been replaced by bridges.

A particular difficult problem occurs in traversing the broad alluvial cones which commonly form at the mouths of desert canyons. A bridge crossing of a well-defined wash near the apex of the cone may be short and simple. Further down the cone, the path of future floods is very uncertain as the easily erodable streambed shifts capriciously. Thus, a much longer structure is required to avoid loss of approach embankment. Training dikes, rock slope protection, check dams and lined channels are often used to reduce the risk of washouts at these locations. Typical storm damage is shown in Figures 3 and 4.

Figure 3. Damage to bridge structure caused by large boulders on collision.

Figure 4. Severe Embankment Erosion

STRUCTURES

Obviously, the magnitude of drainage structures is selected to match site conditions. Culverts of varying sizes are used where they will carry flows and pass anticipated debris. Bridges are required at many locations to attain adequate waterway.

The choice of bridge type for use in the desert as in other locations is primarily one of economics. In California, the most frequently used bridge type for spanning desert washes is a cast-in-place reinforced concrete slab. For conditions that prevail in much of California, cast-in-place concrete structures are usually the most economical choice. However, we have found that precast concrete superstructures become competitive in the more remote locations, particularly where the magnitude of the projects allows repetitive use of similar structural members.

This experience would lead us to believe that some type of prefabricated elements would become an economical choice in desert sites that are very remote from developed areas. Typically, with any of the available prefabricated systems, parts of the structure are manufactured at a central location and transported to the site where these components are then assembled into place. This method of construction reduces the need for skilled labor at the site, the expense of setting up a concrete batch plant and importing cement and suitable aggregates may be avoided, and the demand for water for mixing and

STRUCTURAL SECTION		STR. DEPTH / MAX. SPAN	SPAN RANGE	REMARKS
R C SLAB	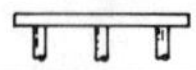	.055	10-40	
R C T-BEAM		.065	30-60	
R C BOX		.055	50-150	
CIP/PS BOX		.045	100-250	
PC/PS SLAB	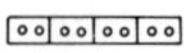	.035	30-60	
PC/PS T, TT, ⊥	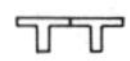	.055	30-120	
PC/PS I	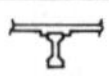	.055	30-120	
PC/PS BOX		.045	30-150	
STRUCT. STEEL I-BEAM	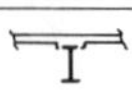	.055	60-300	
STRUCT. STEEL BOX		.050	60-500	
STEEL TRUSS		Varies	350-1500	
CANTILEVERED SEGMENTAL	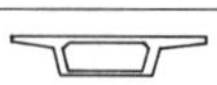	.050	100-500	

Figure 5. Structure Types

curing concrete at a remote site is eliminated. Quality control on the manufacture of prefabricated elements is usually much better in a well-equipped, fully manned factory setting than at a construction site.

Precast concrete members can be made in a variety of shapes and sizes to accommodate a wide range of span lengths. Some of the shapes that have been commonly used in California are illustrated in Figure 5.

The state of Alaska has found the use of single T-girders an economical choice in the remote areas of that state. The girders are tied together with diaphragms spaced at about 25 foot (7.6 meters) centers and the top surfaces of the T's are paved with asphalt concrete to provide a smooth riding surface. The small quantities of cast-in-place concrete required with this system can be mixed and placed manually or with simple equipment.

In California, most precast concrete designs have been based on precast members that are oriented in a longitudinal direction. There is, however, a growing worldwide trend for using transverse precast segments. In this system the entire bridge cross section is cast in lengths of 10 - 15 feet (3 - 5 meters). After being shipped to the site, these short sections are joined together with prestressing steel like a string of beads to form a continuous superstructure. Early applications of this system were typically for long span bridges, but this method of construction has recently been used economically for shorter span bridges in various parts of the world. Bridges of this type were recently used for freeway overcrossing structures near Vail, Colorado.

In the United States, structural steel girders are always fabricated at a central location and transported to the site either as full length members or as shorter segments that can be spliced together at the site. Although a cast-in-place concrete deck is commonly used in conjunction with steel girders, it is possible to avoid cast-in-place concrete either by using an orthotropic steel deck which is an integral part of the girder system or by attaching precast concrete deck slabs to the steel girders.

Larger diameter structural plate culverts are a suitable substitute for conventional bridges under some circumstances. The basic components of these structures are relatively small, easily transported, and can be assembled with small tools without highly skilled labor. Proprietary, arch-shaped structural plate culverts are available in spans up to about 40 feet (12 meters), and they have been used successfully under fill heights up to about 20 feet (6 meters). These structures should be used with caution in desert soils that are corrosive in nature or at sites where excessive loss of metal through abrasion will be a problem. Careful design of the inlet and outlet configurations to avoid excessive erosion is also required for a low maintenance installation.

Although timber may not be readily available as a native material in countries that have a predominantly arid climate, it should not be overlooked as a possible economical material for constructing desert bridges. Timber is light and relatively easy to transport, and can be modified at the site without sophisticated equipment. In addition to ordinary sawn planks and beams, bridge systems utilizing laminated girders and prefabricated deck modules have been used economically in remote areas of this country. Generally, wood requires treatment with a preservative to give it a reasonable life expectancy in any exterior exposure.

The foundation design of a desert structure deserves the same careful attention as the superstructure. Gathering reliable data on soil conditions necessary for design requires a visit to the site with exploratory drilling equipment. This is frequently a problem, because it must be done before construction access roads have been completed, if the contract documents are to reflect the benefit of this information.

An ordinary spread footing foundation is usually the most economical type if it satisfies site conditions. In weak soils or when extensive scour is anticipated, piles of timber, concrete or steel may be used. This adds to the cost because pile driving equipment is generally cumbersome, requires good access roads and must have skilled operators.

In general, prefabrication has been used less frequently for substructure elements than for superstructure elements. Caltrans did a project in San Bernardino County in which all of the substructure was precast. Although this system did not prove to be particularly economical in this application, it did demonstrate the feasibility of precasting all of the substructure elements and assembling them into a sound, durable structure as shown in Figure 6.

For small bridges, precast concrete or steel piling can be extended above the groundline to support the superstructure, thus avoiding a cast-in-place concrete pier. While this is usually an economical solution, it should not be used at sites where the piling would be subject to impact damage from heavy boulders that are rolled downstream during a heavy runoff. In highly corrosive soils concrete piles should provide additional cover over reinforcing steel and steel piling should not be used.

A possible method of providing an economical foundation is to adapt crib wall elements for this purpose. Crib wall construction is relatively simple, fast, and can be done without heavy equipment or highly skilled labor. These walls can be built to provide areas at the top for supporting structural beams. This kind of construction would be suitable for low-clearance bridges in areas of

Figure 6. Construction of Precast Concrete Bridge and Retaining Wall System

low seismicity. It could be expected to perform badly during an earthquake.

Another system of earth stabilization that can be used as an alternate to crib wall construction is reinforced earth. This development is a French patented system which has been utilized extensively in Europe and marketed in the United States by the Reinforced Earth Company of Washington, D.C., [6 & 7].

It utilizes reinforcing elements in soil to increase shear strength and thus permit earthworks with vertical side slopes. The reinforcement can be of any material that will provide the necessary tensile resistance, resistance to pullout and also sufficient service life with corrosion considerations utilizing appropriate factors of safety for both internal and external stability. The reinforcement is attached to either concrete or light gage steel facing elements. Wall heights are controlled by the bearing capacity of the foundation materials and the embedment requirements for stability. Refer to Figure 7. An embankment has also been constructed by Caltrans using discarded automobile tire sidewalls to reinforce the soil and allow steeper fill slope construction.

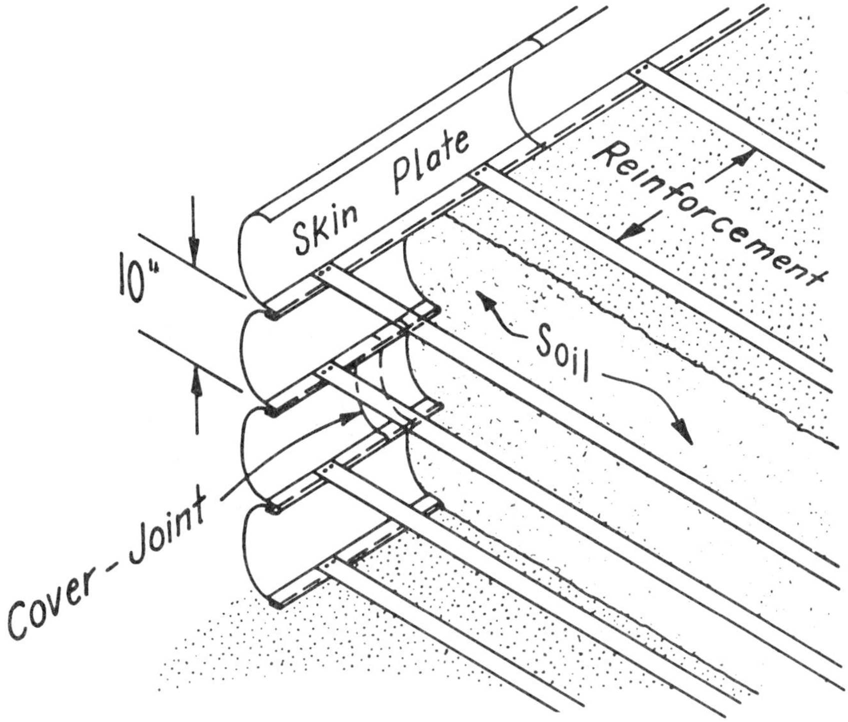

Figure 7. Schematic of Reinforced Earth Fill

REFERENCES

1. Dekema, J. and Kocher, R. F., Sand Hills-Historic Dunes Section of U.S. 80 Being Reconstructed as Freeway, California Highways and Public Works, California Department of Public Works, July-August 1961.

2. California State Department of Public Works, Desert Dune Buggies of Yesteryear - Old Plank Road, Bulletin, California State Public Works, Fall 1972.

3. Bureau of Reclamation, Concrete Manual, Unites States Department of Interior, Seventh Edition, Denver, Colorado, 1963.

4. Kerr, R. C. and Nigra, J. O., Eolian Sand Control, Bulletin of the American Association of Petroleum Geologists, Vol. 36, No. 8, August 1952, pp. 1541-1573.

5. Adhoc Committee of Coachella Valley Associated Governments (CVAG), Blow Sand Control and Protection Plan, Draft Report, Sections 2 and 3, March 1977.

6. Chang, J. C., Earthwork Reinforcement Techniques, Final Research Report (TL-2115-9-74-37), Transportation Laboratory, California Department of Transportation, October 1974.

7. Walkinshaw, J. L., Reinforced Earth Construction, Final Report No. FHWA-DP-18, Region 15 Demonstration Project No. 18, Department of Transportation, Federal Highway Administration, April 1975.

WILDLIFE MANAGEMENT IN THE U.S. DESERTS

Norman S. Smith
Arizona Cooperative Wildlife Research Unit
University of Arizona
Tucson, Arizona 85721

ABSTRACT

The objective of wildlife management is to maintain wildlife and their habitat for the benefit of society. The needs and desires of society have changed from production of game animals to maintaining all species of wildlife. The basic principles of wildlife management do not differ significantly among different biomes. Some of the specific problems of the desert include: water use and development, energy production, urban encroachment and competition from domestic or feral animals. The mission, goals, and programs of the U.S. Fish and Wildlife Service are discussed.

The objective of wildlife management is to maintain viable populations of wild species and their habitats in a manner that is consistent with the needs and desires of society. The determination of the needs and desires of a changing society presents difficulties of wildlife management anywhere, from the deserts to the tropical forests. The difficulties often seem unsurmountable in a fragile environment such as the desert.

THE CHANGING ROLE OF WILDLIFE MANAGEMENT

In the past, the needs of society appeared to be for the production of as many game animals as possible with less attention being directed toward non-game forms of wildlife. While managers strived to maximize production of game animals, many non-game species often benefited as a side effect. Wildlife managers also tried to suppress some species such as predators and other species that were thought to be competitors of desirable species. As our population has grown, our needs and desires as a society have changed. We now look at practically all species as having some merit. Idealistically, we shouldn't categorize wild species as "good or bad" animals. What constitutes "good or bad" animals is the fuel for explosive controversy. The question of how many coyotes are acceptable is an example. The introduction of exotic species to add to our native fauna is another volatile issue.

Curiously, as the population of the U.S. has expanded and the demands we put on the environment to support the population have increased so has the emphasis on maintaining natural areas and all wild species. These two desires often seem incompatible. Apparently, only when the general public becomes aware that the things we used to have are really in jeopardy do we as a society become active in defense of those things. As a result, we now have various well-meaning but rabid factions literally screaming, "We need more development of these lands for production of our necessities" or "We must halt the destruction of our wild ecosystems immediately."

The role of the wildlife managing agencies is to implement the will of the public as it is represented by law or by direction from representative bodies. Often it appears that these wildlife programs are at cross purposes with man's other uses of the land.

BASIC WILDLIFE MANAGEMENT PROCEDURES

Once the public's will has been established and objectives set, the principles by which biologists manage wildlife of the desert are no different from thse guiding managers of more mesic environments.

Wildlife Inventory

At some time, the species of interest should be inventoried or at least listed. The literature of wildlife management is replete with species lists, censuses, indexes of abundance and techniques to obtain and evaluate these data. Even though these data are most basic, many of the methods need refinement. For example, total counts of most species are impossible or at least very difficult to get because of the distribution or behavior of the animals. Most methods developed in the past were for game birds and game mammals and other species were studied more by academic zoologists than by governmental agencies. Thus, management techniques for non-game wildlife are not as complete as those for game species.

Habitat Requirements

Description of the habitat requirements of wildlife is a critical step in wildlife management. As part of most ecological studies of species or groups of species, food, cover and water needs are described. Currently, the trend is toward classification, inventory, and evaluation of habitat for ecological units that meet the needs of all wildlife rather than for single species only. If critical habitats are known, the benefits of any action that jeopardizes them can be weighed against the loss. It is at this point that governmental agencies can forestall or minimize losses of critical habitat because of other human activities. It is now more economical for all concerned to have wildlife considered in advance, rather than changing or stopping the project after it has started.

Habitat Improvement

The inexorable advance of man has altered or destroyed natural habitats. Because of this, wildlife managers expend much energy for the restoration of desirable habitat. Many programs or activities are designed to offset the deleterious effects of: environmental pollutants, introduced or natural diseases, excessive predation brought on by various natural or artificial reasons, increased competition between species (wild, domestic or feral) and disturbance of some critical phases of the life cycle for various wild species.

Regulation of Use

Because of (and, sometimes, in spite of) efforts of wildlife managers, there are often usable surpluses of many wild species, surpluses that can be hunted or harvested without danger to the population. To regulate this harvest, seasons and bag limits and methods of harvest must be established. Enforcement of these regulations is a task using a large portion of the finances available to managing agencies. In addition to the uses that directly remove wildlife, there are other activities that require regulation. The "so-called" non-consumptive uses such as bird-watching, game viewing, wildlife photography and wildlife interpretation all require some sort of regulation if their disturbing effects are to be minimized. People seldom realize that their presence, if uncontrolled, can be detrimental to the well being of wildlife. A few examples are: trampling of vegetation near a nest of a bird may cause the offspring to be more vulnerable to a predator, or to exposure, or causing doves to congregate at a bird feeding station where trichosomiasis is easily transmitted, or recreation near the only available source of water.

Planning

In addition to the general methods of wildlife management, there is the anticipation of future needs of society and how to minimize the impact on wildlife. It is at this point that we could use a crystal ball, tea leaves or some other mystical means for predicting the future. Lacking these, we must use the best educated guesses available. A management tool that is now evolving is the population simulation model and the computer. After making some basic assumptions (which may or may not be true), various parameters such as harvest rate, reproductive rate, mortality rate, etc. are manipulated to view the possible effects on the population. The computer is becoming more and more useful for wildlife management as it is in all other human endeavors.

WILDLIFE MANAGEMENT IN THE DESERT

Specific major problems that relate to wildlife management in the desert of the Southwest include those that involve water use and development, energy production, urban encroachment on wildlife habitat, competition from livestock and feral livestock.

Water Use and Developments

The deserts of the Southwestern U.S. are nice places to live provided there is water. The provision and allocation of water to the rapidly expanding human population of our deserts causes many problems for wildlife. The ground water is being mined at a rate greater than recharge and the level of ground water is lowering. This has had serious effects upon many riparian habitats. Riparian vegetation has been removed in the name of water conservation and justified (rationalized?) by explaining that bare soil or concrete do not transpire water like phreatophytes. Riparian habitat has been inundated by the creation of hydroelectric, flood control and irrigation dams. The interface between the newly created water line and the desert is not stable enough for a new riparian zone to establish. The rising and falling water levels in the reservoirs behind the dams cause numerous headaches for wildlife managers.

The construction of the facilities associated with water developments and transporting systems such as roads, fences, canals, dikes, powerlines, etc. can all have detrimental effects on wildlife. Effects such as impeding animal movements from one part of a traditional range to another part, disruption of microhabitat drainage by roads, canals, etc., accidental drowning in canals or accidental collision of animals and motor vehicles are common examples.

The provision of water to previously unwatered habitats creates new environments at the expense of the old. Some wildlife can adapt to the new and their numbers may increase. Other forms may be obligatory desert dwellers and may diminish along with their habitat. Some mobile species whose numbers are water limited may increase their numbers in response to the new source of water and then begin to detrimentally effect the vegetative components of their habitat. The newly created mesic habitat may allow parasitic diseases to expand their range and into animals previously uninfected. During the course of the conference uncounted references will or have been made about environmental changes pertaining to water. Almost all can affect wildlife.

Urban Encroachment

Urban encroachment upon desert wildlife habitats is a serious problem. As the human population moves into the deserts, the number of interactions between man and wildlife increase, usually to the detriment of the wildlife. Even in some instances where man's intentions have been benevolent, the results may be disastrous. For example, for the last few years, some people living in suburban developments near Tucson, Arizona have encouraged the buildup of coyote populations by feeding them. In March and April, 5 rabid coyotes, a rabid fox and a rabid dog have been killed. By definition, this constituted an epizootic and the County Board of Health requested the U.S. Fish and Wildlife Service to reduce the coyote population in 3 areas around Tucson. Human population centers contiguous to wildlife habitat increase the visitor pressure on the wildlife. The sheer force of numbers disturbing some wild animals will cause them to abandon otherwise apparently suitable habitat. Man's pets frequently prey upon, harass, disturb and transmit diseases to wild animals.

There are vast tracts of land in Arizona where real estate developers have purchased land, laid out plots, built roads, put up street signs, sold a few plots and have gone out of business. The roads remain, criss-crossing the desert, changing water flow patterns sometimes to the detriment of wildlife. Other developments, successful from the real estate view, have thrived and built up human populations always to the detriment of wildlife. Trailer houses mushrooming in the area between Tucson and Phoenix may be a threat to the Harris' hawk which nests in this area.

Livestock

Since the white man first arrived on the scene in our deserts, livestock interests have conflicted with wildlife. Overgrazing of desert ranges has been such a common occurrence in the past that it now seems to be accepted as the correct way to do it because of precedence. Predator control in the American west was initiated for the benefit of the livestock industry. Predator control has caused so many heated arguments between livestockmen, conservationists, hunters, preservationists and politicians that I doubted this controversy could be topped. Now, however, the controversy about feral horses and burros seems destined to rival that of the predator issue. Laws recently passed have made management very difficult on land where feral horses and burros are found. Even National Parks are having difficulty performing their mandated responsibilities because of emotional segments of the public who are opposed to killing the intruders.

Energy Production

Man's need for more energy has been responsible for many wildlife-man conflicts. The effect of dams for hydroelectric power was mentioned previously. Coal mining, oil drilling and extraction of oil from shale, development of solar collecting devices and geothermal development are either creating problems for wildlife management now or they will in the future.

Anytime the landscape is changed, or the use of the land changes, there will be problems for wildlife. To predict the extent of the effect and to minimize or mitigate it, is the challenge that wildlife managers face.

THE U.S. FISH AND WILDLIFE SERVICE

This agency is responsible for the federal stewardship of wildlife. Other federal agencies such as the National Park Service, Forest Service, Bureau of Land Management and Bureau of Indian Affairs, to mention a few, have programs that influence wildlife or its habitat. However, the Fish and Wildlife Service's stated mission ". . . is to assure maximum opportunity for the American people, consistent with their needs and desires, to benefit from fish and wildlife resources as part of man's natural environment" (U.S. Fish & Wildlife Service Program Management Document, 1973).

First among their goals is to assist in the establishment of an environmental ethnic and a sense of responsibility. The means by which this is to be done are: to detect and reduce environmental pollutants, to prevent environmental deterioration, to acquire, protect and manage unique ecosystems with high environmental value to fish and wildlife, to mitigate damage to fish and wildlife habitat caused by man's manipulation, and to prevent the extinction of each of the nation's fish and wildlife and their habitats.

The second goal is to assist and guide the conservation development and management of wildlife resources. This is done by: maintaining the levels and distributional patterns of wildlife so that all Americans have an equal opportunity for recreational benefit, maintaining the diversity of resident and migratory wildlife on Service controlled lands, providing professional service to localities or social groups who benefit economically from the fish and wildlife resources, controlling wildlife populations to prevent self-destruction of their habitat, and reducing adverse effects on the health or enjoyment of man and industry.

A third goal is to improve the American public's understanding, appreciation and use of our wildlife resources. This is done by maintaining a high quality wildlife experience, by providing information about wise use of the resource and by alerting the public to consequences of environmental changes and how the change will affect the well-being of the people.

There are 17 programs in the Fish and Wildlife Service designed to carry out the mission and meet the goals. All are important, but I will select only a few to describe some of the ways they deal with current wildlife management problems in the desert. Programs such as Animal Damage Control, Mammals and Non-migratory Birds, Migratory Birds, Cooperative Fishery and Wildlife Research Units, and Federal Aid are among the more traditional programs and have been and will continue to be central to the operation of the Fish and Wildlife Service. Programs that include Biological Monitoring, Environmental Impact Assessment, Comprehensive Natural Resource Planning, Endangered Species have been instituted more recently.

Because all agencies of the Federal Government are required to ensure an integrated use of natural and social sciences and environmental design arts in planning and in decision making which may have an impact on man's environment, the Fish and Wildlife Service has developed a habitat evaluation procedure to provide a uniform method of determining the impact of water development projects on fish and wildlife and their habitat. These procedures combined with habitat classification and inventory enable the Fish and Wildlife Service to act decisively before habitat is lost rather than react defensively after the fact.

Wildlife is the product of the biotic and abiotic factors of the environment. Man is a monumental component of the biotic element that wields tremendous influence on the other biota and the abiotic as well. Because of this power, he should also be responsible for his actions. All sectors of society, be they federal, international, state, private, or industry should feel and shoulder this responsibility. The existence of this conference is recognition of the responsibility we have for all aspects of our environment.

TWENTY-FIVE YEARS OF DESERT WATER DEVELOPMENT FOR WILDLIFE PURPOSES

Daniel P. Schadle

ABSTRACT

Arizona is the sixth largest state in the United States which contains 113,909 square miles. Elevations within the State vary from 200 ft. in the lower desert areas to 12,611 ft. above sea level in our highest mountain. Arizona's temperatures range from a record high of 128°F. to -33°F. Rainfall extremes range from less than ½ inch in the lower deserts to 25 inches in our higher mountains. Arizona receives 80-85% sunshine.

The southwestern portion of Arizona or approximately one quarter of the land area can be classified as Sonoran Desert Shrub and it is this generally broad area that will be discussed here.

The desert wildlife species which are managed in this habitat include: Desert Mule Deer, Desert Bighorn Sheep, Sonoran Antelope, White-Winged Dove, Mourning Dove, Gambel Quail and the Cottontail Rabbit.

Permanent water has been the limiting factor in this habitat and the management of these wildlife species has consisted of the strategic placement of watering devices for the species with the widest daily cruising radii - this being approximately 6 miles for the Desert Mule Deer. All other species receive direct benefits from this placement.

The Arizona Game and Fish Department has been working on this management problem for the past 25 years. Our first efforts to furnish permanent water for wildlife began with only small game animals in mind. We, more or less, followed in the footsteps of California and their "Gallinaceous Guzzler" type of watering device. It was soon evident that these small-capacity reservoirs and small artificial rainwater collecting surfaces were not adequate for our purpose.

Drastic changes and inovations for this original "guzzler" were made to produce the type of "rainwater catchment" we needed in most situations. The "all purpose" rainwater catchment was designed so that all wildlife would benefit from each installation. In order to produce the right catchment for the right situation, many variations of the original catchment were tried in search of the ideal catchment. Reservoir materials have included concrete, butyl rubber bags, galvanized metal tanks and heavy vinyl bags. Rainwater collecting apron materials have included compacted earth sprayed with asphalt, concrete, butyl rubber sheeting, corrugated metal, asphalt impregnated fiberglass and barren hillsides. Storage reservoirs have ranged in their capacities from a 55 gallon oil drum to 17,000 gallon vinyl liner bags. Through many changes over the years, the Arizona Game and Fish Department has produced what felt to be the most desirable "all purpose" rainwater catchment for areas of extremely low rainfall. This catchment consists of circular corrugated steel rings of varying diameters with vinyl bag liners. The collecting aprons consist of either natural runoff surfaces, asphalt impregnated fiberglass or corrugated metal sheets assembled adjacent to the reservoir. Water is piped to an above-ground concrete trough where water flows are regulated with a float valve. This catchment has provided the Department with the versatility of its prefabrication and its ability to fit into most situations. It has provided us with the management tool which not only assures wildlife water permanency, but also can be used to distribute game and establish resident wildlife populations in areas where only remnant and transient populations once existed. Just as important, we are now providing a growing human population with many hours of recreation from an area where little or no previous recreational value was realized. The Arizona Game and Fish Department will continue to exploit all water harvesting possibilities in these desert areas for the wise harvest of our natural resources.

INTRODUCTION

The State of Arizona is the sixth largest state in the United States, containing a land mass of 113,909 square miles (295,000 km.) (1). Arizona is bordered on the west by California and Nevada, on the south by Mexico, on the east by New Mexico and on the north by Utah. There are three distinct topography areas in the state: a high plateau region in the northeast corner, a mountainous area which runs diagonally through the mid-section to the northwestern tip, and the southwestern district which is divided between desert valleys and low mountain ranges. Arizona's topography ranges from 200 ft. (61 m.) above sea level in the lowest desert areas to Humphrey's Peak near Flagstaff at elevation 12,611 ft. (3,843 m.) above sea level (2).

Arizona's yearly average precipitation ranges from 25 inches (63.5 cm.) in the higher mountain areas to a low of 0.47 inches (1.19 cm.) recorded in 1928 at Yuma, Arizona (3). Much of this precipitation is realized from tropical storms which are of the typical downpour variety, where an entire yearly annual precipitation can be received in a very short period of time. Usually these types of rains result in rapid runoff which causes flash flooding and the formation of permanent washes and arroyos.

Arizona's extremes in temperature range from 33° (5.9°C.) below zero to 127° (53.2°C.) above zero, and it is true that our Grand Canyon State claims sunshine 80-85% of the time (3). This is more sunshine than any other state can claim, and though Arizona could be known as the Sunshine State, she no longer can claim the Baby State title. That was lost when Alaska and Hawaii were admitted to statehood.

MANAGEMENT RESPONSIBILITIES

Under Title 17 (A.R.S.) Arizona Revised Statutes, the Arizona Game and Fish Department has the sole responsibility of managing and protecting all the wildlife species within its boundaries. The Department, through administrative, research and management practices, is pursuing these objectives on a statewide basis. However, because this conference is primarily interested in desert developments, only the management techniques for our desert regions will be discussed here.

SPECIES MANAGEMENT

In a land such as this that can be extremely harsh both in winter and summer, a unique problem is presented for anyone attempting to manage wildlife. Arizona's ecology is without parallel. One might say that Arizona is a biological island situated in a sea of contrast. The wide, often abrupt variations within this island demand a certain unique knowledge and appreciation in order to develop and use them advantageously.

Wildlife management, when all optimum management features are present, can present a rather humdrum situation to the wildlife manager. But, in contrast, if one or more of the basic habitat requirements is missing (i.e. shelter, food or water), recommendations based on field evaluations must be made to determine the limiting factors of the habitat and then the appropriate solutions to remove the deficiencies must be instituted. Unfortunately, in Arizona even in our more choice habitat, there is usually a total lack of one basic requirement - or at best a deficiency. With the shortage of water (a normal situation), it is a constant struggle in Arizona to maintain any species of wildlife. Every attempt to manage the environment, whether for water, forage or cover, produces a decided impact on wildlife, and this must be fully considered in planning for any phase of game management. Providing permanent water in a near waterless habitat presents many unforeseen management problems that are sometimes not recognized initially. We have learned much from such situations. Arizona's water development in our Southwestern Regions has met with long range and outstanding success, indicated not only in increased hunting participation, but also by the recreational involvement in an area which once supported very little activity of this type.

Arizona is indeed fortunate to have a wide diversity of wildlife species which are managed for restoration and recreational purposes. The most important species are listed here: Rocky Mountain Mule Deer (*Odocoileus hemionus macrotis*); Desert Mule Deer (*Odocoileus hemionus canus*); Wapiti (Elk) (*Cervius canadensis*); White-tailed Deer (*Odocoileus cousi*); Bighorn Sheep (*Ovis nelsoni*); Antelope (*Antilocarpa Americana*); Javelina (*Picari angulatus sonoriensis*); Black Bear (*Urus americanus*); Turkey (*Meleagris gallapava merriami*); Cougar (*Felis concolor*); Cottontail Rabbit (*Sylvilagus nuttali*); Tree Squirrel (*Sciurus fremonti*), (*Sciurus alberti*), (*Sciurus kaibaensis*), (*Sciurus apache*), (*Sciurus Arizonensis*); American Buffalo (*Bison Bison*); Gambel quail (*Laphortyx gambeli*); Mearns quail (*Cyrtonyx montezumae mearnsi*); Scaled quail (*Calipepla squamata pallida*); migratory birds, including Mourning dove (*Zenaida macroura*), Whitewinged dove (*Zenaida asiatica*) and the Band-tailed pigeon (*Clumba fasciate*). This list also should include the waterfowl species common to the Pacific flyway.

SOUTHWEST DESERT REGION

The southwestern area of Arizona is typically Sonoran Desert Shrub and it contains the flora and fauna associated with this habitat. Much of the flora of this region is of little or no value to the species which occupy the habitat. However, the key forage species for small and big game are present in sufficient quantities to provide the necessary habitat to produce a resident population. Three large game species - Desert Bighorn, Desert Mule Deer and Javelina have received extensive management in this desert area. The small game and predators common to this type of terrain all received direct benefits from desert management.

The Southwestern desert area in which this extensive management was undertaken must be classified as marginal habitat for most game species, with the exception of the Bighorn sheep and the Gambel quail. Suitable habitat for all species is dependent upon the desert washes or arroyos for most of their food and cover, with permanent water being the limiting factor within this hostile habitat. Seasonal rains and flash floods provide the moisture necessary to produce adequate vegetation to satisfy the requirements of transient game for food and cover. However, the high desert temperatures, without permanent water available, negates the possibility of providing a resident game population. Providing permanent water in a habitat such as this is the obvious solution for the establishment of resident game in these vast dry areas.

Securing suitable and adequate water for the existence of civilization has been one of man's problems for centuries. Entire civilizations have come and gone due to either the abundance or the absence of water. Water harvesting and the storage of collected water has been carried on since 4500 b.c. (1). The pursuit of additional water from our natural sources is more evident in this generation than in any previous generation (5). Water demands in all areas of our nation continue to be of national interest as this country pursues its quest for additional energy.

WATER HARVEST PROGRESS

Arizona's first attempt to supply quail with water resulted in erecting a 55 gallon (208 l.) gas drum perched on the side of the road. A short piece of galvanized pipe was inserted into the barrel as an outlet. This pipe was capped on the end and allowed to drip slowly into a shallow basin. The barrel was plainly designated as "Quail Water" by the wording on the drum. These slowly dripping drums soon became empty and had to be refilled by hauling water, since there was no other water source. Also, the hot radiators of passing motorists constantly depleted this small water supply - usually before the quail could utilize it.

The first "Gallinaceous Guzzler" was established by B. Glading in California in 1947. Arizona's first attempt to produce a type of "Gallinaceous Guzzler" resulted in a 900 gallon (3,406 l.) concrete underground storage tank with a small 8 inch (20.3 cm.) ramp which ran diagonally from the rim of one side of the reservoir to the floor of the opposite side. A 2" x 12" (5.08 x 30.4 cm.) plank was bolted over the ramp opening to suppress evaporation. A small opening was left at one end of the ramp for small game to enter at the top of the ramp. Of course, large game could not utilize this water. The collecting apron of this catchment consisted of cleared and compacted earth sprayed with hot asphalt. As these units became older, the asphalt aprons broke down from weathering and the apron material was eroded into the reservoir. These aprons were eventually replaced with concrete.

It is unfortunate that the early wildlife biologists thought only small game needed supplemental water. The feeling then was that larger game with larger daily cruising radii could and would seek out natural waters. Through continued research, it was soon evident that our larger game animals were in search of and needed this man-produced water. As a result of this, larger concrete catchments with increased storage capacities and larger concrete runoff surfaces were needed. The open drinking ramp was also needed so that all wildlife could utilize the water. At this point, we no longer constructed the small 900 gallon (3,406 l.) capacity concrete catchment which we labeled as the "all purpose" catchment, since no wildlife could be excluded from utilizing the stored water. This unit remained as the standard catchment and this type catchment was constructed in water-deficient areas of the State with only minor changes being applied.

The drier areas of the State were the last areas where attempts to furnish permanent water through catchment construction was tried. One experimental concrete catchment was constructed in the Tank Mountain area of our southwestern desert. This catchment was observed for one year and utilization figures indicated that additional catchments were needed. In the succeeding two years, two additional concrete catchments were constructed, approximately six miles (9.6 km.) apart. Wildlife usage on all three units was extremely high, both with small and large game - mostly desert mule deer and quail. It was necessary to haul water to these units in an effort to maintain water permanency. It was obvious that larger catchments with larger storage and runoff capacities were needed. The 2000 gallon (7,750 l.) storage capacities were enlarged to 2500 gallons (9,463 l.). Apron surfaces were increased to over 2000 sq. ft. (186,000 sq. cm.). Over the next few years this 2500 gallon (9,643 l.) storage was enlarged to 3500 gallons (approximately 13,250 l.). Continued utilization of these catchments by big game throughout the entire hot summer months indicated that we were now managing a resident big game population rather than a transient winter herd. Due to the inconsistency and the general lack of rain in this area, extensive water hauling over rough desert terrain during the hot summer was necessary to keep this resident desert deer herd healthy.

It was again obvious that we needed even larger catchments with larger and more efficient runoff surfaces. A new, all galvanized catchment was designed to answer our needs. The first written specification for this catchment reads as follows:

> "This development consists of a 2000 sq. ft. (186,000 sq. cm.) 29 ga. (0.03 cm.) corrugated steel apron assembled on a natural slope to insure maximum runoff receipts. The apron is fixed into position of 55 concrete footings with ½" x 6" (1.52 x 15.2 cm.) foundation bolts and eight 1" x 3/16" (2.45 x 0.47624 cm.) strap iron lengths extending across the apron surface at 12' (3.65 m.) intervals. The strap iron is drilled and placed over the foundation bolts and secured to the apron with machine nuts. These two operations secure the apron from destruction by high winds. Runoff is collected in a reinforced concrete trough which is constructed at the lower end of the apron. The dimensions of the trough are 19½ ft. (5.95 m.) long x 22" (55.8 cm.) wide x 24" (60.9 cm.) deep. The outlet from this collecting trough is a screened 8" (20 cm.) corrugated culvert which leads directly into a 4000 gallon (15,151 l.) 16 ga. (0.15 cm.) and 18 ga. (0.12 cm.) covered metal reservoir. From this storage tank, water is piped to a concrete wildlife trough where flow is regulated by a float valve."

A rock rubble platform serves as footing support for the circular reservoir which is 10 ft. (3 m.) high x 7¼ ft. (2.2 m.) in diameter.

This became the basic design of our rainwater catchments where annual precipitation ranged from below 4" (10 cm.) to 8" (20 cm.). As with any new product, the search for improvement continued and was directed at this first galvanized catchment. We found that the concrete collecting trough at the lower end of the apron could be replaced with one-half of a 24" diameter (60.9 cm.) galvanized culvert cut longitudinally. This eliminated the need of form setting and mixing concrete for this water-collecting sump. Another factor that became evident was that the electrolytic action of some desert soils was not compatible with unprotected galvanized apron material. Also, the back filling to cover the lower portions of the tanks to prevent vandalism by shooting was abandoned due to the corrosive action of the soil. The extensive use of rust prohibitors was used to combat this problem. We felt that with these changes, we finally had the ideal catchment for this desert habitat. However, this was not the case. Increased utilization in all water-developed areas soon had us scurrying to provide still additional storage at these developed catchment sites. We turned to the installation of a variety of storage tanks which presented some rather grotesque developments. Surplus aluminum aircraft fuel tanks were used by assembling them in tandem to acquire the desired storage capacities. In one case, an iron boiler was also installed at one of our heavily utilized catchments. The square footage of the galvanized aprons was also increased to produce the additional runoff. Over the years, these make-shift, sometimes peculiar looking array of surplus tanks, together with some of our original galvanized tanks, were replaced because of corrosion and leakage. These replacements or redevelopments were made with larger storage tanks and runoff areas. Ideally, the Department was seeking a catchment that was both mobile and prefabricated and which also had a large storage capacity.

At this time, several commercial companies were stressing these use of butyl rubber products to collect rainwater and store it in huge bladder-like bags. The Federal government, through the B.I.A. (Bureau of Indian Affairs), U.S.F.S. (United States Forest Service) and B.L.M. (Bureau of Land Management) began experimenting with these poducts to produce water for livestock and range improvements. The Arizona Game and Fish Department began using this material for wildlife water catchments with somewhat less than successful results. It seems that Arizona's wind, sun and rodents, of which we have plenty, did not provide the proper atmosphere for this material. As of this date, we have replaced or have scheduled to replace all butyl rubber catchment materials.

Still in pursuit of the large storage catchment with mobile and prefabricated properties, we experimented with the "backyard swimming pool" type of reservoir. Not only does this unit have the desired storage capacities which we need, but it can be easily increased and has the features of being able to fit into almost any terrain requirement. This unit can supply 5,000 gallons (18,925 l.), 10,000 gallons (37,853 l.), or 17,000 (64,200 l.) of water. From using this type of reservoir, we developed a galvanized steel ring with a vinyl bag liner catchment reservoir. The vinyl liner is of twenty-seven mil. and is a tough and durable membrane. The capacity of the reservoirs can be increased by the addition of more rings of corrugated metal for support. The diameter of the vinyl liner is ordered to fit the size and height of the ring desired. The reservoir is covered with a wooden support frame which is then covered with either corrugated metal or smooth metal sheeting. Additional runoff into the tanks is made possible through the cover on the reservoir. This reservoir can be attached to our conventional galvanized aprons or can be used with natural runoff from barren hillsides. The water collected in these reservoirs is piped to a wildlife drinking trough where the flow is regulated by a float valve.

We feel that this unit is providing us with the desired features for supplying permanent water in our desert regions. It is true that emergencies will arise which will require water hauling to these units, but our water hauling is continually being reduced as the years pass.

Evaluations on the present water developments continue, and in cooperation with other State and Federal agencies, we are experimenting with additional runoff surfaces. We have tried fiberglass impregnated asphalt and flaked parrafin, which melts in Arizona's hot sun and produces a water-tight runoff surface. Progress in this field is a must and every avenue should be exploited to meet our continually increasing water requirements.

WATER HARVEST FOR BIGHORN SHEEP

Providing water for the Arizona Desert Bighorn Sheep in its own habitat presents a unique and completely different problem.

Due to the encroachment of man, the Desert Bighorn, which once inhabited the entire desert regions, has been forced back to occupy the more desolate and remote mountains of the southwestern desert. To provide water for this species involved developing water in this extremely rough habitat. In short, bring the water to the animal - not merely place the water where the animal can eventually get to it. Due to the lack of any natural springs and seeps, all potential water possibilities must be explored and developed. Usually, these include sealing natural temporary potholes to prevent them from leaking, increasing the holding capacities of a natural pothole by constructing small waterproof dams, constructing silt retention dams to prevent the potholes from silting closed and constructing shade roofs to retard evaporation. The Arizona Desert Bighorn Sheep Society, a group of laymen who have dedicated themselves to the restoration and preservation of this animal, has become deeply involved with these "waterhole projects" and has assisted with much of the menial and arduous work involved in this activity. On occasion, a helicopter is used to aid them in this work. The present program has been active for over seven years and we feel that the efforts expended are providing excellent Bighorn Sheep management.

HABITAT WATER FROM OTHER SOURCES

The southwestern deserts have been the focal point of mineral, gas and oil explorations over the past centuries and through these explorations, many abandoned drill holes and mine shafts have produced underground water sources. We have tapped these sources through the use of windmills and siphons to make this water available to wildlife. Ramps are constructed into shallow excavations, not only to make this water available, but to eliminate a possible death-trap situation. Any springs or seeps that do surface in this desert habitat are improved in order to produce to their fullest capacity.

REFERENCES

(1) Valley National Bank, Arizona Statistical Review, 32nd Edition, Phoenix, Arizona, 1976.

(2) Russo, John P., The Desert Bighorn Sheep in Arizona, Arizona Game and Fish Department, 1956.

(3) Smith, H.V., The Climate of Arizona, Arizona Experiment Station Bulletin 279, University of Arizona, Tucson, 1956.

(4) Arizona Conservation Club, Inc., Natural Resources of Arizona, Arizona Conservation Club, Inc., Tempe, Arizona pp. 86-107, 1950.

(5) U. S. Department of Agriculture, Proceedings of the Water Harvesting Symposium, Phoenix, Arizona, March 26-28, 1974, 1975.

(6) U. S. Department of the Interior, Geothermal Handbook, Page 175, Fish and Wildlife Service, 1976.

ADAPTATIONS USED BY WILDLIFE AND IMPLICATIONS FOR MAN IN DESERT REGIONS

Roger Hungerford[1]

ABSTRACT

Deserts of the world support many forms of free living vertebrates jointly termed wildlife. Individual species and whole classes of wildlife have evolved procedures allowing many of them to find food, exist without free water to drink, regulate their body temperature, reproduce their kind and control their populations within the carrying capacity of their ranges. Wildlife, while seldom found in large numbers in one desert area, exhibits a remarkable diversity in numbers of kinds. The actual mechanisms they possess are also varied and they survive by applying physiological, structural, and behavioral mechanisms in the hottest, driest and most barren regions. Man has learned much and can learn much more from desert wildlife, yet extinction threatens them today more than ever before. Desert soils, plants, and animal ecosystems are linked together in a fragile pattern, easily disrupted by man's actions. Some of the greatest past and present threats to desert wildlife resources of the world include overgrazing by domestic animals, physical disturbance factors, and environmental pollution. Wildlife resources have many present and potential values. Only careful integrated management can help man retain and enhance them for the use of future generations.

INTRODUCTION

The deserts of the world are not devoid of life, they only appear to be. Except in places where man has made extensive changes, most desert regions support several forms of free-living vertebrates (wildlife).

Desert animals are often dispersed - just as desert plants arrange themselves in scattered patterns in order to survive. The low total number of one kind and a

---

[1]University of Arizona.

low vertebrate biomass is normal for arid regions. Often, however, there may be a considerable variety of different kinds existing in a desert. The necessity of adapting to rigorous conditions has led to many unusual reptiles, birds and mammals. Within each of these groups of wildlife, scientists throughout the world have discovered mechanisms which allow these life-forms to exist under seemingly impossible conditions. Perhaps man might learn to use these processes for his own well-being or even for his survival in arid regions.

This paper will attempt to demonstrate that free-living vertebrates should be on our list of capital reserves of renewable natural resources that can be managed in the desert regions of the earth.

Today, some values of wildlife seem nebulous, yet the store of scientific knowledge is growing which can easily lead to more direct benefits from wildlife in the future. Man's lack of confidence that this will happen or his indifference has placed much of the world's desert wildlife in jeopardy. The extinction of species continues, nothing can be more irreversible, and each extinction is a final loss of a renewable resource by mismanagement. The actions of man that endangers desert wildlife are numerous but they can be classified, described, and perhaps a lesson learned from one arid area of the world can be applied in the future to another.

ADAPTATIONS OF DESERT WILDLIFE

All living organisms, including man, must be prepared in several ways to exist for long periods in desert areas. Animals, like desert plants, have developed special adaptations through time which allow them a free-living existence here. No desert animal carries or wears sun glasses, but their adjustments to fill their particular niche in a desert ecosystem may be innovative. These adaptations of desert wildlife can be classified as physical or structural changes, physiological specializations and behavioral adjustments. The literature on desert life reveals many examples of how one or more of these three classes of adaptations allow for survival.

Food and water are essential to survival. Animal life is often faced with either a lack of or only a seasonal abundance of both in arid regions. Plant food sources are frequently restricted to 1) annuals, 2) succulents (like cacti), and 3) desert shrubs (Odum, 1971). The annual plants are important; Walter (1954) found in African deserts a direct linear relationship between total dry matter produced and desert rainfall. He found the net primary productivity of these true deserts to be less than 2000 kg per hectare. This means wildlife must have the mobility to go to where the food or water is or to adopt a stop-and-go existence of some kind. Birds can easily migrate to take advantage of seasonal abundance or avoid places devoid of food or lacking moisture sources. Many mammals including many ungulates have the mobility to drift and even migrate. Stewart (1963) cited an outstanding example in the Arabian oryx (<u>Oryx leuceryx</u>) which travels far and fast over steppes and deep into the sand hills of the Arabian desert in search of food. He stated the oryx is capable of locating, from several miles away, a fall of rain which will quickly result in a flush of fresh plant growth. Jackrabbits (<u>Lepus</u> sp.) demonstrate a similar response to scattered showers in the American southwest, but their mobility is much reduced.

A stop-and-go existence is fairly common in many creatures, often including both physiological and behavioral adjustments. These individuals might be classed as the desert annuals among the vertebrates. Reptiles may often respond by sleeping away periods of food and water deficiency and the associated temperatures inimical to their comfort; small mammals, especially rodents may estivate and even birds may undergo a long sleep and a reduction in their body temperature. Shilov (1968) gives a good review of this adaptation in birds, including the capability of several types that inhabit arid regions of the world.

Faced with a limited list of possible foods in a desert, its wildlife has made interesting adjustments. Some have become generalists like the formerly wild camels (very thorny foods acceptable) and asses (almost anything green seems to be eaten). A few may supplement their diet or even specialize in the less usual, like the juicy parasites on the roots of desert shrubs dug up by the Arabian oryx (Caruthers, 1935). Extremely spiny cacti like _Agave_ sp. and _Opuntia_ sp. are eaten by the collared peccary (_Pecari tajacu_) in the arid southwestern United States and northern Mexico (Eddy, 1961), and a large proportion of Kalahari and Australian desert birds specialize in ground insects (Pionka and Huey, 1971).

Man must carry water with him into the desert, but the wildlife that lives there has made interesting adjustments and indeed many vertebrates literally do not drink water. The existence of these adaptations have fascinated numerous investigators for many years, yet new discoveries from obscure wildlife species may benefit us directly by showing new ways that man himself may someday satisfy his desert thirst.

The list of those vertebrates now known to exist without water is too long to enumerate, but examples by types of wildlife and from different parts of the world include the reptiles and amphibians. Lizards and snakes seem to represent a high degree of adjustment to an absence of drinking water. Some of the mechanisms they use, like elimination of a solid or pulpy mass instead of liquid urine are known, but further research is needed to fully describe all of the mechanisms they use (Chew, 1961).

Most birds are much more mobile than reptiles and many that are strong fliers like the dove or the sandgrouse will fly miles to isolated watering places in order to drink. The sandgrouse may also carry water in its feathers to its nesting young (Cade and Maclean, 1967). Some of the ground dwellers, because of their short flight habit and the long impractical walk to water, do not drink and must satisfy their needs in other ways. Some of the desert-dwelling new-world quails can and do exist and reproduce where no drinking water is present (Hungerford, 1960; Lowe, 1955; Sumner, 1935). They avoid moisture loss by shade or hole-seeking behavior and they select a plant diet containing sufficient moisture. Research is continuing with these and other desert birds to find their other secret adaptations.

Several types of mammals are dwellers of the driest deserts and are known abstainers from drinking water. Rodents like the gerbil, kangaroo rat, pack rat and ground squierrel subsist on food moisture without free water by using all three of the adaptive mechanisms here described. Perhaps the kangaroo rat and the gerbil use one of the most unique mechanisms. Organic compounds in foods contain hydrogen. By oxidizing hydrogen from even dry food substances to form water they can exist. One hundred grams of barley yields only 54 grams of water but a kangaroo rat can exist for five weeks on this small amount of water from food grain if he conserves it. He conserves water by spending much of the day in his burrow where the humidity is high (Schmidt-Nielsen, 1953). MacMillan and Lee (1967) report Australian desert mice use a mechanism already described for reptiles. They owe their water independence to an extreme renal capacity for concentrating urine and to reduce the output of water in the feces. This efficiency of water conservation, physiologically combined with behavioral adaptations, ensures their survival in extreme aridity, diurnal heat, and low rainfall which characterize the Australian desert regions in which they live.

Larger mammals have adapted to existence in several of the world's desert regions. The African eland and oryx can survive indefinitely without drinking water by physiological and behavioral strategems that minimize the amount of water they lose through evaporation (Taylor, 1969). This is also true of the different and nearly extinct Arabian oryx (Stewart, 1963) and it may also be true for other large desert-dwelling mammals. The camel is an exceptional performer in hot desert climates and he has been available for study. His secrets have been more fully explained by published investigations. The camel's desert dwelling procedures include proper insulation, a variable body temperature, a low rate of evapotranspirational loss of water for cooling his body, and the added ability to take on an ample supply of water all at one time (Schmidt-Nielsen, 1964).

A man can die from but one day of exposure to extreme hot desert conditions without water to drink. If so many vertebrate forms can and do exist without free water, perhaps we have more to learn from them.

The physiological processes through which population numbers of desert life are self-regulated is another procedure where man might learn valuable lessons. Annual plants are characteristic of arid regions. Seeds of annuals lie dormant for years, then produce in abundance when rain finally falls. Other forms of life also behave like annuals in the desert. Production of animal life can be futile without the plant resource base to support it. The case for natural self-regulating of populations was brought out by Wynne-Edwards (1965 and Odum (1971:197). They agree that "populations evolve toward self-regulation".

Only one of the several ways in which populations can be self-regulated need to be mentioned here, but this process is especially noticeable in deserts. Ellis and Osborne (1965) found that gibberellins and other plant hormones influence the time of sexual maturity of desert locusts. Sexual maturation and reproduction is triggered at the onset of rains by hormones in plants thus gearing the breeding of the locusts to rainfall. Similarly, in the deserts of the southwestern U.S., quail reproduction is geared to the availability of sources of vitamin A from green plants (Hungerford, 1964). Adequate rainfall is followed by pairing, nesting and the production and survival of large broods of young. If rainfall is inadequate, these birds may wait until the next year to breed. While some of the mechanisms are still a mystery, many desert wildlife species "bloom" when it rains or wait until it does - just like desert wildflowers.

THREATS TO DESERT WILDLIFE

Widespread disruption of habitat is now the main reason for the disappearance of species (Myers, 1977). This overshadows over-hunting or exploitation of wildlife, moving them to a status of minor causes today. Desert habitats are fragile and easy to disrupt. Roads, canals, power transmission of pipe-lines, settlements or increased cultivation, and the need to increase food production for our increasing human population - all degrade habitats. Pollution should also be considered a type of habitat disruption; in some areas it has become so bad that it alone threatens a number of species. Management for wildlife can be combined with technology to mitigate some of these losses of habitat for wildlife.

Many of the world's deserts resisted major invasion by man and his technology. Barriers to their development fell in the last few decades because of resource needs for an expanding human population. Today, most desert areas receive increasing impact and development continues on an unprecedented scale.

Ironically, it is not a form of modern technology that is the greatest current threat to desert habitat, but overgrazing by domestic livestock. Current attempts to settle nomadic peoples often confines their livestock to use only a single area of range continuously and heavily. Darling (1955) suggested that a better form of livestock handling approximates the old nomadic herding and allows each area of land a period of rest following selective use.

Dasmann (1959) among others, has seriously questioned whether livestock have any permanent place on arid rangelands or whether any but the lightest degree of stocking will not eventually lead to serious range damage. We can approach this later question by comparing historical and more recent events.

Overgrazing is perhaps the first way that man began to change the earth's surface. By 3000 b.c., heavy grazing pressure by sheep and goats denuded the Armenian headwater of the Tigris and Euphrates rivers leading to siltation of canals and destruction of farms in the "land of milk and honey" (Dasmann, 1959). Much of the more arid lands of the Mediterranean basin could not stand continuous grazing by herd animals. Plant cover disappeared, erosion followed, and desert spread into former scrub, forest, and steppe. Destruction of past cities and cultures in this, the cradle of civilization, began with overgrazing and this factor alone created some of the deserts of today. Petrides (1956) reports recent destruction in East Africa of vast areas, formerly undeveloped for generations due to the presence of the tsetse fly, carrier of sleeping sickness. Under pressure of increasing native populations, whole ranges have been cleared of big game, the flies irradicated and these places made safe for human habitation. In a remarkable short time, unrestricted cattle grazing by native herds have changed the aspect from lush grassland to barren eroded wastes. On arid ranges in interior Australia, drifting sands and blowing dust in the 1940's marked te eastward march of the desert due to overgrazing by sheep and hordes of European rabbits, both exotic grazing animals brought to the land down under (Pick, 1944). The Papago people of the southwestern desert of the U.S. have possessed cattle for 300 years. Desertification has taken place here, especially since 1880 and today the carrying capacity of these ranges is reduced yearly (Fontana, 1976). Hastings and Turner (1965:289) report the change in Arizona vegetation both on and off the Indian lands are in response to a combination of climatic and cultural stress . . ."that climate and cattle have united to produce it". Gallizioli (1976:162) believes . . ."prolonged range abuse by livestock is far more damaging to wildlife in general than various other insults to which the habitat is subjected". He believes this to be true for other states in the U.S. as well as Arizona, and that some species like the desert bighorn sheep will not tolerate the presence of any cattle within their habitat. The more hot and arid section of the western U.S., on its public lands, and even in national parks and monuments now have two additional grazing animals to consider. Wild horses and burros roam free here, and they have both legal protection (1971 wild horse and burro act) and they have the sympathy of most Americans. Extensive areas of Nevada, the desert portions of the Grand Canyon, and deserts along the lower Colorado River show the result of too many grazing animals, including these two new ones. Destruction of wildlife habitat by overgrazing is not now and perhaps has never been limited to older cultures and it occurs today in developed and developing nations alike.

Scientific range management has increased the production of food in many parts of the protein starved world. More and accelerated efforts in this direction will be needed in more places in the future. It also seems that people management is essential to its full success. Modern grazing systems, new breeds of livestock, planned cropping of wild African ungulates, and increasing total forage production are needed technology. By design, range management maintains habitats for wildlife and retains the other resources of the land. The types of livestock now used by man seem, however, best suited for the more mesic habitats of the world.

Dasmann's (1959) question, if livestock have any permanent place in the most arid lands, is still a legitimate one.

VALUES OF DESERT WILDLIFE

Some of the values of desert wildlife seem dependent upon scientific discovery before a net profit can be realized from them as a resource. This is still true of desert plants, although squash and date palms are not the only ones man now uses. Guyule for rubber and the goatnut of Sonoran deserts for waxes and oils are more recent additions. Sperm whales were formerly the only living source, but now the lowly goatnut shrub also provides a good substitute for fine sperm oil (Stubblefield and Wright, 1976).

Watts (1968:70-71) reviewed some of the economic arguments for grazing wildlife rather than livestock, which has been proven to be profitable both in Texas and in Southern Rhodesia. Some of his arguments which apply to desert situations include:

1. "Native wild game has been selected by nature for eons to withstand extreme conditions and endemic hazards in their native habitats. These adaptations will be lacking in goats, sheep and cattle."
2. "Wild game can remain in excellent condition even under the alternating excessive drought and excessive rain characteristic of African plains, which would soon be lethal to domestic cattle."
3. "Native wild game make better use of incident solar energy because of their extreme diversification and specialization."

Perhaps the best critic of American inability so far to integrate wisdom with knowledge and technology into wise land use planning, was Aldo Leopold (1933:637). He described why much of the arid southwestern U.S. changed as it developed into a weedy waste...

> "because the ecology of the Southwest was set on a hair trigger. Because cows eat brush when the grass is gone, and thus postpone the penalties of over utilization. Because certain grasses when grazed too closely to bear seed stalks, are weakened and give way to inferior grasses and those to inferior weeds, and these to naked earth. Because rain which splatters on vegetated soil stays clear and sinks while rain which splatters upon devegetated soil seals its interstices with colloidal muck and hence most runs away in floods, cutting the heart out of the country as it goes. Are these phenomena any more difficult to foresee than the paths of stars which science deciphers without the error of a single second? Which is more important to the welfare of civilization?"

Deserts too often are a prime example of man's increasing tendency to use the land only for short-range economic benefits. To change our course would not be a retreat from technology, but a redirection of it based on an ecological understanding.

History and climate have played creative roles in determining the character of deserts, its plant and animal life, and its people. Dubos (1970:16) noted that "each place is characterized by a set of attributes that give it uniqueness". He recognized that this "genius of the place" was made up of the physical, biological, social and historic forces all working together. Future desert management should consider the quality element of these habitats and attempt to attain this sense of fitness between man and his surroundings.

Economic arguments alone will yield little immediate results in saving the remaining desert wildlife of the world. Dollar values to others in other places cannot compete with today's immediate local needs for space, scarce water, and other precious resources that must be torn from the desert. Individuals and nations must through knowledge and reason be convinced that free-living vertebrates should not be omitted from our list of capital reserves of renewable natural resources. Each species may become a treasure for future genetic transformations, may yield a secret to allow man's future survival, or merely enrich our lives in a mechanized world. To preserve them we must concentrate on maintaining their habitats. Homeless, they are doomed to extinction even though to us their habitat seems to be a barren desert. In fact, desert vertebrates are a sensitive biological indicator of things going wrong in our world that may be our next concern. We should treasure them as the miner used to care for his caged canary whose death warned him of black-damp in the coal mines of Wales.

REFERENCES

Cade, T. J. and G. L. Maclean. 1967. Transport of water by adult sand-grouse to their young. Condor 69(4):323-343.

Carruthers, D. 1935. Arabian adventure to the great Nafud in quest of the oryx. London.

Chew, R. M. 1961. Water metabolism of desert inhabiting vertebrates. Biol. Rev. 36:1-31.

Darling, F. F. 1955. Pastoralism in relation to populations of men and animals. Oliver and Boyd, Edinburgh. 152 pp.

Dasmann, T. A. 1959. Environmental conservation. J. Wiley & Sons. London. 307 pp.

Dubos, R. J. 1970. The genius of the place. Amer. For. 32:16-19, 61-62.

Eddy, T. A. 1961. Foods and feeding pattern of the collared peccary in southern Arizona. J. Wildl. Manage. 25(3):240-257.

Ellis, P. E. and D. J. Osborne. 1965. Desert locusts: sexual maturation delayed by feeding on senescent vegetation. Science 19:546-547.

Fontana, B. L. 1976. Desertification of Papagueria: Cattle and the Papago. Desertification, Office of Arid Lands, Univ. of Ariz. 125 pp.

Gallizioli, S. 1976. Deadly overgrazing. Defenders 31(3):161-163.

Hastings, J. R. and R. M. Turner. 1965. The changing mile. Univ. Ariz. Press, Tucson. 317 pp.

Hungerford, C. R. 1960. Water requirements of Gambel quail. Trans. N. Am. Wildlife Conf. 25:231-240.

Leopold, A. 1933. The conservation ethic. J. For. 31:634:643.

Lowe, C. H. 1955. Gambel quail and water supply on Tiberon Island, Sonora, Mexico. Condor 57(4):244.

MacMillan, R. E. and A. K. Lee. 1967. Australian mice: Independence of exogenous water. Science 158:383-385.

Myers, N. 1977. Garden of Eden to weed patch. NRDC Newsletter 6(1):1-15.

Odum, E. P. 1971. Fundamentals of ecology. W. B. Saunders, Philadelphia. p. 5-74.

Ohmart, R. D., T. E. Chapman and L. E. McFarland. 1970. Water turnover in roadrunners under different environmental conditions. Auk 87(4):787-793.

Petrides, G. A. 1956. Big game densities and range carrying capacities in East Africa. Trans. N. Am. Wildlife Conf. 21:525-537.

Pianka, E. R. and R. B. Huey. 1971. Bird species density in the Kalahari and the Australian deserts. Koedoe 14:123-129.

Pick, J. H. 1944. Australia's dying heart. Melbourne Univ. Press. 119 pp.

Schmidt-Nielsen, K. and B. Schmidt-Nielsen. 1953. The desert rat. Sci. Amer. 189(1):73-78.

Schmidt-Nielsen, K. 1964. Desert animals: Physiological problem of heat and water. Clarendon Press, Oxford, England.

Shilov, I. A. 1968. Regulyatsiya Tebloobmena U Ptits. Moscow Univ. Press. 279 pp.

Stewart, D. R. M. 1963. The Arabian oryx (Oryx leucoryx, Pallas). East African Wild. J. 1:103-117.

Stubblefield, T. M. and G. W. Wright 1976. Jojoba beats cotton. Prog. Agric. 28(1):14-17.

Sumner, L. E. 1935. A life history of California quail with recommendations for conservation and management. Calif. Fish & Game 21(324):185-342.

Taylor, C. R. 1969. The eland and the oryx. Sci. Amer. 220(1):88-95.

Walter, H. 1954. Le facteue eau dans les regions arides et sa signification pour l'organization de la vegetacion dans les contries subtropecoles. In Les Divisions Ecologiques du Monde. Centre Nationale de la Recherche Scientifique, Paris. p. 27-39.

Watts, K. E. 1968. Ecology and resource management. McGraw-Hill, New York. 450 pp.

DESERT DEVELOPMENT: THE PERSEPCTIVE OF A CALIFORNIA CONSERVATIONIST

Thaddeus C. Trzyna, Ph.D.[1]

By now, I am sure, you have heard a good deal about technical aspects of individual environmental considerations in developing and managing desert regions, such as soil conservation and wildlife protection. I will try to give you a somewhat broader perspective as a social scientist who has worked both professionally and as a citizen activist on local conservation issues in the California desert, as well as - more recently - on promoting better international cooperation on global problems of the environment.

First of all, I would like to give you a brief description of the citizen environmental movement in the United States and internationally, what it stands for, and how it relates to the developing countries. Second, I will give an example from our experience in the California desert of how the misuse of technology can cause severe and long-lasting damage over a very short period of time. Finally, I will make an appeal for preserving a few areas in each of the world's arid regions in as nearly a natural state as possible, pointing to a new means of assistance for doing so.

THE ENVIRONMENTAL MOVEMENT

The great changes that have taken place in the world's social order over the past two hundred years are largely the result of social movements. Even if all of their goals are not achieved, many of their values and parts of their programs are usually accepted into the ever-changing social order. Examples are the international anti-slavery movement of the last century, and the African freedom movement of the 1950s and '60s (1).

---

[1] President, Center for California Public Affairs and Chairman, International Committee of the Sierra Club, P.O. Box 30, Claremont, California 91711, USA.

The environmental movement is one of the newest members of this social phenomenon to emerge on the scene. The values it represents are nothing new: a respect or even reverence for natural resources is implicit in many so-called primitive cultures. In the West, laws regulating environmental quality can be traced back to ancient times. Citizens' groups concerned with conservation, town planning, and pollution control have been active in industrialized countries for a century or more.

What _is_ new and different is that the citizen conservation movement has evolved from being concerned with single resources to realizing their interdependence and the need to view the whole problem of environmental protection in its entirety. We have finally begun to realize, for example, that in urban areas, air pollution cannot be separated from transportation; that flood control must be linked to land use planning; and that unbridled population growth is at the root of many of our environmental ills. The environmental movement has become _ecological_: people concerned about protecting individual resources are looking more and more to see how they relate to the natural, social, and economic system as a whole.

The Movement in North America

A good example of this is the organization that I represent, the Sierra Club. It was founded in San Francisco in 1892, with the main purpose of preserving the natural wonders and recreational resources of California's Sierra Nevada range. Today, it has broadened its concerns to include nearly every sort of environmental problem, from air quality and solid waste management to nuclear power safety, toxic chemicals, preservation of prime agricultural lands, and population. It has taken to heart the words of its founder, naturalist John Muir, who said in the 1880s, "When we try to pick out anything by itself, we find it hitched to everything else in the universe" (2).

The Sierra Club is also an example of how public concern and involvement in environmental issues have mounted in North America, especially over the past fifteen years or so. When I joined the Club as a university student in 1959, it had fewer than 10,000 members and it was still confined mainly to working for protection of wilderness areas in California and elsewhere in the western U.S. Today, it has over 180,000 dues-paying members. It has become a national and even international organization, with several hundred local groups throughout the U.S. and Canada, professionally-staffed offices in ten cities, and an annual budget of $7 million.

Although the Sierra Club is one of the largest and most active groups, there are some 3,000 voluntary environmental organizations in the United States, with several million members. Several hundred of them are national or regional in scope, often specializing in specific problems, but the majority are local groups that work on a broad range of issues in their immediate communities. Most are supported entirely by individual contributions and staffed by volunteers (3). The movement in Canada is similar in structure, and a number of major organizations have active members in both countries.

The activities of citizens' environmental organizations vary, but all of them gather and disseminate information to their members and to public officials and the communications media, and most of them work to pass environmental legislation at the national, state, or local levels of government. Some of them also bring lawsuits against governmental or private acts deemed damaging to the environmenta, and such "public interest" litigation has been an important means of getting long-range environmental quality considerations incorporated into development projects.

What the Movement Stands For

The philosophy behind the environmental movement can be summed up in one word: stewardship. As the late French scholar and futurist Bertrand de Jouvenel put it:

> We esteem ourselves masters of the Earth. But should not an owner be the husbandman of his land? Must he not tend it as he uses it? Shall he not indeed delight in its beauty as well as enjoy its fruit?
>
> "The Lord God planted Man in a garden of delight, to dress it and to tend it" (Genesis 2:15). This I take to mean that the Earth has been given to us for our utility and enjoyment, but also entrusted to our care, that we should be its caretakers and gardeners. (4)

We stand not for mindless opposition to progress, but for opposition to mindless progress. We are for progress that is genuine, that enriches this and future generations. We are for reforestation, for reclamation of derelict lands, for recycling resources, for returning nutrient wastes to the soil, for clean air and water, for conservation of all our limited natural resources, and for better public health.

We are against the kind of mindless "progress" that runs roughshod over human values, that befouls the environment, that destroys our heritage, and defrauds posterity. We are against trading off immediate economic benefits for wasteful depletion of non-renewable resources or irreversible damage to the environment. We are against leaving future generations with a biologically impoverished planet and exhausted natural resources (5).

Outside North America

I am telling you all this because the environmental movement is by no means exclusively a North American phenomenon or a phenomenon of the West. It is found in varying degrees and in different forms in all parts of the world and is gradually taking on a transnational character. It is a force that each of you will most likely have to deal with in your careers, and I think it is important that you understand what it is and what it stands for.

In the other industrialized countries of the West, particularly in northern Europe and in Australia, development of the citizens' environmental movement has paralleled that in the U.S. and Canada, though probably nowhere has it yet had as strong a political influence as in North America.

In the states with centrally-planned economies, citizen conversation groups can also have a significant role. In the Soviet Union, for example, the All-Russia Society for the Conservation of Nature, founded in 1924, has some 11 million members (mostly school children in its youth division). The law of the Russian Soviet Federated Socialist Republic permits it to conduct public conservation inspections and express its position on the most desirable uses for natural resources or areas. The RSFSR Government is required to at least formally review the Society's proposals. As Philip Pryde has noted,

> It goes without saying that public conservation organizations in the Soviet Union... do not enjoy the political freedom or leverage that an organization such as the Sierra Club is able to exercise in the United States.

> Still, the Soviet societies can be effective at the local level, if they so choose, in checking on compliance with existing conservation laws, advancing conservation education, and bringing to the attention of government organs areas where improvements would be possible or desirable. (6)

The citizen environmental movement has touched the countries of the Third World in different ways. In some regions, notably in Latin America and Southeast Asia, small but active indigenous private groups have emerged (for example, the Amigos de la Naturaleza, which has members in 11 South and Central American countries; and in Malaysia). In other areas, for instance, East Africa and India, conservation groups set up by colonials have been taken over by local leaders. In many Third World Countries, however, short-term problems of mere survival have precluded any efforts to organize citizens to work toward longer-range goals of environmental protection, and in some states, any such political activity is discouraged (7).

An International Movement

I mentioned above that the environmental movement has evolved from being concerned with individual resources to recognizing their interdependence and how they relate to the whole. There is another thing that is new and different about the movement: its emergence as an international and global force.

About 1970, at the same time citizen interest began to "take off" in North America and northern Europe, there was increasing concern in the world scientific and political communities that something ought to be done to promote more effective international action on environmental problems. After all, such problems don't respect national boundaries. The main result, as you know, was the UN Conference on the Human Environment at Stockholm in 1972, and the subsequent creation of the United Nations Environment Programme (UNEP).

What is less widely known is that the same concern triggered much greater awareness among many national and international non-governmental organizations (NGOs) of the need to work more actively on transnational and global environment issues. As a result, a number of these groups have been giving more attention to promoting international cooperation. The environmental movement is still mainly confined to individual countries, but it is more and more taking on the distinct characteristics of a global social movement.

Let me again take the example of the Sierra Club, with which I am most familiar. In 1971, the Club's governing board established a formal International Program and opened a professionally-staffed office opposite UN Headquarters in New York. In doing so, the board stated that "not only is the internationalization of the environmental movement essential for man's survival, but that very movement provides the countries of the world with their best chance yet" to improve international cooperation generally.

The Club's international office, now called the Earthcare Center (8), carries on a broad range of activities targeted at increasing the environmental awareness of international and U.S. Government agency officials, other NGOs, and political, business, and citizen leaders. Among its projects have been setting up a multinational coalition to work for protecting the ocean environment at the UN Law of the Sea Conference, organizing a training institute on international environment affairs, and advising the U.S. Government on its participation in the UN Water Conference.

Several other U.S. environmental groups are active in international issues, including Friends of the Earth, National Audubon Society, and Natural Resources Defense Council. We try to work together as much as possible, among ourselves and with kindred groups abroad, and through such world umbrella organizations as the International Union for Conservation of Nature and Natural Resources (IUCN), which has become, in the words of former UNEP Executive Director Maurice Strong, "the nerve center of a world conservation movement" (9).

Cooperation with Developing Countries

Another new thing about the environmental movement, and I hope it is the beginning of a trend, is the idea of citizens' groups in industrialized countries working closely with similar groups in developing countries on problems that concern both of us, or the world community at large.

One example of such cooperation is a major survey of tropical rain forests in Venezuela that the Sierra Club has just completed, working jointly with the Government of Venezuela, the Venezuelan Rural Development Council, and UNEP. The purpose of the survey, which was done by a binational team headed by a professional forester from the Sierra Club (who is also a professor at Cornell University), was to determine the condition of the rainforests and establish guidelines for their use and preservation, both in Venezuela and in other countries with similar resources (10).

I would like to see other citizens' environmental organizations in the developed countries involved in similar projects, and I hope that they, like our Venezuela-U.S. one, will be truly binational in character, with each side learning from the other. None of us has all the answers.

MISUSE OF TECHNOLOGY: THE EXAMPLE OF RECREATIONAL OFF-ROAD VEHICLES ON THE CALIFORNIA DESERT

Developing countries are often sensitive, and understandably so, to what they perceive to be a "holier-than-thou" attitude by environmentalists in industrialized countries. Let me give you an example, then, of how California, which has been called "the world's most advanced industrial society" (11), failed in allowing a new use of technology to cause severe and long-lasting damage over a large area within a relatively short time.

Technology has been the main agent of environmental change over the past two centuries. It has been responsible for many of our environmental problems, but it is also one of the major sources of increased standards of living. Within it are solutions to many - but surely not all - of the problems it creates. I am not speaking against technology, only against the misuse of it.

A striking example of how technology can be misused and cause great damage to the natural environment is the short and sad history of recreational off-road vehicles on the California desert. It may not be directly relevant to your countries, but I think it shows two things: how fragile desert environments are; and how misuse of a new technology - or a new use of technology - can lead to great damage to arid environments. I think it holds a lesson for all of us.

Off-Road Vehicles and the California Desert

The California desert covers some 6.4 million hectares in the south-eastern part of the state (of which about 4.8 million ha. is public land controlled by the federal Bureau of Land Management). It is easily accessible from the Los Angeles metropolitan area, which is the second largest urban region in the U.S., and the world's seventh largest, with over 8 million people.

For many years, the desert has served as an outdoor playground for Southern Californians, who are addicted to various forms of outdoor recreation. Until recently, their activities, mainly confined to weekend camping, walking, and driving over established roads, had relatively little impact on the desert environment.

Beginning in the early 1960's, however, Southern Californians started to buy recreational off-road vehicles in large numbers. These include Jeeps and other four-wheel-drive cars; dune buggies, which are specially-built cars with large tires designed to navigate sand dunes; and motorcycles, including trail bikes, which are built to travel over extremely rough and steep terrain.

The desert was highly attractive to off-road vehicles (ORV) recreationists. There were open spaces unhindered in most places by trees, bushes, or large rocks; there were hills and cliffs that offered challenges to drivers; and most of the land was unregulated public domain.

ORVs soon became not only a means of transportation into previously inaccessible remote areas, but an end in themselves. Tearing across the desert in a state of reckless abandon became a major "sport."

By 1974 (the latest information I could find), there were some 1.2 million registered trail bikes in Southern California, a half-million four-wheel-drive vehicles, and 200,000 dune buggies (12). By now, the numbers are considerably higher.

There are several reasons for the explosion of ORV use on the California desert. They include urban affluence and thus ability to buy the machines, the proximity of the desert to a major population center, availability of leisure time, increased need for individual freedom and self-expression, urban "claustrophobia," a desire to get back to nature coupled with being used to ease of mobility, remnants of a frontier culture, and heavy advertising campaigns which often present ORVs as symbols of virility.

The Damage Done

Whatever the reason, the ORV explosion happened so quickly that government found itself unready and unable -- and in some cases, unwilling -- to act to protect the environment. Laws regulating ORVs were nonexistent or inadequate, and there were no effective means, over such a huge area, of enforcing what laws did exist. Public agencies, including the Bureau of Land Management, yielded to pressure from ORV enthusiasts and lagged on setting up a system of regulation.

The California desert is hardly a wasteland. Robert C. Stebbings, a prominent zoologist, has written that it

> contains a great array of biologic, physical, and historical resources. It is rich in plant and animal life. Many species are highly specialized to withstand extremes

> of aridity, temperature, and light. There are isolated springs, humid mountain-tops, sand dunes, and playas containing living remnants; fish, mammals, amphibians, reptiles, invertebrates, and plants found nowhere else... By desert standards the California desert is lush compared to the Sahara. (13)

The impact of heavy ORV use on the desert over a period of little more than a decade has been severe. In a study of the damage done by cross-desert motorcycle races, the U.S. Geological Survey concluded that a single annual event held over an eight-year period had

> destroyed desert pavement, mature vegetation, and other natural barriers to erosion that required centuries to form within areas now totaling more than 9000 acres [3650 ha.] ...produced mechanical erosion of hundreds of tons of soil and laid bare large areas to future wind and water erosion. Major loss of animal and plant life has occurred and their habitats in tracked areas have been seriously modified. Some archaeological and geological records have been destroyed and others have been greatly damaged. (14)

In another recent study, the U.S. Department of the Interior found that

> Serious harm has been done to the extremely dry soils of the California desert... Desert crusts, pavements, and playas require from thirty years to a century or more to be reestablished after they have been destabilized by ORVs. (15)

Drivers of off-road vehicles have caused great damage to archaeological sites in the California desert, both by driving over them, often unwittingly, and by making sites formerly protected by their sheer remoteness easily accessible. The more noticeable kinds of sites, especially petroglyphs and pictographs, are favorite targets of vandals (16).

The result of ORV misuse, as Stebbins has written, is that a "national heritage of enormous biological, historical, and recreational value is being steadily, and in some areas rapidly, damaged for a special form of desert recreation enjoyed by less than one percent of the American people." (17)

Both the federal and state governments have now taken steps to try to minimize the damage done by ORVs, but enforcement is made difficult by the large area of land and the numbers of vehicles involved. Citizens' environmental groups, including the Sierra Club and the Environmental Defense Fund, have had a big role in encouraging governmental action, and their interest will continue.

The Australian Experience

The problem of recreational off-road vehicles is not limited to the United States. In Australia, for example, a parliamentary committee concluded earlier this year that "the indiscriminate use of off-road vehicles can cause serious damage to the physical environment," and recommended legislation to control their use (18).

In arid lands, which cover 75% of Australia, intensive motorcycle activity has caused gullies over a meter deep in some places, and erosion of soil has resulted in introduction of noxious plant species that compete with the native flora. The committee recommended that arid areas be given the greatest protection against ORVs because they are most vulnerable (19).

According to a Northern Territory conservation official, the use of four-wheel-drive vehicles by Australian Aborigines is causing "extremely significant change" to the natural environment of the 244,000-sq. km. of native reserves in the territory. "Aborigines used to show concern for the environment when they were using their traditional hunting methods, but this appears to have changed now." Use of ORVs for recreational purposes by Australians of European origin has also become a problem in the Northern Territory, and a fine of A$500 has been proposed to protect areas where ORVs can cause extensive soil erosion; mere possession of such a vehicle in the area would result in a fine (20).

The Lesson

The lesson to be drawn from off-road vehicles is, I think, that if a new technology or use of technology is shown, or even reasonably suspected, to cause a serious impact on the environment, then a conservative approach must be taken, even perhaps to the point of prohibition, at least until all of the possible impacts are known (21).

We should take a Napoleonic Code attitude toward such things, whether they are off-road vehicles, a new form of pesticide, or genetic engineering. that is, they should be presumed guilty until proved innocent. It may be that the risks involved are worth the benefits derived (as may be the case with the new genetic research), but we should know what those risks are.

PRESERVATION OF NATURAL AREAS

I would like to close by making a brief appeal to you - because I do not see it listed elsewhere on your agenda - for setting aside a few areas in each of the world's arid regions in as nearly a natural state as possible.

Preservation of natural areas has been a central theme of the environmental movement, and much of our effort in the United States has been devoted to creating and protecting systems of national and state parks and wilderness areas.

The value of preserving natural areas has long been recognized in the international community, and many such areas in desert regions are listed in the United Nations List of National Parks and Equivalent Reserves (22). In fact, three of the ten largest national parks of the world are in arid countries: the Central Kalahari Game Reserve (5.2 million hectares) and the Gemsbok National Park (2.5 million ha.), both in Botswana; and the Ouadi Rime-Ouadi Achim Reserve (4.9 million ha.) in Chad (23).

There are a number of very good reasons for setting aside areas that do not yet bear the marks of man's work, which include their value as cultural and historical heritage, natural laboratories and reference points for science, recreational resources, economic resources for attracting tourists and the money they bring with them, genetic pools, and places to educate children and adults about the natural world. Some of us also have a gnawing feeling that it somehow isn't right to develop everything. As David Brower, president of Friends of the Earth, has put it: "The wild places are where we began. When they end, so do we. We had better not speed their passing."(24)

While there is a broad support for the concept of preserving natural areas, there has been a problem in getting enough funds to manage and protect them once they are designated, especially in the Third World, and thus a reluctance to set aside new reserves.

One answer to this problem, at least for areas of truly global significance, is the World Heritage Fund (properly known as the Fund for the Protection of the World Cultural and Natural Heritage), set up under Unesco as a special trust fund by the World Heritage Convention (25), which was signed in 1972 and came into effect in 1975. The main function of the Fund will be to provide financial help to less-developed countries for protecting natural areas "of special value from the aesthetic or scientific point of view," and also cultural monuments, buildings, and sites.

The environmental movement, especially through the IUCN, has been pressing for such a measure since 1968, and we are now trying to get the governments of industrialized countries to contribute to the World Heritage Fund so it can begin its work.

REFERENCES

(1) For a description of such movements, see Gerald A. Sumida, Transnational Movements and Economic Structures, in C.E. Black and R.A. Falk, eds., The Future of the International Legal Order, vol. IV, The Structure of the International Environment, Princeton University Press, Princeton, 1972.

(2) Sierra Club, 530 Bush Street, San Francisco, California 94108.

(3) Barney, Gerald O., ed., The Unfinished Agenda: The Citizens Policy Guide to Environmental Issues, Crowell, New York, 1977, p. 161.

(4) de Jouvenel, Bertrand, The Stewardship of the Earth, in The Fitness of Man's Environment, Smithsonian Institution Press, Washington, 1968, p. 102.

(5) Paraphrased in part from a recent statement by J. Michael McCloskey, Executive Director, Sierra Club.

(6) Pryde, Philip R., Conservation in the Soviet Union, Cambridge University Press, London, 1972, pp. 20-21.

(7) The World Directory of Environmental Organizations, Second Edition, 1976, edited by the author and Eugene V. Coan, and published for the Sierra Club by Public Affairs Clearinghouse (P.O. Box 30, Claremont, California 91711), lists some 3,500 organizations of all types in nearly every country of the world.

(8) Earthcare Center, Sierra Club, 800 Second Avenue, New York, New York 10017.

(9) IUCN Bulletin, New Series 8, 5 (May 1977), 27. IUCN's secretariat is at 1110 Morges, Switzerland.

(10) Hamilton, Lawrence S., Tropical Rainforest Use and Preservation: A Study of Problems and Practices in Venezuela, Sierra Club, 1976. Spanish-language edition, Conservacion de los Bosques Humidos de Venezuela, issued 1977 and available from CONARE, Apartado 17015, El Conde, Caracas 101, Venezuela.

(11) Bradshaw, Ted K., New Issues for California, the World's Most Advanced Industrial Society, Public Affairs Report (University of California, Berkeley), 17, 4 (August 1976).

(12) Heath, Russell, The Environmental Consequences of the Off-Road Vehicle, with Profiles of the Industry and the Enthusiast, Defenders of Wildlife and Friends of the Earth, Washington, 1974, p. 20.

(13) Stebbins, Robert C., Off-Road Vehicles and the Fragile Desert, The American Biology Teacher, 36, 4 (1974), 204.

(14) Wilshire, H.G., and J.K. Nakata, Off-Road Vehicle Effects on California's Mojave Desert, California Geology, 29, 6 (1976), 131-132.

(15) Draft Environmental Impact Statement, Departmental Implementation of Executive Order 11644 pertaining to Use of Off-Road Vehicles on the Public Lands, U.S. Department of the Interior, Washington, 1976, p. 38.

(16) Drawn from Sylvia Broadbent, Scavengers on Wheels, Sierra Club Bulletin, 59, 4 (1974).

(17) Stebbins, Robert C., Off-Road Vehicles and the Fragile Desert (Part 2), The American Biology Teacher, 36, 5 (1974), 303.

(18) Off-Road Vehicles: Impact on the Australian Environment, Commonwealth House of Representatives, Standing Committee on Environment and Conservation, Canberra, 1977, p. 25.

(19) Ibid., 21.

(20) Centralian Advocate (Alice Springs, Australia), July 29, 1976, 1.

(21) Helpful in preparing these comments on ORVs were the above-cited works of Heath, Stebbins, Wilshire and Nakata, and Broadbent, as well as Off-Road Recreation Vehicles: A Research Summary, 1969-1975, Texas Agricultural Experiment Station, College Station, Texas, 1976; and Submission to the Select Committee [of the Ontario Legislative Assembly] on Motorized Snow Vehicles and All Terrain Vehicles, Federation of Ontario Naturalists, Toronto, 1973. Russell Shay of the Sierra Club staff kindly provided materials and background information.

(22) Published by IUCN (see n. 8), latest edition, 1975.

(23) IUCN Bulletin, New Series 6, 8 (August 1975), 30.

(24) Brower, David, A Time for Sarsaparilla, in Garrett de Bell, ed., The Environmental Handbook, Ballantine Books, New York, 1970, p. 11.

(25) Convention Concerning the Protection of the World Cultural and Natural Heritage, adopted by the General Conference of the United Nations Educational, Scientific and Cultural Organization, Paris, November 16, 1972. For further information, address: World Heritage Committee, Unseco, place de Fontenoy, 75007 Paris, France.

SOME ELEMENTS OF THE DEVELOPMENT AND MANAGEMENT OF DESERTS

Recommendations of the Sacramento Conference

The Conference presented a selection of experiences and technologies from both developed and developing countries. To utilize the information presented and developed in the discussion, it will be necessary for the participants and their associates and their governments to select the strategies and techniques that will best meet the needs of their people.

This report has used the term "desert" in the broad sense of the Conference. It is recognized that both true deserts, with no vegetation whatever, and arid lands are included. Most of the recommendations apply to arid lands, some apply to both.

DESERTIFICATION

Many millions of people are presently being endangered by desertification processes in a region of about one million square kilometers bordering on desert areas. The factors leading to this desertification are fairly well known but not always well understood. These would include:

1. The rapid growth of population which has resulted in increased flocks, overgrazing and depletion of woody vegetation for fuel.
2. Incorrect land and water management and destructive techniques of cultivation and irrigation.
3. Climatic variation and long-range climatic change.
4. Social, political and cultural factors.

Millions of hectares of formerly productive soils are lost annually, some of this loss being irreversible. Obviously steps must be taken to contain and even to reverse these destructive processes. The technologies presented at this conference are partly well known and even if there is still a great deal to be learned about them they could theoretically be put into practice today.

Populations have been growing at a rapid rate so that even where development projects have been successfully implemented, per capita income has not substantially increased and at best, these projects have provided only a partial answer to the problems of desertification.

It seems clear that development programs will have to be accompanied by an intensive program of education and by far-reaching social and political reforms. Changes in custom and tradition where necessary have to be generated from within. The developed nations can help by:

a. Providing capital and technical assistance for improvement of educational facilities and services.
b. Providing credits, subsidies and other incentives for measures leading to better land management and to reduction of population and livestock pressures.
c. Establishing produce and marketing institutions.
d. Aiding in the establishment and operation of local and regional research centers.
e. Aiding in the development of local extention services to assist in the transfer of information developed through research to the farmers and other users.

DESERT DEVELOPMENT AND STRATEGIES

A number of the papers read at the Conference placed heavy emphasis on techniques and technology involving management of some basic components such as water, mineral, energy sources (including petroleum, solar, and geothermal) and the primary and secondary productivity of natural and agriculturally modified ecosystems.

Regional and national planners charged with strategy selection are often faced with simultaneously conflicting situations, especially when, as usual, the financial, labor, material, or other resource is limited. In such situations strategy decisions are often made with the hope of achieving maximal short term returns in order that other strategies may be activated later. However, in most cases two or more strategies are being carried out simultaneously. Even if local governmental agencies are emphasizing only a single strategy (e.g., a cash crop produced for export), local peoples are generally involved in food crop production, animal husbandry and harvesting of native fruits, berries and wildlife species as well as, perhaps, limited mining or home industry.

A number of different strategies are evident in the papers presented but most appear to be centered around the maximal acquisition and most efficient utilization of that basic resource in short supply -- WATER!

Further, most appear to involve the use of water for some agricultural purpose -- either field crops or domestic animals. Specific projects included in the reports that could be emphasized as strategies include the following:

1. Development of coastal zones and regions, utilizing the extra food resources and cheap transportation supplied by the marine interface.
2. Maximize storage of excess waters during wet seasons with inexpensive compartmented reservoirs and similar devices for use in dry seasons.

3. Where transportation to markets and technology exists, grow and export high value crops. In some cases this can include the use of controlled environment chambers (greenhouses).
4. By irrigation (from underground, imported surface waters or local impoundment in wet season) develop, where appropriate, local agriculture including field crops and animal husbandry. Although a desert area does not need to produce its own food, it must produce something resulting in an export surplus, so food can be purchased.
5. Utilize tourism (together with natural areas, game reserves and game farms) as high labor intensive source of foreign exchange.
6. Where possible, utilize extractive industries (mining).
7. In all possible situations (agriculture, mining) accomplish processing with local processing plants and labor, if water requirements are not excessive. The added value and reduced shipping charges will result in added value to local populations.
8. Creation of desert reserves in all areas is recommended to preserve a portion of the natural flora, fauna and land forms. Such reserves are valuable for study, research, and enjoyment and to preserve the indigenous germ plasm. If this is not done many species may become extinct.

WATER RESOURCES AND MANAGEMENT

Deserts and arid lands are by definition deficient in water supplies. Any strategy for management or development, be it only the improvement of life for the existing population, the extension of agriculture and grazing, or mining, petroleum development, industrial development or the encouragement of tourism, must consider the most effective use of available water.

It is likely that a combination of strategies will be employed, depending on the resources, needs, and political attitudes of the country. There sould be a thorough survey of the water supplies available including surface supplies, ground water, geothermal water, rainfall and the possibility of imports. The type and extent of development must depend on the results of such a survey. It may be necessary to adjust quality standards and use patterns to the quality of water actually available. Maximum use must be made of water conservation measures and the reclamation and reuse of waste waters.

Long range planning is essential. For example, in some areas there are large bodies of fossil groundwater. Such water should be used benficially, even though there is no recharge, but there must be a decision as to what rate of use will be most beneficial to the people.

Some techniques to maximize the effective use of available water include improved irrigation techniques, evaporation control, salinity control and drainage, and water harvesting to maximize use of rainfall. Institutional and political solutions must be found, as well as technological answers. The economics of development must be realistic. The goal of long-term water planning should be a safe and reliable supply of water designed to shield the local economy from drought and depletion of available water.

AGRICULTURAL DEVELOPMENT

Cropping, Ranching, and Tree Plantations

Since some desert areas are being invaded by the plow, it seems important that this situation is given a close second look in many developing countries. This calls for a proper resources survey including soil, water and other investigations.

Lands which can sustain limited cultivation must be so identified and other lands should be considered for retirement from cultivation and diversion for other alternative land uses, e.g., ranching, tree plantation, widlife development, recreation, etc.

Grazing and harvesting of trees and other vegetation would have to be limited or the process of degradation would not be reversed.

Adequate means should be used to minimize losses to crops and native vegetation caused by pests.

Low water requirement and drought resistant crops and crop varieties including forage crops deserve special attention where cropping and grazing are considered desirable. Other practices, e.g., soil and water conservation practices, water harvesting, mulch farming, suitable conservation rotations, mixed cropping, etc., may be considered.

Scientists and extention workers should evaluate indigenous and traditional agricultural/farming practices in the region, for easy adoption and quick results. These methods should simultaneously be evaluated in the light of modern knowledge so that further improvement may be possible.

Mixed Farming

Recognizing the place of plant species of grazing value, shrubs and trees and by-products of some field crops as may be grown in the area, animal production in the area would deserve special attention in addition to cropping. Such an approach would provide for year-round work of a productive type to the rural people in many developing countries.

Where data on relative economics of cropping vs. animal husbandry vs. mixed farming is limited, the need to generate it is indicated. Wildlife studies need special consideration. Grazing behavior of mixed herds of domestic animals and also mixed herds having a suitable mix of wildlife with domestic animals calls for systematic studies. Wildlife may be an excellent source of meat and hides, but will not provide milk or wool. Upgrading of animals and appropriate health care have to be provided for.

Grazing and Livestock Management

In many countries, over-grazing of low rainfall desert areas is among the major factors responsible for desertification.

1. Studies on carrying capacity of such areas should be made where such information is lacking so that regulation of grazing may be considered against this background.

2. Unrestricted grazing be discouraged as much as rapidly possible. This could be enforced easily on lands held by the Government. For private lands, suitable education and, if necessary, legislative action may be needed.
3. Instead of numbers of low quality animals, every effort has to be made to build the population of quality animals adapted to the area by reducing unproductive or less productive animals. Cross-breeding to a limited extent and introduction of new breeds/species of animals may be considered in this context.
4. Both immediate and long-range programs for installing watering points have to be formulated and implemented so as to reduce excessive grazing pressures on lands around the existing or newly developed watering points.
5. A suitable program for upgrading of the degraded areas should be developed to improve vegetative cover, reduce the incidence of erosion, and to improve the economy of the areas. Revegetation of degraded areas with natural vegetation of economic value must be considered as a first step. Further improvement through reseeding and management of better types of grasses and other vegetation, both locally available and introduced species/types should be scientifically evaluated.
6. Development and improvement of marketing systems would require close attention in areas where such arrangements are lacking.

Land Policy

Lands in many desert countries tend to be used beyond their capabilities, due to population pressures which leads to land deterioration. Land allotment policy in such areas has to be developed in keeping with the land capability. Lands which cannot sustain cultivation must be adequately safe-guarded by prescribing appropriate land use and needed practices.

The Wildlife Resources of Deserts

One of the most abused resources of the desert areas of the world is the indigenous animal resource. Its potential has, to date, been greatly underrated.

Sizable numbers of human desert dwellers have traditionally, and still do, depend on the wildlife resource as their major and often, only, source of protein. In Botswana, for example, an estimated 45% of the human population depends upon wild animals for protein.

Many desert dwellers produce various items derived from the indigenous animals. Often this may be the only source of cash income. This economic resource requires minimal infrastructural, no physical changes to the region, little or no capital investment, and necessitates no changes in life styles.

We highly recommend that the various wildlife species be investigated further wherever sizable populations currently or formerly existed. This investigation should be set up as a large-scale, long-term, on-the-ground practical trial with careful accounting of all variables.

Not until the results of these trials are available will we know whether, with low management resources and energy input, a higher or equal level of productivity (as compared to domestic stock) can be obtained.

In many cases even with lower productivity, optimal land use may prove wildlife raising as the only long-term sustained yield protein producing system possible. In many such places the domestic animals now present should be removed.

INDUSTRY

Industrial development can be an important factor in the transformation of an underdeveloped area into a modern society. The desert offers several advantages which could be utilized for industrial development such as: wide spaces, a dry climate and, in many cases, easy disposal of waste effluents without ecological deterioration.

The industrial potential of agriculture products could rapidly be tapped. The by-products of regular agricultural crops such as cotton seed and bulbs, molasses, bran and others could be the basis for the manufacture of many products. Desert grown industrial crops such as jojoba, guayule, acacias, etc., could be the basis for many sophisticated chemical products.

Governments can aid in establishing industry in desert areas by:

a. Granting special incentives to investors who establish their manufacturing facilities in these areas.
b. Training workers and technicians.
c. Establishing the necessary infrastructure.
d. Extending aid for marketing and export.
e. Inventorying all available resources.

Countries planning mining of minerals including oil and gas should realize that these are depletable resources whereas desert industrial plants are renewable resources. Thus, careful planning is required to avoid ghost towns and economic disruptions.

The conference recommends that UNITAR and the United Nations Industrial Development Organization (UNIDO), in cooperation with FAO organize a conference to discuss:

Arid and semi-arid zone plants as a source of industrial raw material and its processing possibilities.

Minerals including petroleum and their utilization in deserts.

Types of industries and industrial operations which most efficiently utilize water resources.

Different forms of incentives to encourage the development of the above industries.

COASTAL RESOURCES

In addition to other natural resources in the hinterland of the desert coastal area, maritime oriented activities including off-shore resource exploration would seem to be decisive to the strategy for the development of the coastal desert and/or arid regions of the world.

Traditional human activities oriented towards exploitation of living marine resources, in the coastal desert and/or arid areas would be enhanced if successful mariculture were applied, modern fishing methods and fleets were introduced and rational fish processing and marketing schemes were developed.

Traditional maritime industries such as ship-building, drydock and ship repair and maintenance facilities should be regarded as compatible development schemes for coastal desert and/or arid region.

In the context of the emerging environmental constraints and other limitations to further industrial growth of the coastal areas in developed regions of the world, on one hand, and the need for continued and sustained economic growth for the benefit of mankind on the other, it would seem reasonable that institutional as well as economic policy guidelines for long term development of coastal desert and/or arid regions should be compatible with environmental consideration and consistent with yet to be defined policies for a new economic growth.

TOURISM

Tourism as a means of exploiting and developing the potential of arid and semi-arid areas should be carefully considered. Such developments are embryonic in most desert areas, except in some coastal areas.

The following aspects of the desert have tourism potential:

a. Wildlife (viewing, photo safaris, big game hunting).
b. The unique environment of the desert/observation, detailed examination (study tours) of certain aspects, e.g., wild foods/scenery.
c. Health spas (tourists going to the dry region for health reasons).

The development of mineral spas or general spas is being viewed as a new potential strategy of fostering tourism in deserts.

A careful evaluation of the impact on desert residents, and the relative benefit to them (e.g., employment, investment opportunity) must be made before undertaking tourism development, to assure that there will be substantial benefit to the local people. A substantial infrastructure is needed for tourist development including transportation, food services, health services, and water supply.

ENERGY AND MINERALS

Energy

In desert countries energy is as vital a supply as in non-desert areas. The use of firewood for cooking still is wide-spread in desert areas and is continuing to destroy the remaining trees and bushes, and unless a country can organize on a regular basis the planting of trees or bushes for firewood supply, countries should study carefully the possibilities of replacing the use of firewood with other energy sources.

The introduction of electricity in desert areas in addition to its energy application is vital for communications and many other uses.

Wind power units should be considered in those areas where wind conditions are favorable. Unfortunately precise local wind condition surveys have not been carried out in most areas, and until such surveys are available, the application of wind power cannot be judged.

Solar energy at the present stage of development is useful for water warming, but for electricity generation the costs are too high at present. Geothermal energy is known to exist in many desert areas, and when geothermal energy can be developed it could also be used among many other applications for electricity generation.

If a country possesses sedimentary areas it should, as a matter of course, be a primary task for the country to explore its underground potential, and in such areas there is a possibility of finding fuel of many types from conventional oil, oil shale, tar sands, and gas to coal, lignite, and peat. It is equally obvious that the discovery in a desert country of large deposits of oil and gas or coal cannot only contribute to the local energy supply, but could also provide an important source of income for the country, in addition to local employment. In general, in the energy policy of a desert country the destruction of bushes and remaining forests through use as firewood should gradually be eliminated and any available source of energy discussed above should be introduced and electricity should be made available as soon as possible to the whole population.

MINERALS

It is likely that all desert areas contain mineral resources of one type or the other. There are many mineral resources which are specific to desert areas because they are formed in the deserts, such as saline deposits, including a variety of salts such as potash, as well as other minerals, such as lithium. Depending on the geology of a desert area one can expect to find materials from gold to silver, lead, zinc, and many others, including bauxite, the raw material for aluminum. Unfortunately, mineral exploration requires high skill and risk capital; moreover, many of the desert areas have limited geological information, frequently no detailed geological mapping, and modern geochemical and geophysical surveys covering only very limited areas. It is recommended, therefore, that desert areas consider mining exploration as one of their primary tasks, including training of their own geologists and geochemists and geophysicists so that the government at a fairly early stage will be able to evaluate its own underground resources, and on the basis of this, can make policy decisions on which resources to develop first.

Modern mineral development in desert areas need not to be water intensive, and the present situation where mineral deposits are left undeveloped because of water shortages could in many cases be overcome, by application of dry processing methods. In addition, the reuse and reprocessing of mining water can sharply reduce the total quantity of fresh water which may be required. Furthermore, most mining processes allow the use of low quality water, and so on; and thus, if the mine is within proximity of agricultural areas or urban centers, the mining development, except for the need of the personnel, could be covered by waste water.

The same policy should be applied to the oil industry in desert areas which still use in some countries large quantities of fresh water. Thus, the development of mineral resources in desert areas need not be a burden on the limited water resources of a desert country, but could make a considerable contribution to the development of such a country through the income effect, the employment effect, and supply of energy, and other material also needed in the country.

HEALTH

Introduction

This topic falls into two categories:

1. Maintaining the health, or improving the health, of the resident and immigrant desert population; and
2. Health care as a focus for tourists potential. The inherent dry nature of deserts, it should be noted, means health problems are probably generally of a less chronic nature than in more humid areas.

Recommendations

a. Examination of the traditional desert-peoples diet could reveal foods and means of keeping a desert population healthy;
b. Strategies for bringing health-care to resident desert populations should be further investigated and different country experiences collated. Mobile health services (ranging from flying doctors to barefoot doctor type) is particularly stressed.
c. Tackling nutritional needs and desert-specific health problems and diseases is critical to other developments.

INFRASTRUCTURE

In some desert areas there may be large supplies of ground water or other resources that have not been developed because there is no infrastructure. To make development possible there must be transportation systems (roads, railroads, airports) and communications (telephone or radio). There must be energy resources, a marketing structure and community services. The latter includes health services, education and vocational training, and government functions.

The infrastructure should be self-supporting eventually, but it is probable that outside capital will be needed to provide the initial services so development can begin.

PARTICIPANT LIST

- A -

Dr. Mansur M. Aba-Husayn
Assistant Deputy Minister for Research and Development
Ministry of Agriculture and Water
Riyadh, Saudi Arabia

Adly Abdel-Meguid
Senior Industrial Development Officer
United Nations Industrial Development Organization
United Nations, Room 2766
New York, New York 10017

Dr. Ahmed Gamal Abdel Samie
Vice President
Academy of Scientific Research and Technology
101 Kasr El Eini Street
Cairo, Egypt

Adolfo Aguilar
Center for Third World Economic and Social Studies
Cor. Profirio Diaz 80
Mexico 20, D.F.

Rhousmane Ahmoud
Chef de Service Planification Sectorielle, Ministere du Plan
B. P. 862
Niamey, Niger

Shahid Akbar
Chief (Corporate Planning)
Oil and Gas Development Corporation
Shafi Chambers, Club Road
Karachi-4, Pakistan

Kola Aladejana
Acting Director of Forestry
Federal Department of Forestry
P.M.B. 5011
Ibadan, Nigeria

Omar A. Mughram Al-Ghamdi
Mechanical Engineer, Saline Water Conversion Corporation
P.O. Box 4931
Jeddah, Saudi Arabia

Dr. Abdullah Al-Jahmi
Director General of Plant Production
Ministry of Agriculture
Sidi Misri
Tripoli, Libya

Laika Khidir Al-Jibury
Scientific Research Foundation
Jadiriya
Baghdad, Iraq

Ahmad A. Al-Mansour
P.O. Box 2
Alkhobar, Saudi Arabia

Nasser Mohamed Amer
Director, Irrigation Department
Ministry of Agriculture
P. O. Box No. 1161, Khormksar
Aden, P.D.R. Yemen

Abdullah Abdir-Razik Arar
Regional Land and Water Development Officer
Food and Agriculture Organization
P. O. Box 2223
Cairo, Egypt

Mazhar Aslam
Manager, Groundwater Development
Resources Development Corporation
Dawood Centre
Karachi, Pakistan

Mohammad Osman Atif
Officer in Charge of Personnel Training,
Forestry Department
Ministry of Agriculture
Kabul, Afghanistan

El Hag Makki Awouda
Head, Management and Services Division,
GUM Arabic, Sudan Forest Service
c/o Forestry Department
Khartoum, Sudan

Mesfin Aytenffisu
Regional Manager,
Harar Regional Office
Ethiopian Water Resources Authority
P. O. Box 2
Harar, Ethiopia

Eduardo Azuara Salas
Center for Third World Economic
and Social Studies
Cor. Profirio Diaz 80
Mexico 20, D.F.

- B -

Dr. Omar Ba
General Director
Sahalien Institute, CILSS
c/o I.E.R.
Bamako, Mali

Dr. Amanclyte M. Babayev
Desert Institute
744012 Ashkhabad
U.S.S.R.

E. Bakutina
c/o State Committee for Science &
Technology
11 Gorkey Street
Moscow
U.S.S.R.

Dr. Joseph Bamea
Co-Director of Conference
United Nations Institute of Training
and Research
801 United Nations Plaza
New York, New York 10017

Roshan B. Bhappu
Vice President and General Manager,
Mountain States Research and
Development Co.
P.O. Box 17960
Tucson, Arizona 85731

Michael Bloome
United Nations Institute of Training
and Research
801 United Nations Plaza
New York, New York 10017

Sergio E. Bonilla
Vice President, INIA
Santiago, Chile

Boualem Bousseloub
Resources Planner, California
Department of Water Resources
P.O. Box 388
Sacramento, California 95802

Jakov Bradanovic
Centre for National Resources, Energy
and Transport
United Nations
P.O. Box 20 G.C.P.O.
New York, New York 10017

Dr. Michael Bradley
Associate Professor
Department of Hydrology and Water
Resources
University of Arizona
Tucson, Arizona 85721

Dr. Donat B. Brice
Staff Chemical Engineer
California Department of Water
Resources
P. O. Box 388
Sacramento, California 95802

John E. Bryson
Chairman, California State Water
Resources Control Board
1416 Ninth Street
Sacramento, California 95814

William E. Bullard, Jr.
4931 N.E. Roselawn
Portland, Oregon 97218

Joseph I. Burns
President, World Affairs Council of Sacramento
600 Forum Building
Sacramento, California 95814

William E. Bye
Environmental Engineer
U.S. Environmental Protection Agency
401 "M" Street, S.W.
Washington, D.C. 20460

- C -

Professor Chang Sen-dou
Department of Geography
University of Hawaii
Honolulu, Hawaii 96822

Bertrand H. Chatel
Chief, Technology Applications Section, Office of Science and Technology
United Nations
New York, New York 10017

Dr. C. Brent Cluff
Water Resources Research
University of Arizona
Tucson, Arizona 85721

Dr. E. Lendell Cockrum
Head, Department of Ecology and Evolutionary Biology
University of Arizona
Tucson, Arizona 85721

A. Gene Collins
Project Leader, Petroleum Production Research
U.S. Energy Research and Development Administration
P.O. Box 1398
Bartlesville, Oklahoma 94003

Keith R. Cooley
United States Water Conservation Laboratory
4331 East Broadway
Phoenix, Arizona 85040

Carlos Correa Sanfuentes
Regional Secretariat for Planning and Coordination
Prat 350 La Serena
P.O. Box 387
La Serena, Chile

Kit Cullen
California Office of Economic Opportunity
555 Capitol Mall
Sacramento, California 95814

- D -

Sami R. Daghistani
Institute for Research on Natural Resources
Jederiah
Baghdad, Iraq

Dr. B. C. Dando
Epidemiologist
World Health Organization
Box 54
Gaborone, Botswana

Ali Darag Ali
Ministry of Agriculture and Natural Resources
P.O. Box 199
Khartoum, Sudan

Christian DeClerq
Economist, U.N. Economic Commission for Western Asia
P. B. 4656
Beirut, Lebanon

Dr. K. James DeCook
Associate Hydrologist, Water Resources Research Center
University of Arizona
Tucson, Arizona 85721

Claire T. Dedrick
Secretary for Resources
California Resources Agency
1416 Ninth Street
Sacramento, California 95814

Georges Cohen deLara
Chief, Department of Water Quality, Sogreah, Inc.
6 rue de Lorraine
Grenoble, France

Philippe de Seynes
Director, Programme of Future Studies
United Nations Institute of Training and Research
801 United Nations Plaza
New York, New York 10017

Aliou M. Diallo
Programme Management Office
Sahelian Unit, Office of Technical Co-operation
United Nations
New York, New York 10017

Efren Diaz
Pergolas 192
Xochimilco, Mexico D.F.

John R. Donnell
Chief, Oil Shale Section
Chemical Research Branch
U.S. Geological Survey
P. O. Box 25046; Stop 939
Denver, Colorado 80225

- E -

Lic. Rodolfo Echeverria Zuno
Center for Third World Economic and Social Studies
Cor. Profirio Diaz 80
Mexico 20, D.F.

Stahrl W. Edmunds
Dean, Graduate School of Administration, University of California Riverside
Riverside, California 92521

Dr. William C. Ellet
Director, Marketing Research
Northrop Corporation
1800 Century Park East
Los Angeles, California 90067

Azouz Ennifar
Director, Tunisian National Tourist Office
630 Fifth Avenue
New York, New York 10020

Dr. Mohamed Aly Ezzat
Undersecretary for Desert Irrigation
Ministry for Irrigation
70 Gumhuriya Street
Cairo, Egypt

- F -

Michael W. Fall
Fish and Wildlife Service
U.S. Department of the Interior
Denver Wildlife Research Center
Lakewood, Colorado 80255

Egidio Feliu Acevedo
SERPLAC
Edificio Gobierno Regional
Iquique, Chile

Dr. Elias Fereres-Castiel
Area Soil and Water Specialist
LAWR: Water Science and Engineering
University of California, Davis
Davis, California 95616

Robert F. Fingado
Chief of Implementation, Delta Branch, Central District
California Department of Water Resources
P. O. Box 160088
Sacramento, California 95816

Walter W. Fisher
Metallurgist
Arizona Bureau of Mines
University of Arizona
Tucson, Arizona 85721

Carlton E. Forbes
Chief Engineer, California Department of Transportation
1120 N Street
Sacramento, California 95814

- G -

Kay Gary
University of California Davis
Davis, California 95616

James J. Geraghty
President, Geraghty and Miller, Inc.
P. O. Box 17174
Tampa, Florida 33682

Professor McGuire Gibson
Oriental Institute
University of Chicago
Chicago, Illinois 60637

Herbert W. Greydanus
Chief, Division of Planning
California Department of Water Resources
P. O. Box 388
Sacramento, California 95802

Dr. Joseph G. Gringof
Director, Middle Asian Regional Hydrometeorological Research Institute
Uzbekistan, Tashkent
U.S.S.R.

- H -

Dr. Robert M. Hagan
Professor of Water Science
University of California Davis
Davis, California 95616

Dr. Hassan H. Hajrah
Head of Biology Department
Faculty of Science
King Abdulaziz University
P. O. Box 1540
Jeddah, Saudi Arabia

Walter G. Halset
United Nations International Labour Organization
P. O. Box 1966
Lusaka, Zambia

Dr. Adnan Hardan
Agricultural Consultant
Ministry of Planning
P. O. Box 2018
Baghdad, Iraq

Edward L. Hastey
State Director
U.S. Bureau of Land Management
2800 Cottage Way
Sacramento, California 95825

Thomas J. Henderson
President, Atmospherics, Inc.
5652 E. Dayton
Fresno, California 93727

Lee F. Hermsmier
Agricultural Engineer
Imperial Valley Conservation Research Center
4151 Highway 86
Brawley, California 92227

Lonnie Hiebert
Saline Processors
Box 21
Claremont, California 91711

Diane Hilden
University of California Davis
Davis, California 95616

Dr. Colin W. Holloway
International Union for Conservation of Nature and Natural Resources (IUCN)
1110 Morges
Switzerland

Dr. David Hopcraft
Director, Wildlife Ranching and Research
Athi River, Box 47272
Nairobi, Kenya

H. E. Horton
Assistant Regional Director
U. S. Bureau of Reclamation
2800 Cottage Way
Sacramento, California 95825

Abdoulgader Houdairi
Hydrogeologist
P.O. Box 4411
Tripoli, Libya

Dr. William L. Hughes
Director, Engineering Energy Laboratory
Oklahoma State University
Stillwater, Oklahoma 74074

Dr. John L. Hult
President, Application Concepts and Technology Association
P. O. Box 1731
Santa Monica, California 90406

- I -

El Mahdi H. Ismail
Director, Agricultural and Industrial Section
Ministry of Planning
P. O. Box 600
Tripoli, Libya

Dr. Shawkat Abdel-kader Ismail
Advisory to H.R.H. Governor
Saline Water Conversion Corporation
P. O. Box 4931
Jeddah, Kingdom of Saudi Arabia

- J -

Dr. Merle H. Jensen
Environmental Research Laboratory
University of Arizona
Tucson International Airport
Tucson, Arizona 85706

E. Mostafa Jermoumi
Chef de Service de la Mise en Valeur Agricole de Marrakech
Ministère de l'Agriculture et de la Réforme Agraire
Direction Provinciale de l'Agriculture
B. P. 21
Marrakech, Morocco

Darshan Johal
Centre for Housing, Building and Planning
United Nations
New York, New York 10017

Dr. Jack D. Johnson
Director, Arid Lands Studies
University of Arizona
845 N. Park Avenue
Tucson, Arizona 85719

Professor Arthur W. Jokela
Department of Landscape Architecture
California State Polytechnic University
3801 West Temple Avenue
Pomona, California 91768

Professor Paul H. Jones
Department of Geology
Louisiana State University
Baton Rouge, Louisiana 70803

- K -

Dr. Michel Keita
Institute de Recherche en Sciences Humaines
B. P. 318
Niamey, Niger

Roy J. Kelly
Planning Economist
Metropolitan Water District of Southern California
P. O. Box 54153
Los Angeles, California 90054

Ngare Kessely
Conseiller
Ministère des Affaires Etrangères
N'djamena, Tchad

M. Said Kettanah
Scientific Research Foundation
Jadiriya
Baghdad, Iraq

Dr. Nikolai G. Kharin
Biologist, Desert Institute
Academy of Sciences of Turkmen S.S.R.
GSP 24, 744012 Ashkhabad
U.S.S.R.

Ali Kholdbarin
Expert, Soil Conservation Bureau
Forest and Range Organization
Ministry of Agriculture and Natural Resources
Farahg Onobi-Mansour
Farahnal 25
Tehran, Iran

Robert E. King
Consultant
580 Shore Acres Drive
Mamaroneck, New York 10543

Ira E. Klein
Consulting Geologist
1521 El Nido Way
Sacramento, California 95825

- L -

Frank L. Lambrecht, D.T.M.L.H.
International Health Program, Health Sciences
University of Arizona
Tucson, Arizona 85724

Everard Mervyn Lofting
Engineering-Economics Associates
1700 Solano Avenue
Berkeley, California 94707

Hussain A. K. Lolo
Undersecretary, Ministry of Agriculture
Sanaa, Yemen Arab Republic

Janece Long
Administrative Assistant
California Secretary of State
925 "L" Street, Suite 605
Sacramento, California 95814

Manuel Lopez, Jr.
Regional Director, U.S. Bureau of Reclamation
P. O. Box 427
Boulder City, Nevada 89005

- M -

F. G. "Phil" Macias
Pacific Southwest Planning Officer, U.S. Department of the Interior
450 Golden Gate Avenue
San Francisco, California 94102

Dr. Joel Maltos Romo
Secretaria de Agricultura Zonas Aridas
Apartado Postal 426
Saltillo, Coahuila, Mexico

Billy E. Martin
Regional Director, U. S. Bureau of Reclamation
2800 Cottage Way
Sacramento, California 95825

John W. Masier
Chief, Planning Investigations Branch, San Joaquin District
California Department of Water Resources
P. O. Box 5710
Fresno, California 93755

Dr. W. Gerald Matlock
International Agriculture
University of Arizona
Tucson, Arizona 95721

Melvin Mattson
General Engineer
Membrane Processes Division
Office of Water Research and Technology, U.S. Department of the Interior
Washington, D.C. 20240

Richard L. Maullin
Chairman, California Energy Resources Conservation and Development Commission
1111 Howe Avenue
Sacramento, California 95825

Dr. Fernando Medellin-Leal
Director, Instituto de Investigacion de Zonas Deserticas
Universidad Autonoma de San Luis Potosi
Apartado Postal 458
San Luis Potosi, S.L.P., Mexico

Ing. Salvador Mejia-Acevedo
Center for Third World Economic and Social Studies
Cor. Profirio Diaz 80
Mexico 20, D.F.

Professor Aden B. Meinel
Optical Sciences Center
University of Arizona
Tucson, Arizona 85721

Dominick Mendola
Vice President and Director of Operations
Solar Aqua Systems, Inc.
P. O. Box 88
Encinitas, California 92024

Bashkar P. Menon
Center for Economic and Social Information
United Nations
New York, New York 10017

S. W. O. Menzel
Chief Executive
Reed Irrigation Systems
8400 K Magnolia
Santee, California 92071

Jewell L. Meyer
Area Soil and Water Specialist
University of California
Cooperative Extension
433 County Center, III Court
Modesto, California 95350

Robert W. Miller
Chief, Interstate Planning
California Department of Water Resources
P. O. Box 388
Sacramento, California 95802

Mohamed Khattar Mohamed
General Director for Planning
Secretariat of Municipalities
P. O. Box 2739
Tripoli, Libya

Lucien Monition
Scientific Advisor, Bureau de recherches geologiques et minieres
B. P. 6009
45018 Orleans Cedex, France

Virginia S. Mueller
Vice President, World Affairs Council of Sacramento
Stanford Brothers Warehouse
106 L Street
Sacramento, California 95814

- N -

Stephen J. Nicola
Senior Fishery Biologist
California Department of Fish and Game
1416 Ninth Street
Sacramento, California 95814

F. Niknam
Director General, Soil Conservation Bureau, Forestry and Range Organization
Ministry of Agriculture and Natural Resources
Tehran, Iran

Rose M. Nonini
Special Assistant
California Department of Water Resources
P. O. Box 388
Sacramento, California 95802

- O -

Mohamed J. Omar
Planning Director, Settlement Development Programme
P. O. Box 1407
Mogadishu, Somalia

Dr. James E. Osborn
Chairman, Department of Agricultural Economics
Texas Tech University
P.O. Box 4169
Lubbock, Texas 79409

Dr. B. F. Osorio-Tafall
Director General
Center for Third World Economic and Social Studies
Cor. Profirio Diaz 80
Mexico 20, D.F.

Abdullah Ouderdan
Forest Director
Forestry Department
Kabul, Afghanistan

John Dalmas Owvor Onyango
Deputy Chief Conservator of Forests, Forest Department
P.O. Box 30513
Nairobi, Kenya

- P -

Patricia Paylore
Office of Arid Lands Studies
University of Arizona
845 N. Park
Tucson, Arizona 85719

Professor Zvi Pelah
Head, R. and D. Authority
Ben-Gurion University of the Negev
P. O. Box 1025
Beer-Sheva, Israel

Dr. Ishwar Prakash
Principal Animal Ecologist
Central Arid Zone Research Institute
Jodhpur, India

Dr. Robert M. Pratt
Special Assistant
California Department of Food and Agriculture
1220 N. Street
Sacramento, California 95814

Edgar P. Price
Special Assistant to the Regional Director
U.S. Bureau of Reclamation
2800 Cottage Way
Sacramento, California 95825

- Q -

Dr. William J. Quirk
Deputy Chief Scientist for the Global Atmospheric Research Project
NASA Goddard Institute for Space Studies
2880 Broadway
New York, New York 10027

- R -

Joseph Ramamonjisoa
Directeur du Project de Development de l'Androy
Ministere du Development Rural et de la Reforme Agraire
Anosy
Tananarive, Madagascar

Robin R. Reynolds
Co-Director of Conference and
Deputy Director, California Department of Water Resources
P. O. Box 388
Sacramento, California 95802

Robin R. Reynolds III
California State Water Resources Control Board
1416 Ninth Street
Sacramento, California 95814

Ronald B. Robie
Director, California Department of Water Resources
P. O. Box 388
Sacramento, California 95802

Frank E. Robinson
Water Scientist, Imperial Valley Field Station
University of California
1004 E. Holton Road
El Centro, California 92243

Bruce C. Rogers
Special Assistant
California Secretary of State
925 L Street, Suite 605
Sacramento, California 95814

Dr. Virgilio Roig
Director, Instituto de Investigaciones de Zonas Aridas
Universidad Nacional de Cuyo
Gregorio Torres 2085
Godoycruz
Mendoza, Argentina

Richard E. Rominger
Director, California Department of Food and Agriculture
1220 N Street
Sacramento, California 95814

Dr. Adam Rose
Assistant Professor,
Department of Economics
University of California Riverside
Riverside, California 92521

Professor Boris G. Rozanov
Assistant to the Secretary-General, United Nations Conference on Desertification
U. N. Environment Programme
P. O. Box 30552
Nairobi, Kenya

- S -

Antonio Sainz Unzueta
Jefe Dpto. de Suelos
Ministerio de Asuntos Campesinos y Agropecuarios
Carlos Blanco No 841
Casilla de Correo No 5780
La Paz, Bolivia

Nampaa N. Sanogho
Ingenieur Principal, Direction Nationale des Eaux et Forets
B. P. 275
Bamako, Mali

Dr. Albert Sasson
Division of Ecological Sciences
U.N.E.S.C.O.
Paris 75007, France

Margaret W. Savage
Thermo. Mist Company
9528 Lemoran Avenue
Downey, California 90240

Glenn B. Sawyer
Supvr. Land and Water Use Analyst,
California Department of Water Resources
P. O. Box 388
Sacramento, California 95802

Daniel P. Schadle
Supervisor, Lands, Development and Maintenance
Arizona Game and Fish Department
2222 W. Greenway Road
Phoenix, Arizona 85023

Dr. Joel Schechter
Director, Applied Research Institute
Ben-Gurion University
P. O. Box 1025
Beer-Sheva, Israel

Frank J. Schober, Jr.
Commanding General, California Military Department
P. O. Box 214405
Sacramento, California 95821

Abdul Gasim Seif El Din
Senior Research Officer
GUM Research Division
P. O. Box 302
El Obeid, Sudan

Gonsalo Sevilla Rodriguez
Instituto de Reforma Agraria y Colonizacion
Carrio 1040, Casilla 146-A
Quito, Ecuador

Richard P. Sheldon
Chief Geologist
U. S. Geological Survey
Reston, Virginia 22092

Dr. Marilyn L. Shelton
Assistant Professor of Geography
University of California Davis
Davis, California 95616

Clark W. Shraeder
Sierra Club
Box 621
Broderick, California 95605

Dr. Ousmane Silla
C.I.L.S.S.
B. P. 7049
Ouagadougou, Upper Volta

Dr. Ranbir Singh
Director of Agriculture
Government of Rajasthan
Jaipur, India

Dr. Subrata Sinha
Director, Environmental Geology (W.R.)
Geological Survey of India
C-33, Lajpat Marg
Jaipur 302001, India

George I. Smith
Geologist, U.S. Geological Survey
345 Middlefield Road
Meno Park, California 94025

Professor Norman S. Smith
Arizona Cooperative Wildlife Research Unit
219 Biological Science Building
University of Arizona
Tucson, Arizona 85721

Dr. J. Herbert Synder
Director, Water Resources Center
University of California Davis
Davis, California 95616

Alberto A. Sojit
Project Director
United Nations Development Program/ Inter-American Development Bank UNDP
P. B. O. 197D
Santiago, Chile

William Spaulding, Jr.
Geothermal Advisor, U.S. Fish and Wildlife Service
U.S. Department of the Interior
1730 K Street
Washington, D.C. 20240

Professor Brian Spooner
Department of Anthropology
University of Pennsylvania
Philadelphia, Pennsylvania 19174

Terry G. Spragg
A.C.T.A.
221 Third Place
Manhattan Beach, California 90266

Dr. Preston Stegenga
Director, International Center
California State University, Sacramento
6000 J Street
Sacramento, California 95819

Wesley E. Steiner
Executive Director
Arizona Water Commission
222 North Central Avenue, Suite 800
Phoenix, Arizona 85004

C. L. Stetson
Chief, San Joaquin District
California Department of Water Resources
P. O. Box 5710
Fresno, California 93755

Professor H. H. Stoevener
Department of Agriculture and Resource Economics
Oregon State University
Corvallis, Oregon 97331

- T -

Michael Takla
Project Engineer, Bechtel Inc.
P. O. Box 3965
San Francisco, California 94119

Harold E. Thomas
1339 Portola Road
Woodside, California 94062

A. Toure
Directeur de la Protection et Amélioration Efficace
Espèce Agro.-Pastorale
B. P. 170
Nouackchott, Mauritania,
West Africa

P.A. Towner
Chief Counsel, California Department of Water Resources
P. O. Box 388
Sacramento, California 95802

Dr. Thaddeus C. Trzyna
President, Center for California Public Affairs and
Chairman, Sierra Club International Committee
P. O. Box 30
Claremont, California 91711

George L. Turcott
Associate Director
Bureau of Land Management
U.S. Department of the Interior
Washington, D.C. 20240

- U -

Zafar Uddin
Deputy Director of Research
Agricultural Research Council
F-7/2 Islamabad, Pakistan

- V -

Frank P. Vandemaele
Senior Technical Adviser
United Nations Development Programme
Room DC-2062
New York, New York 10017

- W -

Kenneth Wahl
Water Resources Division
U. S. Geological Survey
855 Oak Grove Avenue
Menlo Park, California 94025

William E. Warner
Consultant
2090 Eighth Avenue
Sacramento, California 95818

Robert O. Watkins
District Director, California Department of Transportation
P. O. Box 847, 500 South Main
Bishop, California 93574

Gary D. Weatherford
Deputy Secretary for Resources
California Resources Agency
1416 Ninth Street
Sacramento, California 95814

Richard A. Weaver
Wildlife Manager, California Department of Fish and Game
1416 Ninth Street
Sacramento, California 95814

Ernest M. Weber
Engineering Geologist
Colorado River Board of California
107 South Broadway, Room 8103
Los Angeles, California 90012

James L. Welsh
Chief, Statewide Planning Branch,
California Department of Water Resources
P. O. Box 388
Sacramento, California 95802

Ruth Welsh
California Department of Water Resources
P. O. Box 388
Sacramento, California 95802

Elizabeth Wily
Ministry of Local Government and Lands
P. Bag 006
Gaborone, Botswana

Dr. Vicente M. Witt
Chief, Environmental Program Support
World Health Organization
Pan American Health Organization
525 Twenty-Third Street, N.W.
Washington, D.C. 20037

- X Y Z -

Dr. James A. Young
Range Scientist
Agricultural Research Service
U.S. Department of Agriculture
Reno, Nevada 89512

INDEX

Volume I: Energy and Minerals includes page 1-254.
Volume II: Agriculture includes pages 255-481.
Volume III: Water includes page 482-946.
Volume IV: Desert Management includes page 947-1396.